The Hypothalamus–Pituitary–Adrenal Axis

Neuroimmune Biology, Volume 7

# Neuroimmune Biology

*Series Editors*

I. Berczi, A. Szentivanyi

# The Hypothalamus–Pituitary–Adrenal Axis

*Volume Editors*

Adriana del Rey
*Institute of Physiology*
*University of Marburg*
*Marburg, Germany*

George P. Chrousos
*University of Athens*
*Athens, Greece*

Hugo O. Besedovsky
*Institute of Physiology*
*University of Marburg*
*Marburg, Germany*

ELSEVIER

AMSTERDAM • BOSTON • HEIDELBERG • LONDON • NEW YORK • OXFORD
PARIS • SAN DIEGO • SAN FRANCISCO • SINGAPORE • SYDNEY • TOKYO

Elsevier
Radarweg 29, PO Box 211, 1000 AE Amsterdam, The Netherlands
Linacre House, Jordan Hill, Oxford OX2 8DP, UK

First edition 2008



**Library of Congress Cataloging-in-Publication Data**
A catalog record for this book is available from the Library of Congress

**British Library Cataloguing in Publication Data**
A catalogue record for this book is available from the British Library

ISBN: 978-0-444-53040-0
ISSN: 1567-7443

For information on all Elsevier publications
visit our website at elsevierdirect.com

Printed and bound in the United Kingdom

Transferred to Digital Print 2011

# Contents

# *List of Contributors*

**Matias Acuña**
*Laboratorio de Fisiología y Biología Molecular, Departamento de Fisiología, Biología Molecular y Celular, Facultad de Ciencias Exactas y Naturales, Universidad de Buenos Aires, Ciudad Universitaria, Pabellón II, C1428EHA, Buenos Aires, Argentina*

**Eduardo Arzt**
*Laboratorio de Fisiología y Biología Molecular, Departamento de Fisiología, Biología Molecular y Celular, Facultad de Ciencias Exactas y Naturales, Universidad de Buenos Aires, Ciudad Universitaria, Pabellón II, C1428EHA, Buenos Aires, Argentina*

**Peter J. Barnes**
*Department of Thoracic Medicine, National Heart and Lung Institute, Dovehouse Street, London, SW3 6LY, UK*

**Albertus Beishuizen**
*Department of Intensive Care, VU University Medical Center, De Boelelaan 1117, 1081 HV Amsterdam, The Netherlands*

**Hugo O. Besedovsky**
*Department of Immunophysiology, Institute of Physiology and Pathophysiology, Medical Faculty, 35037 Marburg, Germany*

**Stefan R. Bornstein**
*III Medical Clinic, University Hospital Carl Gustav Carus, at the Technical University of Dresden, Fetscherstrasse 74 01307 Dresden, Germany*

**Carolina P. Castro**
*Laboratorio de Fisiología y Biología Molecular, Departamento de Fisiología, Biología Molecular y Celular, Facultad de Ciencias Exactas y Naturales, Universidad de Buenos Aires, Ciudad Universitaria, Pabellón II, C1428EHA, Buenos Aires, Argentina*

**George P. Chrousos**
*First Department of Pediatrics, Aghia Sophia Childrens Hospital, University of Athens, Athens, Greece*

**John A. Cidlowski**
*NIEHS, 111 T.W. Alexander Dr. Building 101, Room F-354, Research Triangle Park, Durham, NC 27709, USA*

**Firdaus S. Dhabhar**
*Department of Psychiatry, Program in Immunology & Cancer Center, Stanford University, Grant Building, Lab S-207, 300 Pasteur Drive, MC 5135, Stanford, CA 94305-5135, USA*

**Adrian J. Dunn**
*Pacific Biosciences Research Center and Department of Psychology, University of Hawaii at Manoa Honolulu, Hawaii, USA*

**Ilia J. Elenkov**
*Institute of Neurobiology and Molecular Medicine, Italian National Research Council, Via Fosso del Cavaliere 100, 00133 Rome, Italy*

**Denis P. Franchimont**
*Division of Gastroenterology, Montreal General Hospital, Mcgill University, Montreal, Canada*

**Damiana Giacomini**
*Laboratorio de Fisiología y Biología Molecular, Departamento de Fisiología, Biología Molecular y Celular, Facultad de Ciencias Exactas y Naturales, Universidad de Buenos Aires, Ciudad Universitaria, Pabellón II, C1428EHA, Buenos Aires, Argentina*

**Paul M. Guyre**
*Department of Physiology, Dartmouth Medical School, Lebanon NH 03756-0001, USA*

**Peter Härle**
*Department of Internal Medicine I, University Hospital Regensburg, 93042 Regensburg, Germany*

**Klaus Heeg**
*Dept. of Hygiene and Medical Microbiology, University of Heidelberg, Im Neuenheimer Feld 324, 69120 Heidelberg, Germany*

**Carmen Hernández-López**
*Centre for Microscopy and Cytometry, Complutense University, 28040 Madrid, Spain*

**David Janele**
*Department of Internal Medicine I, University Hospital Regensburg, 93042 Regensburg, Germany*

**Wilfrid Jänig**
*Physiologisches Institut, Christian-Albrechts-Universität zu Kiel, Olshausenstr. 40, Olshausenstr. 40, Germany*

**Eva Jiménez**
*Department of Cell Biology, Faculty of Medicine, Complutense University, 28040 Madrid, Spain*

**T. Bucky Jones**
*The Ohio State University, College of Medicine, Center for Brain and Spinal Cord Repair, 460 W. 12th Avenue, 786 Biomedical Research Tower, Columbus, Ohio 43210, USA*

**Martin Kees**
*Department of Internal Medicine I, University Hospital Regensburg, 93042 Regensburg, Germany*

**Alexander W. Krug**
*Medical Clinic III, Carl Gustav Carus University Hospital, University of Dresden, Dresden, Germany*

**Ivan Lefkovits**
*Research Department of the University Clinics, Physiology Institute of University of Basel, Vesalianum, Vesalgasse 1, CH-4051 Basel, Switzerland*

**Kurt M. Lucin**
*Department of Neurology and Neurological Sciences, Stanford University School of Medicine, 300 Pasteur Drive, Rm 343, MC 5235, Stanford, CA 94305, USA*

**Jeanette I. Webster Marketon**
*Institute for Behavioral Medicine Research and Division of Pulmonary, Allergy, Critical Care and Sleep Medicine, Department of Internal Medicine, The Ohio State University, 2058A Graves Hall, 333 W. 10th Ave, Columbus, OH 43210, USA*

**Allan Munck**
*Emeritus Department of Physiology, Dartmouth Medical School, Lebanon NH 03756-0001, USA*

**Juan J. Muñoz**
*Centre for Microscopy and Cytometry, Complutense University, 28040 Madrid, Spain*

**Alberto C. Nagashima**
*Laboratorio de Fisiología y Biología Molecular, Departamento de Fisiología, Biología Molecular y Celular, Facultad de Ciencias Exactas y Naturales, Universidad de Buenos Aires, Ciudad Universitaria, Pabellón II, C1428EHA, Buenos Aires, Argentina*

**Marcelo Páez-Pereda**
*Max-Planck Institute of Psychiatry, Kraepelinstr. 10, 80804 Munich, Germany*

**Louis Du Pasquier**
*Institute of Zoology and Evolutionary Biology, Vesalgasse 1, CH-4051 Basel, Switzerland*

**Georg Pongratz**
*Department of Internal Medicine I, University Hospital Regensburg, 93042 Regensburg, Germany*

**Phillip G. Popovich**
*The Ohio State University, College of Medicine, Center for Brain and Spinal Cord Repair, 460 W. 12th Avenue, 786 Biomedical Research Tower, Columbus, Ohio 43210, USA*

**Ulrich Renner**
*Max-Planck Institute of Psychiatry, Kraepelinstr. 10, 80804 Munich, Germany*

**Adriana del Rey**
*Institute of Physiology, University of Marburg, Marburg, Germany*

**Rosa Sacedón**
*Department of Cell Biology, Faculty of Medicine, Complutense University, 28040 Madrid, Spain*

**Jürgen Schölmerich**
*Department of Internal Medicine I, University Hospital Regensburg, 93042 Regensburg, Germany*

**Kathleen A. Smoak**
*Laboratory of Signal Transduction, National Institute of Environmental Health Sciences, Department of Health and Human Services, National Institutes of Health, Research Triangle Park, North Carolina 27709, USA*

**Günter K. Stalla**
*Max-Planck Institute of Psychiatry, Kraepelinstr. 10, 80804 Munich, Germany*

**Esther M. Sternberg**
*Section on Neuroendocrine Immunology and Behavior, National Institute of Mental Health, National Institutes of Health, Department of Health and Human Services, 36, Convent Drive MSC-4020, Bldg 36, Room 1A23, Bethesda, MD 20892-4020, USA*

**Rainer H. Straub**
*Laboratory of Neuroendocrinoimmunology, Dept. of Internal Medicine I, University Hospital, 93042 Regensburg, Germany*

**Alberto Varas**
*Department of Cell Biology, Faculty of Medicine, Complutense University, 28040 Madrid, Spain*

**Helmut Vedder**
*Department of General Psychiatry, Psychotherapy and Psychosomatics II Nordbaden Psychiatric Centre, Heidelberger Strasse 1a, 69168 Wiesloch, Germany*

**Istvan Vermes**
*Department of Clinical Chemistry, Medical Spectrum Twente, Hospital Group, P.O. Box 50000, 7500 KA Enschede, The Netherlands*

**Angeles Vicente**
*Department of Cell Biology, Faculty of Medicine, Complutense University, 28040 Madrid, Spain*

**Ilias Vrezas**
*Department of Endocrinology, University Hospital of Duesseldorf, Moorenstr. 5 40225, Duesseldorf, Germany*

**Mark P. Yeager**
*Dartmouth Medical School, Lebanon NH 03756-0001, USA*

**Agustín G. Zapata**
*Department of Cell Biology, Faculty of Biology, Complutense University, 28040 Madrid, Spain*

**Christian G. Ziegler**
*Medical Clinic III, Carl Gustav Carus University Hospital, University of Dresden, Dresden, Germany*

# *Foreword*

Studies on interactions between the immune, endocrine, and nervous systems can no longer be considered as an emerging field in biological and medical sciences. Indeed, in the last 30 years, hundreds of publications dealing with different aspects of this field appeared in prestigious multidisciplinary journals and in journals specialized in immunology, endocrinology, and neurosciences. As it usually occurs, the development of a scientific field is preceded by a stage that can be called "prehistory," in which few primitive facts serve for claims that are not based on scientifically structured hypotheses and concepts. We have to admit that the "prehistoric" period in the field of immune–neuroendocrine interactions was a little bit too long, in which holistic statements such as "the brain and the mind control everything and then ... why not the immune system?" predominated. Furthermore, the experimental evidence available at that time often derived from a primitive technology or from studies restricted to non-adaptive acute hypersensitivity and anaphylactic reactions. The most solid evidence of endocrine effects on immunity was that adrenocortical hormones are anti-inflammatory and can affect immune organs and the distribution of lymphoid cells. Such knowledge led Hans Selye to predict that stress can influence immune process, a prediction that is at present fully confirmed and continues to be investigated in detail.

Since then, an enormous bulk of information has accumulated indicating that the immune and neuroendocrine systems interact and control each other. Unfortunately, for some time, most immunologists were reluctant to accept the existence of a level of neuroendocrine control of immune processes. Retrospectively, this reluctancy is easy to justify since, at that time, it was still necessary to elucidate crucial intrinsic immune mechanisms. The main concern of immunologists was to understand the immune system from "within," with a tendency to ignore that immune cells are also exposed to external signals that affect the system. To look "inside" the immune system yielded formidable results. Now, the structure of the main molecules (e.g., antibodies, T-cell receptor) that recognizes the huge universe of antigens is known. The molecular and genetic bases of the differentiation and diversification of immune cells as well as the types and subtypes of cells that participate in an immune response are largely understood. The biochemical pathways of immune cell activation and how these cells interact and receive information from antigen-presenting cells have also been clarified to a great extent. This knowledge showed that immune cells are extremely complex, and that refined interactions between them constitute the basis of different types of immune responses.

The identification of efficient autoregulatory mechanisms may suggest that, compared to other physiologic systems, the immune system displays a privileged autonomy. However, a physiological immune response in an organism depends on mechanisms that are under neuroendocrine control. The most obvious level of such dependence is the need to control the highly demanding metabolic processes that underlie immune cell functions. Another example of dependence is related to the circulatory systems. Indeed, no effective immune response would be possible if immune cells could not circulate and reach the places where antigens are presented.

Hormones, neurotransmitters, and neuropeptides can also influence more refined mechanisms underlying immune cell activity. Immune cells perform in an environment where all these agents are present and can perceive neuroendocrine signals via specific receptors. There is clear evidence that immune cells can establish close contacts with nerve fibers both in lymphoid organs and in tissues where they meet and recognize antigens, and expand and develop as effector cells. It is also known that certain immune products and some hormones and neurotransmitters shear second intracellular messengers, transcription factors, and post-transcriptional mechanisms, a situation that allows a mutual modulation of their effects at intracellular levels. On the other hand, different types of immune cells express a different number of receptors for a given neuroendocrine agent and this number varies after activation. Thus, the distinct sensitivity of resting and activated cells to these agents contributes to immunospecificity and allows a neuroendocrine control of defined steps of the immune response. It is important to remark here that, under natural conditions, hormones, neurotransmitters, and neuropeptides represent the efferent messengers of complex immune–neuroendocrine regulatory circuits. Indeed, as with other systems under neuroendocrine regulation, the immune system conveys signals to central regulatory agencies that in turn respond and affect the course, development, and termination of the immune response. The first neuroendocrine immunoregulatory circuit was postulated when it was shown that, following antigenic challenge, immune cell products stimulate the hypothalamus–pituitary–adrenal (HPA) axis and that the resulting increase in glucocorticoid output can affect the immune response. The organization of this circuit and its relevance for health and disease is the main topic of this volume.

The first chapter (A. Dunn) provides a general perspective of the content of this volume. Then, we considered it useful to offer an anatomical and physiological description of the HPA axis (H. Vedder) and an analysis of how glucocorticoids can signal during health and disease (K. Smoak and J. Cidlowski). Because of its complementary physiological functioning with the HPA axis, the organization of the sympathetic nervous system is included (W. Jänig). Also a brief description of the innate (K. Heeg) and specific (I. Lefkovits and L. Du Pasquier) immune responses, as well as how the immune system functions in the brain (T. Jones, K. Lucin, and P. Popovich), is presented. The second part of this volume is devoted to effects and mechanisms of action of glucocorticoids on immune processes (P. Guyre, M. Yeager, and A. Munck), including their effects on the developing thymus (R. Sacedón *et al.*). The action of catecholamines (I. Elenkov) on immunity and the dual suppressive and enhancing effects of stress on immune function (F. Dhabhar) are also covered in this part. The third part of this volume deals with afferent signals delivered to the brain, endocrine glands, and the autonomic nerve system during immune responses. How immune signals are processed at brain level and affect the HPA axis (H. Besedovsky and A. del Rey), and the effect of inflammatory mediators on the pituitary (D. Giacomini *et al.*), the adrenal glands (I. Vrezas *et al.*), and the autonomic nervous system (R. Straub *et al.*) are addressed here. The last part covers the clinical relevance of HPA axis–immune interactions, particularly during sepsis and the multiple organ dysfunction syndrome (I. Vermes and A. Beishuizen) and inflammatory disorders (J. Webster Marketon and E. Sternberg). The development of glucocorticoid resistance during inflammatory diseases (D. Franchimont and G. Chrousos) and the role of glucocorticoids in asthma (P. Barnes) are also discussed.

We hope that this volume provides a comprehensive view of the molecular and functional organization of the immune–HPA axis circuit and its clinical relevance. We also expect that it promotes further research to identify other circuits integrating immune and neuroendocrine messengers and to understand their relevance for health and disease. In a broader context, we believe that studies of interactions between the main regulatory and adaptive mechanisms of the organism constitute a serious attempt to face the complexity of biological systems.

Hugo O. Besedovsky, Adriana del Rey

# Preface

Glucocorticoids and catecholamines were proposed to regulate immune and inflammatory reactions over half a century ago. However, these seminal observations were rejected later by the scientific community based on the argument that the anti-inflammatory effect of glucocorticoids was a pharmacologic, rather than a physiologic effect. As for catecholamines, contradictory findings were produced with regard to their immunoregulatory role for decades and consequently this issue was not resolved until recently.

Some 35 years ago, or so, a handful of laboratories started to re-investigate the interaction of the neuroendocrine and immune systems. The role of innervation and neuropeptides and that of hormones and cytokines were all thoroughly investigated with accurate and sensitive scientific methodology. The brain and various neural pathways were also examined for immunoregulatory properties.

In this book, the scientists who initiated the work on the HPA axis and on the autonomic nervous system, and others who joined the field later, present their own findings, interpret them, and integrate the new knowledge into this rapidly growing research discipline. It is apparent from this volume that the science of Neuroendoimmune Biology is very solid today with overwhelming evidence showing that the nervous, endocrine, and immune systems interact continuously with each other and play fundamental roles in the physiology and pathophysiology of higher organisms. Indeed, it is possible to suggest on the basis of the evidence presented in this book that the Nervous, Endocrine, and Immune Systems form the Neuroendoimmune Supersystem, which integrates all the biological functions of higher organisms both in health and disease for their entire life cycle.

Istvan Berczi

# I. PHYSIOLOGY

Edited by A. del Rey, G. P. Chrousos and H. O. Besedovsky


# *The HPA Axis and the Immune System: A Perspective*

ADRIAN J. DUNN

*Pacific Biosciences Research Center and Department of Psychology, University of Hawaii at Manoa Honolulu, Hawaii, USA*

## ABSTRACT

Although it had long been known that steroids secreted by the adrenal cortex exerted important inhibitory effects on cells of the immune system, the concept of bidirectional communication between the immune system and the nervous system became widely accepted when it was recognized that activation of the hypothalamo–pituitary–adrenocortical (HPA) axis associated with stress was not the only effect the central nervous system (CNS) exerted on the immune system, and that factors secreted by immune cells could have important effects on the CNS. Activation of the HPA axis results in elevation of circulating concentrations of corticosteroids which drive them from the thymus and spleen into the periphery and inhibit various immune activities. The immune system can also signal the CNS regarding the presence of foreign antigens in peripheral organs. A key finding was that interleukin-1 (IL-1) was a potent stimulator of HPA axis activity. The major known mechanism by which the immune system signals the brain involves cytokines, which are the major chemical messengers within the immune system. Certain other cytokines [e.g, IL-6, IL-10, and tumor necrosis factor-α (TNF-α)] can activate the HPA axis. Cytokines exert their effects on the CNS via afferents of nerves, such as the vagus, although some direct actions of circulating cytokines on the nervous system may also occur. Immune cells can penetrate the brain under some circumstances. The CNS in turn can influence immune function via the autonomic nervous system and various hormones secreted by the pituitary, the adrenal, and other endocrine organs. Cytokines (especially IL-1) can alter the activity of neurotransmitters in the CNS, most notably norepinephrine and corticotropin-releasing factor (CRF), and can have important effects on behavior.

## 1. INTRODUCTION

In retrospect, it is surprising that the concept of communication between the nervous and the immune systems was resisted so strongly only 25 years ago. It has long been recognized that diseases induced a state in which one did not feel well, indicating that the nervous system was able to detect the pathology, and it should have been evident that the immune system was likely to be involved in detecting infections and other pathological states. Nevertheless, immunologists resisted the concept of nervous system–immune system communication, because it was perceived to indicate nervous system control over the immune system, whereas physiologists recognized that communication between the nervous and the immune systems needed to be bidirectional. Eventually, the concept of communication between the immune and the nervous systems was accepted because of the compelling evidence for interactions between the two systems. A role for the nervous system was indicated by the discovery that glucocorticoids could inhibit immune function, implicating the hypothalamus–pituitary–adrenocortical (HPA) axis which is under the control of the brain. A role for the immune system was initially suggested by Blalock [1] who argued that the immune system could be considered the sixth sense, because it was capable of detecting the presence of pathogens or damaged tissue and communicated this to the brain (see Ref. [2]).

## 2. STRESS AND THE HPA AXIS

Over 70 years ago, Hans Selye perceived a critical role for the adrenal gland in stress based on the profound histological changes that occurred in the adrenal cortex. He subsequently determined that the key mediators were certain steroids synthesized in the adrenal cortex (hence corticosteroids). This complemented Cannon's recognition of the role of increased circulating catecholamines from the sympathetic nervous system and adrenal medulla in preparation for the flight or fight responses associated with stress. The principal steroids were cortisol and corticosterone both of which are produced in most species, but in rodents, corticosterone predominates, whereas cortisol predominates in most other animals. These hormones were subsequently named "glucocorticoids" because of their ability to increase the availability of glucose in the body. These steroids are the end-product of activation of the HPA axis, a three-tier system (Fig. 1) that ultimately results in adrenal corticosteroid secretion. It is a cascade initiated by activation of corticotropin-releasing factor (CRF)-containing neurons, with cell bodies in the paraventricular nucleus (PVN) of the hypothalamus. The CRF is secreted into the portal vessels in the median eminence region of the hypothalamus, and is carried in them to the anterior pituitary gland, where it stimulates the secretion of adrenocorticotrophic hormone (corticotropin, ACTH) into the peripheral circulation. The blood-borne ACTH then directly stimulates the adrenal cortex to synthesize and secrete glucocorticoids into the general circulation (see Fig. 1).

## 3. GLUCOCORTICOID INHIBITION OF IMMUNE FUNCTION

The first evidence indicating interactions between the HPA axis and the immune system was the recognition in the late 1940s that glucocorticoids were immunosuppressive (see review by Claman [3]). One manifestation of this was the involution (i.e., shrinking) of the thymus gland which had been known to occur in stressed animals since the mid-nineteenth century. This effect reflects apoptosis of immature lymphocytes in the thymus, and also that glucocorticoids can cause some lymphocytes to leave the thymus and enter the general circulation, perhaps for immune defense. Glucocorticoids can kill lymphocytes in concentrations, only slightly higher than those observed in stressed animals ($10^{-6}$–$10^{-5}$ M), although it is unlikely that cell death is the major mechanism for the immunosuppression. To this day, glucocorticoids and their analogs are the agents most commonly used therapeutically to induce immunosuppression, for example, following tissue transplantation. Nevertheless, the mechanisms underlying the immunosuppressive effects are not fully understood (see Ref. [4]). Initially, they involve glucocorticoid binding

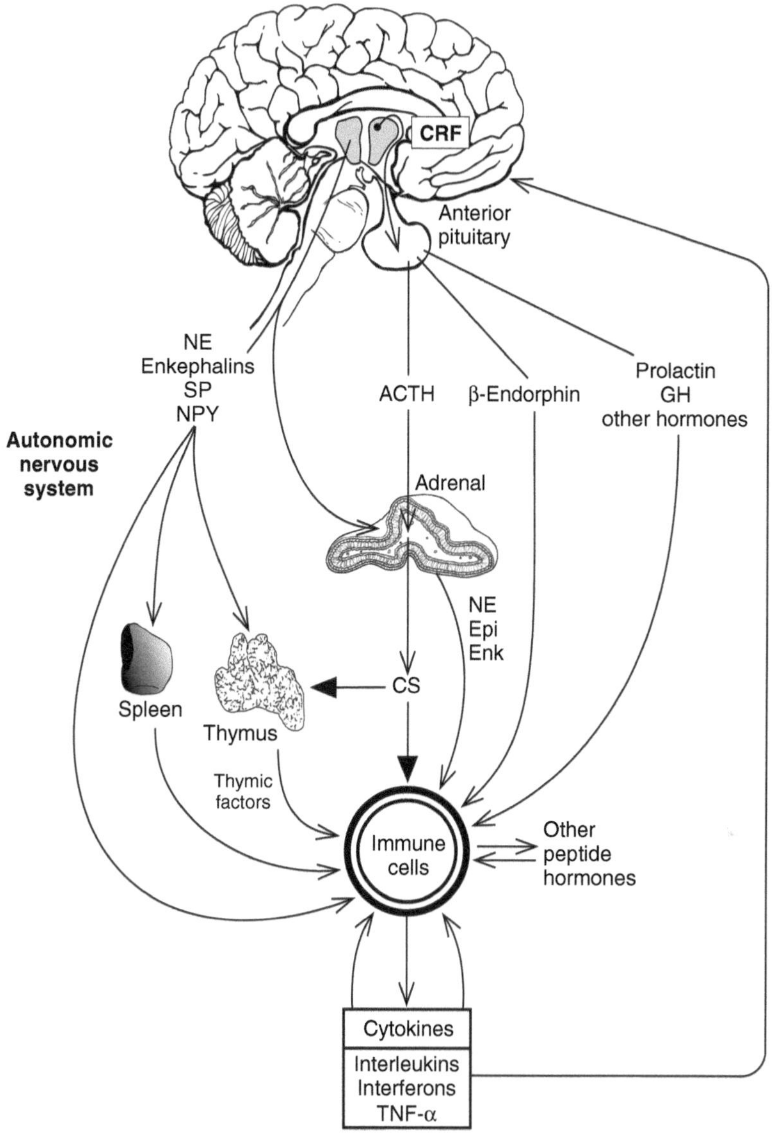


Figure 1. Diagram indicating the relationships between the nervous and the immune systems discussed. ACTH, adrenocorticotrophin; CRF, corticotropin-releasing factor; CS, corticosteroids (cortisol and/or corticosterone); Enk, enkephalins; Epi, epinephrine; GH, growth hormone; NE, norepinephrine; NPY, neuropeptide Y; SP, substance P; TNF-α, tumor necrosis factor-α.

to the glucocorticoid receptor (GR), which indirectly affects the functions of certain intracellular proteins important for immune cell function. A key finding was that glucocorticoid action was associated with the suppression of the function of nuclear transcription factor, NF-κB, by the induction of the inhibitory protein, IκBα [5]. Other proteins such as AP-1 (AP – activator protein) have also been implicated. There are probably multiple mechanisms by which activated GRs inhibit immune function [4].

## 4. IL-1 AND THE HPA AXIS

Evidence for effects of the immune system on the HPA axis derived from the observations that infections were associated with activation of the axis, the accepted indicator of stress (see the detailed review of Besedovsky and del Rey [6]). However, a critical link was established when it was demonstrated that IL-1 was a potent stimulator of the HPA axis, such that the peripheral administration of small doses of IL-1 to rats elevated the plasma concentrations of ACTH and corticosterone [7]. IL-1 is an important cytokine produced by various immune cells (principally macrophages and monocytes) and other tissues in response to a multiplicity of infections, and tissue pathologies. The HPA-activating effect was observed with both purified human IL-1 and synthetic recombinant human IL-1. Together with the observation that the HPA response to Newcastle disease virus (NDV) could be inhibited by the administration of an antibody to IL-1 [7], these observations clearly implicated IL-1 as a mediator of the infection-induced HPA activation. The potent HPA stimulatory activity of IL-1 was subsequently confirmed in many other laboratories and in many other species.

Certain other cytokines have been suggested to be able to activate the HPA axis, specifically IL-6, IL-10, and tumor necrosis factor-α (TNF-α), but none of these is as potent or effective as IL-1.

Needless to say, these observations established a straightforward mechanism by which the immune system could signal the brain. The mechanism was subsequently studied extensively in many laboratories. The results of a large number of studies (see reviews by Besedovsky [8], Turnbull and Rivier [9], and Dunn [10,11]) indicate that IL-1 has the ability to activate CRF-containing cells in the PVN, which is considered the critical initial step in HPA axis activation. In certain conditions, IL-1 may also induce ACTH secretion directly from the anterior pituitary, and/or glucocorticoids directly from the adrenal cortex, although neither of these routes appears to be the primary mechanism under normal physiological conditions [8,10,11].

### 4.1. Mechanisms by which peripherally generated or administered cytokines could act on the brain to activate the HPA axis

Cytokines such as IL-1 have a relatively high molecular weight, large enough that they will not readily penetrate the blood–brain barrier. Thus, it is unlikely that the action of IL-1 is exerted directly on the hypothalamus even though injection of IL-1 into the hypothalamus can activate the HPA axis. So, how does IL-1 induce its effects on the brain? Once again, there appear to be multiple mechanisms by which cytokines can affect the activity of the brain, some of which do not require cytokine penetration of the brain (see reviews by Ericsson et al. [12] and Dunn [13]).

Cytokines can act on brain cells in the circumventricular organs (CVOs), sites at which there is no blood–brain barrier. There is some evidence that IL-1 may act on the median eminence [9], on the organum vasculosum laminae terminalis (OVLT) and the preoptic area [14,15], and in the area postrema [9,12,16]. Several of these CVOs are located in the hypothalamus (the median eminence, OVLT, and preoptic area), and others have direct connections to the hypothalamus. Thus, cytokines may be able to exert effects directly on brain structures close to the PVN.

Cytokines can be transported into the brain to a limited extent using selective uptake systems (transporters), thus bypassing the blood–brain barrier [17]. The capacity of these systems is quite limited, and their functional significance is unclear. The anatomical distribution of the uptake sites has revealed little, but they appear to be important for certain specific functions (see, e.g., Ref. [18]).

Cytokines may act directly or indirectly on peripheral nerves that send afferent signals to the brain. The hypothalamus can be activated indirectly, for example, by the vagus nerve. The vagus contains afferent neurons that project to the brain stem, and which can activate cell bodies of neurons that project to the hypothalamus. Numerous studies have indicated that IL-1 (and endotoxin [lipopolysaccharide, LPS]) can signal the brain by activating such afferents, because lesions of the vagus nerve can prevent various physiological and behavioral responses to intraperitoneally (i.p.) injected LPS [19,20]. Such lesions also affect HPA responses to IL-1 [21,22] and TNF-$\alpha$ [23]. The mechanism for IL-1 (and LPS which induces IL-1 production) appears to be binding to paraganglion cells which in turn interact with the vagal neurons, signaling the brain [20].

Cytokines may act on peripheral tissues inducing the synthesis of lipophilic molecules whose ability to penetrate the brain is not limited by the blood–brain barrier. A major target appears to be brain endothelial cells which bear receptors for IL-1 and LPS. Systemic treatments with LPS and IL-1 are known to induce cyclooxygenase (COX)-2 in brain endothelia [24,25]. COX-2 activation may result in the production of prostaglandin $E_2$ ($PGE_2$) which can freely traverse the blood–brain barrier. $PGE_2$ can induce fever in the anterior hypothalamus [15], and can activate PVN-CRF cells, and thus the HPA axis [26].

Cytokines can be synthesized by immune cells that infiltrate the brain. It is well established that peripheral LPS administration initiates a process that results in macrophages invading the brain and migrating through the brain parenchyma [27]. In this form, the macrophages have the morphology of microglia.

## 4.2. Mechanisms by which IL-1 activates the HPA axis

The mechanisms by which IL-1 activates the HPA axis have been the subject of intense investigation, but are still not completely understood. Most likely this is because there are multiple mechanisms for this response. The biological redundancy undoubtedly reflects the importance of this function of IL-1. It is likely that IL-1 has the potential to act on the brain at all three levels of the HPA axis, the hypothalamus, the anterior pituitary, and the adrenal cortex. Obviously, the mechanism must vary to some extent with the site of IL-1 production *and* the route of administration, as well as with the dose, and possibly with the physiological state of the animal. However, the preponderance of the evidence suggests that under normal physiological circumstances, the principal action of IL-1 on the HPA axis involves hypothalamic CRF. The details of the critical studies have been reviewed earlier [9–11] and are addressed briefly below.

### *4.2.1. IL-1 action on the adrenal cortex*

Some early studies indicated that IL-1 could exert a direct action on adrenal glands *in vitro*; however, in most cases very high doses of IL-1 were necessary [28], and the effects were very dependent on the nature of the preparation studied (e.g., Ref. [29]), and could not always be replicated in other laboratories (e.g., Ref. [30]). A direct effect on the adrenal cortex is unlikely to explain completely the normal *in vivo* elevation of plasma concentrations of corticosterone, because IL-1 administration also elevates plasma ACTH in both rats and mice [7,31]. Also, IL-1 failed to induce increases in plasma corticosterone in hypophysectomized rats [29] and mice [31]. Moreover, the IL-1-induced increases in plasma ACTH and corticosterone were largely prevented by *in vivo* pretreatment with antibodies to CRF [31–34] and were substantially diminished in CRF-knockout mice [35].

#### *4.2.2. IL-1 action on the anterior pituitary*

The fact that hypophysectomy prevented the ACTH and corticosterone responses to IL-1 indicates a critical role for the pituitary in the HPA response to IL-1. A direct action of IL-1 on pituitary cells *in vitro* has been reported, but the data are complex and conflicting (see Ref. [10]). Generally, prolonged incubations are necessary to observe such effects. However, a direct pituitary effect of IL-1 appears to be excluded as the normal physiological mechanism because lesions of the PVN largely prevented the ACTH and corticosterone responses to IL-1 [36,37], and, as indicated above, pretreatment with antibody to CRF or knocking out the gene for CRF prevented the IL-1-induced increases in plasma ACTH and glucocorticoids. That prolonged incubation of both adrenocortical and adenohypophyseal cells increases their sensitivity to IL-1 may indicate that the ability of IL-1 to elevate circulating glucocorticoids is so critical for the organism that when higher components of the HPA axis fail to function properly, downstream organs can develop or assume that function.

#### *4.2.3. IL-1 action on the hypothalamus*

Hypothalamic CRF appears to be critical for the elevations of plasma ACTH and corticosterone induced by peripherally administered IL-1, because complete mediobasal hypothalamic deafferentation prevented the HPA response to i.p.-injected IL-1 [38]. This conclusion was supported by the observation that lesions of the PVN prevented the increases in plasma ACTH and corticosterone induced by i.p. IL-1 [36,37]. It is relevant that IL-1 increases the electrophysiological activity of CRF neurons *in vivo* [39,40]. IL-1 also stimulates CRF release from hypothalamic slabs *in vitro* [41], and activates the HPA axis when injected directly into the hypothalamus [38]. CRF is implicated by the observation that peripherally administered IL-1 elevates concentrations of CRF in portal blood [33], and depletes CRF from the median eminence [32], the latter suggesting increased release of CRF. Moreover, pretreatment of rats [32–34] and mice [31] with an antibody to CRF prevented the increases in plasma ACTH and corticosterone induced by IL-1. In addition, CRF-knockout mice showed only a minuscule increase in plasma corticosterone after IL-1 administration [35].

The *in vivo* evidence summarized above strongly favors a role for hypothalamic CRF as the major mechanism for the action of peripheral IL-1 in normal healthy animals. The evidence for direct actions on the pituitary and adrenal glands derives largely from *in vitro* experiments and is therefore susceptible to artifact. However, in studies on mice treated with antibody to CRF, there were small increases in plasma corticosterone following i.p. IL-1 [31], and very small, but statistically significant, increases were also observed in CRF-knockout mice [35]. This suggests that when the functions of higher levels of the HPA axis are impaired, the pituitary and/or adrenal cortex may gain the ability to respond to IL-1 and mount a glucocorticoid response. It should also be noted that IL-1- and LPS-induced secretion of ACTH and corticosterone can be observed in very young rats [42] and mice [43] at a time when HPA responses to stressors are minimal. Thus the ability of IL-1 to induce glucocorticoid secretion appears ontogenetically very early in life.

#### *4.2.4. Involvement of norepinephrine in IL-1-induced activation of the HPA axis*

Peripheral administration of IL-1β activates brain noradrenergic neurons, especially those in the hypothalamus [44,45]. The activation must occur in the nucleus tractus solitarius of the brain stem, the site of origin of ascending noradrenergic neurons that innervate the hypothalamus, including the PVN. The activation may be local via the area postrema, or indirectly via vagal afferents from the periphery [16,46]. Very high correlations are observed between the increases of noradrenergic activity in the mouse hypothalamus and HPA activation induced by IL-1 (as well as LPS, NDV, and influenza virus) [47]. Similar close correlations are also observed between hypothalamic norepinephrine (NE) release measured by *in vivo* microdialysis and plasma corticosterone following intravenous (i.v.) or i.p. injection of IL-1 into freely moving rats (see below). The relationship between hypothalamic NE and the HPA axis is consistent with the evidence for noradrenergic activation of PVN-CRF neurons [48,49]. Lesioning of this projection in the rat markedly decreased the magnitude of the HPA response (see below). The fact that the COX inhibitor, indomethacin, inhibits the noradrenergic response to IL-1 with equivalent reductions in plasma ACTH and corticosterone [50,51] bolsters the argument that this noradrenergic activation drives the HPA activation.

However, studies of the effects of adrenergic receptor antagonists suggest a more complex relationship. Rivier et al. [52] failed to find any effect of the β-adrenergic antagonist, propranolol, or the $\alpha_1$-adrenergic antagonist, prazosin, or both drugs combined on the HPA activation by IL-1 in rats. We have made similar observations in mice; propranolol had no effect even at high doses, although a small inhibition of the increase in plasma corticosterone was observed at a high dose (1 mg/kg) of prazosin [47]. This effect of prazosin was not enhanced by propranolol. However, when the ventral noradrenergic ascending bundle or the PVN of rats was lesioned with 6-hydroxydopamine (6-OHDA), the IL-1-induced increase in plasma corticosterone was markedly decreased when the depletions of NE in the PVN exceeded 70% [53]. Curiously, however, in mice 6-OHDA depleted whole brain NE by 96% or more, but had little effect on the plasma corticosterone response to i.p. IL-1β although there were small but statistically significant reductions in two of the six replicate experiments [54].

Studies using microdialysis for NE while monitoring HPA axis activation by measuring plasma ACTH and corticosterone have also indicated very close correlations between ACTH and corticosterone [22,51,55]. Subdiaphragmatic vagotomy which lesions the vagal afferents that project to the nucleus tractus solitarius also inhibited the decrease in hypothalamic NE observed in response to IL-1 [21]. Others have shown that a subdiaphragmatic vagotomy substantially reduced the ACTH and corticosterone response to IL-1 [56,57]. Interestingly, in studies in which the release of hypothalamic NE was measured by microdialysis while simultaneously monitoring HPA activation by measuring ACTH and corticosterone in plasma drawn from i.v. catheters, we found that subdiaphragmatic vagotomy in rats prevented the IL-1-induced increase in microdialysate NE from the medial hypothalamus but only reduced the increases in plasma ACTH and corticosterone [22]. Thus the noradrenergic activation does not appear to be essential for the HPA activation. The results also suggest that IL-1 can activate the HPA axis by at least two mechanisms: a vagal route involving activation of hypothalamic NE [22], and an independent route involving COX [51], possibly in the cerebral vasculature [58]. Subdiaphragmatic vagotomy of mice also decreased the noradrenergic and plasma ACTH and corticosterone responses to IL-1 and LPS, although the effects were markedly smaller than we observed in rats [59]. Interestingly, the combination of subdiaphragmatic vagotomy and indomethacin treatment completely inhibited the responses in plasma ACTH and corticosterone in both rats and mice (Abstract from the Society for Neuroscience as reference [59A]).

#### *4.2.5. The involvement of COX*

It has long been known that COX enzymes are involved in the IL-1-induced activation of the HPA axis. Several studies have indicated that various COX inhibitors inhibited the elevation of plasma ACTH and corticosterone following IL-1 administration. However, there was very little inhibition by indomethacin of the elevation of plasma corticosterone when IL-1 was injected into mice intraperitoneally [50]. Also, the early phase of the response to i.p. IL-1 was inhibited by COX inhibitors, whereas the later phase was not [50]. This provided good evidence that the effect of IL-1 depended upon its route of injection, and indicated that more than one mechanism is involved in the HPA responses to i.p. IL-1.

### Interleukin-6

Peripheral administration of IL-6 activates the HPA axis, but maximal activations are not achieved and the doses of IL-6 required are substantially higher than those of IL-1 [60,61]. Nevertheless, IL-6 appears to be able to assume CRF-like activity (stimulating pituitary ACTH secretion) in CRF-knockout mice challenged with turpentine [62] or infected with murine cytomegalovirus [61]. It may also assume ACTH-like activity, inducing glucocorticoid release from the adrenals [61,62], although some ACTH is necessary to prime the adrenal glands.

### Interleukin-10

IL-10 is a cytokine associated with inflammation. Its principal function seems to be limiting the magnitude of immune responses; it is also known as the cytokine synthesis inhibitory factor. IL-10 has been found in the anterior pituitary and in the hypothalamus, and it has been shown that IL-10 can stimulate CRF production in the hypothalamus, and that of ACTH in the pituitary [63]. Curiously, IL-10 inhibits 3-β-hydroxysteroid dehydrogenase, an enzyme critical for glucocorticoid synthesis in the adrenal gland, and IL-10 receptors have been found in the adrenal fasciculata, the region of the adrenal gland largely responsible for glucocorticoid synthesis. This may explain why IL-10-knockout mice exhibit abnormally high basal concentrations of corticosterone and an enhanced response to stress [63].

### Tumor necrosis factor-α

TNF-α also stimulates the HPA axis [64,65]; however, in common with IL-6, it is far less potent than IL-1 and the response is less prolonged. This effect of TNF-α is largely dependent on CRF [37,66]. COX appears to be involved because the HPA activation is inhibited by indomethacin [67].

### Interferons

Type 1 interferons (IFN-α and IFN-β) have frequently been shown to activate the HPA axis in humans (e.g., Ref. [68]). However, IFN administration to rats and mice failed to activate the HPA axis in our studies, and one study reported inhibition (see Ref. [11]).

### Other cytokines

There is also evidence that IL-11, IL-12, and leukemia inhibitory factor (LIF), granulocyte-macrophage colony-stimulating factor (GM-CSF), and stem cell factor (SCF) can also modulate the activity of the HPA axis (see chapter by Besedovsky and del Rey in this volume).

### The role of CRF, ACTH, and endorphins in the immune system

In an early study, Blalock [69] reported the presence of pro-opiomelanocortin (POMC, the precursor to ACTH and β-endorphin) in lymphocytes and macrophages. Subsequently,

Smith et al. [70] showed that lymphocytes can bear receptors for ACTH. However, the presence of functional POMC in lymphocytes has been questioned [71] (see review by Sharp and Linner [72]).

ACTH can apparently stimulate cyclic AMP production in lymphocytes by increasing the influx of calcium ions [73]. The biological significance of this effect of ACTH is not known.

There is evidence that CRF can be synthesized in certain cells of the immune system, and that immune cells bear receptors than can respond to CRF [74,75]. The closely related peptide, urocortin, has also been identified in lymphocytes [76], but these authors found no evidence for the expression of CRF. CRF is also involved in inflammation induced by subcutaneous administration of carrageenin [77].

It is well established that opiate receptors are present on lymphocytes. All three major opiate receptors have been identified: μ-, κ-, and δ-receptors [78]. Their function is not clear, but they could provide a mechanism for regulation of immune function.

Endogenous opiates (primarily β-endorphin) produced by immune cells have been implicated in the antinociception associated with inflammation. The production of these opiates is thought to provide some antinociceptive effects on the pain experienced during tissue inflammation. CRF appears to be involved in this response which involves CRF receptors on immune cells [74,79]. Interestingly, IL-1 may also be involved because it can stimulate β-endorphin production and induce antinociception [74]. Both molecules can trigger the release of endorphins from immune cells.

## 5. CONCLUSIONS

The activation of the HPA axis by IL-1 is consistent with the revisionist hypothesis of Munck and Guyre that a major function of the elevation of circulating glucocorticoids is to limit the body's reactions to stress. Munck and Guyre [80] noted the paradox that whereas the stress-induced elevation in circulating glucocorticoids was initially thought to enhance the organism's resistance to stress (e.g., by increasing available glucose), moderate-to-high circulating concentrations of glucocorticoids tended to suppress many of the defense mechanisms. They proposed that the "stress-induced increases in glucocorticoid levels protect not against the source of stress itself but rather against the body's normal reactions to stress, preventing those reactions from overshooting and themselves threatening homeostasis." This concept was invoked because of the anti-inflammatory properties of glucocorticoids which could not be reconciled with a defensive role. Thus, according to the Munck and Guyre hypothesis, the elevation of glucocorticoids associated with immune activation served to restrain the immune response, and thus limit the extent of the immune response to prevent inappropriate tissue damage. Nevertheless, glucocorticoids do facilitate the acute responses in stress, for example, enabling an increased availability of glucose, fuel for fighting or fleeing. In this context, Munck and Guyre noted that the suppressive effects of glucocorticoids tended to be delayed, and thus did not impair the acute responses in stress. Thus corticosteroids may affect the immune system in different ways at different times during stress.

Stress is not necessarily immunosuppressive. Acute stress can actually enhance certain immune responses [81], although immunosuppression is commonly observed with chronic stress [82]. Moreover, stress-related immunosuppression is not exclusively mediated by glucocorticoids. It can occur in adrenalectomized [83,84] and hypophysectomized animals [84,85]. There is good evidence for the involvement of the autonomic nervous system, especially the catecholamines [86], and of opioid peptides (see, e.g., Moynihan [82]).

In summary, the major effects of the nervous system on the immune system are exerted by glucocorticoids and other components of the HPA axis, apparently to prevent a long-term hyperactivity of the immune system. The immune system in turn activates the HPA axis via several different cytokines, the most important of which appears to be IL-1 although IL-6 can assume this role. Catecholamines are also involved in modulating immune function. The balance between the nervous and the immune systems forms a complex system that reacts effectively to environmental challenges and thus preserves homeostasis.

## REFERENCES

1. Blalock JE. The immune system as a sensory organ. J Immunol 1984;132:1067–70.
2. Blalock JE, Smith EM. Conceptual development of the immune system as a sixth sense. Brain Behav Immun 2007;21:23–33.
3. Claman HN. Corticosteroids and lymphoid cells. N Eng J Med 1972;287:388–97.
4. De Bosscher K, Vanden Berghe W, Haegeman G. Mechanisms of anti-inflammatory action and of immunosuppression by glucocorticoids: negative interference of activated glucocorticoid receptor with transcription factors. J Neuroimmunol 2000;109:16–22.
5. Auphan N, DiDonato JA, Rosette C, Helmberg A, Karin M. Immunosuppression by glucocorticoids: inhibition of NF-kappa B activity through induction of I kappa B synthesis. Science 1995;270:286–90.
6. Besedovsky HO, del Rey A. Physiology of psychoneuroimmunology: a personal view. Brain Behav Immun 2007;21:34–44.
7. Besedovsky HO, del Rey A, Sorkin E, Dinarello CA. Immunoregulatory feedback between interleukin-1 and glucocorticoid hormones. Science 1986;233:652–4.
8. Besedovsky HO, del Rey A. Immune-neuro-endocrine interactions: facts and hypotheses. Endocr Rev 1996;17:64–102.
9. Turnbull AV, Rivier C. Regulation of the hypothalamic-pituitary-adrenal axis by cytokines: actions and mechanisms of action. Physiol Rev 1999;79:1–71.
10. Dunn AJ. Cytokine activation of the HPA axis. Ann NY Acad Sci 2000;917:608–17.
11. Dunn AJ. Cytokine activation of the hypothalamo-pituitary-adrenal axis. In Handbook of Stress and the Brain Part 2: Stress: Integrative and Clincial Aspects. Steckler T, Kalin N, Reul JMHM, Eds; Amsterdam: Elsevier, 2005; pp. 157–74.
12. Ericsson A, Ek M, Wahlstrom I, Kovacs K, Liu C, Hart R et al. Pathways and mechanisms for interleukin-1 mediated regulation of the hypothalamo-pituitary-adrenal axis. In Stress: Molecular Genetic and Neurobiological Advances. McCarty R, Aguilera G, Sabban E, Kvetnansky R, Eds; New York: Gordon and Breach Science Publishers, 1996; pp. 101–20.
13. Dunn AJ. Mechanisms by which cytokines signal the brain. In Neurobiology of the Immune System, International Review of Neurobiology. Clow A, Hucklebridge F, Eds; San Diego: Academic Press, 2002; pp. 43–65.
14. Katsuura G, Arimura A, Koves K,Gottschall PE. Involvement of organum vasculosum of lamina terminalis and preoptic area in interleukin 1β-induced ACTH release. Am J Physiol 1990;258: E163–71.
15. Blatteis CM, Sehic E. Fever: how may circulating pyrogens signal the brain. News Physiol Sci 1997;12:1–9.
16. Goehler LE, Erisir A, Gaykema RPA. Neural-immune interface in the rat area postrema. Neuroscience 2006;140:1415–34.
17. Banks WA, Kastin AJ, Broadwell RD. Passage of cytokines across the blood-brain barrier. Neuroimmunomodulation 1995;2:241–8.
18. Banks WA, Farr SA, La Scola ME, Morley JE. Intravenous human interleukin-1α impairs memory processing in mice: dependence on blood-brain barrier transport into posterior division of the septum. J Pharmacol Exp Ther 2001;299:1–6.

19. Watkins LR, Maier SF, Goehler LE. Cytokine-to-brain communication: a review & analysis of alternative mechanisms. Life Sci 1995;57:1011–026.
20. Goehler LE, Gaykema RPA, Hansen MK, Anderson K, Maier SF, Watkins LR. Vagal immune-to-brain communication: a visceral chemosensory pathway. Auton Neurosci 2000;85:49–59.
21. Fleshner M, Goehler LE, Hermann J, Relton JK, Maier SF, Watkins LR. Interleukin-1β induced corticosterone elevation and hypothalamic NE depletion is vagally mediated. Brain Res Bull 1995;37(6):605–10.
22. Wieczorek M, Dunn AJ. Effect of subdiaphragmatic vagotomy on the noradrenergic and HPA axis activation induced by intraperitoneal interleukin-1 administration in rats. Brain Res 2006;1101:73–84.
23. Fleshner M, Goehler LE, Schwartz BA, McGorry M, Martin D, Maier SF et al. Thermogenic and corticosterone responses to intravenous cytokines (IL-1β and TNF-α) are attenuated by subdiaphragmatic vagotomy. J Neuroimmunol 1998;86:134–41.
24. Quan N, Whiteside M, Herkenham M. Cyclooxygenase 2 mRNA expression in rat brain after peripheral injection of lipopolysaccharide. Brain Res 1998;802:189–97.
25. Cao C, Matsumura K, Yamagata K, Watanabe Y. Involvement of cyclooxygenase-2 in LPS-induced fever and regulation of its mRNA by LPS in the rat brain. Am J Physiol 1997;282:R1712–25.
26. Rivest S. How circulating cytokines trigger the neural circuits that control the hypothalamic-pituitary-adrenal axis. Psychoneuroendocrinology 2001;26:761–88.
27. Quan N, Stern EL, Whiteside MB, Herkenham M. Induction of pro-inflammatory cytokine mRNAs in the brain after peripheral injection of subseptic doses of lipopolysaccharide in the rat. J Neuroimmunol 1999;93:72–80.
28. Roh MS, Drazenovich KA, Barbose JJ, Dinarello CA, Cobb CF. Direct stimulation of the adrenal cortex by interleukin. Surgery 1987;1:140–46.
29. Gwosdow AR, Kumar MSA, Bode HH. Interleukin 1 stimulation of the hypothalamic-pituitary-adrenal axis. Am J Physiol 1990;258:E65–70.
30. Cambronero JC, Rivas FJ, Borrell J, Guaza C. Is the adrenal cortex a putative site for the action of interleukin-1. Horm Metab Res 1992;24:48–9.
31. Dunn AJ. Role of cytokines in infection-induced stress. Ann N Y Acad Sci 1993;697:189–202.
32. Berkenbosch F, van Oers J, del Rey A, Tilders F, Besedovsky H. Corticotropin-releasing factor-producing neurons in the rat activated by interleukin-1. Science 1987;238:524–6.
33. Sapolsky R, Rivier C, Yamamoto G, Plotsky P, Vale W. Interleukin-1 stimulates the secretion of hypothalamic corticotropin-releasing factor. Science 1987;238:522–4.
34. Uehara A, Gottschall PE, Dahl RR, Arimura A. Interleukin-1 stimulates ACTH release by an indirect action which requires endogenous corticotropin releasing factor. Endocrinology 1987;121:1580–2.
35. Dunn AJ, Swiergiel AH. Behavioral responses to stress are intact in CRF-deficient mice. Brain Res 1999;845:14–20.
36. Rivest S, Rivier C. Influence of the paraventricular nucleus of the hypothalamus in the alteration of neuroendocrine functions induced by intermittent footshock or interleukin. Endocrinology 1991;129:2049–57.
37. Kovács KJ, Elenkov IJ. Differential dependence of ACTH secretion induced by various cytokines on the integrity of the paraventricular nucleus. J Neuroendocrinol 1995;7:15–23.
38. Ovadia H, Abramsky O, Barak V, Conforti N, Saphier D, Weidenfeld J. Effect of interleukin-1 on adrenocortical activity in intact and hypothalamic deafferentated male rats. Exp Brain Res 1989;76:246–9.
39. Saphier D. Neurophysiological and endocrine consequences of immune activity. Psychoneuroendocrinology 1989;14:63–87.
40. Besedovsky HO, del Rey A. Physiological implications of the immune-neuro-endocrine network. In. Psychoneuroimmunology. Ader R, Felten D, Cohen N, Eds; New York: Academic Press, 1991; pp. 589–608.
41. Tsagarakis S, Gillies G, Rees LH, Besser M, Grossman A. Interleukin-1 directly stimulates the release of corticotrophin releasing factor from rat hypothalamus. Neuroendocrinology 1989;49:98–101.

42. Levine S, Berkenbosch F, Suchecki D, Tilders FJH. Pituitary-adrenal and interleukin-6 responses to recombinant interleukin-1 in neonatal rats. Psychoneuroendocrinology 1994;19:143–54.
43. del Rey A, Furukawa H, Monge-Arditi G, Kabiersch A, Voigt K-H, Besedovsky HO. Alterations in the pituitary-adrenal axis of adult mice following neonatal exposure to interleukin-1. Brain Behav Immun 1996;10(3):235–48.
44. Dunn AJ. Systemic interleukin-1 administration stimulates hypothalamic norepinephrine metabolism parallelling the increased plasma corticosterone. Life Sci 1988;43:429–35.
45. Kabiersch A, del Rey A, Honegger CG, Besedovsky HO. Interleukin-1 induces changes in norepinephrine metabolism in the rat brain. Brain Behav Immun 1988;2:267–74.
46. Schiltz JC, Sawchenko PE. Signaling the brain in systemic inflammation: The role of perivascular cells. Front Biosci 2003;8:s1321–9.
47. Dunn AJ, Wang J-P, Ando T. Effects of cytokines on cerebral neurotransmission: Comparison with the effects of stress. Adv Exp Med Biol 1999;461:117–27.
48. Al-Damluji S. Adrenergic mechanisms in the control of corticotrophin secretion. J Endocrinol 1988; 119:5–14.
49. Plotsky PM, Cunningham ET, Widmaier EP. Catecholaminergic modulation of corticotropin-releasing factor and adrenocorticotropin secretion. Endocr Rev 1989;10:437–58.
50. Dunn AJ, Chuluyan H. The role of cyclo-oxygenase and lipoxygenase in the interleukin-1-induced activation of the HPA axis: dependence on the route of injection. Life Sci 1992;51:219–25.
51. Wieczorek M, Dunn AJ. Relationships among the behavioral, noradrenergic and pituitary-adrenal responses to interleukin-1 and the effects of indomethacin. Brain Behav Immun 2006;20:477–87.
52. Rivier C. Influence of immune signals on the hypothalamic-pituitary axis of the rodent. Frontiers Neuroendocrinol 1995;16:151–82.
53. Chuluyan H, Saphier D, Rohn WM, Dunn AJ. Noradrenergic innervation of the hypothalamus participates in the adrenocortical responses to interleukin-1. Neuroendocrinology 1992;56:106–11.
54. Swiergiel AH, Dunn AJ, Stone EA. The role of cerebral noradrenergic systems in the Fos response to interleukin-1. Brain Res Bull 1996;41:61–4.
55. Smagin GN, Swiergiel AH, Dunn AJ. Peripheral administration of interleukin-1 increases extracellular concentrations of norepinephrine in rat hypothalamus: comparison with plasma corticosterone. Psychoneuroendocrinology 1996;21:83–93.
56. Gaykema RPA, Dijkstra I, Tilders FJH. Subdiaphragmatic vagotomy suppresses endotoxin-induced activation of hypothalamic corticotropin-releasing hormone neurons and ACTH secretion. Endocrinology 1995;136(10):4717–20.
57. Kapcala LP, He JR, Gao Y, Pieper JO, DeTolla LJ. Subdiaphragmatic vagotomy inhibits intra-abdominal interleukin-1β stimulation of adrenocorticotropin secretion. Brain Res 1996;728:247–54.
58. Dunn AJ, Swiergiel AH, Zhang H, Quan N. The hypophagia Induced by interleukin-1 and lipopolysaccharide is associated with induction of cyclooxygenase-2 in brain endothelia. Neuroimmunomodulation 2006;13:96–104.
59. Wieczorek M, Pournajafi-Nazarloo H, Swiergiel AH, Dunn AJ. Physiological and behavioral responses to interleukin-1β and LPS in vagotomized mice. Physiol Behav 2005;84:500–11.

59A. Dunn AJ, Wieczorek M, Swiergiel AH. The vagus nerve and cyclooxygenase both participate in interleukin-1-induced activation of brain noradrenergic neurons and the HPA axis in mice and rats. Soc Neurosci Abs 2007;733:8.

60. Wang JP, Dunn AJ. Mouse interleukin-6 stimulates the HPA axis and increases brain tryptophan and serotonin metabolism. Neurochem Int 1998;33:143–54.
61. Silverman MN, Miller AH, Biron CA, Pearce BD. Characterization of an interleukin-6 and adrenocorticotropin dependent, immune-to-adrenal pathway during viral infection. Endocrinology 2004;145:3580–9.
62. Venihaki M, Dikkes P, Carrigan A, Karalis KP. Corticotropin-releasing hormone regulates IL-6 expression during inflammation. J Clin Invest 2001;108:1159–66.
63. Smith EM, Tyu H, Hughes TK. Interleukin-10 and the hypothalamic-pituitary-adrenal axis. In. Cytokines, Stress and Immunity, 2nd Edn. Plotnikoff N, Faith R E, Murgo A J, Good R A, Eds. Boca Raton, FL: CRC Press 2007:169–92.

64. Sharp BM, Matta SG, Peterson PK, Newton R, Chao C, McAllen K. Tumor necrosis factor-α is a potent ACTH secretagogue: comparison to interleukin-1β. Endocrinology 1989;124:3131–3.
65. Ando T, Dunn AJ. Mouse tumor necrosis factor-α increases brain tryptophan concentrations and norepinephrine metabolism while activating the HPA axis in mice. Neuroimmunomodulation 1999;6:319–29.
66. Bernardini B, Kamilaris TC, Calogero AE, Johnson EO, Gomez MT, Gold PW et al. Interactions between tumor necrosis factor-α, hypothalamic corticotropin-releasing hormone, and adrenocorticotropin secretion in the rat. Endocrinology 1990;126:2876–81.
67. Sharp BM, Matta SG. Prostaglandins mediate the adrenocorticotropin response to tumor necrosis factor in rats. Endocrinology 1993;132:269–74.
68. Cassidy EM, Manning D, Byrne S, Bolger E, Murray F, Sharifi N et al. Acute effects of low-dose interferon-alpha on serum cortisol and plasma interleukin-6. J Psychopharmacol 2002;16:230–4.
69. Blalock JE, Harbour-McMenamin D, Smith EM. Peptide hormones shared by the neuroendocrine and immunologic systems. J Immunol 1985;135:858–61s.
70. Smith EM, Brosnan P, Meyer WJ, Blalock JE. An ACTH receptor on human mononuclear leukocytes. N Engl J Med 1987;317:1266–9.
71. Van Woudenberg AD, Metzelaar MJ, Van Der Kleij AAM, De Wied D, Burbach JPH, Wiegant VM. Analysis of proopiomelanocortin (POMC) messenger ribonucleic acid and POMC-derived peptides in human peripheral blood mononuclear cells: no evidence for a lymphocyte-derived POMC system. Endocrinology 1993;133(5):1922–33.
72. Sharp B, Linner K. What do we know about the expression of proopiomelanocortin transcripts and related peptides in lymphoid tissue? Endocrinology 1993;133(5):1921–2.
73. Clarke BL. Calcium uptake by ACTH-stimulated lymphocytes: what is the physiological significance. Adv Neuroimmunol 1995;5:275–81.
74. Schäfer M, Carter L, Stein C. Interleukin 1β and corticotropin-releasing factor inhibit pain by releasing opioids from immune cells in inflamed tissue. Proc Natl Acad Sci USA 1994;91:4219–23.
75. Tsagarakis S, Grossman A. Corticotropin-releasing hormone: interactions with the immune system. Neuroimmunomodulation 1994;1:329–34.
76. Perrin MH, Vale WW. Corticotropin releasing factor receptors and their ligand family. Ann NY Acad Sci 1999;885:312–28.
77. Karalis K, Sano H, Redwine J, Listwak S, Wilder RL, Chrousos GP. Autocrine or paracrine inflammatory actions of corticotropin-releasing hormone in vivo. Science 1991;254:421–3.
78. Sharp BM. Multiple opiate receptors on immune cells modulate intracellular signaling. Brain Behav Immun 2006;20:9–14.
79. Schafer M, Mousa SA, Stein C. Corticotropin-releasing factor in antinociception and inflammation. Eur J Pharmacol 1997;323:1–10.
80. Munck A, Guyre P. Glucocorticoid physiology, pharmacology and stress. Adv Exp Med Biol 1986;196:81–96.
81. Dhabhar FS, McEwen BS. Acute stress enhances while chronic stress suppresses cell-mediated immunity in vivo: a potential role for leukocyte trafficking. Brain Behav Immun 1997;11:286–306.
82. Moynihan JA. Mechanisms of stress-induced modulation of immunity. Brain Behav Immun 2003;17:S11–16.
83. Keller SE, Weiss JM, Schleifer SJ, Miller NE, Stein M. Stress-induced suppression of immunity in adrenalectomized rats. Science 1983;221:1301–4.
84. Rabin BS, Cunnick JE, Lysle DT. Stress-induced alteration of immune function. Prog NeuroEndocrinImmunol 1990;3:116–24.
85. Keller SE, Schleifer SJ, Liotta AS, Bond RN, Farhoody N, Stein M. Stress-induced alterations of immunity in hypophysectomized rats. Proc Natl Acad Sci USA 1988;85:9297–301.
86. Moynihan J, Kruszewska B, Madden K, Callahan T. Sympathetic nervous system regulation of immunity. J Neuroimmunol 2004;147:87–90.

# *Physiology of the Hypothalamic–Pituitary–Adrenocortical Axis*

HELMUT VEDDER

*Department of Psychiatry and Psychotherapy, Philipps-University of Marburg, Marburg, Germany*

## ABSTRACT

The hypothalamic–pituitary–adrenocortical (HPA) axis represents one of the adaptational systems of the body with the function of adjusting the organism to challenges of homeostasis, the so-called stress response. Besides circadian and ultradian fluctuations in the activity of the HPA axis, also stressors can induce activation of this system. These stressors may include physical external or internal threats such as immune activation, pain, and exposure to heat and cold. The most potent stressors are, however, those psychological situations, either real or imagined, in which the HPA axis is activated, which subsequently may facilitate adaptation to the stressor. The HPA axis consists of both static (cell and tissues) and dynamic secretable (hormones and other modulators) elements which are components of the stress reaction. They react in a uniform manner under conditions of activation, although with marked quantitative differences. Following an increased secretion of corticotropin-releasing hormone (CRH) and arginine vasopressin (AVP) within the brain, the activation of the pituitary gland leads to the secretion of adrenocorticotropic hormone (ACTH), which in turn induces the release of corticosteroid hormones from the adrenal cortex. Multiple mediators also affect the HPA axis at all levels of biological organization, leading to distinct reactions of the key elements, to adequately respond to the type of stressor. The overall biological and psychological response of the organism under conditions of stress was termed "general adaptation syndrome." If these responses persist or are inadequate, physiological immune, metabolic, and cardiovascular functions may be compromised. As a consequence, the individual may become more vulnerable to stress-related somatic and also neuropsychiatric disorders. Moreover, pathological reactions within the HPA system such as neuroendocrine disturbances and regulatory dysfunctions with a disinhibition and dysregulation of HPA axis functions may lead to inherent disease symptoms within the HPA system.

## ABBREVIATIONS

| | |
|---|---|
| ACTH | adrenocorticotropic hormone |
| AVP | arginine vasopressin |
| BNST | bed nucleus of the stria terminalis |
| CNS | central nervous system |
| CRF | corticotropin-releasing factor |
| CRH | corticotropin-releasing hormone |
| GR | glucocorticoid receptor |

| | |
|---|---|
| HPA | hypothalamic–pituitary–adrenocortical axis |
| IL-1β | Interleukin-1β |
| IL-6 | Interleukin-6 |
| LPS | lipopolysaccharide |
| MR | mineralocorticoid receptor |
| NE | norepinephrine |
| PVN | paraventricular nucleus |
| SCN | suprachiasmatic nucleus |

## 1. INTRODUCTION

### 1.1. The HPA axis represents one of the mediators of the stress reaction

Living organisms are continuously in dynamic contact with their environment. Because biological organisms are dynamic systems with regard to their basic structures and functions, they have the ability to alter their metabolism and behavior in order to react and to adapt to a changing environment – although within limits. In response to external and internal stimuli, termed "stressors," the organism is (1) able to react and to adapt to a given stimulus via an alteration of inherent biological and/or behavioral responses; (2) able to evade the stimulus via a flight reaction, seek confrontation by fighting or use freezing as a survival strategy [1,2]; or (3) unable to survive the stimulus and die. The latter condition denotes the failure of the homeostatic mechanisms, whereas the former conditions induce altered patterns of homeostasis in the organism.

In more complex biological systems, including man, a multiorgan system reacts and adapts to a variety of stimuli/stressors in order to maintain homeostasis. Every cell and tissue is functionally flexible with regard to its functions for the final goal of surviving and adapting to gain a better use of the environmental resources for the organism. This process continuously follows the rules of evolution. Multiorgan systems have developed coordinating systems to orchestrate the homeostatic reactions of the diverse tissues when a stress response is initiated. This process includes the hypothalamic–pituitary–adrenocortical (HPA) system, the so-called stress axis, and the accompanying reactions of the sympatho-adrenomedullary system and other tissues and body systems as the main players in this reaction [3].

Hans Selye defined "stress" as the "unspecific response of the body to every demand" and uncovered the important function of the corticosteroid hormones and the coordinated set of reactions during the stress response. He called this reaction the "general adaptation syndrome" [2,4]. In contrast to the "general adaptation syndrome," a more localized stress reaction has been termed "local adaptation syndrome." The type of stress that evokes a successful or even beneficial adaptation has been described as "eustress," whereas stress resulting in an unsuccessful adaptation reaction was termed "disstress" by Hans Selye [2] (Fig. 1).

The basic concepts of this syndromic reaction are still valid, although they have been enriched by a large amount of additional data during the last century [2,4–6]. These data show that every player within the HPA axis is involved in multiple aspects and reactions to optimally adapt the organism to its physical, biological, and psychosocial environment. Moreover, the HPA axis and its components not only facilitate adaptation of the organism as a whole to the challenges of daily life, but also specifically affect single responses of the body, with regard to either local reactions in separate tissues and organs, metabolic pathways in cells and tissues, or more complex systems such as immune reactions and complex brain functions such as cognition, emotion, and behavior.

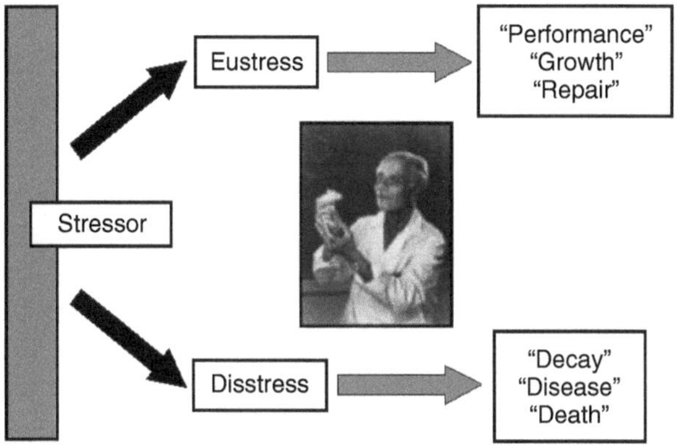


Figure 1. Reaction pathways to a stressor, leading either to "eustress" or to "disstress" of the biological unit and denoting the consequences. (The inset shows Hans Selye holding a rat, who described the different types of stress responses in tissues and organisms.)

## 2. STATIC AND DYNAMIC COMPONENTS OF THE HPA AXIS

On a simplified structural basis, the HPA axis ("stress axis") is formed by static and dynamic components ("static" denotes a localized position in the body on a complex cellular basis in contrast to "dynamic" elements, which are soluble compounds acting at multiple sites of the organism). Static components consist of different cellular body systems and tissues and are in constant interaction with dynamic elements like hormones and other dynamic components inside and outside the HPA axis (Fig. 2).

The structural description of only a few key elements in the central nervous system (CNS) affecting the HPA axis represents a very limited approach to the complex functions of the "information collecting and integrating" brain. For example, the gathering and processing of information, including the function of association, constitutes one of the most refined senses of higher organisms, reaching levels beyond the physical environment. Different perceptions from the past and the present are integrated by the organism via mnemonic and cognitive processes to a model that finally influences the coordinated action of the HPA system under normal and pathological conditions. Quantitatively, it may be assumed that low-level stressors only affect integrated

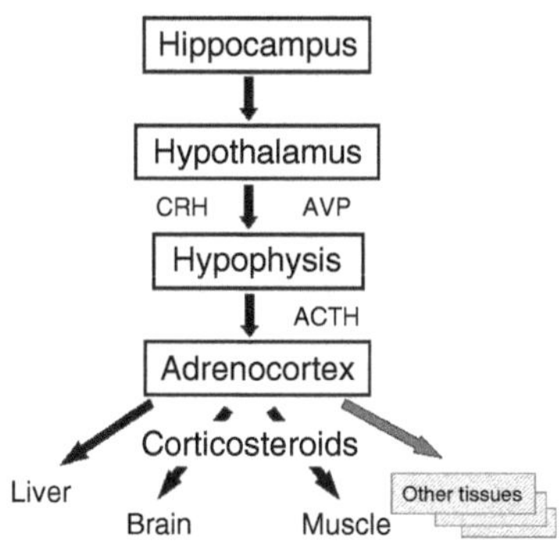


Figure 2. Components of the hypothalamic–pituitary–adrenocortical axis. Boxes denote static components, text within the arrows shows soluble molecules/dynamic compounds.

circadian functions of the HPA without pronounced alterations in the activity of the HPA. If the "stress level" due to internal or external factors rises, a more powerful and distinct reaction of the HPA axis is induced. This may represent the result of a cognitive or brain-related stimulus, but may also occur as the result of an immune reaction or a disturbance in other body functions including malfunctions of the HPA system itself.

Up to now, the hippocampal formation is regarded as the first and most complex integrating part of the body influencing the activation of more downstream components of the axis [7,8]. Other brain areas that contribute to this process include the amygdala, the brainstem, the prefrontal cortex, and the bed nucleus of the stria terminalis (BNST) [9–14]. Following activation of the hippocampal formation, a specific reaction is evoked in distinct locations of the hypothalamus. One of the key players in this reaction is the corticotropin-releasing hormone (CRH; or corticotropin-releasing factor, CRF) system with its dynamic players CRH, the urocortins I, II, and III [15,16], and the corresponding receptor proteins. Interestingly, these substances are not only mediators of the neuroendocrine system, but are also involved in arousal, emotionality, aversive processes, anxiety, and also anxiolysis [15,16] leading to an integrative modulation of psychophysical responses of the organism.

Under normal conditions, the hippocampal formation exerts a blocking/inhibiting action on the HPA system via the hypothalamic centers of the axis [8,9,17,18]. If this functional inhibition of the hippocampal formation disappears, a consecutive *activation* of the HPA axis occurs, leading to a regulatory dysfunction with increased secretion of the different dynamic elements, including corticosteroid hormones, and to enhanced and often also continuous effects of these players in the organism, with the possibility of detrimental effects.

At the hypothalamic level, the medial parvocellular and magnocellular divisions of the lateral paraventricular nucleus (PVN) of the hypothalamus are critically involved in the activity of the HPA system. Neurons of this nucleus synthesize CRH and arginine vasopressin (AVP) and project to the median eminence [10,19]. The activity of the PVN can be directly inhibited by gamma amino butyric acid (GABA)-ergic neurons of the bed nucleus of the stria terminalis (BNST), the preoptic area, and the hypothalamus. In contrast, glutamate is able to activate the neuroendocrine cells via hypothalamic and brainstem projections to the PVN [10]. Additionally, these neurons interact with norepinephrine (NE)-containing neurons in the brainstem, resulting in reciprocal interactions between the central NE and CRH systems, probably even in the manner of a "feed-forward" loop [20]. Furthermore, serotonin and acetylcholine are also involved in the regulation of the HPA axis on different levels including the hypothalamus [9,13] and also with direct and indirect effects such as the different types of actions of serotonin on corticosteroid receptors.

In addition to its neuroendocrine functions, AVP is also involved in specific social behavior functions [21] and other functions such as the water balance of the body [22], suggesting pleiotropic effects of this peptide similar to CRH and the urocortins.

Moreover, CRH and AVP are not only acting as neurotransmitters in a direct synaptic action in the hypothalamus, but are also secreted in a circadian and pulsatile manner into a specific part of the blood system, the so-called hypophyseal portal circulatory system [23]. Thus, these peptides are secreted into blood vessels, outside the blood–brain barrier, and gain access to the pituitary and the periphery via this pathway. In the pituitary, they mainly act on a specific cell population, the corticotrophic cells of the anterior pituitary. These cells release adrenocorticotropic hormone (ACTH) in a highly complex process and specific pulsatile manner [24] and represent the anatomical end-point of the brain-related part of the HPA axis. Functionally, all central – and also a number of peripheral – influences, such as immune factors and the feedback actions of the corticosteroids, are integrated by corticotrophic cells, rendering them

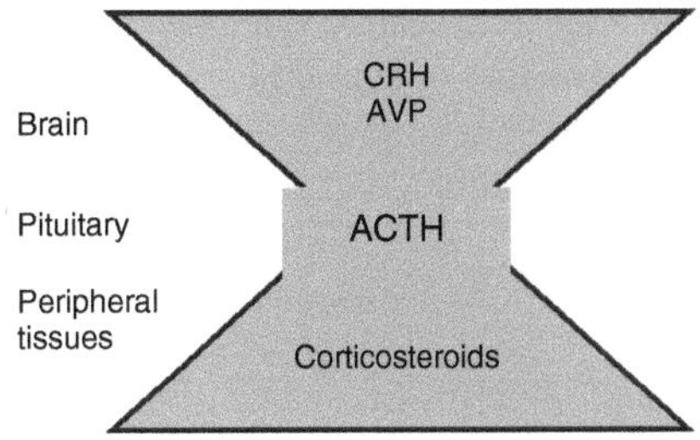


Figure 3. The pituitary, "bottleneck" of the hypothalamic–pituitary–adrenocortical axis, and the central role of adrenocorticotrophic hormone (ACTH) as the mediator between the brain/CRH and AVP and the periphery/the corticosteroid hormones.

the "functional bottle-neck" of this important pathway from the CNS to the adrenocortex and the tissues of the body [25] (Fig. 3).

The HPA "pathway" from the hippocampus to the hypothalamus and to the pituitary is *not* a simple structural and uniform line of functioning in one direction, but consists of a variety of elements feeding into the "hormonal effector system of the brain," the pituitary gland. Numerous other neuropeptides, immune mediators, cellular components, and also the biochemical and genetic setting within the factor-producing cells influence the secretion of the dynamic key components of the HPA system, such as CRH, ACTH, and AVP. Only during the last decade, an important function of immune mediators such as interleukin-1 (IL-1), interleukin-6 (IL-6), and tumor necrosis factor-α has emerged. These factors affect the HPA axis at both cellular and systemic levels [26–28]. This leads to a functional connection between the main mediators of both systems, the corticosteroids and the cytokines, termed the "cytokine-HPA axis feedback circuit" [29].

Thus, the corticotroph cells of the pituitary are one of the most highly integrating parts of the axis, powerfully affecting the activity of the adrenocortex, the next structural downstream element of the system. The adrenal cortex is a part of the peripheral body and is also affected by numerous other factors, including cytokines and other hormones. Thus, the release of the dynamic components cortisol or corticosterone (and of other steroids and compounds) from the adrenal and adrenocortical cells is not only regulated by ACTH. For example, in humans, IL-6 and other immune factors affect and regulate the release of cortisol from adrenocortical cells [30–32]. These data again underscore the pleiotropic nature of the inputs into the HPA axis, even at this most downstream localized level of the system.

Corticosteroid hormones are able to affect nearly every tissue and cell of the body. Remarkably, the blood–brain barrier does not represent a major obstacle for these lipophilic hormones, which are even efficiently concentrated and taken up by highly specific binding sites, the mineralocorticoid receptors (MRs) [33,34] in the hippocampus, the "first structural key element of the HPA axis" (see above). Additionally, binding sites with lower affinity for cortisol/corticosterone, the glucocorticoid receptors (GRs), are found in nearly all brain areas including the hippocampus and also in most cells of the body [33,35]. Moreover, a variety of other interacting mechanisms act on the GR- and MR-mediated regulation of cellular and tissue events, leading to a complex network of effects of corticosteroids on body functions [35]. Thus, even at intracellular levels, these systems allow a tight adaptation of the evoked cell and tissue functions and the necessary biological, behavioral, cognitive, and psychosocial reactions of the adapting organism. This is supported by data on the behavioral effects of cortisol in humans, affecting numerous emotional and cognitive processes including attention, perception, and memory [36,37].

Overall, these actions and also the neuroendocrine feedback effects of corticosteroids on the elements of the HPA axis show that the system is not only acting unidirectionally, but is also retrogradely influenced by the evoked reactions, tightly modulated, although adapting in a more or less open circuit, and affected by numerous external and internal influences. On the other side, the dynamic key elements ACTH, AVP, CRH, and cortisol form part of a more closed feedback circuit, setting the internal framework and most likely also some set-points for the evoked reactions [35].

An even more complex view of the HPA axis has developed in the last decades: It has been shown that other biological mechanisms such as immune reactions and functions of the autonomic nervous system interact in a closely integrated manner with this system, giving rise to a most complex network, which finally integrates all reactions of the body, although with an ever-changing hierarchy of the reactions due to the varying daily and situational demands.

## 3. CIRCADIAN RHYTHMICITY OF THE HPA AXIS

Under physiological conditions, the activity of the HPA axis, as reflected by cortisol secretion, shows a circadian pattern with large interindividual differences (e.g., see Ref. [38]). A refined stochastic model on diurnal cortisol patterns indicated distinct aspects of diurnal cortisol secretion: In the early morning, before wakening, cortisol levels begin to rise from very low levels and show a distinct peak around wakening time. Levels then decrease and show another– albeit smaller– maximum after noon, dropping during the afternoon and the first part of the night to very low levels in the early morning hours [38]. Thus, the circadian pattern is related to the rest–activity cycle of the organism. This is also observed in the rat, in which peak levels are detected in the late evening, when the activity of the animal increases, and a nadir is detected during the start of the resting phase in the early morning.

The circadian pattern of cortisol secretion develops in humans within the first months of life and is initiated in close relationship with the circadian sleep–wake rhythm [39]. In the brain, CRH shows a corticosteroid-independent pattern of daily fluctuations [40], representing probably one of the initiating pathways of the circadian rhythmicity of the HPA axis. Additional data indicate that other inputs are mediated, for example, by the effect of light on the eyes via the suprachiasmatic nucleus (SCN). The increased activity in the SCN leads to an enhanced secretion of CRH, which subsequently activates the HPA axis [41]. Besides the HPA activity, numerous other body functions and hormones show circadian fluctuations [42], embedding the HPA system into a complex network of daily oscillations within the organism, even on an ultradian level. Interestingly, also the stress response differs depending on the activity of the HPA axis and, indirectly, on the time of the day. This is most likely due to a differential occupation of the corticosteroid receptors, for example the MRs and GRs. Under conditions of low circulating corticosteroids, MRs, the high-affinity-type receptors, are occupied by the ligand and subsequently affect cellular functions. MRs are predominantly localized in the hippocampal formation of the brain, and MR functions influence the basal and low activity of the HPA during the afternoon, the evening, and the first part of the night [35,43]. Under conditions of increased corticosteroid secretion and an increased activity of the HPA axis, the GRs throughout the body are occupied, leading to a different pattern of cellular activation/ inactivation and subsequently also to specifically altered tissue responses of the organism [35]. As mentioned before, the components of the HPA system may be differentially activated by other systems, such as the immune system, in relation to the circadian activity levels, resulting in a coordinated interaction between the various systems [44].

The extent of an acute stress response may be dampened during periods when the occupation/activation of cellular systems via the GRs is increased. For example, during the morning hours, a decreased response to an activation of the HPA might occur due to the already occupied GRs and consecutive ceiling effects [35,43]. During periods of lower occupation of GRs in the evening hours and during the night, a more pronounced influence of external stressors on body functions is possible.

Physiologically, this diurnal pattern corresponds to the resting phases of the organism in the afternoon and during the night, when circumstances with a low likelihood of occurring stressors are established by the organism via its self-initiated influences on the surrounding conditions and the environment (resting and sleeping conditions). This pattern corresponds to a primary and "daily-evoked mild endogenous and physiological stress reaction" from the morning hours to the early afternoon, when the body system works in an activating tone, including an activation of the HPA axis.

If an increased "load with external and internal stressors" occurs during this part of the day, a dampening of the extent of the HPA activation due to ceiling effects should be detectable, whereas an increased reactivity to external stressors should be evident in the afternoon and the night due to the low HPA activity during this time period.

This reaction pattern has been demonstrated during mild chronic stress [45] and also following a standardized immune stimulation of the HPA axis by bacterial lipopolysaccharide (LPS; endotoxin) in the morning and in the evening hours [46]. Subsequently, stressors induce increased hormonal responses at times when the internal activity of the stress axis is low [46], thus disturbing more profoundly the circadian rhythmicity of the HPA system.

At present, there are only few data on the "reserve potential" of the HPA axis relating the coordinated regulation of circadian rhythmicity and the extent of an additional activation by an increased stressor load. For example, during normal aging, the inhibiting tone of the hippocampal system may gradually diminish due to a morphological decrease in the static components, such as the neuronal cells of the hippocampal formation. This leads to an altered circadian rhythmicity, for example a flattening of the diurnal amplitude [47], and most likely to a decreased response to external and internal stressors due to an already increased tone of the HPA system [48]. This may result from structural and functional changes of the HPA axis with aging, leading to a prolongation of the cortisol secretion and an insufficient feedback regulation, most likely within the hippocampal formation [48–51]. These alterations and subsequent dysfunctions may then contribute to neuroendocrine disturbances in a number of disorders and diseases such as cardiovascular diseases, diabetes, hypertension, and major depression [52,53].

## 4. PARADIGMATIC ACTIVATION OF THE HPA AXIS

### 4.1. Psychological activation of the HPA axis

It is well known that psychological stressors may induce an activation of the HPA axis as indicated by an increased secretion of cortisol [54–56]. Several studies have shown that novelty, predictability, anticipation of negative consequences, controllability, and ego involvement are critical factors that mediate the extent of the activation of the HPA axis. These data have been replicated in animals with regard to novelty effects [57,58] and in man [59,60] as well as in other species and for other variables under different conditions. Besides these stressor-related factors, individual variables, such as a chronic type of the stressor or emotional reactions, are relevant for the HPA activation. Other studies have also introduced and connected more specific personality-related factors, such as the ability to cope with a stressful situation, with the extent of the neuroendocrine reaction, and the secretion of cortisol [61,62].

In addition to corticosteroid secretion, effects of stress on different brain systems and especially on the CRH system have been examined, suggesting that chronic psychological influences may lead to a specific activation of the CRH system itself, even in a perpetuating manner [63].

A large number of other psychological paradigms have been used to induce the activation of the HPA axis including social separation [57], psychosocial stress [64], and high and low emotional events [65]. Even more sophisticated stressors such as $CO_2$ inhalation, which subsequently induce anxiety and increase emotional pressure [66,67], have been employed.

### 4.2. Neuroendocrine activation of the HPA axis

Neuroendocrine activation of the HPA axis by external application of AVP, ACTH, and CRH or analog substances represents endocrine challenges for the diagnosis of endocrine malfunctions (e.g., see Refs [68–71]). More refined applications to test feedback mechanisms of the HPA axis use the combined application of dexamethasone, a synthetic glucocorticoid, and CRH (e.g., see Ref. [72]). Due to the relevance of stressors for the clinical incidence of affective disorders, these tests have been widely used for the detection of neuroendocrine disturbances in psychiatric disorders (e.g., see Refs [73–76]).

The time frame of ACTH and cortisol secretion is well coordinated. After stimulation with human CRH, there is a nearly instant release of both, ACTH and cortisol (e.g., see Ref. [77]), clearly supporting an almost instantaneous reaction which affects nearly all body systems.

### 4.3. Immune activation of the HPA axis

Bacterial LPS has also been used to activate the HPA axis in humans (for example and review, see Ref. [78]). Comparison of a neuroendocrine (CRH) and an immune (LPS) activation showed that the neuroendocrine response occurs about 2 h later and with a somewhat different time frame after LPS application, indicating that secondarily induced factors mediate the neuroendocrine activation and the subsequent hormonal response [77]. Further detailed studies underscored this finding showing that the complex network of reactions is also influenced by other factors, for example the effects of the cytokines interleukin-6 (IL-6) and interleukin-1β (IL-1β) at virtually all levels of the HPA axis [32,78–80].

Thus, the effects of different psychosocial stressors, of hormones such as CRH and AVP, as well as of other stressors such as an immune stimulation may induce a clearly detectable activation of the HPA axis, although with different characteristics and time kinetics.

## 5. INTERACTION OF THE HPA AXIS WITH OTHER SYSTEMS OF THE BODY

It is well known that an increased stress response including the activation of the HPA axis leads to alterations in basal body functions such as cardiovascular adaptation, changes in metabolic and immune functions as well as cognitive and behavioral alterations like changes in sleep and activity patterns (Fig. 4). For example, the basal relationship between the resting/activity cycle and the circadian HPA activity underscores this interaction already under physiological conditions. Infused glucocorticoids raise blood pressure and increase the reactivity of tissues to NE directly affecting the blood pressure and also glucose tolerance. Moreover, glucocorticoids or an activation of the HPA axis may be involved in the development of obesity [81–84]. Recently, these data led to the hypothesis that stress-induced disturbances of the HPA axis also contribute to the incidence of type 2 diabetes [85–87].

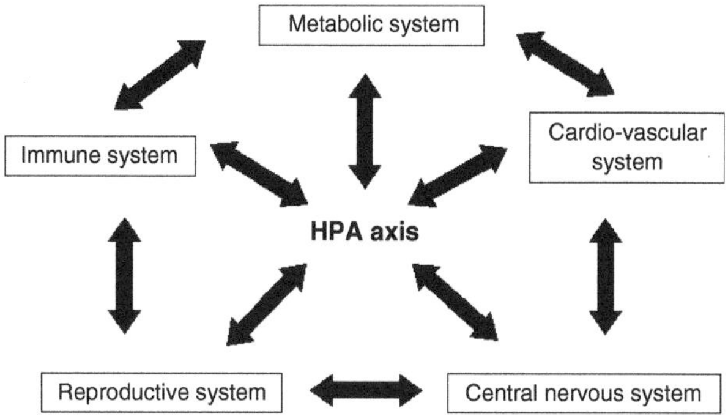


Figure 4. Central role of the hypothalamic–pituitary–adrenocortical axis (HPA axis) in the coordination of the homeostatic systems of the body.

Numerous other reports have shown that stress experience alters the sleep–wake cycle [88–90], pointing to a tight relationship between HPA axis activity and the sleep–wakefulness cycle [88,91]. In this regard, sleep seems to represent a major pacemaker for the initiation of diurnal processes including the secretion of several hormones [42]. Remarkably, not only sleep in general depends on HPA activity but different sleep elements are selectively influenced by HPA components such as ACTH and cortisol. ACTH increases slow-wave sleep and paradoxical sleep [92] Cortisol and other synthetic glucocorticoids also alter slow-wave sleep and therefore critically affect sleep architecture [93].

Recently, the neuroendocrine–immune network has been extended to sleep regulation (e.g., see Refs [94,95]), linking the networks and mediators to specific alterations in sleep parameters.

With focus on behavioral aspects, it has been shown that cortisol affects cognitive processing and specific emotions such as fear and anxiety [96,97]. Additionally, it has been demonstrated that CRH induces specific alterations in anxiety-related behavior, other behavioral responses, and cognitive deficits [98].

Thus, these data show that activation of the HPA system leads to alterations in a large number of body functions including metabolic, cardiovascular, and behavioral aspects as well as specific brain functions such as cognition and memory.

## 6. REGULATION AND CHRONIC ACTIVATION OF THE HPA AXIS

As described above, the activation and adaptation of the HPA axis is tightly regulated at different levels by several feedback circuits [18,99] (Fig. 5). These circuits consist of time-related reactions such as a fast feedback response via circulating glucocorticoids on the secretion of CRH and ACTH [18,99] probably even on the biosynthesis and secretion of the glucocorticoids in the adrenocortical tissue itself [100]. GRs and MRs have been implicated in different types of responses with the GR mediating a fast response at different levels of the HPA axis via a negative influence on the HPA activity [101]. MRs have been implicated in the regulation of the set-point of the axis within a longer time frame (days and weeks), setting the basal activity and subsequently also the reactivity pattern of the axis (see page 22) [43,102].

Therefore, mechanisms such as fast cellular responses are influenced via the fast effects of glucocorticoids on energy metabolism [103], calcium metabolism, and protein processing [104] and protein secretion. Moreover, they are accompanied by responses occurring within longer

**HPA axis activity – interactive factors**

1. Fast cellular feedback mechanisms (Calcium, ATP, others)
2. Genomic mechanisms (corticosteroid receptors)
3. Adaptation to transmitter responses (serotonin, GABA, NMDA, others)
4. Metabolic mechanisms (energy metabolism, liver metabolism, muscle metabolism, others)
5. Individual behavioral adaptations (anxiety, cognition, psychosocial behavior, others)
6. Influences of environmental factors (physical and sociocultural factors, e.g., more or less "stressful environmental," etc.)

Figure 5. Modulatory factors of the activity of the hypothalamic–pituitary–adrenocortical axis (HPA axis) from the cellular level (1) up to sociocultural factors (6).

time periods such as genomic responses, affecting the protein equipment of a cell and other cellular processes within the relevant tissues and feedback loops. Longer feedback loops consist of adaptational effects within the HPA itself such as the influences of immune mediators and neurotransmitters on the activity of the HPA axis, also most likely including effects on the activity of the MR system within the HPA axis [43,102].

Further loops are represented by the adaptation of behavior such as the stress-coping strategies or memory storage [105], the fight or flight reaction itself, and the alteration in the environment of the individual with regard to physical or psychosocial stressors, leading to very long-loop feedback reactions to the responses of the HPA axis.

If an adequate termination of the reactions of the HPA axis fails, a prolonged secretion of mediators impairs cellular adaptational events. This may result in cell and tissue damage, including alterations of elements of the HPA system such as the hippocampus [104]. A disinhibition of the activity of the HPA system may occur under certain conditions of aging [51,534], demonstrating that a *prolonged*, not an increased, activation of the HPA system with a consecutively altered secretion of cortisol as well as other dynamic components of the axis represents the common neuroendocrine alteration.

In summary, the activity of the HPA axis is regulated by numerous and different types of feedback mechanisms that either inhibit or enforce the secretory drive of the neuroendocrine mediators or affect the basal activity of the cellular networks, constantly challenging the adaptational processes of the organism and improving homeostasis in a permanently changing environment.

## 7. COMPREHENSIVE ASPECTS

Organisms are complex biological systems based on static, dynamic, and homeostatic conditions. All principles work together to ensure the survival of biological components from the cellular level to the whole organism and even further to the optimization of psychosocial, cultural, and environmental structures. Stress continuously influences and modulates these systems by initiating adaptational responses, either at a general or at a local level. The HPA axis represents one of the major components of these adaptation mechanisms integrating both psychosocial and physical influences on the organism, and contributing to improve the adaptation to the environment, to a more effective use of the resources, and to optimize survival. When a successful adaptation to given conditions cannot be achieved, even under conditions of recruitment of compensating systems such as the components of the HPA axis, disorders and diseases occur and symptoms of partial failure are observed. Death is the consequence of a total failure of adaptation, if the biological unit, which may be a cell, a tissue, an organism, or even a whole species, has no adequate resources to withstand the intensity and the type of the stressor.

## ACKNOWLEDGMENTS

I thank J.-C. Krieg for a number of helpful comments as well as for his continuous support. The editorial help of Mrs A. Tittmar and Mrs B. Schmalz is gratefully acknowledged. Special thanks are given to E.R. deKloet for his valuable comments and additional substantial input to the contribution.

## REFERENCES

1. Cannon WB. The wisdom of the body. New York: Norton 1932.
2. Selye H. Confusion and controversy in the stress field. J Human. Stress 1975;1(2):37–44.
3. deKloet ER. Hormones and the stressed brain. Ann N Y Acad Sci 2004;1018:1–15.
4. Selye H. Forty years of stress research: principal remaining problems and misconceptions. Can Med Assoc J 1976 3;115(1):53–6.
5. Chrousos GP. Stressors, stress, and neuroendocrine integration of the adaptive response. The 1997 Hans Selye Memorial Lecture. Ann N Y Acad Sci 1998;851:311–35.
6. Charmandari E, Tsigos C, Chrousos G. Endocrinology of the stress response (1). Annu Rev Physiol 2005 17;67:259–84.
7. Jacobson L, Sapolsky R. The role of the hippocampus in feedback regulation of the hypothalamic-pituitary-adrenocortical axis. Endocr Rev 1991;12(2):118–34.
8. Herman JP, Cullinan WE, Young EA, Akil H, Watson SJ. Selective forebrain fiber tract lesions implicate ventral hippocampal structures in tonic regulation of paraventricular nucleus corticotropin-releasing hormone (CRH) and arginine vasopressin (AVP) mRNA expression. Brain Res 1992 2;592(1–2):228–38.
9. Herman JP, Prewitt CM, Cullinan WE. Neuronal circuit regulation of the hypothalamo-pituitary-adrenocortical stress axis. Crit Rev Neurobiol 1996;10(3–4):371–94.
10. Herman JP, Mueller NK, Figueiredo H. Role of GABA and glutamate circuitry in hypothalamo-pituitary-adrenocortical stress integration. Ann N Y Acad Sci 2004;1018:35–45.
11. de Kloet ER, Vreugdenhil E, Oitzl MS, Joels M. Brain corticosteroid receptor balance in health and disease. Endocr Rev 1998;19(3):269–301.
12. Herman JP, Ostrander MM, Mueller NK, Figueiredo H. Limbic system mechanism of stress regulation: hypothalamo-pituitary-adrenocortical axis Prog Neuropsychopharmacol. Biol Psychiatry 2005;29(8):1201–13.
13. Lowry C. Functional subsets of serotonergic neurones: implications for control of the hypothalamic-pituitary-adrenal axis. J Neuroendocrinol 2002;14(11):911–23.
14. Forray MI, Gysling K. Role of noradrenergic projections to the bed nucleus of the stria terminalis in the regulation of the hypothalamic-pituitary-adrenal axis. Brain Res Brain Res Rev 2004; 47(1–3):145–60.
15. Suda T, Kageyama K, Sakihara S, Nigawara T. Physiological roles of urocortins, human homologues of fish urotensin I, and their receptors. Peptides 2004;25(10):1689–701.
16. Heinrichs SC, Koob GF. Corticotropin-releasing factor in brain: a role in activation, arousal, and affect regulation. J Pharmacol Exp Ther 2004;311(2):427–40. Epub 2004 Aug 05.
17. Calogero AE, Galluci WT, Gold PW, Chrousos GP. Multiple feedback regulatory loops upon rat hypothalamic corticotrophin-releasing hormone secretion. J Clin Invest 1988;88:767–74.

18. Keller-Wood ME, Dallman MF. Corticosteroid inhibition of ACTH secretion. Endocr Rev 1984;5(1):1–24.
19. Engelmann M, Landgraf R, Wotjak CT. The hypothalamic-neurohypophysial system regulates the hypothalamic-pituitary-adrenal axis under stress: an old concept revisited. Front Neuroendocrinol 2004;25(3–4):132–49.
20. Dunn AJ, Swiergiel AH, Palamarchouk V. Brain circuits involved in corticotropin-releasing factor-norepinephrine interactions during stress. Ann N Y Acad Sci 2004;1018:25–34.
21. Goodson JL, Bass AH. Social behavior functions and related anatomical characteristics of vasotocin/vasopressin systems in vertebrates. Brain Res Brain Res Rev 2001;35(3):246–65.
22. Hertz L, Chen Y, Spatz M. Involvement of non-neuronal brain cells in AVP-mediated regulation of water space at the cellular, organ, and whole-body level. J Neurosci Res 2000;62(4):480–90.
23. Koenig JI. Pituitary gland: neuropeptides, neurotransmitters and growth factors. Toxicol Pathol 1989;17(2):256–65.
24. Gudmundsson A, Carnes M. Pulsatile adrenocorticotropic hormone: an overview. Biol Psychiatry 1997 1;41(3):342–65.
25. Dallman MF, Akana SF, Cascio CS, Darlington DN, Jacobson L, Levin N. Regulation of ACTH secretion: variations on a theme of B. Recent Prog Horm Res 1987;43:113–73.
26. Turnbull AV, Rivier CL. Regulation of the hypothalamic-pituitary-adrenal axis by cytokines: actions and mechanisms of action. Physiol Rev 1999;79(1):1–71.
27. Schöbitz B, Sutanto W, Carey MP, Holsboer F, de Kloet ER. Endotoxin and interleukin 1 decrease the affinity of hippocampal mineralocorticoid (type I) receptor in parallel to activation of the hypothalamic-pituitary-adrenal axis. Neuroendocrinology 1994;60(2):124–33.
28. Dunn AJ. Cytokine activation of the HPA axis. Ann N Y Acad Sci 2000;917:608–17.
29. Besedovsky HO, del Rey A. The cytokine-HPA axis feed-back circuit. Z Rheumatol 2000;59 (Suppl 2:II):26–30.
30. Marx C, Ehrhart-Bornstein M, Scherbaum WA, Bornstein SR. Regulation of adrenocortical function by cytokines – relevance for immune-endocrine interaction. Horm Metab Res 1998;30(6–7):416–20.
31. Mastorakos G, Chrousos GP, Weber JS. Recombinant interleukin-6 activates the hypothalamic-pituitary-adrenal axis in humans. J Clin Endocrinol Metab 1993;77(6):1690–4.
32. Päth G, Scherbaum WA, Bornstein SR. The role of interleukin-6 in the human adrenal gland. Eur J Clin Invest 2000;30(Suppl 3):91–5.
33. de Kloet ER, Reul JM, Sutanto W. Corticosteroids and the brain. J Steroid Biochem Mol Biol 1990 20;37(3):387–94.
34. de Kloet ER, Sutanto W, Rots N, van Haarst A, van den Berg D, Oitzl M, van Eekelen A, Voorhuis D. Plasticity and function of brain corticosteroid receptors during aging. 1. Acta Endocrinol (Copenh) 1991;125(Suppl 1):65–72.
35. de Kloet ER, Derijk R. Signaling pathways in brain involved in predisposition and pathogenesis of stress-related disease: genetic and kinetic factors affecting the MR/GR balance. Ann N Y Acad Sci 2004;1032:14–34.
36. Erickson K, Drevets W, Schulkin J. Glucocorticoid regulation of diverse cognitive functions in normal and pathological emotional states. Neurosci Biobehav Rev 2003;27(3):233–46.
37. Bemelmans KJ, Goekoop JG, de Rijk R, van Kempen GM. Recall performance, plasma cortisol and plasma norepinephrine in normal human subjects. Biol Psychol 2003;62(1):1–15.
38. Brown EN, Meehan PM, Dempster AP. A stochastic differential equation model of diurnal cortisol patterns. Am J Physiol Endocrinol Metab 2001 Mar;280(3):E450–61.
39. de Weerth C, Zijl RH, Buitelaar JK. Development of cortisol circadian rhythm in infancy. Early Hum Dev 2003;73(1–2):39–52.
40. Kwak SP, Morano MI, Young EA, Watson SJ, Akil H. Diurnal CRH mRNA rhythm in the hypothalamus: decreased expression in the evening is not dependent on endogenous glucocorticoids. Neuroendocrinology 1993;57(1):96–105.
41. Daikoku S, Yokote R, Aizawa T, Kawano H. Light stimulation of the hypothalamic neuroendocrine system. Arch Histol Cytol 1992;55(1):67–76.

42. Czeisler CA, Klerman EB. Circadian and sleep-dependent regulation of hormone release in humans. Recent Prog Horm Res 1999;54:97–130; discussion 130–2.
43. Deuschle M, Weber B, Colla M, Muller M, Kniest A, Heuser I. Mineralocorticoid receptor also modulates basal activity of hypothalamus-pituitary-adrenocortical system in humans. Neuroendocrinology 1998;68(5):355–60.
44. de Kloet ER, Oitzl MS, Schöbitz B. Cytokines and the brain corticosteroid receptor balance: relevance to pathophysiology of neuroendocrine-immune communication. Psychoneuroendocrinology 1994;19(2):121–34.
45. Dallman MF, Akana SF, Bhatnagar S, Bell ME, Strack AM. Bottomed out: metabolic significance of the circadian trough in glucocorticoid concentrations. Int J Obes Relat Metab Disord 2000; 24(Suppl 2):S40–6.
46. Pollmächer T, Mullington J, Korth C, Schreiber W, Hermann D, Orth A, Galanos C, Holsboer F. Diurnal variations in the human host response to endotoxin. J Infect Dis 1996; 174(5):1040–5.
47. Deuschle M, Gotthardt U, Schweiger U, Weber B, Körner A, Schmider J, Standhardt H, Lammers CH, Heuser I. With aging in humans the activity of the hypothalamus-pituitary-adrenal system increases and its diurnal amplitude flattens. Life Sci 1997;61(22):2239–46.
48. de Kloet ER, Sutanto W, Rots N, van Haarst A, van den Berg D, Oitzl M, van Eekelen. Plasticity and function of brain corticosteroid receptors during aging. 1. Acta Endocrinol (Copenh) 1991;125 (Suppl 1):65–72.
49. Björntorp P. Alterations in the ageing corticotropic stress-response axis. Novartis Found Symp 2002;242:46–58; discussion 58–65.
50. Sapolsky R, Armanini M, Packan D, Tombaugh G. Stress and glucocorticoids in. Endocrinol Metab Clin North Am 1987;16(4):965–80.
51. Miller DB, O'Callaghan JP. Effects of aging and stress on hippocampal structure and function. Metabolism 2003;52(10 Suppl 2):17–21.
52. McEwen BS. Sex, stress and the hippocampus: allostasis, allostatic load and the aging process. Neurobiol Aging 2002;23(5):921–39.
53. Traustadottir T, Bosch PR, Matt KS. Gender differences in cardiovascular and hypothalamic-pituitary-adrenal axis responses to psychological stress in healthy older adult men and women. Stress 2003;6(2):133–40.
54. Kirschbaum C, Hellhammer DH. Salivary cortisol in psychobiological research: an overview. Neuropsychobiology 1989a;22:150–69.
55. Biondi M, Picardi A. Psychological stress and neuroendocrine function in humans: the last two decades of research. Psychotherapy and. Psychosomatics 1999;68:114–50.
56. Mason JW. A review of psychoendocrine research on the pituitary-adrenal cortical system. Psychosom Med 1968;30(Suppl):576–607.
57. Hennessy MB. Hypothalamic-pituitary-adrenal responses to brief social separation. Neurosci Biobehav Rev 1997;21(1):11–29.
58. Weinberg J, Wong R. Adrenocortical responsiveness to novelty in the hamster. Physiol Behav 1986;37(5):669–72.
59. Gunnar MR, Connors J, Isensee J. Lack of stability in neonatal adrenocortical reactivity because of rapid habituation of the adrenocortical response. Dev Psychobiol 1989;22:221–3.
60. Gunnar MR, Hertsgaard L, Larson M, Rigatusco J. Cortisol and behavioural responses to repeated stressors in the human newborn. Dev Psychobiol 1991;24:487–505.
61. Kirschbaum C, Bartussek D, Strasburger CJ. Cortisol responses to psychological stress and correlations with personality traits. Pers Individ Dif 1992;13(12):1353–7.
62. van Eck M, Berkhof H, Nicolson N, Sulon J. The effects of perceived stress, traits, mood states, and stressful daily events on salivary cortisol. Psychosom Med 1996;58:432–46.
63. Makino S, Hashimoto K, Gold PW. Multiple feedback mechanisms activating corticotropin-releasing hormone system in the brain during stress. Pharmacol Biochem Behav 2002;73(1): 147–58.

64. Kudielka BM, Schommer NC, Hellhammer DH, Kirschbaum C. Acute HPA axis responses, heart rate, and mood changes to psychosocial stress (TSST) in humans at different times of day. Psychoneuroendocrinology 2004;29(8):983–92.
65. Miller DB, O'Callaghan JP. Neuroendocrine aspects of the response to stress. Metabolism 2002;51(6 Suppl 1):5–10.
66. Argyropoulos SV, Bailey JE, Hood SD, Kendrick AH, Rich AS, Laszlo G, Nash JR, Lightman SL, Nutt DJ. Inhalation of 35% $CO_2$ results in activation of the HPA axis in healthy volunteers. Psychoneuroendocrinology 2002;27(6):715–29.
67. Kaye J, Buchanan F, Kendrick A, Johnson P, Lowry C, Bailey J, Nutt D, Lightman S. Acute carbon dioxide exposure in healthy adults: evaluation of a novel means of investigating the stress response. J Neuroendocrinol 2004;16(3):256–64.
68. Nye EJ, Grice JE, Hockings GI, Strakosch CR, Crosbie GV, Walters MM, Jackson RV. Comparison of adrenocorticotropin (ACTH) stimulation tests and insulin hypoglycemia in normal humans: low dose, standard high dose, and 8-hour ACTH-(1-24) infusion tests. J Clin Endocrinol Metab 1999;84(10):3648–55.
69. Clark PM, Neylon I, Raggatt PR, Sheppard MC, Stewart PM. Defining the normal cortisol response to the short Synacthen test: implications for the investigation of hypothalamic-pituitary disorders. Clin Endocrinol (Oxf) 1998;49(3):287–92.
70. Dullaart RP, Pasterkamp SH, Beentjes JA, Sluiter WJ. Evaluation of adrenal function in patients with hypothalamic and pituitary disorders: comparison of serum cortisol, urinary free cortisol and the human-corticotrophin releasing hormone test with the insulin tolerance test. Clin Endocrinol (Oxf) 1999 Apr;50(4):465–71.
71. Loriaux DL, Nieman L. Corticotropin-releasing hormone testing in pituitary disease. Endocrinol Metab Clin North Am 1991;20(2):363–9.
72. Reimondo G, Paccotti P, Minetto M, Termine A, Stura G, Bergui M, Angeli A, Terzolo M. The corticotrophin-releasing hormone test is the most reliable noninvasive method to differentiate pituitary from ectopic ACTH secretion in Cushing's syndrome. Clin Endocrinol (Oxf) 2003;58(6):718–24.
73. Gold PW, Kling MA, Khan I, Calabrese JR, Kalogeras K, Post RM, Avgerinos PC, Loriaux DL, Chrousos GP. Corticotropin releasing hormone: relevance to normal physiology and to the pathophysiology and differential diagnosis of hypercortisolism and adrenal insufficiency. Adv Biochem Psychopharmacol 1987;43:183–200.
74. Holsboer F. Psychiatric implications of altered limbic-hypothalamic-pituitary-adrenocortical activity. Eur Arch Psychiatry Neurol Sci 1989;238(5–6):302–22.
75. Stratakis CA, Chrousos GP. Neuroendocrinology and pathophysiology of the stress system. Ann N Y Acad Sci 1995 Dec 29;771:1–18.
76. Claes SJ. Corticotropin-releasing hormone (CRH) in psychiatry: from stress to psychopathology. Ann Med 2004;36(1):50–61.
77. Schreiber W, Pollmächer T, Fassbender K, Gudewill S, Vedder H, Wiedemann K, Galanos C, Holsboer F. Endotoxin– and corticotropin-releasing hormone-induced release of ACTH and cortisol. A comparative study in men. Neuroendocrinology 1993;58(1):123–8.
78. Beishuizen A, Thijs LG. Endotoxin and the hypothalamo-pituitary-adrenal (HPA) axis. J Endotoxin Res 2003;9(1):3–24.
79. Del Rey A, Besedovsky HO. Metabolic and neuroendocrine effects of pro-inflammatory cytokines. Eur J Clin Invest 1992;22(Suppl 1):10–15.
80. Vedder H, Schreiber W, Yassouridis A, Gudewill S, Galanos C, Pollmächer T. Dose-dependence of bacterial lipopolysaccharide (LPS) effects on peak response and time course of the immune-endocrine host response in humans. Inflamm Res 1999;48(2):67–74.
81. Chalew S, Nagel H, Shore S. The hypothalamic-pituitary-adrenal axis in obesity. Obes Res 1995;3(4):371–82.
82. Pasquali R, Vicennati V. Activity of the hypothalamic-pituitary-adrenal axis in different obesity phenotypes. Int J Obes Relat Metab Disord 2000;24(Suppl 2):S47–9.

83. Björntorp P, Rosmond R. Neuroendocrine abnormalities in visceral obesity. Int J Obes Relat Metab Disord 2000;24(Suppl 2):S80–5.
84. Mastorakos G, Zapanti E. The hypothalamic-pituitary-adrenal axis in the neuroendocrine regulation of food intake and obesity: the role of corticotropin releasing hormone. Nutr Neurosci 2004; 7(5–6):271–80.
85. Chan O, Inouye K, Riddell MC, Vranic M, Matthews SG. Diabetes and the hypothalamo-pituitary-adrenal (HPA) axis. Minerva Endocrinol 2003;28(2):87–102.
86. Prestele S, Aldenhoff J, Reiff J. The HPA-axis as a possible link between depression, diabetes mellitus and cognitive dysfunction. Fortschr Neurol Psychiatr 2003;71(1):24–36.
87. Rosmond R. Stress induced disturbances of the HPA axis: a pathway to Type 2 diabetes. Med Sci Monit 2003;9(2):RA35–9.
88. Born J, Fehm HL. Hypothalamus-pituitary-adrenal activity during human sleep: a coordinating role for the limbic hippocampal system. Exp Clin Endocrinol Diabetes 1998;106(3):153–63.
89. Cespuglio R, Marinesco S, Baubet G, Bonnet C, el Kafi B. Evidence for a sleep-promoting influence of stress. Adv Neuroimmunol 1995;5:145–54.
90. Opp MR. Corticotropin-releasing hormone involvement in stressor-induced alterations in sleep and in the regulation of waking. Adv Neuroimmunol 1995;5:127–43.
91. Steiger A. Sleep and the hypothalamo-pituitary-adrenocortical system. Sleep Med Rev 2002;6(2):125–38.
92. Chastrette N, Cespuglio R, Jouvet M. Proopiomelanocortin POMC-derived peptides and sleep in the rat: Part 1. Hypnogenic properties of ACTH derivatives. Neuropeptides 1990;15:61–74.
93. Young AH, Sharpley AL, Campling GM, Hockney RA, Cowen PJ. Effects of hydrocortisone on brain 5-HT function and sleep. J Affect Disord 1994;32(2):139–46.
94. Marshall L, Born J. Brain-immune interactions in sleep. Int Rev Neurobiol 2002;52:93–131.
95. Schuld A, Haack M, Hinze-Selch D, Mullington J, Pollmächer T. Experimental studies on the interaction between sleep and the immune system in humans. Psychother Psychosom Med Psychol 2005;55(1):29–35.
96. Ohl F, Michaelis T,Vollmann-Honsdorf GK, Kirschbaum C, Fuchs E. Effect of chronic psychosocial stress and long-term cortisol treatment on hippocampus-mediated memory and hippocampal volume: a pilot-study in tree shrews. Psychoneuroendocrinology 2000;25(4):357–63.
97. Müller MB, Landgraf R, Keck ME. Vasopressin, major depression, and hypothalamic-pituitary-adrenocortical desensitization. Biol Psychiatry 2000 15;48(4):330–3.
98. Steckler T, Holsboer F. Corticotropin-releasing hormone receptor subtypes and emotion. Biol Psychiatry 1999 1;46(11):1480–508.
99. Calogero AE, Gallucci WT, Chrousos GP, Gold PW. Interaction between GABAergic neurotransmission and rat hypothalamic corticotropin-releasing hormone secretion in vitro. Brain Res 1988 25;463(1):28–36.
100. Brentano ST, Picado-Leonard J, Mellon SH, Moore CCD, Miller WL. Tissue-specific, cAMP-induced, and phorbolester repressed expression from the human P450c17 promoter in mouse cells. Mol Endocrinol 1990;4:1972–9.
101. de Kloet ER, Oitzl MS, Joels M. Functional implications of brain corticosteroid receptor diversity. Cell Mol Neurobiol 1993;13(4):433–55.
102. Oitzl MS, van Haarst AD, de Kloet ER. Behavioral and neuroendocrine responses controlled by the concerted action of central mineralocorticoid (MRS) and glucocorticoid receptors (GRS). Psychoneuroendocrinology 1997;22(Suppl 1):S87–93.
103. Martens ME, Peterson PL, Lee CP. In vitro effects of glucocorticoid on mitochondrial energy metabolism. Biochim Biophys Acta 1991 17;1058(2):152–60.
104. Sapolsky RM. Potential behavioral modification of glucocorticoid damage to the hippocampus. Behav Brain Res 1993 30;57(2):175–82.
105. de Kloet ER, Sutanto W, Rots N, van Haarst A, van den Berg D, Oitzl M, van Eekelen A, Voorhuis D. Plasticity and function of brain corticosteroid receptors during aging. Acta Endocrinol (Copenh) 1991;125(Suppl 1):65–72.

*The Hypothalamus–Pituitary–Adrenal Axis*
Edited by A. del Rey, G. P. Chrousos and H. O. Besedovsky



# *Glucocorticoid Signaling in Health and Disease*

KATHLEEN A. SMOAK and JOHN A. CIDLOWSKI

*Laboratory of Signal Transduction, National Institute of Environmental Health Sciences, Department of Health and Human Services, National Institutes of Health, Research Triangle Park, North Carolina, USA*

## ABSTRACT

Glucocorticoids are among the most widely prescribed drugs for chronic inflammatory and immune diseases. They act by binding to the glucocorticoid receptor (GR) that, upon activation, translocates to the nucleus and either stimulates or inhibits gene expression. GR inhibition of many proinflammatory response genes occurs through the induction of anti-inflammatory genes via direct DNA binding and/or through the repression of proinflammatory transcription factors such as nuclear factor-κB (NF-κB) or activator protein-1 (AP-1). In this chapter, we discuss the molecular mechanisms involved in GR inhibition of inflammatory responses as well as the adverse side effects of long-term glucocorticoid administration in humans.

## 1. INTRODUCTION

During an inflammatory response, cytokines produced by activated lymphocytes and macrophages stimulate the hypothalamo–pituitary–adrenal (HPA) axis to synthesize and secrete glucocorticoids from the adrenal cortex. The released glucocorticoids subsequently exert anti-inflammatory effects on many cell types including T cells, macrophages, eosinophils, neutrophils, mast cells, and endothelial and epithelial cells, thereby creating a classical endocrine feedback loop that quells the inflammatory response. Glucocorticoids accomplish these actions primarily through the interruption of proinflammatory, cytokine-mediated signaling pathways as well as through induction of apoptosis in certain cells of the immune system. The molecular mechanisms behind these glucocorticoid-mediated anti-inflammatory effects were, however, unknown to clinicians who first utilized them in the treatment of disease.

Glucocorticoid use in the clinic was made possible by the efforts of Kendall and Reichstein who independently extracted Compound E, now known as cortisone (an endogenous glucocorticoid), from the adrenal cortices of cattle and sheep in 1935 [1]. At the time, large-scale production of cortisone was originally undertaken by pharmaceutical firms in order to find an effective treatment for Addison's disease, a condition of hypocortisolism caused by adrenal insufficiency [1]. In the meantime, rheumatologist Philip Hench noticed that many of his patients exhibited certain symptoms, such as listlessness and hypotension, which were common to those suffering from Addison's disease. This observation was one of many which prompted

Hench and his colleagues to bid for the small amount of commercially available Compound E for use in clinical trials. In 1949, they administered Compound E to a bed-ridden patient suffering from rheumatoid arthritis and observed for the first time the potent anti-inflammatory effects of glucocorticoids as the patient subsequently got out of bed and attempted to dance [2]! A year later, other clinicians reported successful use of cortisone for the treatment of asthma [3]. Kendall, Reichstein, and Hench were awarded the Nobel Prize in 1950 in Physiology or Medicine for their respective contributions to the field of adrenal steroid hormones [1]. However, the enthusiasm felt by these investigators was quickly dampened by the realization that glucocorticoids also have systemic side effects that limited their usefulness. This finding made understanding the mechanisms behind both the beneficial and the detrimental effects of glucocorticoids of primary importance.

Now, over 50 years later, synthetic glucocorticoids are used not only for the treatment of rheumatoid arthritis and asthma, but also for many other immune and inflammatory diseases, including multiple sclerosis, inflammatory bowel disease, and allergy. They are also used in combination therapy for treatment of certain malignancies primarily due to their role in immune cell apoptosis and for their ability to palliate the side effects resulting from other chemotherapeutic agents. Much progress has been made in understanding the signaling of these hormones, most important of which was the discovery of the glucocorticoid receptor (GR), which functions as a ligand-dependent transcription factor and mediates both the anti-inflammatory and the side effects related to glucocorticoid therapy. In this chapter, we review the mechanisms of glucocorticoid signaling and discuss the beneficial and deleterious outcomes of glucocorticoid use in health and disease.

## 2. THE GLUCOCORTICOID RECEPTOR

Glucocorticoids and their receptors are necessary for life, as suggested by studies from GR-deficient mice which show severe abnormalities and die shortly after birth [4]. GR is a member of the nuclear receptor superfamily of transcription factors that includes receptors for mineralocorticoids, progesterone, androgen, estrogen, and thyroid hormones, vitamin D, and retinoic acid [5]. The two most widely known isoforms of human GR, GRα and GRβ, are generated by alternative splicing of a single gene (Fig. 1). GRα is the classic receptor that binds hormone and activates glucocorticoid responsive genes. GRβ differs from GRα in its C-terminus, does not bind glucocorticoids, and until recently, was thought to be incapable of binding ligand. It is now known that GRβ binds to the glucocorticoid antagonist, RU-486, but that it is only transcriptionally active in the absence of this ligand [6]. While the function of GRβ is unclear, it may act as a dominant negative regulator of GRα-mediated transactivation [7–9] and play a role in glucocorticoid resistance. Increased expression of GRβ has been reported in glucocorticoid-resistant asthma [10], rheumatoid arthritis [11], and ulcerative colitis [12]. Interestingly, both GRα and GRβ mRNAs can undergo alternative translation initiation, each generating an additional eight receptor isoforms identified to date [13,14]. While GRβ isoforms have not yet been investigated, GRα isoforms have recently been found to exhibit differing transcriptional activities as well as distinct tissue-specific distribution patterns, which may explain both the cell type-specific effects of glucocorticoids and possibly glucocorticoid resistance as well [15]. GRα is expressed in virtually every cell type in mammals, and will be the focus of this chapter.

The GRα protein consists of three domains whose functions are maintained if expressed independently of the full-length protein (Fig. 1). The N-terminal transactivation domain

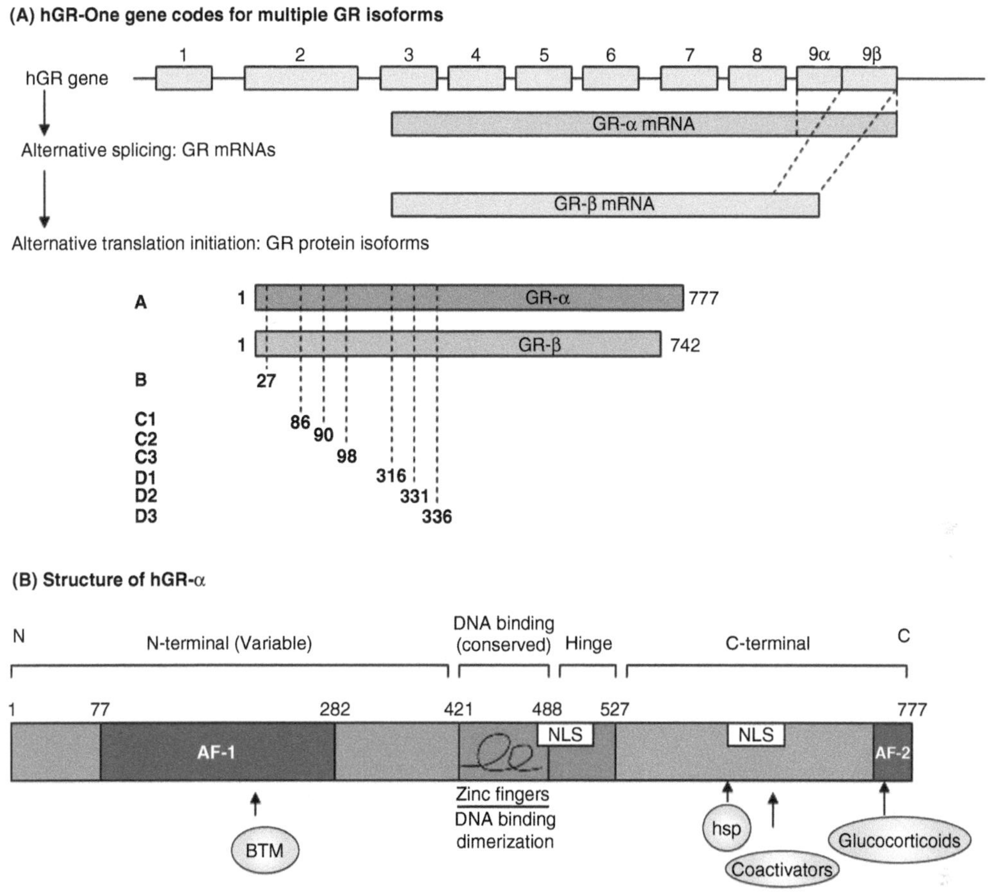


Figure 1. Genomic and protein structure of glucocorticoid receptor (GR). (A) A single GR gene codes for multiple GR isoforms. Alternative splicing of exon 9 generates GRα and GRβ mRNAs. For both GRα and GRβ transcripts, an additional eight protein isoforms are produced by alternative translation initiation: GR-A, GR-B, GR-C1, GR-C2, GR-C3, GR-D1, GR-D2, and GR-D3. Methionine residues corresponding to the specific isoform are shown in bold. (B) Organization of the GRα protein. The GRα protein contains three major domains, shown here from left to right. The N-terminal transactivation domain consists of the activation function (AF)-1 activation domain, which is required for transcriptional enhancement and association with the basal transcription machinery (BTM). The central, DNA-binding domain is highly conserved and contains two zinc finger regions critical for receptor dimerization and target binding. The C-terminal hormone-binding domain serves as a binding site for heat-shock proteins and coactivators, and contains nuclear localization signals (NLS), as well as the ligand-dependent AF-2 domain.

includes the AF-1 (AF – activation function) activation domain required for transcriptional enhancement and association of the receptor with basal transcription factors [16]. The central DNA-binding domain is highly conserved among nuclear receptors. It consists of two zinc finger regions which are critical for receptor dimerization and target binding. The C-terminal domain is the site of hormone binding, and also serves as a binding site for heat-shock proteins (hsp). This hormone-binding domain contains nuclear localization signals (NLSs), as well as the ligand-dependent AF-2 domain [17,18].

Due to their lipophilic nature, glucocorticoids readily diffuse through the plasma membrane by a passive process and bind to the intracellular GRα, which appears to be sequestered in the cytoplasm as part of an inactive complex (Fig. 2) consisting of hsp90, hsp70, hsp56, hsp40, a low molecular weight protein (p23), and several immunophilins [19,20]. Upon binding to its

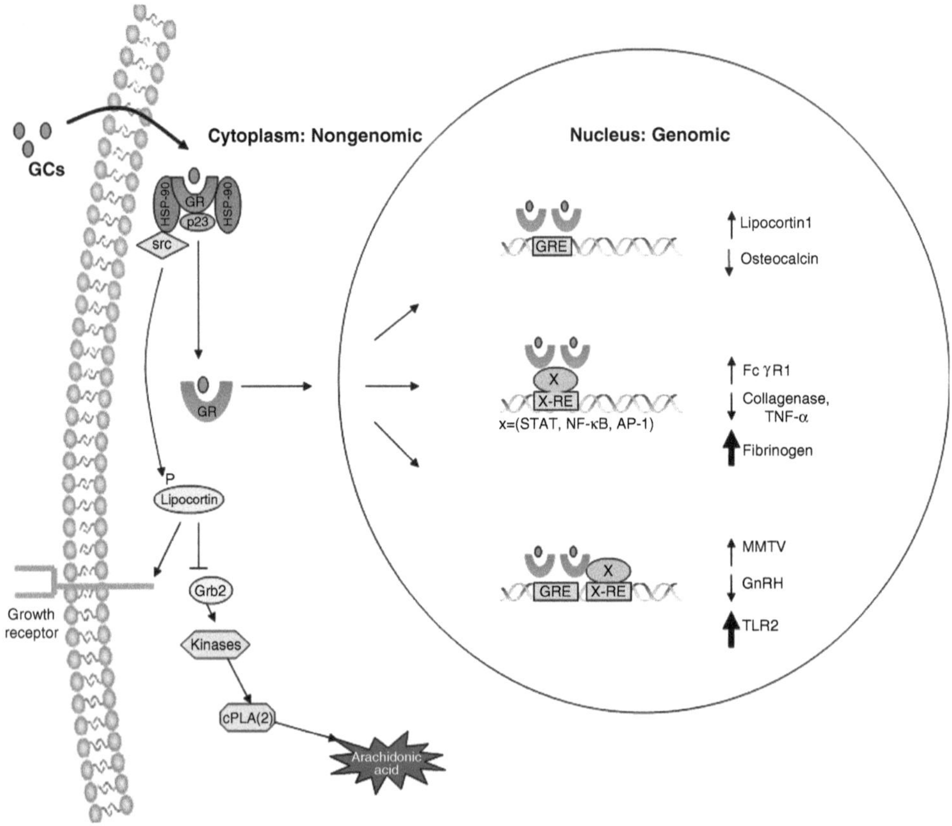


Figure 2. Glucocorticoid receptor (GR)α nongenomic and genomic mechanisms. The inactive glucocorticoid receptor is sequestered in the cytoplasm complexed with chaperones until it becomes an activated transcription factor upon binding to ligand. At this point, GRα dissociates from the multimeric protein complex and can either signal through nongenomic pathways in the cytoplasm or through genomic mechanisms after translocating to the nucleus. Nongenomic: Following treatment with glucocorticoids, SRC kinase is rapidly released from the hsp90 complex. It then phosphorylates lipocortin 1, which subsequently competes with Grb2, blocking epidermal growth factor (EGF)-stimulated cytosolic phospholipase A ($PLA_2$) activation and arachidonic acid (AA) release. Genomic: The ligand-receptor complex can stimulate or inhibit transcriptional responses by the following: (1) Directly binding to DNA at the glucocorticoid response element (GRE) or the negative GRE (nGRE): GRE indicates either GRE or negative GRE. For example, GRα binding to the lipocortin GRE or the osteocalcin nGRE results in activated or repressed gene transcription, respectively. (2) Protein–protein interactions: GRα interaction with transcription factors (X), such as signal transduction and activator of transcription (STATs), nuclear factor (NF)-κB, or activator protein (AP)-1 can either enhance or repress their activity on responsive genes (X-RE). Synergistic activation, denoted by the thicker arrow, has been demonstrated in certain STAT-activated genes such as fibrinogen, resulting from physical interaction between GRα and STATs. (3) GRα interactions with both DNA and other transcription factors: For example, GRα interaction with Smad3 can result in enhanced transcription of the MMTV promoter through the GRE, while GRα tethering to OCT-1 bound to a negative GRE represses transcription of gonadotropin-releasing hormone (GnRH). Synergistic regulation of many genes, such as Toll-like receptor (TLR)-2 occurs through GRα interactions with both STATs and GREs.

ligand (glucocorticoids), GRα undergoes a change in conformation, resulting in its dissociation from the multimeric complex, the exposure of its NLSs, and its rapid translocation to the nucleus. Here, the transcriptionally competent (activated) receptor complex modulates (stimulates or inhibits) transcriptional responses of inflammatory genes by binding directly to DNA and/or by protein–protein interactions between GRα and other transcription factors that regulate proinflammatory genes (Fig. 2). These mechanisms, as well as potential nongenomic and post-transcriptional actions of GRα, culminate in multiple effects on both the adaptive and the innate immune systems.

## 3. GR BINDS DIRECTLY TO GLUCOCORTICOID RESPONSE ELEMENTS

GRα activates transcription by binding as a homodimer to glucocorticoid response elements (GREs) on DNA. The consensus GRE sequence is the palindromic 15-bp motif AGAACAnnnTGTTCT, where the 3′-half-site region is highly conserved, while the 5′-half-site region is more variable. In this regard, a number of different genes have been described where only a GRE half-site sequence was found to be sufficient for glucocorticoid signaling; however, accessory factors or additional GRE half-sites are often involved in this regulation [21]. Both the number of GREs and their relative proximity to the TATA box appear to be determinants of the glucocorticoid inducibility of gene expression [22,23].

Binding of GRα to a GRE results in a conformational change in GRα that promotes the recruitment of multiple coactivators to the GR–DNA complex [24,25]. Some of these coactivators, such as CBP/p300, p/CAF, and SRC-1, contain histone acetylase (HAT) activity, which appears to be critical for remodeling of chromatin structure and full manifestation of glucocorticoid effects. Acetylation of histones results in nucleosomal rearrangement and DNA unwinding, which allows the basal transcription machinery (BTM) access to the promoter. Nucleosomal arrangement and promoter access can also be accomplished via other, non-HAT-containing cofactors including ATP-dependent chromatin-remodeling factors such as SWI/SNF [26]. Glucocorticoids are known to increase the synthesis of certain anti-inflammatory proteins, including lipocortin-1, serum leukoprotease inhibitor, clara cell protein 10 (CC10), interleukin-1 (IL-1) receptor antagonist, IL-10, neural endopeptidase, and mitogen-activated protein kinase (MAPK) phosphatase-1 (MKP-1) [24,25] presumably through GRE-dependent gene activation (Fig. 2). However, activation of these genes likely comprises only a portion of the anti-inflammatory action of glucocorticoids.

GRα can also repress genes by directly binding to a negative GRE (nGRE) (Fig. 2). nGRE sequences exhibit some similarity to GRE sequences but are much more variable, resulting in several mechanisms of nGRE-mediated repression. For example, many mechanisms have been described for GRα downregulation of the adrenocorticotropic hormone (ACTH) precursor gene, pro-opiomelanocortin (POMC). First, GRα binding to the nGRE may inhibit transcription directly through steric hindrance as a result of its close proximity to the TATA box and transcription start site [27]. Notably, GRα binding to the osteocalcin gene is similarly regulated by GRα [28,29] (Fig. 2). In the case of POMC, however, this nGRE region also overlaps with the binding site for Nurr77, which is known to be involved in promoter activation. Thus it has also been postulated that GRα binding antagonizes transcriptional activation of POMC through displacement of Nur77. Finally, GRα-mediated repression of POMC may occur through a tethering mechanism which does not require GRα binding to DNA at all [30]. This tethering mechanism has been described for a number of GRα interactions with transcription factors (below) and has also been described for GRα downregulation of gonadotropin-releasing hormone (GnRH) [31], where GRα binds to octamer-binding protein 1 (OCT-1) which then binds to the nGRE of GnRH. Recently, GRα was found to bind directly to an nGRE found in the Fas ligand promoter [32]; however, it should be noted that very few glucocorticoid-regulated inflammatory genes utilize nGREs as the primary mechanism for repression.

## 4. GR INTERACTIONS WITH TRANSCRIPTION FACTORS

The most potent anti-inflammatory effects of glucocorticoids appear to result from protein–protein interactions between GRα and other transcription factors, particularly nuclear factor-κB (NF)-κB and AP-1 (Fig. 3). This mechanism leads to repression of the production of a number of cytokines that are relevant to inflammatory diseases, including tumor necrosis factor-α (TNF-α), granulocyte-macrophage colony-stimulating factor (GM-CSF), IL-1β, IL-2, IL-3, IL-6, IL-8, and IL-11 [33]. Additionally, several chemokines including eotoxin, MIP, and RANTES, enzymes such as inducible nitric oxide synthase (iNOS) and cyclooxygenase (COX)-2, and adhesion molecules including intercellular adhesion molecule (ICAM)-1 and vascular cell adhesion molecule (VCAM)-1 are also regulated in this manner.

### 4.1. Nuclear factor-κB

NF-κB is a ubiquitous transcription factor whose involvement in inflammatory disorders is well established in both *in vitro* and *in vivo* experimental systems [26]. The NF-κB heterodimer consisting of the p65/p50 subunits is the predominant combination involved in transcriptional activation processes. In the inactive form, NF-κB is maintained in the cytoplasm through its interaction with the inhibitor, IκB [34]. Stimulation of the cell with inflammatory stimuli (such as cytokines or lipopolysaccharide [LPS]) leads to activation of IκB kinase (IKK), which then phosphorylates IκB, resulting in the dissociation of the IκB–NF-κB complex. The released NF-κB translocates to the nucleus, where it activates transcription of a plethora of inflammatory genes, including cytokines such as TNF-α, IL-1β, and GM-CSF, and enzymes associated with the synthesis of inflammatory mediators such as COX-2 and iNOS [33]. Glucocorticoids may repress NF-κB activity through induction of IκB expression by binding to a GRE in its promoter (Fig. 3) [26]. However, this glucocorticoid-mediated mechanism of NF-κB repression is limited to only a few cell types and is not thought to be a major contributor to the GRα anti-inflammatory response, but rather a component of resetting the steady state of NF-κB inhibition.

The primary mechanism by which GRα represses NF-κB transcriptional activity is by physically interacting with the p65 subunit (Fig. 3). GRα may accomplish these inhibitory actions by sequestering NF-κB and preventing its binding to DNA [35,36], or by preventing transmission of an effective signal to other NF-κB responsive genes. For example, a recent report finds that NF-κB-driven genes activated by Toll-like receptor (TLR)-4 or TLR-9 are inhibited by GRα disruption of p65/interferon-regulatory factor (IRF)-3 complexes [37]. GRα also induces the expression of another inhibitor of NF-κB, glucocorticoid-induced leucine zipper (GILZ), which binds to NF-κB, blocking inflammatory signaling relevant in asthma [38,39]. Interestingly, NF-κB has also been shown to negatively regulate GRα-mediated transcription [40,41]. This mutual antagonism may require multiple domains of GRα [35].

Glucocorticoids may also suppress NF-κB transactivation by modifying the BTM at inflammatory gene promoters (Fig. 3). For example, GRα has been shown to interfere with the phosphorylation of the RNA Pol II C-terminal tail, possibly through a serine-2

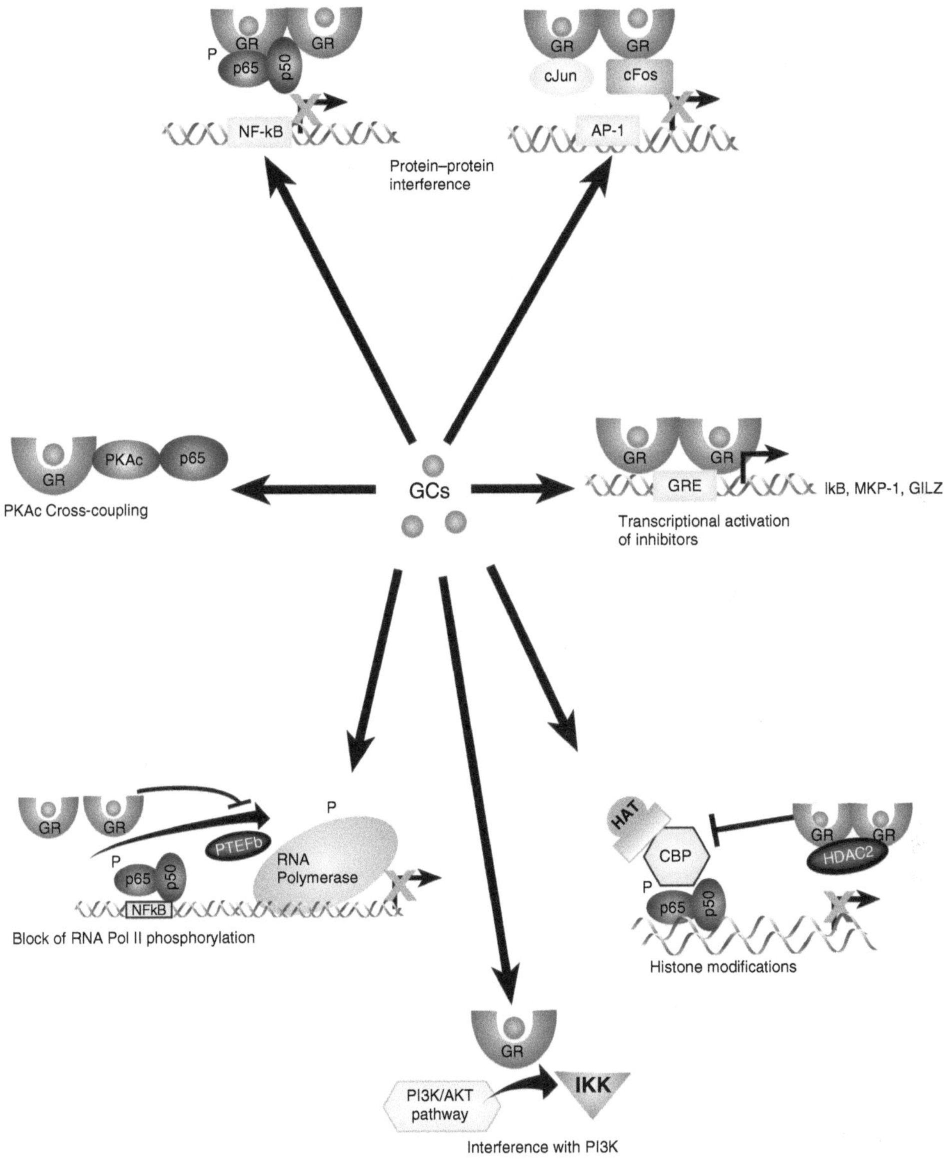


Figure 3. Glucocorticoid receptor (GR)α-mediated repression of nuclear factor (NF)-κB and activator protein (AP)-1 signaling. (1) Protein–protein interactions. GRα interacts with the p65 subunit of NF-κB to repress NF-κB-regulated gene expression and interacts with c-Jun to repress AP-1-regulated gene transcription. (2) GRα and p65 compete for the cytoplasmic catalytic subunit of protein kinase A (PKA) (PKAc), which has been reported to play a role in both GRα-mediated inhibition of NF-κB and p65-mediated inhibition of GRα. (3) In certain cell types, GRα induces IκB, which may lead to the sequestration of NF-κB in the cytoplasm. GRα induces glucocorticoid-induced leucine zipper (GILZ) and mitogen-activated protein kinase (MAPK) phosphatase (MKP)-1, which have repressive effects on NF-κB, AP-1, and MAPK pathways. (4) GRα interferes with the phosphorylation of the C-terminal domain (CTD) of RNA Pol II, thereby disrupting NF-κB-stimulated transcriptional elongation. (5) In the nucleus, GRα may recruit histone deacytelase (HDAC)-2 to the p65 CBP HAT complex. (6) In the cytoplasm, GRα interferes with the phosphatidylinositol 3-kinase (PI3K) pathway (which modulates IκB kinase [IKK] activity), resulting in reduced NF-κB activity.

phosphatase or a serine-2 kinase inhibitor [42]. Additional mechanisms by which GRα may repress NF-κB-mediated gene activation include histone modifications [43] or chromatin remodeling [44,45]. GRα-mediated deacetylation of histones results in increased tightening of DNA around histones, reduced access of transcription factors to their binding sites, and repression of inflammatory genes. GRα associates with histone deacytelase (HDAC)-2 *in vivo* and represses p65 HAT activity [43] (Fig. 3). This mechanism is thought to explain the observation that overexpression of HDAC-2 in glucocorticoid-insensitive alveolar macrophages from patients with COPD is able to restore glucocorticoid sensitivity [46]. Finally, some of the rapid anti-inflammatory effects of glucocorticoids may be due to interference with phosphatidylinositol 3-kinase (PI3K) modulation of IKK activity (Fig. 3): such interference would result in reduced NF-κB activity [47]. Additionally, protein kinase A (PKA) has been proposed to mediate the cross-repression of GRα and NF-κB in the cytoplasm [48] (Fig. 3). Therefore, multiple pathways and/or cell type-specific mechanisms are likely involved in the reciprocal antagonism between GRα and NF-κB.

### 4.2. Activator protein-1

The activator protein-1 (AP-1) complex is also involved in upregulating the expression of many cytokine genes and tissue-destructive enzymes such as collagenase [49]. AP-1 is a member of the basic region-leucine zipper (bZIP) family of DNA-binding proteins, and is comprised of a Jun family member (c-Jun, v-Jun, Jun-B, or Jun-D) homodimerized with another Jun protein or heterodimerized with a Fos family member (c-Fos, Fos-B, Fra-1, or Fra-2). While the c-Fos/c-Jun heterodimer is ubiquitous within the cell, inflammatory cytokines have been shown to stimulate the dimerization via c-Jun phosphorylation by the c-Jun N-terminal kinase (JNK) [49], a member of the MAPK protein family.

GRα can repress AP-1 through some of the same mechanisms discussed above for NF-κB, including direct protein–protein interactions between GRα and the c-Jun subunit which also results in a reciprocal antagonism of transcription [50] (Fig. 3). GRα induces GILZ, which also binds to c-Fos and c-Jun, preventing AP-1 from binding to DNA [51] (Fig. 3). Glucocorticoids have also been suggested to negatively regulate AP-1 through a nongenomic mechanism involving inhibition of the activation/phosphorylation of JNK, and subsequent phosphorylation of c-Jun, thereby repressing AP-1 activity [52]. In this regard, a recent report suggests that this action may be mediated by GRα binding directly to JNK, preventing its phosphorylation/activation by MAP kinase kinase 7 (MKK7) [53]. Therefore, GRα repression of AP-1 resembles that of NF-κB and likely occurs through more than one specific pathway, and the mechanisms behind the mutual antagonism between these two transcription factors remain to be elucidated.

### 4.3. Other transcription factors

It is well documented that transrepression of NF-κB and AP-1 is a primary mechanism whereby glucocorticoids exert their anti-inflammatory effects (Fig. 3); however, GRα can interact with a host of other transcription factors which are activated through various proinflammatory signaling cascades. For example, GRα has been reported to interact with signal transduction and activator of transcription (STATs), interferon-regulatory factors (IRFs), and Sma-MAD

Table 1 Other functional glucocorticoid receptor (GR): transcription factor interactions

| TF | Effect on gene | Mechanism of action | Reference |
|---|---|---|---|
| STATs | | | |
| −1 | ↑ Fc γR1 (no GRE) | GR effect requires IFN-γ-activated STAT1 and PU.1 | [54] |
| −3 | ↑↑ α2-macroglobulin (no GRE) | GR interacts with IL-6-activated STAT3 | [55] |
| | ↑↑ fibrinogen (no GRE) | GR associates with IL-6-activated STAT3 | [56] |
| | ↑ MMTV promoter | GR associates with STAT3 | [57] |
| | ↓ heme oxygenase-1 promoter | GR abrogates IL-6-mediated binding of STAT-3 to DNA | [58] |
| −4 | ↓ IRF-1, IFN-δ | GR inhibits IL-12- or IFN-α-activated STAT-4 phosphorylation | [59] |
| −5 | ↓ a GRE-containing promoter | GR interacts with STAT5 and binds to DNA independent of GRE | [60] |
| | ↑↑ β-casein (half-site GRE) | GR interacts with prolactin activated STAT5 | [61] |
| | ↓ MMTV promoter | GR interacts with STAT5 | [62] |
| −6 | ↑↑ TLR-2 (half-site GRE) | GR effect requires TNF-α-activated STAT5 | [63] |
| | ↑↑ β-casein | GR effect requires IL-4-activated STAT6 | [64] |
| IRF-3 | ↓ IP-10 | GR disrupts p65/IRF-3 interactions | [65] |
| | ↓ IFN-β | GR competes with IRF-3 for GRIP | |
| Smads | ↑ MMTV promoter | GR interacts with Smad 3 | [66] |
| | ↓ Type-1 plasminogen activator inhibitor | GR interacts with Smad 3 | [67] |
| NF-AT | ↓ IL-4 promoter | GR interacts with NF-ATc | [68] |
| NF-IL6 | ↑↑ α 1-acid glycoprotein | GR directly interacts with C/EBP-β | [69] |
| (C/EBP-β) | ↓ IL-1β | GR inhibits DNA binding of C/EBP-β | [70] |
| OCT−1/−2 | ↑↑ MMTV | GR interacts with the POU domain of OCT | [71] |
| | ↓ GnRH | GR tethers itself to OCT at the nGRE | [31] |

↑↑ indicates cooperative or synergistic effects.

(Smads), which are stimulated through Janus Kinase (JAK), TLR, and transforming growth factor β (TGF-β) pathways, respectively. These and other GRα:protein interactions lead to repressive, stimulating, or even synergistic effects on gene transcription and will be described here briefly (see Table 1 for more examples).

STATs are activated by cytokines through the JAK pathway and may regulate numerous genes involved in host defense and innate immunity. Physical interactions between GRα and STAT-3 [55,56], STAT-5 [61,60,64,72], and STAT-6 [64] have been reported to synergistically induce certain STAT-activated genes, such as TLR-2 [63] or fibrinogen [56]. While STATs have also been shown to reciprocally modulate glucocorticoid-responsive genes, the result is not always transcriptional enhancement [62], and the same glucocorticoid-responsive element can be regulated differently depending on the type of STAT [62,57]. A recent report identified a member of the protein inhibitor of activated STAT (PIAS) family of proteins to interact directly with GRα, although the effects of this interaction on gene transcription remain to be determined [73]. Understanding the mechanisms behind glucocorticoid inhibition of STAT signaling will be imperative as these transcription factors have recently been considered to be new targets for asthma therapy [74].

In addition to JAK/STAT signaling, GRα can also modulate TLR signaling. As mentioned above, GRα and TNF-α can synergistically induce TLR-2 mRNA and protein [63]. Alternatively, GRα can repress TLR-signaling pathways through competitive or direct interactions with IRFs. For example, TLR-4 signaling through MyD88 is inhibited by GRα disruption of p65/IRF-3 complexes, which in turn represses IRF-3-dependent activation of IP-10 [37]. A similar mechanism of GRα transrepression was proposed for IRF-7 in TLR-9-dependent activation of target genes. More recently, the GRα coactivator, GRIP, has been reported to compete with IRF-3 in TLR-3 signaling of interferon-β, resulting in inhibition of gene transcription [65].

GRα also interacts with the Smad family of transcription factors, which are activated through the TGF-β-signaling pathway. TGF-β is an important cytokine with pleiotropic roles in cell differentiation, extracellular matrix production, as well as immune and inflammatory responses. Upon TGF-β stimulation, Smads have been found to interact with GRα either to enhance glucocorticoid induction of the mouse mammary tumor virus (MMTV LTR) promoter (Fig. 2) [66] or conversely to inhibit gene transcription of the type-1 plasminogen activator inhibitor (PAI-1) gene [67].

## 5. NONTRANSCRIPTIONAL MECHANISMS

A growing body of evidence suggests that certain rapid effects of glucocorticoids (occurring within seconds) are mediated by actions which are independent of transcription, and thus termed nongenomic. Some of these actions involve interactions with second messenger pathways in the cytoplasm, such as PKA or PI3K (described above and in Fig. 3). In addition, glucocorticoids inhibit epidermal growth factor (EGF)-stimulated cytosolic phospholipase $A_2$ ($cPLA_2$) activation and arachidonic acid (AA) release through a glucocorticoid-dependent, but transcription-independent, mechanism. The mechanism of this inhibition appears to be mediated by SRC kinase, which is rapidly released from the hsp90 complex following treatment with the synthetic glucocorticoid, dexamethasone, and which subsequently phosphorylates lipocortin 1. This action activates lipocortin 1 to compete with Grb2 (an EGF receptor adaptor protein), blocking recruitment of signaling factors to the EGF receptor [75,76] (Fig. 2). Glucocorticoids have also been shown to activate PI3/Akt signaling through a non-nuclear mechanism involving interaction with the regulatory subunit of PI3K, p85α, leading to endothelial nitric oxide release which has cardioprotective effects [77] as well as protective effects in ischemic brain injury [78]. Finally, it is possible that some nongenomic actions of glucocorticoids are mediated by a distinct membrane receptor which has different hormone-binding properties compared to the classical cytoplasmic receptor [79,80]; however, the anti-inflammatory actions of this membrane receptor are yet to be determined.

Glucocorticoids also increase cytokine mRNA turnover by acting at a post-transcriptional level through the adenine–uridine-rich regions of their 3́UTR. One proposed mechanism by which glucocorticoids accomplish these actions is through activation of ribonucleases which target mRNAs for degradation, as demonstrated for the interferon-β gene [81]. Tristetraprolin, an RNA-binding protein which recruits ribonucleases to the 3́UTR of cytokine mRNAs, is induced by dexamethasone, and mediates glucocorticoid suppression of TNF-α by decreasing mRNA half-life [82]. Glucocorticoids have also been shown to inhibit the p38 MAPK pathway, which stabilizes proinflammatory gene mRNAs (such as COX-2, TNF-α, IL-6, and IL-8) as well as negatively regulates innate immune response genes (such as TLR-2). One mechanism by which glucocorticoids suppress the p38 pathway is through transcriptional induction of MKP-1 [24,83] (Fig. 3). MKP-1 has been shown to prevent the phosphorylation/activation of JNK, ERK, and p38 [26]. Thus, dexamethasone inhibits p38 or JNK by inducing MKP-1, resulting in destabilization of inflammatory genes, such as COX-2 [83], TNF-α, and IL1-α/β mRNAs [84] while enhancing TLR-2 expression [63,85]. Glucocorticoids have also been shown to decrease the half-life of other mRNAs, such as IL-5 and monocyte chemoattractant protein (MCP)-1; however, the mechanism behind these effects is currently unknown [86,87].

## 6. GLUCOCORTICOID EFFECTS ON THE IMMUNE RESPONSE

Part of the anti-inflammatory action of glucocorticoids results from their ability to induce apoptosis, or programmed cell death, in immune cells, which would otherwise produce cytokines, chemokines, and adhesion molecules (Fig. 4). Glucocorticoids trigger apoptosis in many

| $\uparrow$ Apoptosis | $\downarrow$ Proinflammatory mediators | Effect of cytokine modulation |
|---|---|---|
|  T-lymphocyte | IL-5<br>MCP-2, MCP-3 | Inhibits eosinophil chemotaxis<br>Inhibits monocyte, dendritic, and natural killer cell chemotaxis |
|  Mast cell | IL-5<br>KC<br>IL-4, ICAM-1 | Inhibits eosinophil chemotaxis<br>Inhibits neutrophil recruitment<br>Induces mast cell apoptosis |
| 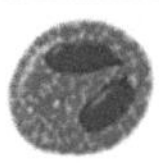 Eosinophil | MCP-1<br>IL-8 | Inhibits mononuclear recruitment and mast cell activation<br>Inhibits chemotaxis of leukocytes and lymphocytes |
| 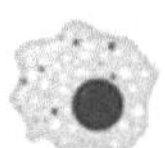 Macrophage | Integrin adhesion molecules<br>IL-12<br>TNF-α, IL-1β, GMCSF | Increases phagocytic clearance of neutrophils<br>Shift from $T_H1$ to $T_H2$<br>Decrease in chemotaxis/inflammation |
| 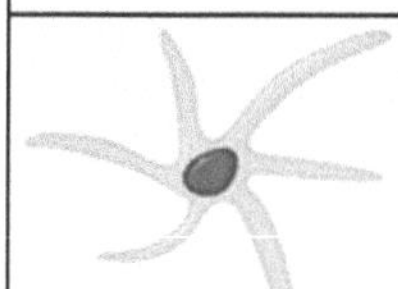 Dendritic cell | IL-12<br>TNF-α, IL-1β | Shift from $T_H1$ to $T_H2$<br>Decrease in inflammation, dendritic cell maturation |

Figure 4. Effect of glucocorticoids on immune cells. Glucocorticoids induce apoptosis in T lymphocytes [93], mast cells [94], eosinophils [95], macrophages [96], and dendritic cells [97]. They also suppress the production of a number of cytokines, chemokines, and adhesion molecules, having effects on phagocytosis, chemotaxis, and T-cell differentiation. IL, interleukin; MCP, monocyte chemoattractant protein; KC, cytokine-induced neutrophil chemoattractant; ICAM, intercellular adhesion molecule; TNF, tumor necrosis factor; GM-CSF = granulocyte macrophage colony-stimulating factor.

inflammatory cell types including mast cells, eosinophils, macrophages, dendritic cells, and T cells, but importantly, they protect against apoptotic stimuli in resident uncompromised cells, such as those of epithelial origin [88,89]. While the exact mechanism behind this cell type-specific effect has not been fully elucidated, it is believed that GRα exerts these effects through gene regulation, including both transactivation and transrepressive mechanisms, which leads to caspase and endonuclease activation and ultimately cell death [90]. Direct interactions between GRα, NF-κB, and AP-1 have been reported as causative events in glucocorticoid-mediated apoptosis [91]. In addition, glucocorticoids regulate transcription of several members of Bcl-2 family, which may be pro- or anti-apoptotic, and which may further undergo post-translational modifications, such as phosphorylation, modulating their activity [92]. The recent identification of different GRα isoforms which regulate unique sets of genes and have tissue-specific distribution [13] may also provide insight into glucocorticoid-mediated cell type-specific apoptosis.

Glucocorticoids not only induce apoptosis but also clear the body of apoptotic debris (which would otherwise cause an inflammatory reaction) by increasing the phagocytotic capacity of macrophages. The effect of dexamethasone on macrophage cytoskeleton and integrin signaling appears to be responsible for augmented phagocytosis of apoptotic neutrophils [98]. Furthermore, glucocorticoids inhibit immune cell trafficking by reducing the expression of cell adhesion molecules, such as VCAM [99], ICAM, or E-selectin [100]. Glucocorticoids also hinder eosinophil activity by decreasing their production and secretion of MCP-1 and IL-8 chemokines [101] and downregulating the eosinophil attractant, IL-5, in other cell types [102,103]. Dexamethasone inhibits mast cell-regulated skin infiltration of neutrophils mainly by attenuating CXC chemokine (KC) secretion and negatively regulating T-cell expression of MCP-2 and MCP-3, potentially affecting monocyte, dendritic cell, and natural killer cell chemotaxis as well [104] (Fig. 4). Certain components of innate immunity, such as acute phase and complement proteins, collectins, and surfactants are enhanced, rather than inhibited, by glucocorticoids [105]. One example is TLR-2, which is upregulated in a synergistic manner by GRα, STAT, and NF-κB as well as MKP-1. It has been proposed that glucocorticoids prime and enhance the innate immune response while repressing part of the adaptive immune response [106].

Glucocorticoid modulation of cytokines has been shown to regulate adaptive immunity by affecting the differentiation of naïve T helper cells and shifting the response from a T helper ($T_H1$) type to a $T_H2$ type. The mechanism of this $T_H1$ to $T_H2$ shift is largely through hindering the production of the $T_H1$ cell-inducing cytokine IL-12 by antigen-presenting cells [107]. Inhibiting the synthesis of $T_H1$ cytokines in turn permits the production of $T_H2$ cytokines [108]. While $T_H1$ cells are primarily involved with cellular immunity, $T_H2$ cells produce cytokines which regulate humoral immunity by promoting B-cell differentiation into antibody-secreting cells. Impaired cellular immunity may explain why people suffering from stress, which produces elevated levels of glucocorticoids, have an increased susceptibility to the common cold or reactivation herpes simplex virus [109,110].

## 7. SIDE EFFECTS OF GLUCOCORTICOID THERAPY

The discovery of the anti-inflammatory effects of glucocorticoids led to their use in the treatment of inflammatory disease; however, prolonged therapy has revealed several debilitating side effects. The most common effects of long-term glucocorticoid therapy are disorders

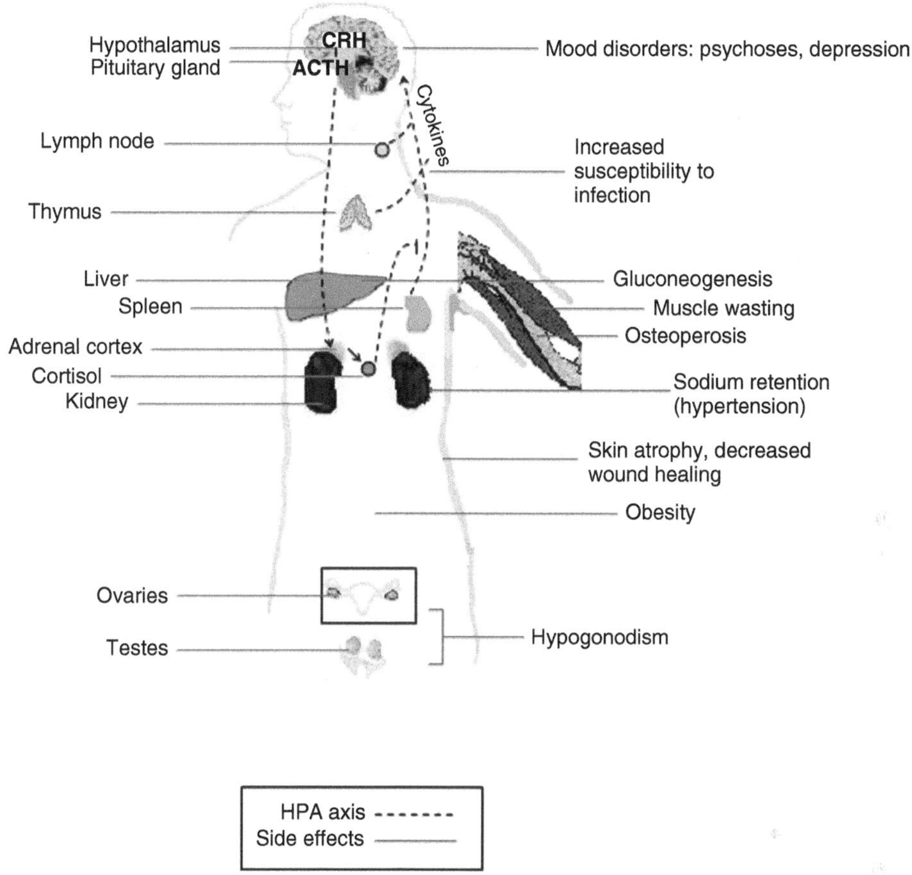


Figure 5. The hypothalamic–pituitary–adrenal (HPA) axis and the side effects of glucocorticoid therapy. The HPA axis is stimulated by cytokines that induce the expression and release of corticotropin-releasing hormone (CRH). CRH subsequently increases the production of adrenocorticotropic hormone (ACTH), which stimulates the cells of the adrenal cortex to synthesize and secrete glucocorticoids. Glucocorticoids in turn inhibit the inflammatory response, thus forming a negative feedback loop. A number of systemic side effects occur as a result of prolonged use of glucocorticoids and resulting HPA axis deregulation.

associated with hyperactivity of the HPA axis, such as diabetes, osteoporosis, and obesity (Fig. 5). A number of the genes thought to mediate these side effects contain GREs; however, GRα:protein interactions and nongenomic pathways have also been implicated. For example, many patients with Cushing's syndrome (hypercortisolism) exhibit increased abdominal fat. Numerous reports *in vitro* have found that glucocorticoids are necessary for fat cell differentiation, and a recent study indicates that GRα may potentiate preadipocyte differentiation through a nongenomic mechanism [111]. Glucocorticoids also stimulate gluconeogenesis in the liver, leading to hyperglycemia and often diabetes. GRα activates the transcription of numerous genes encoding enzymes involved in gluconeogenesis, such as tyrosine aminotransferase [112,113], aspartate aminotransferase [114], glucose-6-phosphatase [115], and phosphoenolpyruvate carboxykinase [116] through GREs or a combination of GREs and other transcription factors.

Long-term topical treatment of glucocorticoids causes thinning of epidermis and an inhibition of wound healing. Type 1 collagen, the major component in skin, is enhanced through TGF-β signaling of PAI-1 gene, and this signaling is inhibited through protein–protein interaction between GRα and Smad3 [67]. In addition, keratins, a family of epidermal genes, are repressed by GRα binding as four monomers to keratin nGREs or by GRα blocking of AP-1-mediated keratin gene transcription [117,118]. Skin atrophy in basal keratinocytes and wound healing in activated keratinocytes correlate with glucocorticoid repression of these keratin genes [118]. Wound healing is also repressed by glucocorticoids through the suppression of cytokines involved in the remodeling of bone and angiogenesis [119].

Glucocorticoids also promote osteoporosis by increasing bone absorption and decreasing bone formation, and patients receiving prolonged steroid treatment have a greater incidence of rib and vertebral fractures. Several bone matrix proteins are regulated by GRα in the osteoblast, namely collagen type 1 reduction (mentioned above) and osteocalcin. GRα binding to the osteocalcin nGRE prevents binding of the transcription factor, TATA box-binding protein (TBP) whose binding site overlaps with the nGRE [28,29]. The use of inhaled or systemic steroids has also been linked to growth suppression in children. The growth plate, the structure responsible for longitudinal growth, becomes thinner as a result of decreased chondrocyte proliferation and increased apoptosis. One mechanism by which dexamethasone both induces apoptosis and decreases proliferation in chondrocytes is through inhibition of Akt phosphorylation and therefore inhibition of the PI3K/Akt-signaling pathway [120].

Psychiatric illnesses, such as mood swings, depression, and psychosis (hallucinations and delusions), have been reported in patients receiving glucocorticoid therapy. These disorders are thought to occur in part through deregulation of the serotonin (5-HT) system. Direct inhibition of basal levels of the neuronal serotonin receptor gene (5-HT1A) occurs by GRα binding to a novel nGRE element involving two directly repeated GRE half-sites spaced by a hexamer [121]. Glucocorticoids have also been reported to repress NF-κB-mediated induction of 5-HT1A via two NF-κB elements in the promoter [122].

## 8. PERSPECTIVES

Glucocorticoids have remained unrivaled as anti-inflammatory agents since the 1940s when they were first put into use. Over the years, basic research has illuminated a number of transcriptional and nontranscriptional mechanisms which GRα utilizes to mediate these effects. Transcriptional mechanisms include GRα binding directly to GREs, other transcription factors, or a combination of both GREs and transcription factors. Nontranscriptional mechanisms include nongenomic actions in the cytoplasm as well as post-transcriptional effects on the stability of mRNAs. These signaling mechanisms lead to a suppression of immune responses at multiple levels, such as apoptosis, chemotaxis, and T-cell differentiation. However, because these mechanisms govern physiologic as well as inflammatory signaling, severe adverse effects accompany any beneficial therapeutic outcomes. It has become generally accepted that most of the anti-inflammatory effects of glucocorticoids are mediated by GRα–protein interaction (transrepression), while the systemic side effects of glucocorticoids are primarily mediated by GRE binding. This profile has led to the development of selective GR modulators that separate activation of transcription from transrepression. These selective, or "dissociated," glucocorticoids have met with some success *in vivo* in animals, but are not currently available for patients as they are generally less effective. Ongoing studies have not ruled out the effects of nonsteroidal, plant-derived GR modulators, such as Compound A, which have been shown to inhibit NF-κB without activating GRE-driven side effects [123,124]. Based on the knowledge gathered from the study of glucocorticoid-signaling pathways, MAPK inhibitors, NF-κB inhibitors, and HAT inhibitors have all been in development for the treatment of chronic inflammatory diseases [125–128]. Interestingly, steroid-sensitive and -resistant patients exhibit a different cellular activation pattern of NF-κB, AP-1, and MAPKs and may provide a diagnostic tool for early recognition of steroid resistance, thereby possibly preventing the side effects associated with prolonged or ineffective steroid treatment [129–133].

## ACKNOWLEDGMENT

We thank Robert Oakley for careful review and editing of this manuscript.

## REFERENCES

1. Glyn J. The discovery and early use of cortisone. J R Soc Med 1998;91(10):513–17.
2. Hench PKE, Slocumb CH, Polley HF. The effect of a hormone of the adrenal cortex (17-Hydroxy-11-Dehydrocorticosterone: Compound E) and of pituitary adrenocorticotropic hormone on rheumatoid arthritis. Rochester, MN. Proc Staff Meet Mayo Clin 1949;24(8):187–97.
3. Carryer HM, Koelsche GA, Prickman LE, Maytum CK, Lake CF, Williams HL. The effect of cortisone of bronchial asthma and hay fever occurring in subjects sensitive to ragweed pollen. J Allergy 1950;21(4):282–7.
4. Cole TJ, Blendy JA, Monaghan AP, Krieglstein K, Schmid W, Aguzzi A et al. Targeted disruption of the glucocorticoid receptor gene blocks adrenergic chromaffin cell development and severely retards lung maturation. Genes Dev 1995;9(13):1608–21.
5. Robinson-Rechavi M, Carpentier AS, Duffraisse M, Laudet V. How many nuclear hormone receptors are there in the human genome. Trends Genet 2001;17(10):554–6.
6. Lewis-Tuffin LJ, Jewell CM, Bienstock RJ, Collins JB, Cidlowski JA. The human glucocorticoid receptor {beta} (hGR{beta}) Binds RU-486 and is transcriptionally active. Mol Cell Biol 2007;27(6):2266–82.
7. Charmandari E, Chrousos GP, Ichijo T, Bhattacharyya N, Vottero A, Souvatzoglou E et al. The human glucocorticoid receptor (hGR) beta isoform suppresses the transcriptional activity of hGRalpha by interfering with formation of active coactivator complexes. Mol Endocrinol 2005;19(1):52–64.
8. Oakley RH, Jewell CM, Yudt MR, Bofetiado DM, Cidlowski JA. The dominant negative activity of the human glucocorticoid receptor beta isoform. Specificity and mechanisms of action. J Biol Chem 1999;274(39):27857–66.
9. Yudt MR, Jewell CM, Bienstock RJ, Cidlowski JA. Molecular origins for the dominant negative function of human glucocorticoid receptor beta. Mol Cell Biol 2003;23(12):4319–30.
10. Leung DY, Hamid Q, Vottero A, Szefler SJ, Surs W, Minshall E et al. Association of glucocorticoid insensitivity with increased expression of glucocorticoid receptor beta. J Exp Med 1997;186(9):1567–74.
11. Chikanza IC. Mechanisms of corticosteroid resistance in rheumatoid arthritis: a putative role for the corticosteroid receptor beta isoform. Ann N Y Acad Sci 2002;966:39–48.
12. Honda M, Orii F, Ayabe T, Imai S, Ashida T, Obara T et al. Expression of glucocorticoid receptor beta in lymphocytes of patients with glucocorticoid-resistant ulcerative colitis. Gastroenterology 2000;118(5):859–66.

13. Lu NZ, Cidlowski JA. Translational regulatory mechanisms generate N-terminal glucocorticoid receptor isoforms with unique transcriptional target genes. Mol Cell 2005;18(3):331–42.
14. Yudt MR, Cidlowski JA. Molecular identification and characterization of a and b forms of the glucocorticoid receptor. Mol Endocrinol 2001;15(7):1093–103.
15. Lu NZ, Cidlowski JA. Glucocorticoid receptor isoforms generate transcription specificity. Trends Cell Biol 2006;16(6):301–7.
16. Kumar R, Thompson EB. Transactivation functions of the N-terminal domains of nuclear hormone receptors: protein folding and coactivator interactions. Mol Endocrinol 2003;17(1):1–10.
17. Coghlan MJ, Elmore SW, Kym PR, Kort ME. The pursuit of differentiated ligands for the glucocorticoid receptor. Curr Top Med Chem 2003;3(14):1617–35.
18. Kumar R, Thompson EB. The structure of the nuclear hormone receptors. Steroids 1999;64(5):310–19.
19. Dittmar KD, Banach M, Galigniana MD, Pratt WB. The role of DnaJ-like proteins in glucocorticoid receptor.hsp90 heterocomplex assembly by the reconstituted hsp90.p60.hsp70 foldosome complex. J Biol Chem 1998;273(13):7358–66.
20. Dittmar KD, Demady DR, Stancato LF, Krishna P, Pratt WB. Folding of the glucocorticoid receptor by the heat shock protein (hsp) 90-based chaperone machinery. The role of p23 is to stabilize receptor.hsp90 heterocomplexes formed by hsp90.p60.hsp70. J Biol Chem 1997;272(34):21213–20.
21. Schoneveld OJ, Gaemers IC, Lamers WH. Mechanisms of glucocorticoid signalling. Biochim Biophys Acta 2004;1680(2):114–28.
22. Wieland S, Schatt MD, Rusconi S. Role of TATA-element in transcription from glucocorticoid receptor-responsive model promoters. Nucleic Acids Res 1990;18(17):5113–18.
23. Jantzen HM, Strahle U, Gloss B, Stewart F, Schmid W, Boshart M et al. Cooperativity of glucocorticoid response elements located far upstream of the tyrosine aminotransferase gene. Cell 1987;49(1):29–38.
24. Kassel O, Sancono A, Kratzschmar J, Kreft B, Stassen M, Cato AC. Glucocorticoids inhibit MAP kinase via increased expression and decreased degradation of MKP-1. EMBO J 2001;20(24):7108–16.
25. Barnes PJ. Anti-inflammatory actions of glucocorticoids: molecular mechanisms. Clin Sci (Lond) 1998;94(6):557–72.
26. De Bosscher K, Vanden Berghe W, Haegeman G. The interplay between the glucocorticoid receptor and nuclear factor-kappa B or activator protein-1. Molecular mechanisms for gene repression. Endocr Rev 2003;24(4):488–522.
27. Drouin J, Sun YL, Chamberland M, Gauthier Y, De Lean A, Nemer M et al. Novel glucocorticoid receptor complex with DNA element of the hormone-repressed POMC gene. EMBO J 1993;12(1):145–56.
28. Morrison N, Eisman J. Role of the negative glucocorticoid regulatory element in glucocorticoid repression of the human osteocalcin promoter. J Bone Miner Res 1993;8(8):969–75.
29. Meyer T, Carlstedt-Duke J, Starr DB. A weak TATA box is a prerequisite for glucocorticoid-dependent repression of the osteocalcin gene. J Biol Chem 1997;272(49):30709–14.
30. Philips A, Maira M, Mullick A, Chamberland M, Lesage S, Hugo P et al. Antagonism between Nur77 and glucocorticoid receptor for control of transcription. Mol Cell Biol 1997;17(10):5952–9.
31. Chandran UR, Warren BS, Baumann CT, Hager GL, DeFranco DB. The glucocorticoid receptor is tethered to DNA-bound Oct-1 at the mouse gonadotropin-releasing hormone distal negative glucocorticoid response element. J Biol Chem 1999;274(4):2372–8.
32. Novac N, Baus D, Dostert A, Heinzel T. Competition between glucocorticoid receptor and NFkappaB for control of the human FasL promoter. FASEB J 2006;20(8):1074–81.
33. Almawi WY, Melemedjian OK. Molecular mechanisms of glucocorticoid antiproliferative effects: antagonism of transcription factor activity by glucocorticoid receptor. J Leukoc Biol 2002;71(1):9–15.
34. Ghosh S, Karin M. Missing pieces in the NF-kappaB puzzle. Cell 2002;109(Suppl):S81–96.
35. McKay LI, Cidlowski JA. Cross-talk between nuclear factor-kappa B and the steroid hormone receptors: mechanisms of mutual antagonism. Mol Endocrinol 1998;12(1):45–56.

36. McKay LI, Cidlowski JA. Molecular control of immune/inflammatory responses: interactions between nuclear factor-kappa B and steroid receptor-signaling pathways. Endocr Rev 1999;20(4):435–59.
37. Ogawa S, Lozach J, Benner C, Pascual G, Tangirala RK, Westin S et al. Molecular determinants of crosstalk between nuclear receptors and Toll-like receptors. Cell 2005;122(5):707–21.
38. Eddleston J, Herschbach J, Wagelie-Steffen AL, Christiansen SC, Zuraw BL. The anti-inflammatory effect of glucocorticoids is mediated by glucocorticoid-induced leucine zipper in epithelial cells. J Allergy Clin Immunol 2007;119(1):115–22.
39. Marco BD, Massetti M, Bruscoli S, Macchiarulo A, Virgilio RD, Velardi E et al. Glucocorticoid-induced leucine zipper (GILZ)/NF-kappa B interaction: role of GILZ homo-dimerization and C-terminal domain. Nucleic Acids Res 2006;35(2):517–28.
40. McKay LI, Cidlowski JA. CBP (CREB binding protein) integrates NF-kappaB (nuclear factor-kappaB) and glucocorticoid receptor physical interactions and antagonism. Mol Endocrinol 2000;14(8):1222–34.
41. Tao Y, Williams-Skipp C, Scheinman RI. Mapping of glucocorticoid receptor DNA binding domain surfaces contributing to transrepression of NF-kappa B and induction of apoptosis. J Biol Chem 2001;276(4):2329–32.
42. Nissen RM, Yamamoto KR. The glucocorticoid receptor inhibits NFkappaB by interfering with serine-2 phosphorylation of the RNA polymerase II carboxy-terminal domain. Genes Dev 2000;14(18):2314–29.
43. Li J, Lin Q, Yoon HG, Huang ZQ, Strahl BD, Allis CD et al. Involvement of histone methylation and phosphorylation in regulation of transcription by thyroid hormone receptor. Mol Cell Biol 2002;22(16):5688–97.
44. Fletcher TM, Ryu BW, Baumann CT, Warren BS, Fragoso G, John S et al. Structure and dynamic properties of a glucocorticoid receptor-induced chromatin transition. Mol Cell Biol 2000;20(17):6466–75.
45. Belikov S, Gelius B, Almouzni G, Wrange O. Hormone activation induces nucleosome positioning in vivo. EMBO J 2000;19(5):1023–33.
46. Ito K, Yamamura S, Essilfie-Quaye S, Cosio B, Ito M, Barnes PJ et al. Histone deacetylase 2-mediated deacetylation of the glucocorticoid receptor enables NF-kappaB suppression. J Exp Med 2006;203(1):7–13.
47. Leis H, Page A, Ramirez A, Bravo A, Segrelles C, Paramio J et al. Glucocorticoid receptor counteracts tumorigenic activity of Akt in skin through interference with the phosphatidylinositol 3-kinase signaling pathway. Mol Endocrinol 2004;18(2):303–11.
48. Doucas V, Shi Y, Miyamoto S, West A, Verma I, Evans RM. Cytoplasmic catalytic subunit of protein kinase A mediates cross-repression by NF-kappa B and the glucocorticoid receptor. Proc Natl Acad Sci USA 2000;97(22):11893–8.
49. Johnson GL, Lapadat R. Mitogen-activated protein kinase pathways mediated by ERK, JNK, and p38 protein kinases. Science 2002;298(5600):1911–12.
50. Schule R, Rangarajan P, Kliewer S, Ransone LJ, Bolado J, Yang N et al. Functional antagonism between oncoprotein c-Jun and the glucocorticoid receptor. Cell 1990;62(6):1217–26.
51. Mittelstadt PR, Ashwell JD. Inhibition of AP-1 by the glucocorticoid-inducible protein GILZ. J Biol Chem 2001;276(31):29603–10.
52. Caelles C, Gonzalez-Sancho JM, Munoz A. Nuclear hormone receptor antagonism with AP-1 by inhibition of the JNK pathway. Genes Dev 1997;11(24):3351–64.
53. Bruna A, Nicolas M, Munoz A, Kyriakis JM, Caelles C. Glucocorticoid receptor-JNK interaction mediates inhibition of the JNK pathway by glucocorticoids. EMBO J 2003;22(22):6035–44.
54. Aittomaki S, Pesu M, Groner B, Janne OA, Palvimo JJ, Silvennoinen O. Cooperation among Stat1, glucocorticoid receptor,PU. 1 in transcriptional activation of the high-affinity Fc gamma receptor I in monocytes. J Immunol 2000;164(11):5689–97.

55. Takeda T, Kurachi H, Yamamoto T, Nishio Y, Nakatsuji Y, Morishige K et al. Crosstalk between the interleukin-6 (IL-6)-JAK-STAT and the glucocorticoid-nuclear receptor pathway: synergistic activation of IL-6 response element by IL-6 and glucocorticoid. J Endocrinol 1998;159(2):323–30.
56. Fuller GM, Zhang Z. Transcriptional control mechanism of fibrinogen gene expression. Ann N Y Acad Sci 2001;936:469–79.
57. Zhang Z, Jones S, Hagood JS, Fuentes NL, Fuller GM. STAT3 acts as a co-activator of glucocorticoid receptor signaling. J Biol Chem 1997;272(49):30607–10.
58. Lavrovsky Y, Drummond GS, Abraham NG. Downregulation of the human heme oxygenase gene by glucocorticoids and identification of 56b regulatory elements. Biochem Biophys Res Commun 1996;218(3):759–65.
59. Franchimont D, Galon J, Gadina M, Visconti R, Zhou Y, Aringer M et al. Inhibition of Th1 immune response by glucocorticoids: dexamethasone selectively inhibits IL-12-induced Stat4 phosphorylation in T lymphocytes. J Immunol 2000;164(4):1768–74.
60. Stocklin E, Wissler M, Gouilleux F, Groner B. Functional interactions between Stat5 and the glucocorticoid receptor. Nature 1996;383(6602):726–8.
61. Stoecklin E, Wissler M, Moriggl R, Groner B. Specific DNA binding of Stat5, but not of glucocorticoid receptor, is required for their functional cooperation in the regulation of gene transcription. Mol Cell Biol 1997;17(11):6708–16.
62. Biola A, Lefebvre P, Perrin-Wolff M, Sturm M, Bertoglio J, Pallardy M. Interleukin-2 inhibits glucocorticoid receptor transcriptional activity through a mechanism involving STAT5 (signal transducer and activator of transcription 5) but not AP-1. Mol Endocrinol 2001;15(7):1062–76.
63. Hermoso MA, Matsuguchi T, Smoak K, Cidlowski JA. Glucocorticoids and tumor necrosis factor alpha cooperatively regulate Toll-like receptor 2 gene expression. Mol Cell Biol 2004;24(11):4743–56.
64. Moriggl R, Berchtold S, Friedrich K, Standke GJ, Kammer W, Heim M et al. Comparison of the transactivation domains of Stat5 and Stat6 in lymphoid cells and mammary epithelial cells. Mol Cell Biol 1997;17(7):3663–78.
65. Reily MM, Pantoja C, Hu X, Chinenov Y, Rogatsky I. The GRIP1:IRF3 interaction as a target for glucocorticoid receptor-mediated immunosuppression. EMBO J 2006;25(1):108–17.
66. Aurrekoetxea-Hernandez K, Buetti E. Transforming growth factor beta enhances the glucocorticoid response of the mouse mammary tumor virus promoter through Smad and GA-binding proteins. J Virol 2004;78(5):2201–11.
67. Song CZ, Tian X, Gelehrter TD. Glucocorticoid receptor inhibits transforming growth factor-beta signaling by directly targeting the transcriptional activation function of Smad3. Proc Natl Acad Sci USA 1999;96(21):11776–81.
68. Chen R, Burke TF, Cumberland JE, Brummet M, Beck LA, Casolaro V et al. Glucocorticoids inhibit calcium- and calcineurin-dependent activation of the human IL-4 promoter. J Immunol 2000;164(2):825–32.
69. Nishio Y, Isshiki H, Kishimoto T, Akira S. A nuclear factor for interleukin-6 expression (NF-IL6) and the glucocorticoid receptor synergistically activate transcription of the rat alpha 1-acid glycoprotein gene via direct protein-protein interaction. Mol Cell Biol 1993;13(3):1854–62.
70. Waterman WR, Xu LL, Tetradis S, Motyckova G, Tsukada J, Saito K et al. Glucocorticoid inhibits the human pro-interleukin lbeta gene (ILIB) by decreasing DNA binding of transactivators to the signal-responsive enhancer. Mol Immunol 2006;43(7):773–82.
71. Prefontaine GG, Lemieux ME, Giffin W, Schild-Poulter C, Pope L, LaCasse E et al. Recruitment of octamer transcription factors to DNA by glucocorticoid receptor. Mol Cell Biol 1998;18(6):3416–30.
72. Wyszomierski SL, Yeh J, Rosen JM. Glucocorticoid receptor/signal transducer and activator of transcription 5 (STAT5) interactions enhance STAT5 activation by prolonging STAT5 DNA binding and tyrosine phosphorylation. Mol Endocrinol 1999;13(2):330–43.
73. Tirard M, Jasbinsek J, Almeida OF, Michaelidis TM. The manifold actions of the protein inhibitor of activated STAT proteins on the transcriptional activity of mineralocorticoid and glucocorticoid receptors in neural cells. J Mol Endocrinol 2004;32(3):825–41.

74. Roth M, Black JL. Transcription factors in asthma: are transcription factors a new target for asthma therapy. Curr Drug Targets 2006;7(5):589–95.
75. Croxtall JD, Choudhury Q, Flower RJ. Glucocorticoids act within minutes to inhibit recruitment of signalling factors to activated EGF receptors through a receptor-dependent, transcription-independent mechanism. Br J Pharmacol 2000;130(2):289–98.
76. Croxtall JD, van Hal PT, Choudhury Q, Gilroy DW, Flower RJ. Different glucocorticoids vary in their genomic and non-genomic mechanism of action in A549 cells. Br J Pharmacol 2002;135(2):511–19.
77. Hafezi-Moghadam A, Simoncini T, Yang Z, Limbourg FP, Plumier JC, Rebsamen MC et al. Acute cardiovascular protective effects of corticosteroids are mediated by non-transcriptional activation of endothelial nitric oxide synthase. Nat Med 2002;8(5):473–9.
78. Limbourg FP, Huang Z, Plumier JC, Simoncini T, Fujioka M, Tuckermann J et al. Rapid nontranscriptional activation of endothelial nitric oxide synthase mediates increased cerebral blood flow and stroke protection by corticosteroids. J Clin Invest 2002;110(11):1729–38.
79. Evans SJ, Searcy BT, Moore FL. A subset of kappa opioid ligands bind to the membrane glucocorticoid receptor in an amphibian brain. Endocrinology 2000;141(7):2294–300.
80. Powell CE, Watson CS, Gametchu B. Immunoaffinity isolation of native membrane glucocorticoid receptor from S-49++ lymphoma cells: biochemical characterization and interaction with Hsp 70 and Hsp 90. Endocrine 1999;10(3):271–80.
81. Peppel K, Vinci JM, Baglioni C. The AU-rich sequences in the 3′ untranslated region mediate the increased turnover of interferon mRNA induced by glucocorticoids. J Exp Med 1991;173(2): 349–55.
82. Smoak K, Cidlowski JA. Glucocorticoids regulate tristetraprolin synthesis and posttranscriptionally regulate tumor necrosis factor alpha inflammatory signaling. Mol Cell Biol 2006;26(23):9126–35.
83. Lasa M, Abraham SM, Boucheron C, Saklatvala J, Clark AR. Dexamethasone causes sustained expression of mitogen-activated protein kinase (MAPK) phosphatase 1 and phosphatase-mediated inhibition of MAPK p38. Mol Cell Biol 2002;22(22):7802–11.
84. Abraham SM, Lawrence T, Kleiman A, Warden P, Medghalchi M, Tuckermann J et al. Antiinflammatory effects of dexamethasone are partly dependent on induction of dual specificity phosphatase 1. J Exp Med 2006;203(8):1883–9.
85. Shuto T, Imasato A, Jono H, Sakai A, Xu H, Watanabe T et al. Glucocorticoids synergistically enhance nontypeable Haemophilus influenzae-induced Toll-like receptor 2 expression via a negative cross-talk with p38 MAP kinase. J Biol Chem 2002;277(19):17263–70.
86. Poon M, Liu B, Taubman MB. Identification of a novel dexamethasone-sensitive RNA-destabilizing region on rat monocyte chemoattractant protein 1 mRNA. Mol Cell Biol 1999;19(10):6471–8.
87. Staples KJ, Bergmann MW, Barnes PJ, Newton R. Evidence for post-transcriptional regulation of interleukin-5 by dexamethasone. Immunology 2003;109(4):527–35.
88. Amsterdam A, Tajima K,Sasson R. Cell-specific regulation of apoptosis by glucocorticoids: implication to their anti-inflammatory action. Biochem Pharmacol 2002;64(5–6):843–50.
89. Herr I, Gassler N, Friess H, Buchler MW. Regulation of differential pro- and anti-apoptotic signaling by glucocorticoids. Apoptosis 2006;12(2):271–91.
90. Distelhorst CW. Recent insights into the mechanism of glucocorticosteroid-induced apoptosis. Cell Death Differ 2002;9(1):6–19.
91. Greenstein S, Ghias K, Krett NL, Rosen ST. Mechanisms of glucocorticoid-mediated apoptosis in hematological malignancies. Clin Cancer Res 2002;8(6):1681–94.
92. Huang ST, Cidlowski JA. Phosphorylation status modulates Bcl-2 function during glucocorticoid-induced apoptosis in T lymphocytes. FASEB J 2002;16(8):825–32.
93. Herold MJ, McPherson KG, Reichardt HM. Glucocorticoids in T cell apoptosis and function. Cell Mol Life Sci 2006;63(1):60–72.
94. Yoshikawa H, Tasaka K. Suppression of mast cell activation by glucocorticoid. Arch Immunol Ther Exp (Warsz) 2000;48(6):487–95.

95. Druilhe A, Letuve S, Pretolani M. Glucocorticoid-induced apoptosis in human eosinophils: mechanisms of action. Apoptosis 2003;8(5):481–95.
96. Schmidt M, Pauels HG, Lugering N, Lugering A, Domschke W, Kucharzik T. Glucocorticoids induce apoptosis in human monocytes: potential role of IL-1 beta. J Immunol 1999;163(6):3484–90.
97. Boor PP, Metselaar HJ, Mancham S, Tilanus HW, Kusters JG, Kwekkeboom J. Prednisolone suppresses the function and promotes apoptosis of plasmacytoid dendritic cells. Am J Transplant 2006;6(10):2332–41.
98. Giles KM, Ross K, Rossi AG, Hotchin NA, Haslett C, Dransfield I. Glucocorticoid augmentation of macrophage capacity for phagocytosis of apoptotic cells is associated with reduced p130Cas expression, loss of paxillin/pyk2 phosphorylation, and high levels of active Rac. J Immunol 2001;167(2):976–86.
99. Atsuta J, Plitt J, Bochner BS, Schleimer RP. Inhibition of VCAM-1 expression in human bronchial epithelial cells by glucocorticoids. Am J Respir Cell Mol Biol 1999;20(4):643–50.
100. Cronstein BN, Kimmel SC, Levin RI, Martiniuk F, Weissmann G. A mechanism for the antiinflammatory effects of corticosteroids: the glucocorticoid receptor regulates leukocyte adhesion to endothelial cells and expression of endothelial-leukocyte adhesion molecule 1 and intercellular adhesion molecule 1. Proc Natl Acad Sci USA 1992;89(21):9991–5.
101. Miyamasu M, Misaki Y, Izumi S, Takaishi T, Morita Y, Nakamura H et al. Glucocorticoids inhibit chemokine generation by human eosinophils. J Allergy Clin Immunol 1998;101(1 Pt 1):75–83.
102. Schramm R, Thorlacius H. Neutrophil recruitment in mast cell-dependent inflammation: inhibitory mechanisms of glucocorticoids. Inflamm Res 2004;53(12):644–52.
103. Sewell WA, Scurr LL, Orphanides H, Kinder S, Ludowyke RI. Induction of interleukin-4 and interleukin-5 expression in mast cells is inhibited by glucocorticoids. Clin Diagn Lab Immunol 1998;5(1):18–23.
104. Pype JL, Dupont LJ, Menten P, Van Coillie E, Opdenakker G, Van Damme J et al. Expression of monocyte chemotactic protein (MCP)-1, MCP-2, and MCP-3 by human airway smooth-muscle cells. Modulation by corticosteroids and T-helper 2 cytokines. Am J Respir Cell Mol Biol 1999;21(4):528–36.
105. Schleimer RP. Glucocorticoids suppress inflammation but spare innate immune responses in airway epithelium. Proc Am Thorac Soc 2004;1(3):222–30.
106. Galon J, Franchimont D, Hiroi N, Frey G, Boettner A, Ehrhart-Bornstein M et al. Gene profiling reveals unknown enhancing and suppressive actions of glucocorticoids on immune cells. FASEB J 2002;16(1):61–71.
107. Blotta MH, DeKruyff RH, Umetsu DT. Corticosteroids inhibit IL-12 production in human monocytes and enhance their capacity to induce IL-4 synthesis in CD4+ lymphocytes. J Immunol 1997;158(12):5589–95.
108. Agarwal SK, Marshall GD, Jr. Dexamethasone promotes type 2 cytokine production primarily through inhibition of type 1 cytokines. J Interferon Cytokine Res 2001;21(3):147–55.
109. Cohen S, Tyrrell DA, Smith AP. Psychological stress and susceptibility to the common cold. N Engl J Med 1991;325(9):606–12.
110. Sainz B, Loutsch JM, Marquart ME, Hill JM. Stress-associated immunomodulation and herpes simplex virus infections. Med Hypotheses 2001;56(3):348–56.
111. Wiper-Bergeron N, Wu D, Pope L, Schild-Poulter C, Hache RJ. Stimulation of preadipocyte differentiation by steroid through targeting of an HDAC1 complex. EMBO J 2003;22(9):2135–45.
112. Grange T, Roux J, Rigaud G, Pictet R. Cell-type specific activity of two glucocorticoid responsive units of rat tyrosine aminotransferase gene is associated with multiple binding sites for C/EBP and a novel liver-specific nuclear factor. Nucleic Acids Res 1991;19(1):131–9.
113. Rigaud G, Roux J, Pictet R, Grange T. In vivo footprinting of rat TAT gene: dynamic interplay between the glucocorticoid receptor and a liver-specific factor. Cell 1991;67(5):977–86.
114. Beurton F, Bandyopadhyay U, Dieumegard B, Barouki R, Aggerbeck M. Delineation of the insulin-responsive sequence in the rat cytosolic aspartate aminotransferase gene: binding sites for hepatocyte nuclear factor-3 and nuclear factor I. Biochem J 1999;343(Pt 3):687–95.

115. Yoshiuchi I, Shingu R, Nakajima H, Hamaguchi T, Horikawa Y, Yamasaki T et al. Mutation/polymorphism scanning of glucose-6-phosphatase gene promoter in noninsulin-dependent diabetes mellitus patients. J Clin Endocrinol Metab 1998;83(3):1016–19.
116. Ruppert S, Boshart M, Bosch FX, Schmid W, Fournier RE, Schutz G. Two genetically defined trans-acting loci coordinately regulate overlapping sets of liver-specific genes. Cell 1990;61(5):895–904.
117. Dostert A, Heinzel T. Negative glucocorticoid receptor response elements and their role in glucocorticoid action. Curr Pharm Des 2004;10(23):2807–16.
118. Radoja N, Komine M, Jho SH, Blumenberg M, Tomic-Canic M. Novel mechanism of steroid action in skin through glucocorticoid receptor monomers. Mol Cell Biol 2000;20(12):4328–39.
119. Anstead GM. Steroids, retinoids, and wound healing. Adv Wound Care 1998;11(6):277–85.
120. Chrysis D, Zaman F, Chagin AS, Takigawa M, Savendahl L. Dexamethasone induces apoptosis in proliferative chondrocytes through activation of caspases and suppression of the Akt-phosphatidylinositol 3′-kinase signaling pathway. Endocrinology 2005;146(3):1391–7.
121. Ou XM, Storring JM, Kushwaha N, Albert PR. Heterodimerization of mineralocorticoid and glucocorticoid receptors at a novel negative response element of the 5-HT1A receptor gene. J Biol Chem 2001;276(17):14299–307.
122. Wissink S, Meijer O, Pearce D, van Der Burg B, van Der Saag PT. Regulation of the rat serotonin-1A receptor gene by corticosteroids. J Biol Chem 2000;275(2):1321–6.
123. De Bosscher K, Vanden Berghe W, Beck IM, Van Molle W, Hennuyer N, Hapgood J et al. A fully dissociated compound of plant origin for inflammatory gene repression. Proc Natl Acad Sci USA 2005;102(44):15827–32.
124. De Bosscher K, Vanden Berghe W, Haegeman G. Cross-talk between nuclear receptors and nuclear factor kappaB. Oncogene 2006;25(51):6868–86.
125. Barnes PJ. Targeting histone deacetylase 2 in chronic obstructive pulmonary disease treatment. Expert Opin Ther Targets 2005;9(6):1111–21.
126. Barnes PJ. Transcription factors in airway diseases. Lab Invest 2006;86(9):867–72.
127. Catley MC, Sukkar MB, Chung KF, Jaffee B, Liao SM, Coyle AJ et al. Validation of the anti-inflammatory properties of small-molecule IkappaB Kinase (IKK)-2 inhibitors by comparison with adenoviral-mediated delivery of dominant-negative IKK1 and IKK2 in human airways smooth muscle. Mol Pharmacol 2006;70(2):697–705.
128. Cosio BG, Tsaprouni L, Ito K, Jazrawi E, Adcock IM, Barnes PJ. Theophylline restores histone deacetylase activity and steroid responses in COPD macrophages. J Exp Med 2004;200(5):689–95.
129. Adcock IM, Lane SJ, Brown CR, Lee TH, Barnes PJ. Abnormal glucocorticoid receptor-activator protein 1 interaction in steroid-resistant asthma. J Exp Med 1995;182(6):1951–8.
130. Bantel H, Domschke W, Schulze-Osthoff K, Kaskas B, Gregor M. Abnormal activation of transcription factor NF-kappaB involved in steroid resistance in chronic inflammatory bowel disease. Am J Gastroenterol 2000;95(7):1845–6.
131. Bantel H, Schmitz ML, Raible A, Gregor M, Schulze-Osthoff K. Critical role of NF-kappaB and stress-activated protein kinases in steroid unresponsiveness. FASEB J 2002;16(13):1832–4.
132. Lane SJ, Adcock IM, Richards D, Hawrylowicz C, Barnes PJ, Lee TH. Corticosteroid-resistant bronchial asthma is associated with increased c-fos expression in monocytes and T lymphocytes. J Clin Invest 1998;102(12):2156–64.
133. Sousa AR, Lane SJ, Cidlowski JA, Staynov DZ, Lee TH. Glucocorticoid resistance in asthma is associated with elevated in vivo expression of the glucocorticoid receptor beta-isoform. J Allergy Clin Immunol 2000;105(5):943–50.

*The Hypothalamus–Pituitary–Adrenal Axis*
Edited by A. del Rey, G. P. Chrousos and H. O. Besedovsky



# *Organization of the Sympathetic Nervous System: Peripheral and Central Aspects*

WILFRID JÄNIG

*Physiologisches Institut, Christian-Albrechts-Universität zu Kiel, Kiel, Germany*

## ABSTRACT

The sympathetic nervous system is involved in many autonomic regulations leading to homeostasis and adaptation of the homeostatic regulations to the internal and external demands on the body. The basic peripheral building blocks of these regulations are the final sympathetic pathways, consisting of populations of pre- and postganglionic neurons that are functionally defined according to the effector cells they supply (vascular smooth muscle cells, non-vascular smooth muscle cells, secretory epithelia, etc.).

Each group of target cells is innervated (and regulated) by one (rarely two) sympathetic final pathways. Neurons of a specific, functionally defined sympathetic pathway exhibit a typical discharge pattern, which is the result of integrative processes in distinct neural circuits of the spinal cord, brain stem, hypothalamus, and telencephalon.

Impulse activity in preganglionic neurons is transmitted in the sympathetic ganglia to postganglionic neurons of the same functional type. There is no communication between functionally different sympathetic pathways in the autonomic ganglia. Activity in sympathetic postganglionic axons is transmitted to the effector cells largely via anatomically and functionally defined neuroeffector junctions.

Spinal cord, brain stem, and hypothalamus contain functionally distinct neural circuits that are connected to the final sympathetic pathways. Only a few of these central circuits have been explored and can be described in detail. The lowest level of central integration occurs in the spinal cord. Spinal circuits are probably integrated in every sympathetically mediated regulation.

The lower brain stem (pons and medulla oblongata) contains the neural circuits involved in (1) homeostatic cardiovascular regulation and its integration with the regulation of respiration, (2) regulation of body core temperature, (3) regulation of pelvic organs, and (4) regulation of the gastrointestinal tract (including regulation of food intake and metabolism).

Neural circuits in spinal cord and lower brain stem, which represent the different types of homeostatic regulation, are integral components of complex regulations represented in the hypothalamus and mesencephalon, which include neuroendocrine systems and somato-motor systems and constitute elementary behaviors. The telencephalon (neocortex and limbic system) adapts these functions to the external state of the organism.

The sympathetic nervous system is involved in the regulation of protection of body tissues against external and internal threatening events. This function is closely related to the neural control of the immune system. During fast defense (confrontational defense, flight, quiescence), the body is prepared for protection and during slow defense the organism switches to recuperation. Both involve the sympatho-neural systems and the sympatho-adrenal (SA) system and are related to the neural control of inflammation and sensitivity of nociceptors (hyperalgesia). The sympathetically mediated protective functions are controlled by neural circuits in the spinal cord, brain, and hypothalamus. These in turn are potentially under telencephalic control.

**Keywords:** Sympathetic nervous system, functional organization, peripheral organization, central organization, signal transmission in peripheral pathways, neuro–immune interaction, body protection and sympathetic nervous system, inflammation and sympatho-adrenal system, nociceptor sensitization and sympatho-adrenal system, hyperalgesia and inflammation, telencephalic control

## 1. INTRODUCTION

All living organisms interact continuously with their environment. They receive multiple signals from the environment via sensory systems and respond by way of their somato-motor systems. Both sensory processing and motor actions are entirely under control of the central nervous system (CNS). The brain contains the representations of extracorporeal space and somatic body domains. It generates complex motor commands on the basis of these central representations and the afferent feedback from the body and the environment leading to movements of the body in its environment against internal and external forces. The tools for performing these actions are the effector machines, the skeletal muscles, and their controlling somato-motoneurons. The body's motor activity is only possible when its internal milieu is controlled to keep the component cells, tissues, and organs (including the brain and skeletal muscles) maintained in an optimal state for their functioning. This enables the organism to adjust its performance to the varying internal and external demands placed on the organism.

The control of the internal milieu is also exerted by the brain by acting on many peripheral target tissues (smooth muscle cells of various organs, cardiac muscle cells, exocrine glands, endocrine cells, metabolic tissues, immune cells, etc.). The *efferent signals* from the brain to the periphery of the body by which this control is achieved are *neural* (by the autonomic nervous systems) and *hormonal* (by the neuroendocrine systems). The *afferent signals* from the periphery of the body to the brain are neural, hormonal (e.g., hormones from endocrine organs including those in the gastrointestinal tract, cytokines from the immune system, leptin from adipocytes), and physicochemical (e.g., blood glucose level, body core temperature, etc.).

The maintenance of physiological parameters like concentration of ions, blood glucose, arterial blood gases, body core temperature, and so on in a narrow range is called *homeostasis*. The concept of homeostasis has been formulated by Walter B. Cannon [1,2] based on the idea of the fixity of the internal milieu of the body, which has been formulated first by Claude Bernard [3]. The process of maintaining stability of the internal milieu during changes in the body and in the environment requires systems that have a large range of activity such as the cardiovascular system, the thermoregulatory system, the metabolic system (involving the gastrointestinal tract and endocrine systems releasing insulin, glucagon, leptin, and thyroxin), and the immune system. The adaptation of the parameters of the internal milieu in response to internal and environmental challenges (exercise, hunger, temperature load, and physical threat) is called *allostasis* [4].[1]

[1] These adaptive allostatic regulations are physiologically rapidly mobilized during internal or environmental perturbations and then turned off when not needed any longer. Allostatic responses maintained active over long periods of time result in "wear and tear" of the mechanisms involved including neuronal ones. This is called *allostatic load*. The consequences of allostatic load may lead to various types of diseases, such as hypertension, myocardial infarction, obesity, diabetes, atherosclerosis, and metabolic syndrome [5–7].

By analogy with the organization of the somato-motor system and the sensory systems in the brain, it is natural to accept the concept that autonomic and endocrine homeostatic regulations are also represented in the brain (i.e., in the hypothalamus, brain stem, and spinal cord) and that these regulations are under the control of the telencephalon and integrated there with the somato-motor and sensory representations. Thus, the brain contains autonomic "sensorimotor programs" for the coordinated regulation of the internal environment of the body's tissues and organs and sends efferent commands to the peripheral target tissues through the autonomic and endocrine routes.

Autonomic regulation of body functions requires the existence of specific neuronal pathways in the periphery and a specific organization of neural circuits connected to these pathways in the CNS, otherwise it would be impossible to understand the rapid adjustments of the body in higher vertebrates during diverse behaviors. This implies that the various autonomic systems must be centrally integrated and must have multiple distinct peripheral pathways. These pathways are defined according to the function they induce in the target cells they innervate. The effector cells of the autonomic nervous system are diverse while those of the somatic efferent system are not. Thus, the autonomic nervous system is the major efferent component of the peripheral nervous system that outweighs in its diversity of function and size the somatic efferent pathways.

Based on this reasoning, I will summarize three groups of experimental data:

1. The sympathetic nervous system is organized in the periphery in many functionally and anatomically separate pathways that transmit the centrally generated impulses faithfully to the effector cells. The same applies to the parasympathetic nervous system. This will not be described further here [8].
2. The peripheral sympathetic pathways are connected to reflex circuits in spinal cord, brain, and hypothalamus that are characteristic for each peripheral sympathetic pathway. Autonomic regulations that are represented in the spinal cord, brain, and hypothalamus and are exerted via the peripheral sympathetic pathways are closely integrated with the neuroendocrine and somato-motor systems.
3. The sympathetic and sympatho-adrenal (SA) systems are involved in the regulation of the immune system, inflammation, and hyperalgesia. In these functions, they are important in the regulation of protection of the body against injury from outside as well as from inside of the body.

Details have been described and reviewed in the literature (see Refs [8–23] for groups 1 and 2 and Refs [24–26] for group 3).

## 2. FUNCTIONAL ORGANIZATION OF SYMPATHETIC PATHWAYS

### 2.1. Definitions

Langley [27] originally proposed the generic term "autonomic nervous system" to describe the innervation of virtually all tissues and organs except striated muscle fibers. Langley's division of the autonomic nervous system into the sympathetic, parasympathetic, and enteric nervous systems is now universally applied. The definition of the *sympathetic* and *parasympathetic nervous systems* is primarily *anatomical* (the thoracolumbar system or sympathetic system; the craniosacral or parasympathetic system). The *enteric nervous system* is intrinsic to the wall of the gastrointestinal tract and consists of interconnecting plexuses along its length [28,29].

In the definition of the terms sympathetic and parasympathetic, afferent neurons are not included. About 85% of the axons in the vagus nerves and up to 50% of those in the splanchnic nerves (greater, lesser, least, lumbar and pelvic) are afferent and are called spinal or vagal visceral afferents. They come from sensory receptors in the internal organs and have their cell bodies in the ganglia of the 9th and 10th cranial nerves and in the dorsal root ganglia of the spinal segments corresponding to the autonomic outflow. Sometimes thoracolumbar and sacral afferents are labeled "sympathetic" or "parasympathetic," but this nomenclature is misleading. This somewhat strict separation does not preclude that visceral afferents are important in most autonomic reflexes and regulations [8,30–35].

The sympathetic and parasympathetic systems consist of two populations of neurons in series which are connected synaptically. The cell bodies of the final sympathetic and parasympathetic neurons are grouped in *autonomic ganglia*. Their axons are unmyelinated and project from these ganglia to the target organs. These neurons are called *ganglion cells* or *postganglionic neurons*. The cell bodies of the *preganglionic* neurons lie in the spinal cord and brain stem. They send axons from the CNS into the ganglia and form synapses on the dendrites and cell bodies of the postganglionic neurons. Their axons are myelinated as well as unmyelinated.

### 2.2. Reflex patterns as functional markers

Many individual sympathetic pre- and postganglionic neurons exhibit spontaneous activity and/or can be activated or inhibited reflexly by appropriate physiological stimuli. This has been shown in anesthetized cats (and for some systems in rats) for neurons of the lumbar sympathetic outflow to skeletal muscle, skin, and pelvic viscera [9,10,12,36–39] and for neurons of the thoracic sympathetic outflow to the head and neck [40,41], as well as in awake humans for the sympathetic outflow to skeletal muscles and skin [16,42,43]. The reflexes observed correspond to the effector responses that are induced by changes in activity of these neurons. The reflex patterns elicited by physiological stimulation of various afferent input systems are characteristic and therefore represent physiological "fingerprints" for each sympathetic pathway. Figure 1 illustrates reflex responses in some major classes of sympathetic neurons in the anesthetized cat:

- Reflex patterns in muscle and visceral vasoconstrictor neurons consist of inhibition by arterial baroreceptors but excitation by arterial chemoreceptors, cutaneous nociceptors, and spinal visceral nociceptors (Fig. 1$A_1$, $A_2$, and B).
- Most cutaneous vasoconstrictor (CVC) neurons are inhibited by stimulation of cutaneous nociceptors of the distal extremities, spinal visceral afferents, arterial chemoreceptors, and central warm-sensitive neurons in the spinal cord and hypothalamus (Fig. 1$A_1$, $A_2$, $A_3$, $A_4$, and B).
- Sudomotor neurons are activated by stimulation of Pacinian corpuscles in skin and by some other afferent stimuli (Fig. 1$A_3$).
- Motility-regulating neurons innervating pelvic organs are excited or inhibited by stimulation of sacral afferents from the urinary bladder, hindgut, or anal canal but are not affected by arterial baroreceptor activation. Functionally, different types of motility-regulating neurons can be discriminated by way of their reflex pattern.
- Neurons firing only during inspiration, which innervate effectors of the head, are excited by noxious stimulation and stimulation of chemoreceptors, most of them not being under baroreceptor control (Fig. 1B).
- Some types of sympathetic neurons are only activated in special functional conditions: pilomotor neurons, vasodilator neurons to skeletal muscle or skin [44], neurons innervating the pineal gland, lipomotor neurons innervating brown adipose tissue in the rat [23,45], pupillomotor neurons,

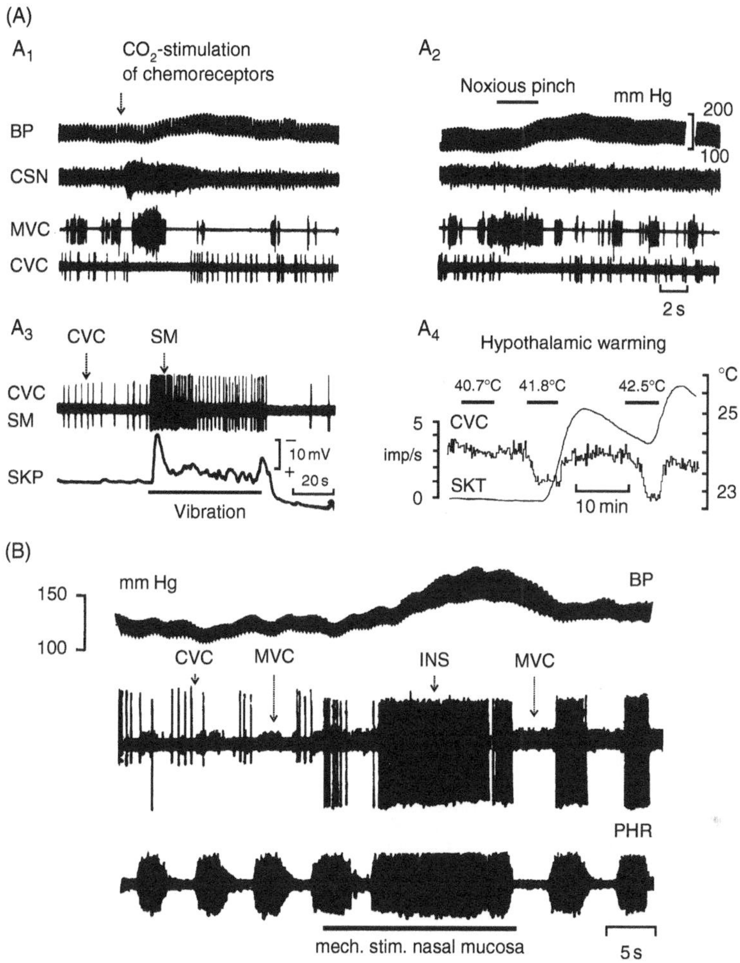


Figure 1. Reflex patterns in sympathetic neurons. (A) Reflexes in muscle (MVC) and cutaneous (CVC) vasoconstrictor and sudomotor (SM) neurons recorded from postganglionic axons in anesthetized cats. ($A_1$) Stimulation of the carotid chemoreceptors by a bolus injection of $CO_2$-enriched saline (at arrow) activated the MVC neurons and inhibited the CVC neurons (recorded simultaneously). Increased afferent activity in the carotid sinus nerve (CSN) was monitored. The increase of blood pressure evoked by chemoreceptor stimulation led to a baroreceptor-mediated inhibition of MVC activity following chemoreceptor reflex activation but not of CVC activity. ($A_2$) Stimulation of cutaneous nociceptors by pinching the ipsilateral hindpaw (indicated by bar) also excited the MVC neurons and inhibited the CVC neurons. ($A_3$) Simultaneous recordings of the activity in a single CVC neuron (small signal) and in a single SM neuron (large signal) and the skin potential (SKP) from the central paw pad. Stimulation of Pacinian corpuscles in the hindpaws by vibration excited the SM neuron and inhibited the CVC neuron. SM activation was correlated with the changes in skin potential. ($A_4$) Inhibition of CVC neurons to warming of the anterior hypothalamus. Note the increase of skin temperature (SKT) on the central paw pad following depression of CVC activity. (B) Reflexes in a CVC neuron, an inspiratory (INS) sympathetic neuron, and a MVC neuron to mechanical (noxious) stimulation of the nasal mucosa in the cat. Simultaneous recording from three preganglionic axons in a strand isolated from the cervical sympathetic trunk in an anesthetized cat. Note that this stimulus (1) activated the INS neuron but only during phrenic nerve discharge (PHR), (2) inhibited the CVC neuron, which was otherwise only active during expiration, and (3) excited the MVC neuron, which was active in inspiration and expiration before the stimulus. The reflexes in these neurons outlasted the stimulus. The increase in blood pressure (BP) was correlated with the continuous MVC discharge during and after the stimulus. Data for $A_1$–$A_3$ from Jänig and Kümmel, unpublished; $A_4$ from Ref. [51]; B modified from Ref. [48].

neurons innervating cells of the adrenal medulla-releasing adrenaline [45], and neurons innervating pelvic erectile tissue and generating dilation of this tissue [46,47]).

- Regulation of respiration and regulation of peripheral sympathetic pathways are closely integrated in the medulla oblongata. This integration is reflected in respiratory rhythmicity of the activity in sympathetic neurons. The rhythmicity is not uniform but varies in its pattern (respiratory activity profile) according to the function of the sympathetic neurons [14,16,48–51]. Figure 1B illustrates an example. The muscle vasoconstrictor (MVC) neurons discharge in inspiration and expiration. The CVC neurons discharge either only in expiration or only in inspiration or do not exhibit respiratory modulation of their activity. The inspiratory neurons discharge only in inspiration. Sudomotor neurons discharge in postinspiration. Most motility-regulating neurons exhibit no respiratory modulation in their activity.

So far, more than 10 different functional groups of postganglionic and preganglionic sympathetic neurons have been identified in the lumbar sympathetic outflow to skin, skeletal muscle, and viscera and in the thoracic sympathetic outflow to head and neck of the cat (see Fig. 2 for

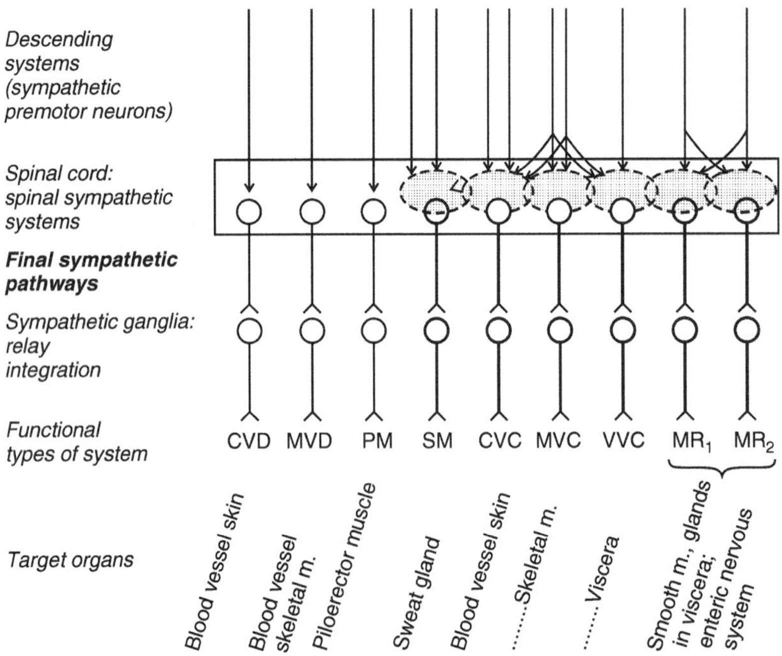


Figure 2. Lumbar sympathetic systems supplying skeletal muscle and skin of the hindlimb and pelvic organs (including distal colon) in the cat. The preganglionic neurons of these systems are largely located in the lumbar segments L1–L5 and project in the lumbar sympathetic trunk distal to paravertebral ganglion L5 or in the lumbar splanchnic nerves and the hypogastric nerves. The first group projects (with a few exceptions related to pelvic organs) to postganglionic neurons innervating somatic tissues and the second group to postganglionic neurons innervating pelvic organs and colon. The second group includes other sympathetic pathways associated with the internal sexual organs and probably other target cells in the pelvic organs that are not included in the figure. Preganglionic neurons are associated with spinal circuits (shaded areas) and are under multiple supraspinal control via descending systems (see Fig. 5). Sympathetic neurons with spontaneous activity are in bold. CVC, cutaneous vasoconstrictor; CVD, cutaneous vasodilator; MR, motility-regulating; MVC, muscle vasoconstrictor; MVD, muscle vasodilator; PM, pilomotor; SM, sudomotor; VVC, visceral vasoconstrictor. Modified from Ref. [12].

the lumbar sympathetic outflow to skin, skeletal muscle, and pelvic organs). The same types of reflex patterns have been observed in both preganglionic as well as postganglionic neurons. Most neurons in several of these pathways (e.g., the vasoconstrictor, sudomotor, motility-regulating pathways) have ongoing activity, whereas in four pathways the neurons are normally silent (e.g., the pilomotor and vasodilator pathways). It is likely that other target cells are innervated by other functionally distinct groups of sympathetic neurons which have not been studied so far. These sympathetic neurons innervate, for example, the kidney (blood vessels and juxtaglomerular cells), the urogenital tract (the urinary bladder, vas deferens, erectile tissue, and glandular tissue), the hindgut, the spleen (immune tissue), the heart, the brown adipose tissue [23,45], and so on. To emphasize, most of the data have been obtained under standardized experimental conditions in anesthetized cats. In humans, this type of standardized experimentation is not possible. Furthermore, no direct recording from autonomic preganglionic neurons and from autonomic neurons innervating viscera and head can be made in humans. However, using microneurographic recordings from bundles with few or single postganglionic axons in human skin and muscle nerves, it has clearly been shown that MVC, CVC, and sudomotor neurons have distinct reflex patterns and that there is also evidence for the existence of sympathetic vasodilator neurons supplying skin and skeletal muscle in humans [15,16,42,43].

Several systematic studies have been made on the functional properties of parasympathetic pre- and postganglionic neurons [11,15]. There are good reasons to assume that the principle of organization into functionally discrete pathways is the same as in the sympathetic nervous system, the main difference being that some targets of the sympathetic system are widely distributed throughout the body (e.g., blood vessels, sweat glands, erector pili muscles, and fat tissue), whereas the targets of most parasympathetic pathways are more restricted [8,52–54].

## 2.3. Transmission of signals in peripheral sympathetic pathways

### *2.3.1. Autonomic ganglia*

A major function of the peripheral ganglia is to distribute the centrally integrated signals by connecting each preganglionic axon with several postganglionic neurons. The extent of divergence varies significantly, the ratio of pre- to postganglionic axons being, in pathways such as in the ciliary ganglion to the iris and ciliary body, as low as 1:4 and in others, such as in the superior cervical ganglion with many vasoconstrictor neurons, as high as 1:150. However, it is clear that limited divergence and much divergence, respectively, are not characteristics of the parasympathetic or sympathetic systems [55]. Probably, by analogy with somatic motor units, limited divergence is common in pathways to small targets with discrete functions (e.g., in autonomic pathways to the inner muscles of the eye), whereas widespread divergence is a feature of pathways to anatomically extensive effectors that act more or less simultaneously (e.g., in vasoconstrictor pathways) [8,11,56].

### *2.3.2. Sympathetic paravertebral ganglia*

Within sympathetic paravertebral ganglia (in the sympathetic chains), ganglionic neurons have uniform properties. Each convergent cholinergic preganglionic axon produces excitatory post-synaptic potentials by activating nicotinic receptor channels. The amplitude of the potentials varies between inputs, ranging from a few millivolts (subthreshold weak synaptic inputs) to

10–40 mV (suprathreshold strong synaptic inputs). In most cases, one or two strong preganglionic synaptic inputs have, like the endplate potential at the skeletal neuromuscular junction, a high safety factor and always initiate an action potential. Thus, the ganglion cell relays the incoming CNS-derived signals in only a few of its preganglionic inputs [57,58]. The function of weak preganglionic synaptic inputs generating small subthreshold potentials in paravertebral ganglia is not clear (Fig. 3A).

### 2.3.3. *Sympathetic prevertebral ganglia*

In prevertebral (sympathetic) ganglia, postganglionic neurons, at least in experimental animals, do not have uniform properties. Three broad groups differ electrophysiologically (by the $K^+$ channels that control excitability), morphologically (by their size and dendritic branching), and neurochemically (by their neuropeptide content) [59]. Two groups, like paravertebral neurons, have suprathreshold synaptic connections with one or two preganglionic axons that determine the firing pattern of these neurons. The mode of synaptic transmission in the third group is different. These neurons receive weak preganglionic synaptic inputs that only activate them by summation. However, they also receive many cholinergic nicotinic synaptic inputs from intestinofugal neurons of the enteric nervous system that can be activated by mechanical stimulation of the gastrointestinal tract. Summation of synaptic potentials from peripheral and preganglionic inputs is necessary to initiate their discharge. These postganglionic neurons also receive synaptic input from collaterals of visceral primary afferent neurons (that have their cell bodies in the dorsal root ganglia) and depolarize slowly when these inputs are activated at high frequency. The slow responses are generated by the neuropeptide substance P released from the afferent collaterals. These prevertebral neurons therefore depend on temporal and spatial integration of incoming synaptic signals and may establish peripheral (extraspinal) reflexes (Fig. 3B).

### 2.3.4. *Parasympathetic ganglia*

The structure of many parasympathetic ganglion cells, with few dendrites, is simpler than that of sympathetic neurons. The preganglionic input is correspondingly simple, often consisting of a single suprathreshold input. However, some parasympathetic ganglia in the body trunk contain, in addition to postganglionic neurons, neurons that behave as primary afferent neurons or interneurons, that is, these ganglia have the potential for reflex activity independent of the CNS, like the enteric nervous system (intracardiac ganglia [60]; see also Ref. [61]).

The pelvic or hypogastric plexuses contain the neurons that innervate the pelvic organs. Some of these ganglion cells are noradrenergic and are innervated by lumbar sympathetic preganglionic axons, others are cholinergic and receive sacral parasympathetic inputs. A small proportion of pelvic neurons receive synaptic connections from both sympathetic preganglionic neurons projecting through the hypogastric nerves and parasympathetic preganglionic neurons projecting through the pelvic nerves [62].

### 2.3.5. *Transmission of signals at the autonomic neuroeffector junctions*

In peripheral tissues, the effects of activity in autonomic nerve terminals on autonomic effector cells are complex and may depend on the release of several different compounds and on the presence and distribution of the receptors in the effector membranes for these compounds. Anatomical investigations of neuroeffector junctions at arterioles, veins, pacemaker cells of the

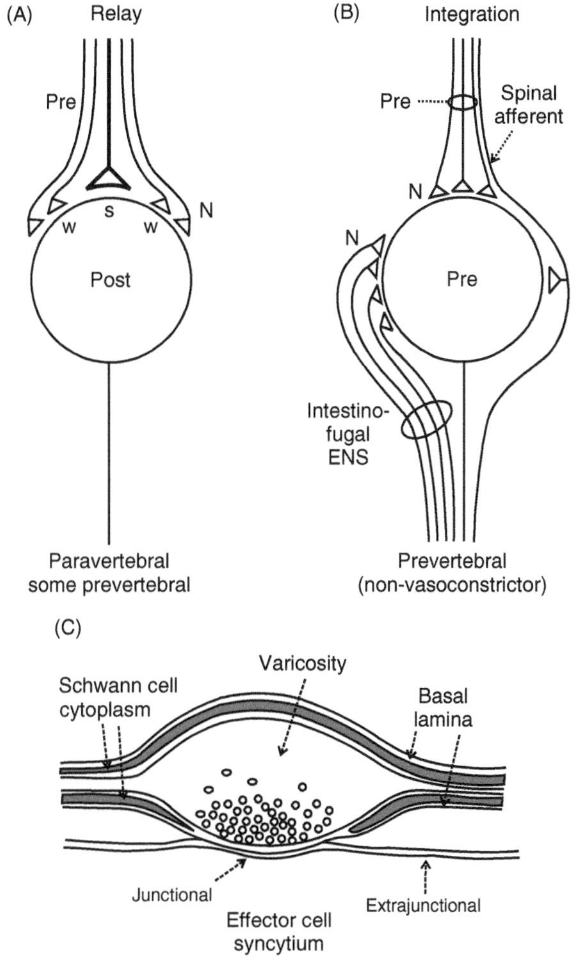


Figure 3. Signal transmission in peripheral autonomic pathways. (A) Relay functions of autonomic ganglion cells. Postganglionic neuron with one (and sometimes two) strong (s; suprathreshold) preganglionic synaptic input (generating postsynaptic potentials of 10–30 mV) and several weak (w; subthreshold) synaptic inputs (generating synaptic potentials of a few millivolts). This connectivity occurs in almost all neurons in paravertebral sympathetic ganglia and some neurons of prevertebral ganglia. These connections mainly function to transmit the activity from pre- to postganglionic neurons. (B) Relay and integrative functions of autonomic ganglion cells in prevertebral ganglia. Postganglionic neuron with weak synaptic inputs from both preganglionic neurons and interneurons of the enteric nervous system (ENS; intestino-fugal neurons) and also from collaterals of spinal visceral afferents. The first two are nicotinic cholinergic (N). The afferent collaterals are peptidergic and use substance P (SP) as transmitter. These postganglionic neurons innervate neurons of the enteric nervous system and possibly other target organs in the viscera. They fire only after integration of several subthreshold cholinergic inputs and/or a slow depolarization [8,56,57]. (C) Simplified scheme of neuroeffector transmission to autonomic target cells (arterioles, heart, non-vascular smooth muscle cells, and secretory cells). Subjuntional (-synaptic) receptors mediate the effect of transmitter released by the nerve terminals during excitation under physiological conditions. The cell surface at which this nerve-effector communication occurs is $\leq 1\%$ of the total cells surface. These subsynaptic receptors are ligand coupled or second messenger coupled to cellular effectors (e.g., ionic channels). Extrasynaptic receptors for the transmitter are either different from the subsynaptic ones and/or are coupled by different intracellular second-messenger pathways to the cellular effectors. The function of the extrajunctionally located receptors is unclear for most innervated effector cells. Small vesicles containing the transmitter are located close to the synaptic cleft. Large vesicles are also present in many varicosities. In these vesicles, neuropeptides are co-localized with "classical" transmitter. Large vesicles are not located close to the synaptic cleft. The physiological role of the neuropeptides is, in most cases, unclear. Modified from Ref. [8].

heart, and longitudinal muscle of the gastrointestinal tract have demonstrated that varicosities of autonomic nerve fibers that are not surrounded by Schwann cells form close synaptic contacts with the effector cells [63,64]. These structures are the morphological substrate for the precise transmission of the centrally generated signals in the postganglionic neurons to the effector cells (Fig. 3C).

Classically, chemical transmission at these neuroeffector junctions is based on the release of one of the "conventional" transmitters, acetylcholine or noradrenaline. However, it is now clear that several chemical substances, often contained within individual autonomic neurons, can be released by action potentials and can have multiple actions on effector tissues [65,66]. The compounds that may be involved are nitric oxide (NO), ATP, or a neuropeptide [e.g., vasoactive intestinal peptide (VIP), neuropeptide Y (NPY), galanin (GAL), and others]. Immunohistochemistry has revealed the presence of many peptides although only a few of these have been demonstrated to modify function after release from nerve terminals in vivo (e.g., NPY or VIP).

Most sympathetic postganglionic axons release noradrenaline, but sympathetic sudomotor and muscle vasodilator axons are cholinergic. Cholinergic sympathetic muscle vasodilator neurons have been shown to exist in some mammalian species (for review, see Refs [67,68]). Whether they exist in humans is a controversial issue [44,69]. Most but not all nerve-mediated effects can be antagonized by blockade of adrenoceptors or muscarinic acetylcholine receptors. All parasympathetic neurons are cholinergic, that is, release acetylcholine on stimulation [62]. However, not all effects of stimulating parasympathetic nerves are blocked by muscarinic antagonists. This clearly implies that other transmitters and/or other receptors are involved.

Responses of tissues to nerve-released noradrenaline or acetylcholine usually only follow repetitive activation of many axons. High-frequency stimuli, particularly in bursts, may produce effector responses due to the concomitant release of a neuropeptide. Alternatively, when the effects of nerve activity are not blocked completely by an adrenoceptor or muscarinic antagonist at a concentration that entirely abolishes the response to exogenous transmitter, it may not necessarily be the case that a transmitter other than acetylcholine or noradrenaline is involved. Although the effects of exogenously applied substances that have putative transmitter function on cellular functions are known for many tissues, the consequences of activation of postjunctional receptors by neurally released transmitters have rarely been investigated. When they have, the mechanisms of neuroeffector transmission have been found to be diverse involving a range of cellular events [54]. One important concept that has emerged is that the cellular mechanisms utilized by an endogenously released transmitter are often not the same as when this transmitter substance or its analog is applied exogenously. The endogenously released transmitter acts primarily or exclusively on the subjunctional receptors (which cover <1% of the surface of the effector cells), whereas the exogenously applied transmitter acts on the extrasynaptically located receptors (Fig. 3C; [64]).

## 2.4. Conclusions

The experimental studies on the sympathetic systems show (Fig. 4) the following:

1. The peripheral sympathetic nervous system is composed of many functionally distinct subsystems (sympathetic pathways) characterized by their reflex patterns. These reflex patterns are the result of integrative processes in spinal cord, brain stem, and hypothalamus.

2. Functionally similar preganglionic and postganglionic neurons are synaptically connected in the autonomic ganglia, with no "cross-talk" between functionally different types of peripheral pathways. The centrally generated reflex patterns are transmitted through the autonomic ganglia without distortion. In prevertebral sympathetic ganglia, the central messages may be modulated by extraspinal synaptic inputs from the enteric nervous system in pathways being involved in regulation of motility or secretion of the gastrointestinal tract.
3. The messages in these functional pathways are transmitted to the autonomic effector cells by distinct neuroeffector mechanisms. This has clearly been shown for arterioles, veins, the heart, and the longitudinal musculature of the gastrointestinal tract.

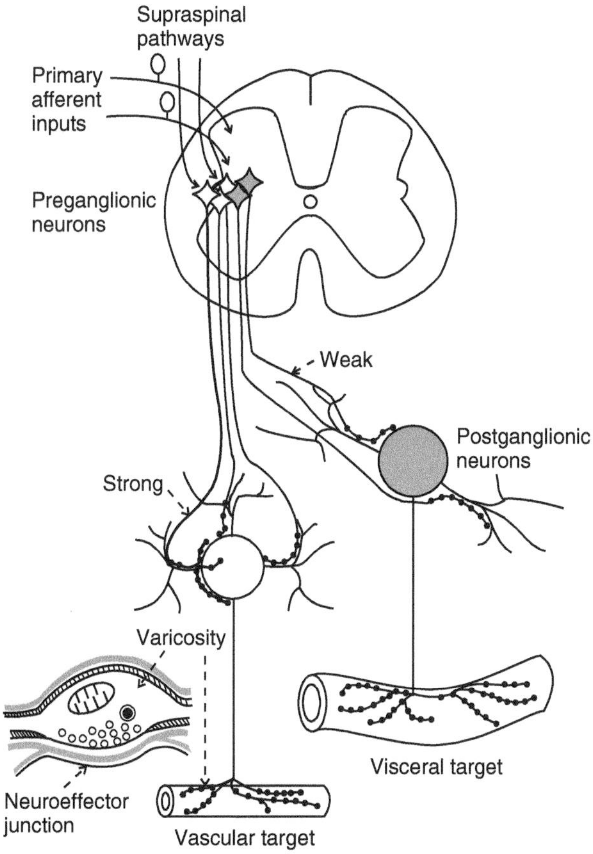


Figure 4. Organization of the sympathetic nervous system into functional pathways. Separate functional pathways extend from the CNS to the effector organs. Preganglionic neurons located in the spinal cord (intermediate zone of the thoraco-lumbar spinal cord) integrate signals descending from brain stem and hypothalamus and arising from primary afferent neurons. This integration involves several classes of autonomic interneurons in the spinal cord. The preganglionic neurons project to peripheral ganglia and converge onto postganglionic neurons. Some preganglionic inputs to postganglionic neurons are always suprathreshold (or strong). Others are subthreshold (weak) and must summate to generate an action potential (see Fig 3A, B). The postganglionic axons form multiple neuroeffector junctions with their target cells. Many varicosities of the terminal axons that contain the synaptic vesicles with the transmitter(s) form close contacts with the target cells (neuroeffector junctions; see Fig. 3B). For details of this concept, see Ref. [8]. With permission from Ref. [52].

## 3. CENTRAL ORGANIZATION OF SYMPATHETIC SYSTEMS

### 3.1. Spinal cord

Sympathetic preganglionic neurons, autonomic interneurons, and populations of primary afferent neurons form spinal autonomic reflex circuits that are integrated in the regulation of activity in preganglionic neurons by supraspinal centers. The spinal cord is an integrative organ in its own right and determines various components of the discharge pattern in the spinal autonomic pathways when the spinal cord is intact. Important constituents in this integration are groups of spinal interneurons that are function-specific and have been postulated on the basis of the study of the reflex patterns in sympathetic neurons (see Section 2.2 and Fig. 1). However, no spinal autonomic interneuron has up to date been functionally identified by direct recording, that is no interneuron has been characterized by its functional types of synaptic input and by the functional type(s) of preganglionic neurons it forms synapses with [70–73].

Neurons projecting from brain stem or hypothalamus to the spinal cord (sympathetic premotor neurons) may synapse with the autonomic interneurons or with the preganglionic neurons. The supraspinal sympathetic premotor neurons are located in the rostral ventrolateral medulla (RVLM), in the caudal raphe nuclei of the medulla oblongata, in the A5 area of the caudal ventrolateral pons, in the lateral hypothalamus, and in the paraventricular nucleus (parvocellular parts) of the hypothalamus (Fig. 5). The primary excitatory transmitter of the sympathetic premotor neurons seems to be glutamate; gamma-aminobutyric acid (GABA) is the primary inhibitory transmitter (in spinal interneurons and also in some groups of sympathetic premotor neurons) to sympathetic preganglionic neurons [74]. The functions of the monoamines in the sympathetic premotor neurons [adrenaline in the sympathetic C1 premotor neurons of the RVLM, noradrenaline in the A5 neurons, serotonin in the sympathetic raphe premotor neurons] and of the neuropeptides (e.g., corticotropin-releasing hormone, enkephalin, substance P, thyreotropin-releasing hormone, oxytocin, and vasopressin) regulating sympathetic final pathways remain unclear.

Analysis of the discharge patterns in the autonomic neurons under standardized experimental conditions leads to the following conclusions [8,12]: (1) The spinal cord contains distinct autonomic reflex pathways that are integrated with supraspinal reflex pathways during regulation of the autonomic target organs. (2) The discharge pattern in the different types of sympathetic neuron consists of components that are associated with integration in the spinal cord as well as in lower brain stem, upper brain stem, and hypothalamus. (3) Reflex integration in the spinal cord is related to distinct afferent inputs from skin, viscera, and deep somatic structures. (4) Signals arising in supraspinal systems are integrated with those from spinal circuits at the preganglionic neuron. (5) Spinal circuits are important in the coordination of somato-motor and autonomic functions.

Thus, spinal circuits, spinal afferent inflows, and descending influences from brain stem and hypothalamus always work together in the integrative activity of the preganglionic sympathetic neurons. By analogy with the somato-motor systems, it is hypothesized that spinal autonomic interneurons and preganglionic neurons constitute spinal autonomic motor programs that are integrated in the normal regulation of autonomic target organs. Depending on function, the sympathetic pathways are under predominant control of the lower brain stem (e.g., muscle and visceral vasoconstrictor pathways innervating resistance vessels, the cardiomotor pathway), of the hypothalamus (e.g., CVC pathways and the pathway to the brown adipose tissue (in rodents), both probably via the caudal raphe nuclei (see section 3.2), or of the circuits in the spinal cord (e.g., the final sympathetic motility-regulating and secretomotor pathways).

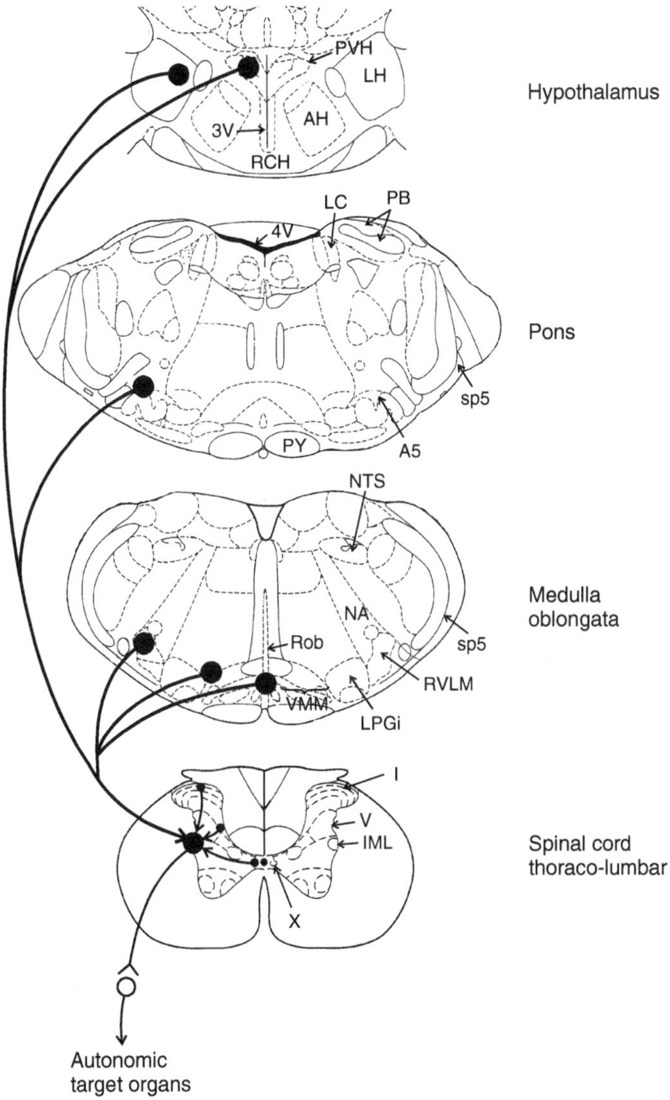


Figure 5. Sympathetic premotoneurons in brain stem and hypothalamus, spinal segmental interneurons, and propriospinal neurons projecting to sympathetic preganglionic neurons. Sympathetic premotor neurons project to preganglionic neurons and to spinal autonomic interneurons. They are located in the RVLM, in the caudal raphe nuclei of the medulla oblongata [raphe magnus, pallidus, and obscurus (Rob)], in the area A5 of the caudal ventrolateral pons, in the lateral hypothalamus (LH), and in the paraventricular nucleus of the hypothalamus (PVH). Autonomic interneurons are located in laminae I, II, V, VI, IX, and X of the spinal cord. Propriospinal neurons are located in laminae I, V, VII, and X of the cervical spinal segments C1–C6, in the lateral funiculus, and in the lateral spinal nucleus of the cervical segments C1–C4 (not shown here). Transverse sections according to Ref. [76]. AH, anterior hypothalamus; LC, locus ceruleus; LPGi, lateral paragigantocellular nucleus; NA, nucleus ambiguus; NTS, nucleus tractus solitarii; PB, parabrachial nuclei; PY, pyramidal tract; RCH, retrochiasmatic area; sp5, spinal trigeminal tract; VMM, ventromedial medulla (includes the gigantocellular reticular nucleus alpha and ventral and the parapyramidal nucleus [76]); 3V, 3rd ventricle; 4V, 4th ventricle. Modified from Refs [70,71,77].

However, in *all* spinal autonomic systems, it seems that the spinal component is essential for this integration because it sets the excitability of the preganglionic neurons and/or shapes the supraspinal signals according to "spinal autonomic programs" [8,12,75].

### 3.2. Lower brain stem

Homeostatic regulations of arterial blood pressure, respiration, and gastrointestinal functions are represented in the lower brain stem. These regulations require a temporally precise coordination and adaptation to somatic body functions and are therefore closely integrated. This integration is reflected in the anatomy and physiology of the neural substrates of these homeostatic regulations in the lower brain stem. Included in this integration are the spinal autonomic circuits and the peripheral autonomic pathways (in addition to the enteric nervous system):

- Neurons being involved in regulation of arterial blood pressure (regulation of heart and peripheral resistance blood vessels) and respiration are situated in rostrocaudally organized columns of neurons in the ventrolateral medulla oblongata (VLM) which extend from the facial nucleus to about 10 mm caudal to the obex. The VLM includes: (1) The ventral respiratory groups of neurons; (2) the rostral ventrolateral medulla (RVLM) containing sympathetic premotor neurons and associated interneurons; and (3) various parts of the caudal ventrolateral medulla (CVLM) that contain excitatory and inhibitory interneurons and mediate several types of cardiovascular reflexes. The VLM is closely associated with parasympathetic preganglionic cardiomotor and bronchomotor neurons in the external formation of the nucleus ambiguus [17,78].
- The RVLM is a sympathetic cardiovascular premotor nucleus mediating homeostatic reflexes to cardiovascular sympathetic preganglionic neurons (muscle, visceral, renal vasoconstrictor neurons, and cardiomotor neurons). Spontaneous activity in the neurons of these cardiovascular pathways is also mediated by the RVLM. It originates in neuronal networks associated with the RVLM and possibly in the premotor neurons themselves [18,79,80].
- Distinct arterial baroreceptor reflexes exist to sympathetic cardiovascular neurons. These baroreceptor reflex pathways consists of four synapses between the baroreceptor input to the nucleus tractus solitarii (NTS) and the preganglionic neurons in the spinal cord: Excitatory interneurons in the NTS project to inhibitory interneurons in the CVLM, these project to the premotor neurons in the RVLM. The transmitter of the inhibitory interneuron is GABA; the transmitter at the other synapses is glutamate. All synapses of the baroreceptor reflexes are under modulatory influence from other nuclei in the lower and upper brain stem, hypothalamus, and telencephalon. The baroreceptor reflex pathway to the parasympathetic cardiomotor neurons in the external formation of the nucleus ambiguus is disynaptic, the transmitter being glutamate in both synapses [18,79].
- Reflexes elicited in sympathetic cardiovascular preganglionic neurons by stimulation of arterial chemoreceptors are mediated via the NTS and the respiratory network and independent of the respiratory network [22].
- Reflexes in sympathetic cardiovascular neurons related to cardiac afferents, lung afferents, or afferents from the gastrointestinal tract are also mediated by the NTS and by sympathetic premotor neurons in the RVLM. However, their central pathways are unknown.
- Reflexes in sympathetic cardiovascular neurons generated by stimulation of somatic and spinal visceral afferent neurons may also be mediated by the RVLM. Again their central pathways are unknown.
- The caudal raphe nuclei of the medulla oblongata (raphe magnus and pallidus) contain sympathetic premotor neurons to preganglionic CVC neurons or to preganglionic lipomotor

neurons (supplying brown adipose tissue in the rat). These premotor neurons are involved in thermoregulation and in regulation of energy balance [23,45,81,82] and are under powerful control of the hypothalamus. Other sympathetic premotor neurons functionally related to the heart or kidney may also be located in the caudal raphe nuclei.

- Sympathetic neurons exhibit respiratory rhythmicity in their activity. The respiratory pattern in cardiovascular autonomic neurons is probably mediated by the "common cardiorespiratory network" in the VLM. This network represents a sensory-motor program that closely coordinates regulation of respiration and regulation of arterial blood pressure under all physiological conditions. The respiratory modulation in the other types of sympathetic neurons, for example, being involved in thermoregulation, regulation of the gastrointestinal tract, or regulation of pelvic organs, and so on, is generated by other mechanisms, probably also in the medulla oblongata [14,16,83,84].

### 3.3. Upper brain stem and hypothalamus

The circuits in the medulla oblongata are under control of the upper brain stem, hypothalamus, and telencephalon. They are neural building blocks of regulations represented in the supramedullary brain centers. Table 1 summarizes the data about functions of the hypothalamus (and upper brain stem) in which the autonomic systems are involved (see Ref. [85]). These complex integrative functions include somato-motor, neuroendocrine, and autonomic components. The important point to be made here is that these precisely organized complex regulations, which are represented in the hypothalamus and upper brain stem, require

1. anatomically and functionally precisely organized peripheral autonomic pathways innervating various target organs (shaded column in Table 1) and
2. precisely organized central autonomic circuits in the spinal cord and lower brain stem.

An important function in which hypothalamus and mesencephalon are involved and which very much seems to depend on the functioning sympathetic systems (in addition to the neuroendocrine systems) is the protection of body tissues during acute and chronic pain and stress; this includes the neuronal control of the immune system (shaded vertical column in Table 1, see contributions to this volume). Some aspects of these protective functions will be discussed in the next section.

### 3.4. Conclusions

The experimental studies of the central organization of sympathetic systems show, corresponding to the functionally distinct peripheral autonomic pathways, the following [8]:

1. The spinal cord contains many autonomic circuits that are characterized by their primary afferent inputs and the final spinal autonomic pathways they are connected to. Several distinct types of autonomic interneurons have to be postulated in the spinal cord.
2. The brain stem and hypothalamus contain several types of sympathetic premotor neurons that connect the supraspinal autonomic centers with the spinal autonomic circuits and form synapses with the preganglionic neurons and/or the spinal autonomic interneurons.
3. The medulla oblongata and caudal pons contain complex circuits that are involved in homeostatic regulation of the cardiovascular system (blood pressure), respiratory system,

Table 1 Properties of integrative functions of the hypothalamus

| Function | Behavior | Nuclei in hypothalamus | Afferent feedback: neural, hormonal, cytokines | Autonomic systems | Endocrine systems, hormones |
|---|---|---|---|---|---|
| Thermoregulation | Thermoregulatory behavior | Preopt. region, ant./posterior hypothalamus, OVLT (pyrogenic zone, fever) | Peripheral thermoreceptors, central thermosens. (preopt.), cytokines (fever) | SyNS (skin [CVC,SM]), (BAT [rodents]) | TRH/Thyr (anterior pituitary) |
| Reproduction, sexual behavior | Sexual behavior, sexual orientation | MPNl (male; human dimorphic) VMvl (female), TU, PM | Afferents from sexual organs, afferents from other sensory systems | SyNS (thor.-lumb.), PaNS (sacral) (sexual organs) | GnRH, FSH/LH (anterior pituitary) |
| Volume-, osmoregulation (fluid homeostasis) | Drinking behavior, thirst | N. paraventr./supraopt., R. preopt. med., AVPV, OVLT, SFO | Osmorecept. in OVLT and liver, volume rec. ri. atrium (vagal), angiotensin II via SFO | NTS, SyNS (kidney) | Vasopressin (posterior pituitary) |
| Regul. of nutrition, Regul. of metabolism | Nutritive behavior, hunger/ satiety | N. arcuatus, N. paraventr., N. ventromed. (regulation of insulin secretion) | Vagal afferents and hormones from GIT (CCK, ghrelin, GLP-1, insulin, glucagon), leptin from BAT, insulin, nutritive signals | Enteric nervous system, PaNS (DMV), SyNS (BAT[rodents], WAT | Insulin, glucagon, orexin, leptin |
| Temporal organization of body functions | Sleep-waking behavior, circadian/endogenous rhythm of body functions | N. suprachiasmaticus, SBPV | Afferents from retina (retino-hypothalamic tract) | SyNS, PaNS, SyNS to gl. pinealis | Melatonin (gl. pinealis) |
| Body protection (acute, e.g. during pain and stress) | Defense behavior (fight, flight, quiescence) | AHN, VMH, PM; lateral and ventrolateral PAG | Nociceptive afferents (from body surface, deep somatic, viscera) | SyNS, PaNS (cardiovascular system [CMN,MVC,VVC, etc.]) | CRH/ACTH (anterior pituitary gland), adrenaline (SA system) |
| Immune defense | Defense of toxic substances and situations (sickness behavior) | N. paraventr. hypothalami, other | Cytokines | SyNS (to immune tissue) | CRH/ACTH (anterior pituitary gland), adrenaline (SA system) |

AH, anterior hypothalamus; AHN, anterior hypothalamic n.; AVPV, anteroventral periventricular n.; BAT, brown adipose tissue; CMN, cardiomotor neurons (symp., parasymp.); CRH/ACTH, corticotropin-RH/adrenocorticotropic hormone; CVC, cutaneous vasoconstrictor neurons; DMV, dorsal motor nucleus of the vagus; FSH/LH, follicle-stimulating hormone/luteinizing hormone; GIT, gastrointestinal tract; GnRH, gonadotropine RH; MPNl, lateral part of the medial preoptic nucleus; MVC, muscle vasoconstrictor neurons; N., nucleus; NTS, nucleus tractus solitarii; OVLT, organum vasculosum laminae terminalis (osmosensors); PAG, periaqueductal grey; PaNS, parasympathetic nervous system; PM, premammillary n.; R., regio; RH, releasing hormone; SA system, sympathoadrenal system (adrenal medulla); SBPV, subparaventricular zone; SFO, subfornical organ (angiotensin sensitivity); SM, sudomotor neurons; SyNS, sympathetic nervous system; thor.-lumb., thoraco-lumbar; TRH/Thyr, thyreotropin RH/thyroxin; TU, tuberal n.; VMvl, ventrolateral part of the ventromedial nucleus; VVC, visceral vasoconstrictor neurons; WAT, white adipose tissue.

The specific autonomic systems that are mainly involved in these functions are shown in the shaded vertical column. Functions related to protection of the body are shown in the horizontal dotted area. Data after Ref [8].

and gastrointestinal tract and in the integration of these three regulations. Some of the underlying circuits (e.g., those related to arterial baroreceptor and chemoreceptor afferents, and to respiration) have been worked out.

4. Hypothalamus and mesencephalon organize somato-motor systems, autonomic systems, neuroendocrine systems, and afferent systems to complex functions (stereotyped behaviors) that are important for survival of the individual and the species. Integrated in these functions are the homeostatic regulations organized in the lower brain stem. The organizing principles and details about the neural circuits in the hypothalamus and mesencephalon have been little explored.

## 4. SYMPATHETIC NERVOUS SYSTEM AND BODY PROTECTION

Responses of the organism during pain and stress, whether elicited by external or internal stimuli, are integral components of an adaptive biological system and important for the organism to function in the confines of a dynamic and frequently challenging and dangerous environment (see Ref. [86]). These responses consist of autonomic, neuroendocrine, and somato-motor responses, which include the appropriate sensory perceptions and felt emotions. They serve to adapt organ functions to the changing behavior and the behavior to changing environments. The integrated responses displayed by the organism are states that are represented in the brain (brain stem, hypothalamus, limbic system, and neocortex) of the organism. Perception of sensations, experience of emotions, autonomic responses, endocrine responses, and somato-motor responses occur principally in parallel and are therefore parallel read-outs of these central representations. These central representations obtain continuous neuronal afferent, hormonal, and humoral signals monitoring the state of the different tissues. The adaptation of the immune system, control of inflammation, and control of hyperalgesia by the CNS are integral components in this scenario and require sympathetic systems which function in a differentiated way. During real or impending tissue damage, this integrated protective system organized by the brain is strongly activated leading to illness responses including pain and hyperalgesia (Fig. 6). Part of this protective system, promoting tissue repair and recuperation, is the bidirectional functioning immune–brain system in which the brain is continuously influenced by signals from the immune system (cytokines) and the brain modulates the reactivity of the immune system, probably also via the sympatho-neural system (see Section 4.1).

This bidirectional brain–immune system is slow and furthers recuperation, that is, tissue healing. Regulation of pain and hyperalgesia are integral components of the fast defense system (fight and flight) and the slow (recuperative) defense system. During *fast defense*, organized by the hypothalamo-mesencephalic system, fast analgesia, mobilization of energy, activation of various sympathetic channels (including the SA system), and activation of the hypothalamo–pituitary–adrenal (HPA) axis occur. This fast defense is activated from the periphery by stimulation of nociceptors and accompanied by an increased vigilance and alertness. During *slow defense*, the organism switches to recuperation and healing of tissues. It is characterized by pain and hyperalgesia, which keep the organism in a state of quiescence and rest. This slower defense system is activated by peripheral signals from the immune system via afferent (e.g., vagal) neurons or by cytokines via the circumventricular organs. The involvement of cytokines in sensitization of nociceptors during inflammation, part of it mediated by the terminals of sympathetic fibers, and the slow change of sensitivity of nociceptors linked to the activity of the SA system may be components of the slow defense system [87].

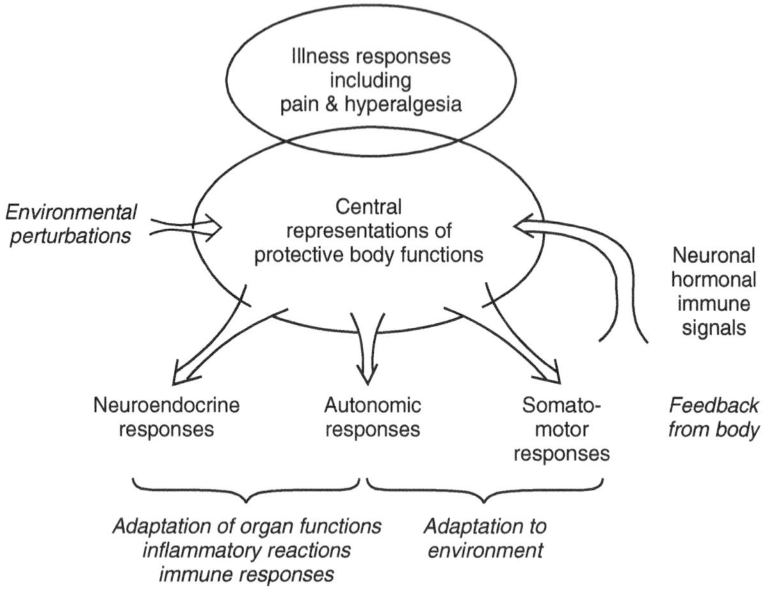


Figure 6. Protective mechanisms of the body regulated by the brain: inputs, central representations, and outputs. Activation of the central circuits leads to neuroendocrine, autonomic, somatomotor, and illness responses (which includes pain and hyperalgesia). The afferent feedback from the body tissues is neural and hormonal and is mediated by cytokines from the immune system. The central circuits in the spinal cord, brain stem, and hypothalamus are adapted by the telencephalon to the environmental situations and to the situations in the body. Modified after Ref. [93].

The neurobiological basis of fast defense behaviors during pain and stress, which are integrated in the lateral and ventrolateral columns of the periaqueductal grey (confrontational defense, flight, quiescence), has been described elsewhere [88–92].

### 4.1. Control of the immune system by the sympathetic nervous system

Anatomical, physiological, pharmacological, and behavioral experiments on animals support the notion that the sympathetic nervous system can influence the immune system and therefore control protective mechanisms of the body at the cellular level (see Refs [94–98]). Control of the immune system by the sympathetic nervous system would mean that the telencephalon is principally able to influence, via the hypothalamus, immune responses. This is an attractive idea and, based on clinical and experimental observations, is propagated by several groups without necessary experimental basis (see Refs [94,95,99]). As far as mechanisms underlying neuroimmune interaction and psychoneuroimmunology (i.e., the reciprocal interaction between brain and immune system) are concerned, the reader is referred to Ader [100].

The mechanisms by which the brain modulates the immune system via the sympathetic outflow remain largely unsolved [100–102]. This has conceptual and methodical reasons. In view of the functional specificity of the peripheral sympathetic pathways as outlined above, the key question to be asked is: Is the immune system supplied by a sympathetic pathway which is distinct from other sympathetic pathways (see Sections 2.2–2.4) and specifically mediates an immune–modulatory effect?

The involvement of the sympathetic nervous system in the modulation of the immune system by the CNS has never been addressed in this way. However, several observations support the idea that the neural communication from the brain to the immune system occurs via the sympathetic nervous system, that is, that it is mediated by a peripheral pathway that is *functionally distinct from all other sympathetic systems (the vasoconstrictor systems, etc.) and under control of the hypothalamus*:

- Primary and secondary lymphoid tissues are innervated by postganglionic noradrenergic sympathetic neurons. Varicosities of the sympathetic terminals can be found in close proximity with T lymphocytes and macrophages (see Refs [94,98]) as described for other sympathetic target cells.
- The spleen of the cat is innervated by sympathetic postganglionic neurons that are numerically, relative to the weight of the organs, three times as large as the sympathetic innervation of the kidneys [103].
- The sympathetic innervation of the spleen is functionally different from that of the kidney: Sympathetic neurons innervating the kidney behave like "classical" vasoconstrictor neurons [10,104–106], yet many sympathetic neurons innervating the spleen are not under control of the arterial baroreceptors and show distinct reflexes to stimulation of afferents from the spleen and the gastrointestinal tract, which are different from those in vasoconstrictor neurons [107,108]. These results suggest that many sympathetic neurons innervating the spleen have a function other than to elicit vasoconstriction. This function may be related to the immune system.
- Functional studies performed on the spleen of rodents have shown that (for review, see Ref. [96], and references herein) (1) surgical and chemical sympathectomy alters the splenic immune responses (e.g., increase of natural killer cell cytotoxicity, lymphocyte proliferation responses to mitogen stimulation, and production of interleukin-1β). (2) Stimulation of the splenic nerve reduces the splenic immune responses. (3) Lesions or stimulation of distinct hypothalamic sites or microinjection of cytokines at distinct hypothalamic sites activates some splenic immune responses. These changes are no longer present after denervation of the spleen. (4) Neural activity in the splenic nerve is affected by these central manipulations, and changes in this activity are correlated with the changes of the splenic immune responses. For example, activity in sympathetic neurons to the spleen elicited by interventions at the hypothalamus (in particular the ventromedial nucleus of the hypothalamus) is significantly correlated with suppression of natural killer cell cytotoxicity in the spleen. This suppression is mediated by β-adrenoceptors [109,110]. A *hypothalamo-sympathetic neural system that controls the immune system* has been postulated [96].
- The skin is innervated by sympathetic vasoconstrictor, sudomotor, pilomotor, and possibly vasodilator neurons, which can functionally be recognized (see Section 2.2). It is, however, obvious from neurophysiological studies that there are many sympathetic neurons projecting in skin nerves that do not exhibit spontaneous and reflex activity and the function of which is unknown. Is it possible that these postganglionic neurons do not innervate the "classical" sympathetic target cells but are associated with the skin–immune system? [111–114]

Using classical neurophysiological recordings in vivo from sympathetic neurons innervating spleen, kidney, skin, skeletal muscle, and so on, it should be possible to discriminate neurons innervating the immune tissue from other functional types of sympathetic neuron (e.g., vasoconstrictor neurons to skin, skeletal muscle, kidney, or spleen). This is exemplified

Table 2 Examples of function-specific sympathetic pathways to different targets within some organs with emphasis on immune tissue

| Organ | Target cells | Function |
|---|---|---|
| Spleen | BV, *IC* | VC, *IR* |
| Hairy skin | $BV_{skin}$, *IC* | VC (VD?), *IR* |
| Hairless skin | $BV_{skin}$, SG, *IC* | VC (VD?), SM, *IR* |
| Kidney | BV, JGA | VC, renin release |
| Skel. muscle | $BV_{muscle}$ | VC (VD?) |

BV, blood vessel; VC, vasoconstriction; VD, vasodilatation; *IC*, immune cells; *IR*, immune response; SG, sweat gland; SM, sudomotor response; JGA, juxtraglomerular cells.
Refs [7,9,28,45].

in Table 2 that shows the target cells in particular organs that are innervated and possibly controlled by sympathetic neurons and the functions of these neurons. One sympathetic channel in three of these organs projects potentially to the immune tissue and possibly regulates the immune response.

Thus, in analogy to what has been described in Sections 2.2 and 2.3 above, it should theoretically be possible to characterize the sympathetic neurons innervating lymphoid tissues by using stimuli that are adequate to elicit immune responses and to assign to these sympathetic neurons characteristic functional markers (for review and references, see Ref. [96]). This idea leads to the formulation of *two alternative testable hypotheses*:

1. Neurons of sympathetic pathways are functionally specific for the immune tissues and can be characterized by distinct reflex patterns elicited in these neurons by physiological stimuli that are related to the immune system and therefore related to defense and protection of the organism.
2. The alternative hypothesis would be that reflex responses in sympathetic neurons that modulate immune responses are found indiscriminately in all populations of sympathetic neurons; these responses would therefore not be functionally specific for the lymphoid tissue. This could mean that more or less *all* sympathetic noradrenergic pathways have, in addition to their specific target-organ related functions, a general function that is related to defense and protection of the tissues.

## 4.2. Control of inflammation and hyperalgesia by the brain involving the sympatho-adrenal system

Inflammation and hyperalgesia following tissue trauma are protective reactions which further healing. Mechanisms of inflammation are commonly considered to be confined to the periphery involving immune-competent and related inflammatory cells as well as vascular cells. The main mechanism of hyperalgesia during inflammation in this view is confined to the sensitization of nociceptors by inflammatory mediators leading to central changes (central sensitization) and appropriate protective behavior. However, using animal models of experimental inflammation of the knee joint synovia and mechanical hyperalgesic behavior, it has recently been shown that both inflammation and nociceptor sensitization are potentially under the control of the SA system, implying that the brain can influence both via this system. Furthermore, both seem to be dependent on the innervation of the tissue (synovia, skin) by sympathetic postganglionic neurons in an unprecedented way.[2] Although both animal models (bradykinin-induced plasma extravasation in the synovia of the rat knee and bradykinin-induced mechanical hyperalgesic behavior) are rather artificial and specific and although we have up to now no human models and no clinical situation that correspond to these animal models, it is of interest to learn from these animal models the ways by which the neuroendocrine systems can principally influence peripheral mechanisms of inflammation and nociceptor sensitization.

Noxious stimulation leads to depression of the *experimental inflammation.* This depression is, depending on the mode of afferent stimulation, mediated either by the SA system [118,119] or by the HPA system [122,123]. Thus, the experimental inflammatory process in the synovia is potentially under the control of both the HPA system and the SA system.[3] Here, only experiments involving the SA system will be summarized. Noxious stimulation (e.g., of skin or viscera) activates the SA system, which releases adrenaline which in turn leads to a depression of bradykinin-induced plasma extravasation in the synovia. This decrease of plasma extravasation is abolished after denervation of both adrenal medullae. It is not generated by a decrease of blood flow through the synovia (due to constriction of blood vessels) because synovial plasma extravasation produced by platelet activating factor is *not* reduced by activation of the SA system [118]. The nociceptive neuroendocrine negative feedback circuit involving the SA system is based on reflex pathways in the spinal cord and brain stem [118,119]. It has been shown that the sensitivity of these reflex control pathways can be changed by activity in vagal afferents from the abdominal viscera. Vagal afferent activity leads to inhibition of these reflex pathways and removal of activity in the vagal afferents [by surgical interruption of the subdiaphragmatic vagus nerves or of the celiac branches of the vagus nerves (which innervate the distal duodenum, ileum, jejunum, and proximal colon) [120]] to central disinhibition and therefore to an enhancement of the nociceptive–neuroendocrine negative feedback control. The inhibition, maintained by activity in the vagal afferents, is operating at the level of the spinal cord (see Fig. 7; [118,119,121,124]).

---

[2](1) Bradykinin-induced plasma extravasation of the knee joint synovia, which occurs at the venular side of the vascular bed, is reduced to about 30% after surgical sympathectomy (removal of the postganglionic neurons) but not after decentralization of the postganglionic neurons (cutting the preganglionic axons). Thus, this synovial plasma extravasation is dependent on the innervation of the synovia by sympathetic postganglionic axons but not on activity (action potentials) in these axons and not on release of noradrenaline. The idea is that the varicosities of the sympathetic postganglionic axons mediate directly or indirectly the effect of bradykinin (see [115,116]). Details about the underlying mechanism of this novel function of postganglionic sympathetic terminals have to be worked out. (2) Bradykinin-induced mechanical hyperalgesic behavior tested with the Randall-Sellito test is also dependent on the innervation of the stimulated skin by sympathetic neurons but *not* on the activity in the sympathetic neurons [comparison of sympathectomized hindlimb with sympathetically decentralized (preganglionic axons sectioned) hindlimb] [117]. Thus, sensitization of cutaneous nociceptors to mechanical stimulation by intracutaneous bradykinin appears to be mediated, at least in part, by the sympathetic terminals. Details about the mechanism underlying this unprecedented effect of postganglionic axons have to be worked out.

[3] Continuous transcutaneous electrical stimulation of cutaneous C-fibers in the plantar skin at low stimulation frequencies ($\leq$4 Hz) leads to depression of bradykinin-induced plasma extravasation which is preferentially mediated by the reflex activation of the HPA axis [26,122,123]. Quasi-natural stimulation of plantar cutaneous nociceptors by the neurotoxin capsaicin (extracted from the paprika) injected into the skin also leads to depression of bradykinin-induced plasma extravasation in the synovia; however, this depression is mediated by activation of the sympatho-adrenal system [118,119]. This differential activation of both neuroendocrine systems generated by the two modes of afferent stimulation is interesting but puzzling. The mechanisms underlying this differential activation are unknown and are presently explored [26].

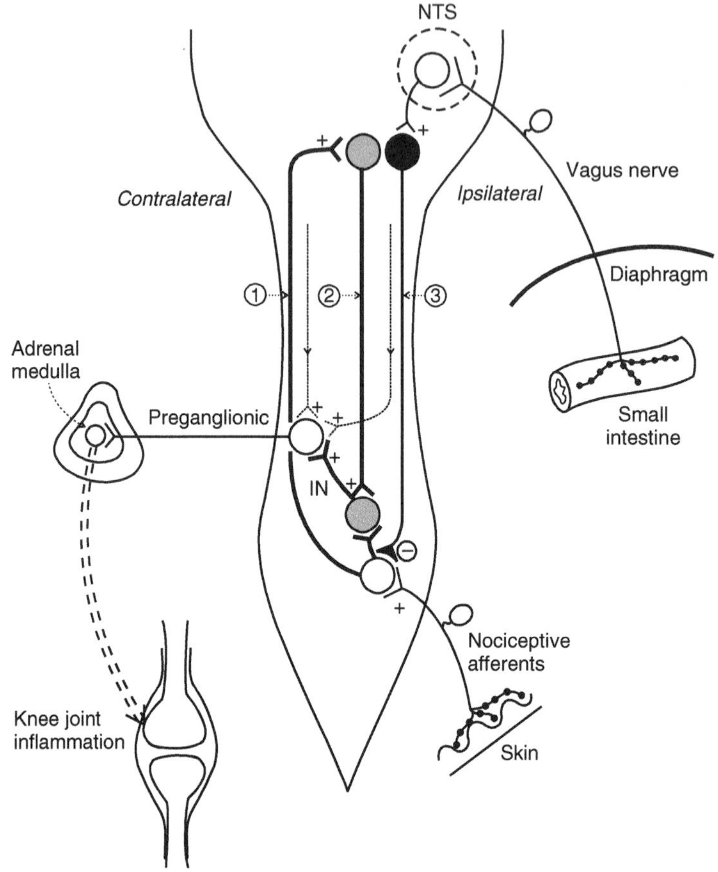


Figure 7. Schematic diagram showing the proposed neural circuits in spinal cord and brain stem that modulate experimental inflammation in the rat knee joint via the SA system (adrenal medulla). Experimental inflammation was generated by perfusion of the knee joint with saline containing the inflammatory mediator bradykinin (160 ng/ml; $1.51 \times 10^{-7}$ M) in the rat. Bradykinin generates plasma extravasation at the venules of the synovia into the knee joint cavity[4]. Stimulation of cutaneous nociceptors by capsaicin (CAP) leads to depression of inflammation by activation of preganglionic neurons innervating the adrenal medullae via a spinal and a spino–bulbo–spinal excitatory circuit (grey neurons). The ascending limb of this spino–bulbo–spinal reflex loop projects through the dorsolateral funiculus (DLF) of the spinal cord contralateral to the nociceptive input (①). The descending limb of this reflex loop projects through the dorsal quadrants (②). The spino–bulbo–spinal and spinal reflex circuits are inhibited by activity in abdominal vagal afferents from the small intestine that is exerted at the level of the spinal cord (black neuron). The descending limb of this inhibitory pathway projects through the DLF ipsilateral to the nociceptive input (③). Dotted thin lines: Axons of sympathetic premotor neurons in the brain stem (see Fig. 5) that project through the dorsolateral funiculi of the spinal cord to the preganglionic neurons of the adrenal medulla. For details, see text. IN, interneuron; NTS, nucleus tractus solitarii; +, excitation; −, inhibition. Modified from Refs [118,123].

[4] Bradykinin (BK)-induced plasma extravasation was determined in the knee joint of anesthetized rats. Skin overlying the knee was excised to expose the joint capsule. 30-gauge inflow and 25-gauge outflow needles were inserted into the knee joint cavity for administration of perfusion fluid (0.9% saline at 250 μL/min without bradykinin or with 160 ng/ml bradykinin). Rats were then given an intravenous injection of Evans blue dye (50 mg/kg). Perfusion fluid was collected every 5 min for up to 120 min, and dye concentration was determined spectrophotometrically at 620 nm. The absorbance at this wave length is linearly related to the dye concentration and therefore to the degree of plasma extravasation of the synovia. Following collection of the first three samples to establish baseline plasma extravasation levels, BK and other substances are added to the perfusing fluid [118,119,122,123].

Bradykinin-induced *mechanical hyperalgesic behavior* is enhanced and baseline paw withdrawal threshold to mechanical stimulation reduced in rats in which abdominal vagal afferents have been interrupted. The vagal afferents involved are those that innervate the small intestine (i.e., project through the celiac branches of the abdominal vagus nerves). This enhancement is generated by activation of the SA system. After bilateral denervation of the adrenal medullae (cutting the sympathetic preganglionic axons projecting through the adrenal nerves to the adrenal medullae), the hyperalgesic behavior either does not develop following vagotomy or reverses in vagotomized animals [92,117,125]. These results imply that there exists a vago-sympathetic reflex pathway that controls the activity in the preganglionic neurons innervating the adrenal medullae, that activity in vagal afferents inhibits this pathway and therefore the preganglionic neurons, and that removal of the vagal activity activates the SA system (Fig. 8). Activation of the adrenal medullae leads to release of adrenaline [126] and subsequently to

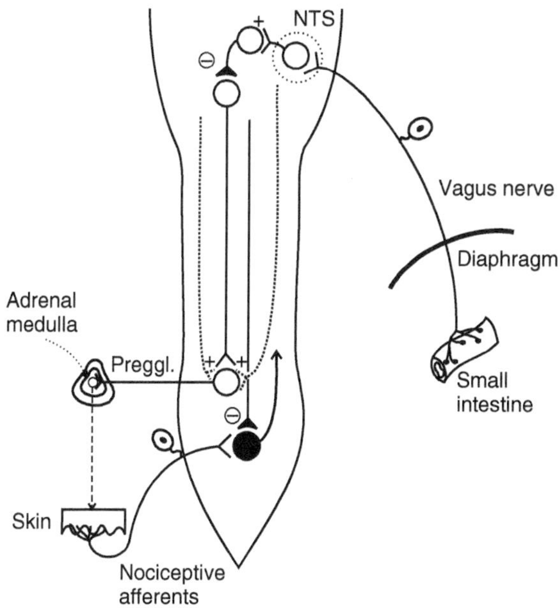


Figure 8. Schematic diagram showing the proposed neural circuits in spinal cord and brain stem that modulate nociceptor sensitivity via the sympatho-adrenal system (adrenal medulla).[5] Sensitivity of cutaneous nociceptors for mechanical stimulation is modulated by adrenaline released by the adrenal medullae. Activation of the adrenal medullae increases the sensitivity of the nociceptors. Activity in preganglionic neurons innervating the adrenal medullae also depends on activity in vagal afferents from the small intestine, which has an inhibitory influence on the central pathways to these preganglionic neurons. Thus, interruption of the vagal afferents leads to activation of the adrenal medullae. It is hypothesized that these neuronal (reflex) circuits in the brain stem are under the control of upper brain stem, hypothalamus, and telencephalon. Dotted thin lines: Axons of sympathetic premotor neurons in the brain stem that project through the dorsolateral funiculi of the spinal cord to the preganglionic neurons of the adrenal medullae (see Fig. 5). For details see text. NTS, nucleus tractus solitarii; +, excitation; −, inhibition. Modified from Refs [117,125].

---

[5] Decrease of paw-withdrawal threshold to mechanical stimulation of the dorsum of the rat hind paw induced by bradykinin (bradykinin-induced behavioral mechanical hyperalgesia) was measured in non-anesthetized rats. Cutaneous nociceptors on the dorsum of the paw were stimulated by a linearly increasing mechanical force. Paw-withdrawal threshold is defined as the mean minimum force (in grams) at which the rat withdraws its paw. Bradykinin (0.1–1000 μg) was injected in a volume of 2.5 μL of saline into the dermis of the skin.

sensitization of cutaneous nociceptors for mechanical stimulation. This sensitization has a *long time course*: It develops slowly over 7 days following activation of the adrenal medullae and reverses slowly following denervation of the adrenal medullae. Continuous infusion of adrenaline into a rat via a minipump over up to 2 weeks generates the same effect (enhancement of bradykinin-induced mechanical hyperalgesic behavior). Enhancement of this behavior following subdiaphragmatic vagotomy is prevented by continuous infusion of a $\beta_2$-adrenoceptor blocker via a minipump for up to 2 weeks [126]. Thus, adrenaline acts possibly on $\beta_2$-adrenoceptors in the skin which in turn generates a sensitization of nociceptors with a slow time course. Adrenaline most likely does not act directly on the nociceptors but on other

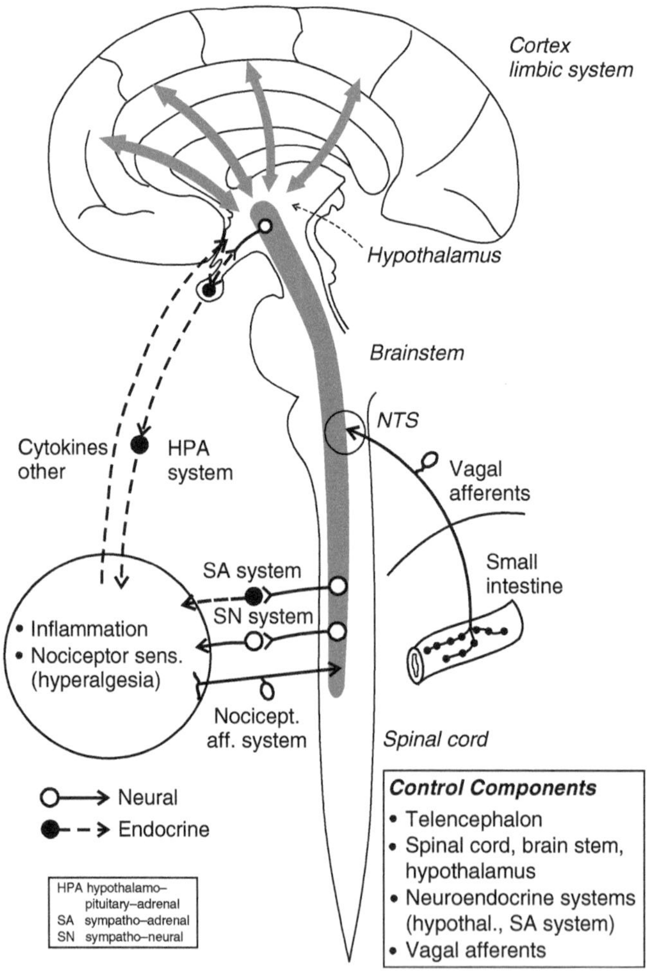


Figure 9. The spinal cord, brain stem, and hypothalamus contain neuronal circuits that control nociceptor sensitivity and inflammation in the periphery of the body via the SA system and the hypothalamo–pituitary–adrenal (HPA) system (shaded area). Feedback information from the peripheral inflammatory process occurs via nociceptive primary afferent neurons and cytokines. The central circuits linked to the SA system and the HPA system are modulated by activity in vagal afferents probably innervating the small intestine (and here the gut-associated lymphoid tissue). The telencephalon controls the inflammation and sensitivity of nociceptors via the circuits in the spinal cord, brain stem, and hypothalamus (see shaded double arrows). The sympatho-neural (SN) system is involved too (see Footnote 1 on page 75). Modified from Ref. [24].

cells – possibly immune system cells such as macrophages, mast cells, or keratinocytes – which then release substances that generate the sensitization, this in particular because prostaglandin $E_2$-induced mechanical hyperalgesic behavior is not changed after vagotomy [117].

This is an entirely new form of nociceptor sensitization involving the SA system. Its peripheral mechanisms are unknown. No recordings from cutaneous nociceptive primary afferent neurons testing sensitization to long-term activation of the adrenal medullae (or to long-term action of adrenaline infused into the rat) have been performed so far.

Overall, these findings mean that inflammatory processes and sensitivity of nociceptors can be changed via the SA system (adrenal medullae) by interventions at remote body domains (here the visceral body domain and activity in vagal afferents). The changes are mediated by reflex circuits in the spinal cord and brain stem (see Figs 7 and 8), which connect the nociceptive afferent inputs and the vagal afferent inputs (via the NTS) with the sympathetic preganglionic neurons innervating the adrenal medullae. It is postulated that the reflex circuits consist of excitatory spinal and spino–bulbo–spinal pathways associated with the nociceptive afferent input and inhibitory bulbo–spinal pathways associated with the vagal afferent input (Figs 7 and 8 [118,119]). Details of these reflex circuits have to be worked out. Experimental investigations performed on rats show that sympathetic preganglionic neurons innervating cells of the adrenal medulla releasing adrenaline are connected to distinct neuronal circuits in the spinal cord and lower brain stem that are different from those connected to preganglionic neurons innervating cells of the adrenal medullae-releasing noradrenaline and preganglionic neurons innervating postganglionic neurons that supply resistance vessels in skeletal muscle or viscera [23,127].

It is not far-fetched to suggest that the multiple reflex circuits connected to the adrenomedullary cells that release adrenaline and the HPA axis, which also forms a nociceptive–neuroendocrine reflex circuit in the control of inflammation [24,25,123], are under the control of hypothalamus and telencephalon. Thus, these basic reflex circuits connected to both neuroendocrine systems could be "used" by higher brain centers to control both inflammation and nociceptors sensitization (Fig. 9). The detailed peripheral and central mechanism by which this occurs has to be worked out.

### 4.3. Conclusions

Based on experiments conducted on rats, several new testable hypotheses are emerging in the field of neural and neuroendocrine control of protection of body tissues (Fig. 9):

1. The SA system has an unprecedented role in the control of sensitivity of nociceptors and inflammation. This control is independent of the control of blood flow through the tissues.
2. The HPA system may also have an unprecedented role in the control of inflammation.
3. In the control of inflammation, both the SA system and the HPA axis form nociceptive–neuroendocrine feedback circuits with the nociceptive primary afferents that have distinct pathways in the spinal cord, brain stem, and hypothalamus.
4. The neural circuits in the spinal cord, brain stem, and hypothalamus that are connected to the SA system and the HPA axis are controlled by vagal afferents that innervate the small intestine. These afferents may be associated with the largest immune system in the body [the gut-associated lymphoid tissue (GALT)].
5. The telencephalon modulates the neural circuits in the spinal cord, brain stem, and hypothalamus associated with the control of nociceptor sensitization and inflammation and controls in this way cellular protective processes in the body.

## REFERENCES


1. Cannon WB. Organization for physiological homeostasis. Physiol Rev 1929;9:399–431.
2. Cannon WB. The Wisdom of the Body, 2nd revised and enlarged edition. New York: Norton, 1939.
3. Bernard C. Leçons sur la Physiologie et la Pathologie du Système Nerveux [On the Physiology and Pathophysiology of the Nervous System]. Paris: Bailliere, 1858.
4. Sterling P, Eyer J. Allostasis: a new paradigm to explain arousal pathology. In Handbook of Life Stress, Cognition and Health. Fisher S, Reason J, Eds; New York: Wiley, 1988; pp. 629–49.
5. McEwen BS. Protective and damaging effects of stress mediators. N Engl J Med 1998;338: 171–9.
6. McEwen BS. Protective and damaging effects of stress mediators: central role of the brain. Prog Brain Res 2000;122:25–34.
7. Schulkin J. Rethinking Homeostasis. Allostatic Regulation in Physiology and Pathophysiology. Cambridge, MA: The MIT press, 2003.
8. Jänig W. The Integrative Action of the Autonomic Nervous System. Neurobiology of Homeostasis. Cambridge, NY: Cambridge University Press, 2006.
9. Jänig W. Organization of the lumbar sympathetic outflow to skeletal muscle and skin of the cat hindlimb and tail. Rev Physiol Biochem Pharmacol 1985;102:119–213.
10. Jänig W. Pre- and postganglionic vasoconstrictor neurons: differentiation, types, and discharge properties. Annu Rev Physiol 1988;50:525–39.
11. Jänig W. Ganglionic transmission in vivo. In Autonomic Ganglia. McLachlan EM, Ed.; Switzerland: Harwood Academic Publishers, 1995; pp. 349–95.
12. Jänig W. Spinal cord reflex organization of sympathetic systems. Prog Brain Res 1996a;107:43–77.
13. Jänig W. The autonomic nervous system and its coordination by the brain. In Handbook of Affective Sciences, Part II. Autonomic Psychophysiology. Davidson RJ, Scherer K, Goldsmith HH, Eds; Oxford, NY: Oxford University Press, 2003; pp. 135–86.
14. Häbler HJ, Jänig W, Michaelis M. Respiratory modulation of activity in sympathetic neurones. Prog Neurobiol 1994;43:567–606.
15. Jänig W, Häbler HJ. Organisation of the autonomic nervous system: structure and function. In Handbook of Clinical Neurology 74, The Autonomic Nervous System, Part I: Normal Functions. Appenzeller O, Ed.; Amsterdam: Elsevier, 1999; pp. 1–52.
16. Jänig W, Häbler HJ. Neurophysiological analysis of target-related sympathetic pathways. From animal to human: similarities and differences. Acta Physiol Scand 2003;177:255–74.
17. Dampney RA. Functional organization of central pathways regulating the cardiovascular system. Physiol Rev 1994;74:323–64.
18. Spyer KM. Central nervous mechanisms contributing to cardiovascular control. J Physiol (Lond) 1994;474:1–19.
19. Guyenet PG, Koshiya N, Huangfu D, Baraban SC, Stornetta RL, Li Y-W. Role of medulla oblongata in generation of sympathetic and vagal outflows. Prog Brain Res 1996;107:127–44.
20. Kirchheim HR, Just A, Ehmke H. Physiology and pathophysiology of baroreceptor function and neuro-hormonal abnormalities in heart failure. Basic Res Cardiol 1998;93(Suppl 1):1–22.
21. Folkow B. Perspectives on the integrative function of the "sympatho-adrenomedullary system". Auton Neurosci 2000;83:101–15.
22. Guyenet PG. Neural structures that mediate sympathoexcitation during hypoxia. Respir Physiol 2000;121:147–62.
23. Morrison SF. Differential control of sympathetic outflow. Am J Physiol Regul Integr Comp Physiol 2001;281:R683–98.
24. Jänig W. Autonomic nervous system and pain. In Basbaum AI, Kaneko A, Shepherd GM, Westheimer G (eds) The Senses: A Comprehensive Reference, Vol. 5: Pain. Bushnell MC, Basbaum AI, Eds; San Diego: Academic Press, 2008; pp. 193–225.
25. Jänig W, Levine JD. Autonomic-endocrine-immune responses in acute and chronic pain. In Wall and Melzack'S Textbook of Pain, 5 Edn. McMahon SB, Koltzenburg M, Eds; Amsterdam Edinburgh: Elsevier Churchill Livingstone, 2006; pp. 205–18.

26. Jänig W, Chapman CR, Green PG. Pain and body protection: sensory, autonomic, neuroendocrine and behavioral mechanisms in the control of inflammation and hyperalgesia. In Proceedings of the 11th World Congress on Pain. Flor H, Kalso E, Dostrovsky JO, Eds; Seattle: IASP Press, 2006; pp. 331–47.
27. Langley JN. The Autonomic Nervous System. Part I. Cambridge: W. Heffer, 1921.
28. Furness JB, Costa M. The Enteric Nervous System. London: Churchill Livingston, 1987.
29. Furness JB. The Enteric Nervous System. Oxford: Blackwell Science Ltd, 2006.
30. Ritter S, Ritter RC, Barnes CD. Neuroanatomy and Physiology of Abdominal Vagal Afferents. Boca Raton, FL: CRC Press, 1992.
31. Undem B, Weinreich D (Eds). Advances in Vagal Afferent Neurobiology. Boca Raton: CRC Press, 2005.
32. Jänig W, Koltzenburg M. Pain arising from the urogenital tract. In Nervous Control of the Urogenital System. Maggi CA, Ed.; Chur, Switzerland: Harwood Academic Publisher, 1993;pp. 523–76.
33. Cervero F. Sensory innervation of the viscera: peripheral basis of visceral pain. Physiol Rev 1994;74:95–138.
34. Jänig W. Neurobiology of visceral afferent neurons: neuroanatomy, functions, organ regulations and sensations. Biol Psychol 1996;42:29–51.
35. Jänig W. Vagal afferents and visceral pain. In Advances in Vagal Afferent Neurobiology. Undem B, Weinreich D, Eds; Boca Raton: CRC Press, 2005; pp. 461–89.
36. Jänig W, McLachlan EM. Organization of lumbar spinal outflow to distal colon and pelvic organs. Physiol Rev 1987;67:1332–404.
37. Jänig W, Schmidt M, Schnitzler A, Wesselmann U. Differentiation of sympathetic neurones projecting in the hypogastric nerves in terms of their discharge patterns in cats. J Physiol (Lond) 1991;437:157–79.
38. Häbler HJ, Jänig W, Krummel M, Peters OA. Respiratory modulation of the activity in postganglionic neurons supplying skeletal muscle and skin of the rat hindlimb. J Neurophysiol 1993;70:920–30.
39. Häbler HJ, Jänig W, Krummel M, Peters OA. Reflex patterns in postganglionic neurons supplying skin and skeletal muscle of the rat hindlimb. J Neurophysiol 1994;72:2222–36.
40. Boczek-Funcke A, Dembowsky K, Häbler HJ, Jänig W, McAllen RM, Michaelis M. Classification of preganglionic neurones projecting into the cat cervical sympathetic trunk. J Physiol (Lond) 1992;453:319–39.
41. Boczek-Funcke A, Dembowsky K, Häbler HJ, Jänig W, Michaelis M. Spontaneous activity, conduction velocity and segmental origin of different classes of thoracic preganglionic neurons projecting into the cat cervical sympathetic trunk. J Auton Nerv Syst 1993;43:189–200.
42. Wallin BG, Fagius J. Peripheral sympathetic neural activity in conscious humans. Annu Rev Physiol 1988;50:565–76.
43. Wallin BG. Intraneural recordings of normal and abnormal sympathetic activity in humans. In Autonomic Failure. Mathias CJ, Bannister R, Eds; Oxford, NY: Oxford University Press, 2002; pp. 224–31.
44. Joyner MJ, Halliwill JR. Sympathetic vasodilatation in human limbs. J Physiol (Lond) 2000;526: 471–80.
45. Morrison SF. RVLM and raphe differentially regulate sympathetic outflows to splanchnic and brown adipose tissue. Am J Physiol 1999;276:R962–73.
46. De Groat WC.Neural control of the urinary bladder and sexual organs. In Autonomic Failure. Mathias CJ, Bannister R, Eds; Oxford: Oxford University Press, 2002; pp. 151–65.
47. Jänig W. (1996). Behavioral and neurovegetative components of reproductive functions. In Comprehensive Human Physiology, Vol. 2. Greger R, Windhorst U, Eds; Berlin, Heidelberg: Springer-Verlag, 1996; pp. 2253–63.
48. Boczek-Funcke A, Häbler HJ, Jänig W, Michaelis M. Respiratory modulation of the activity in sympathetic neurones supplying muscle, skin and pelvic organs in the cat. J Physiol (Lond) 1992;449:333–61.

49. Boczek-Funcke A, Dembowsky K, Häbler HJ, Jänig W, Michaelis M. Respiratory related activity patterns in preganglionic neurones projecting into the cat cervical sympathetic trunk. J Physiol (Lond) 1992;457:277–96.
50. Häbler HJ, Bartsch T, Jänig W. Rhythmicity in single fiber postganglionic activity supplying the rat tail. J Neurophysiol 1999;81:2026–36.
51. Grewe W, Jänig W, Kümmel H. Effects of hypothalamic thermal stimuli on sympathetic neurones innervating skin and skeletal muscle of the cat hindlimb. J Physiol (Lond) 1995;488:139–52.
52. Jänig W, McLachlan EM. Characteristics of function-specific pathways in the sympathetic nervous system. Trends Neurosci 1992;15:475–81.
53. Jänig W, McLachlan EM. Specialized functional pathway are the building blocks of the autonomic nervous system. J Auton Nerv Syst 1992;41:3–13.
54. Jänig W, McLachlan EM. Neurobiology of the autonomic nervous system. In Autonomic Failure. Mathias CJ, Bannister R, Eds; Oxford, NY: Oxford University Press, 2002; pp. 3–15.
55. Wang FB, Holst MC, Powley TL. The ratio of pre- to postganglionic neurons and related issues in the autonomic nervous system. Brain Res Rev 1995;21:93–115.
56. McLachlan EM (Ed.). Autonomic Ganglia. Luxembourg: Harwood Academic Publishers, 1995.
57. McLachlan EM, Davies PJ, Häbler HJ, Jamieson J. On-going and reflex synaptic events in rat superior cervical ganglion cells. J Physiol (Lond) 1997;501:165–81.
58. McLachlan EM, Häbler HJ, Jamieson J, Davies PJ. Analysis of the periodicity of synaptic events in neurones in the superior cervical ganglion of anaesthetized rats. J Physiol (Lond) 1998;511:461–78.
59. Boyd HD, McLachlan EM, Keast JR, Inokuchi H. Three electrophysiological classes of guinea pig sympathetic postganglionic neurone have distinct morphologies. J Comp Neurol 1996;369:372–87.
60. Edwards FR, Hirst GD, Klemm MF, Steele PA. Different types of ganglion cell in the cardiac plexus of guinea-pigs. J Physiol (Lond) 1995;486:453–71.
61. Mawe GM. Prevertebral, pancreatic and gallbladder ganglia: non-enteric ganglia that are involved in gastrointestinal function. In Autonomic Ganglia. McLachlan EM, Ed.; Luxembourg: Harwood Academic Publishers, 1995; pp. 397–444.
62. Keast JR. Pelvic ganglia. In Autonomic Ganglia. McLachlan EM, Ed.; London: Harwood Academic Publishers, 1995; pp. 445–79.
63. Hirst GDS, Bramich NJ, Edwards FR, Klemm M. Transmission at autonomic neuroeffector junctions. Trends Neurosci 1992;15:40–6.
64. Hirst GD, Choate JK, Cousins HM, Edwards FR, Klemm MF. Transmission by post-ganglionic axons of the autonomic nervous system: the importance of the specialized neuroeffector junction. Neurosci 1996;73:7–23.
65. Furness JB, Morris JL, Gibbins IL, Costa M. Chemical coding of neurons and plurichemical transmission. Annu Rev Pharmacol Toxicol 1989;29:289–306.
66. Morris JL, Gibbins IL. Co-transmission and neuromodulation. In Autonomic Neuroeffector Mechanisms. Burnstock G, Hoyle CHV, Eds; Chur, Switzerland: Harwood, 1992; pp. 33–119.
67. Uvnäs B. Sympathetic vasodilator system and blood flow. Physiol Rev 1960;40(Suppl 4):68–75.
68. Bolme B, Novotny J, Uvnäs B, Wright PG. Species distribution of sympathetic cholinergic vasodilator nerves in skeletal muscle. Acta Physiol Scand 1970;78:60–4.
69. Dietz NM, Rivera JM, Eggener SE, Fix RT, Warner DO, Joyner MJ. Nitric oxide contributes to the rise in forearm blood flow during mental stress in humans. J Physiol (Lond) 1994;480:361–68.
70. Strack AM, Sawyer WB, Platt KB, Loewy AD. CNS cell groups regulating the sympathetic outflow to adrenal gland as revealed by transneuronal cell body labeling with pseudorabies virus. Brain Res 1989;491:274–96.
71. Jansen AS, Wessendorf MW, Loewy AD. Transneuronal labeling of CNS neuropeptide and monoamine neurons after pseudorabies virus injections into the stellate ganglion. Brain Res 1995;683:1–24.
72. Cabot JB, Alessi V, Carroll J, Ligorio M. Spinal cord lamina V and lamina VII interneuronal projections to sympathetic preganglionic neurons. J Comp Neurol 1994;347:515–30.

73. Cabot JB. Some principles of the spinal organization of the sympathetic preganglionic outflow. Prog Brain Res 1996;107:29–42.
74. Llewellyn-Smith IJ, Minson JB, Pilowsky PM, Arnolda LF, Chalmers JP. The one hundred percent hypothesis: glutamate or GABA in synapses on sympathetic preganglionic neurons. Clin Exp Hypertens 1995;17:323–33.
75. Weaver LC, Polosa C. Spinal cord circuits providing control of sympathetic preganglionic neurones. In Central Nervous Control of Autonomic Functions. Jordan D, Ed.; Amsterdam: Harwood Academic Publishers, 1994; pp. 29–61.
76. Paxinos G, Watson C. The Rat Brain. San Diego: Academic Press, 1998.
77. Jansen AS, Loewy AD. Neurons lying in the white matter of the upper cervical spinal cord project to the intermediolateral cell column. Neuroscience 1997;77:889–98.
78. Rekling JC, Feldman JL. PreBotzinger complex and pacemaker neurons: hypothesized site and kernel for respiratory rhythm generation. Annu Rev Physiol 1998;60:385–405.
79. Guyenet PG. Role of the ventral medulla oblongata in blood pressure regulation. In Central Regulation of Autonomic Functions. Loewy AD, Spyer KM, Eds; Oxford: Oxford University Press, 1990; pp. 145–67.
80. Lipski J, Lin J, Teo MY, van Wyk M. The network vs pacemaker theory of the activity of RVL presympathetic neurons – a comparison with another putative pacemaker system. Auton Neuroscience 2002;98:85–9.
81. Nagashima K, Nakai S, Tanaka M, Kanosue K. Neuronal circuitries involved in thermoregulation. Auton Neurosci 2000;85:18–25.
82. Rathner JA, Owens NC, McAllen RM. Cold-activated raphe-spinal neurons in rats. J Physiol 2001;535:841–54.
83. Richter DW, Spyer KM. Cardiorespiratory control. In Central Regulation of Autonomic Function. Loewy AD, Spyer KM, Eds; Oxford: Oxford University Press, 1990; pp. 189–207.
84. Richter DW, Spyer KM, Gilbey MP, Lawson EE, Bainton CR, Wilhelm Z. On the existence of a common cardiorespiratory network. In Cardiorespiratory and Motor Coordination. Koepchen HP, Huopaniemi T, Eds; Berlin, Heidelberg, New York: Springer, 1991; pp. 118–30.
85. Card JP, Swanson LW, Moore RY. The hypothalamus: an overview of regulatory systems. In Fundamental Neuroscience. Squire LR, Bloom F, McConnell SR, Roberts JL, Spitzer NC, Zigmond MJ, Eds; San Diego: Academic Press, 2003; pp. 897–909.
86. Brown MR, Koob GF, Rivier C (Eds). Stress. Neurobiology and Neuroendocrinology. New York, Basel, Hongkong: Marcel Dekker, 1991.
87. Maier SF, Watking LR. Cytokines for psychologists: implications of bidirectional immune-to-brain communication for understanding behaviour, mood, and cognition. Psychol Rev 1998;105:83–107.
88. Bandler R, Carrive P, Zhang SP. Integration of somatic and autonomic reactions within the midbrain periaqueductal grey: viscerotopic, somatotopic and functional organization. Prog Brain Res 1991;87:269–305.
89. Bandler R, Shipley MT. Columnar organization in the midbrain periaqueductal gray: modules for emotionel expression. Trends Neurosci 1994;17:379–89.
90. Bandler R, Keay KA. Columnar organization in the midbrain periaqueductal gray and the integration of emotional expression. Prog Brain Res 1996;107:285–300.
91. Bandler R, Price JL, Keay KA. Brain mediation of active and passive emotional coping. Prog Brain Res 2000;122:333–49.
92. Jänig W, Khasar SG, Levine JD, Miao FJ-P. The role of vagal visceral afferents in the control of nociception. Prog Brain Res 2000;122:273–87.
93. Jänig W, Häbler H-J. Sympathetic nervous system, contribution to chronic pain. Prog Brain Res 2000;129:453–70.
94. Besedovsky HO, del Rey A. Immune-neuroendocrine circuits: integrative role of cytokines. Front Neuroendocrinol 1992;13:61–94.
95. Besedovsky HO, del Rey A. Immune-neuroendocrine interactions: facts and hypotheses. Endocr Rev 1995;17:64–102.

96. Hori T, Katafuchi T, Take S, Shimizu N, Niijima A. The autonomic nervous system as a communication channel between the brain and the immune system. Neuroimmunomodulation 1995;2: 203–15.
97. Madden KS, Felten DL. Experimental basis for neural-immune interactions. Physiol Rev 1995;75: 77–106.
98. Madden KS, Sanders K, Felten DL. Catecholamine influences and sympathetic modulation of immune responsiveness. Annu Rev Pharmacol Toxicol 1995;35:417–48.
99. Schedlowski M, Tewes U. Psychoneuroimmunology. An Interdisciplinary Introduction. Boston, NY: Kluwer Academic Plenum Publishers, 1999.
100. Ader A (Ed.). Psychoneuroimmunology. Vols 1 and 2. Amsterdam: Elsevier, 2007.
101. Ader A, Cohen N. Psychoneuroendocrinology: conditioning and stress. Annu Rev Physiol 1993;44:53–85.
102. Saphier D. Psychoimmunology: the missing link. In Hormonally Induced Changes in Mind and Brain. Schulkin J, Ed.; San Diego, Academic Press, 1993; pp. 191–224.
103. Baron R, Jänig W. Sympathetic and afferent neurons projecting in the splenic nerve of the cat. Neurosci Lett 1988;94:109–13.
104. Dorward PK, Burke SL, Jänig W, Cassell J. Reflex responses to baroreceptor, chemoreceptor and nociceptor inputs in single renal sympathetic neurones in the rabbit and the effects of anaesthesia on them. J Auton Nerv Syst 1987;18:39–54.
105. DiBona GF, Kopp UC. Neural control of renal function. Physiol Rev 1997;77:75–197.
106. Kopp UC, Di Bona GF. The neural control of renal function. In The Kidney: Physiology and Pathophysiology. Seldin G, Giebisch G, Eds; New York: Raven Press, 2000; pp. 981–1006.
107. Meckler RL, Weaver LC. Characteristics of ongoing and reflex discharge of single splenic and renal sympathetic postganglionic fibres in the cat. J Physiol (Lond) 1988;396:139–53.
108. Stein RD, Weaver LC. Multi- and single-fibre mesentric and renal sympathetic responses to chemical stimulation of intestinal receptors in cats. J Physiol (Lond) 1988;396:155–72.
109. Katafuchi T, Take S, Hori T. Roles of sympathetic nervous system in the suppression of cytotoxicity of splenic natural killer cells in the rat. J Physiol (Lond) 1993;465:343–57.
110. Okamoto S, Ibaraki K, Hayashi S, Saito M. Ventromedial hypothalamus supresses splenic lymphocyte activity through sympathetic innervation. Brain Res 1996;39:308–13.
111. Bos JD. Skin immune system. Boca Raton: CRC Press, 1989.
112. Bos JD, Kapsenberg ML. The skin immune system. Its cellular constituents and their interactions. Immunol Today 1986;7:235–40.
113. Edelson RF, Fink JM. The immunologic function of skin. Sci Am 1985;252(6):34–41.
114. Williams IR, Kupper TS. Immunity at the surface: homeostatic mechanisms of the skin immune system. Life Sci 1996;58:1485–507.
115. Miao FJ-P, Jänig W, Levine JD. Role of sympathetic postganglionic neurons in synovial plasma extravasation induced by bradykinin. J Neurophysiol 1996;75:715–24.
116. Miao FJ-P, Green PG, Coderre TJ, Jänig W, Levine JD. Sympathetic-dependence in bradykinin-induced synovial plasma extravasation is dose-related. Neurosci Lett 1996;205:165–8.
117. Khasar SG, Miao FJ-P, Jänig W, Levine JD. Modulation of bradykinin-induced mechanical hyperalgesia in the rat by activation in abdominal vagal afferents. Eur J Neurosci 1998;10:435–44.
118. Miao FJ-P, Jänig W, Levine JD. Nociceptive-neuroendocrine negative feedback control of neurogenic inflammation activated by capsaicin in the rat paw: role of the adrenal medulla. J Physiol (Lond) 2000;527:601–10.
119. Miao FJ-P, Jänig W, Jasmin L, Levine JD. Spino-bulbo-spinal pathway mediating vagal modulation of nociceptive-neuroendocrine control of inflammation in the rat. J Physiol 2001;532:811–22.
120. Berthoud HR, Neuhuber WL. Functional and chemical anatomy of the afferent vagal system. Auton Neurosci 2000;85:1–17.
121. Miao FJ-P, Jänig W, Levine JD. Vagal branches involved in inhibition of bradykinin-induced synovial plasma extravasation by intrathecal nicotine and noxious stimulation in the rat. J Physiol (Lond) 1997;498:473–81.

122. Green PG, Miao FJ-P, Jänig W, Levine JD. Negative feedback neuroendocrine control of the inflammatory response in rats. J Neurosci 1995;15:4678–86.
123. Green PG, Jänig W, Levine JD. Negative feedback neuroendocrine control of inflammatory response in the rat is dependent on the sympathetic postganglionic neuron. J Neurosci 1997;17:3234–8.
124. Miao FJ-P, Jänig W, Green PG, Levine JD. Inhibition of bradykinin-induced plasma extravasation produced by noxious cutaneous and visceral stimuli and its modulation by vagal activity. J Neurophysiol 1997;78:1285–92.
125. Khasar SG, Miao FJ-P, Jänig W, Levine JD. Vagotomy-induced enhancement of mechanical hyperalgesia in the rat is sympathoadrenal-mediated. J Neurosci 1998;18:3043–9.
126. Khasar SG, Green PG, Miao FJ-P, Levine JD. Vagal modulation of nociception is mediated by adrenomedullary epinephrine in the rat. Eur J Neurosci 2003;17:909–15.
127. Morrison SF, Cao WH. Different adrenal sympathetic preganglionic neurons regulate epinephrine and norepinephrine secretion. Am J Physiol Regul Integr Comp Physiol 2000;279:R1763–75.

*The Hypothalamus–Pituitary–Adrenal Axis*
Edited by A. del Rey, G. P. Chrousos and H. O. Besedovsky



# *The Innate Immune System*

KLAUS HEEG

*Department of Hygiene and Medical Microbiology, Institute of Hygiene, University of Heidelberg, Germany*

## ABSTRACT

The innate immune system recognizes infectious danger and launches an immediate anti-infective response. It relies on recognition systems (receptors) that were selected during phylogeny and which in part are shared with non-vertebrate organisms. Receptors of innate immunity are encoded within the germ-line and do not undergo somatic recombination. Humoral as well as cellular mechanisms are involved, which interact at multiple levels and are expressed on immune and non-immune cells to ensure an appropriate and efficient defense response. Innate and adaptive immunity are interconnected. Activation of innate immune cells is a necessary prerequisite for induction of adaptive immune responses. Multiple effector mechanisms of adaptive immunity ultimately rely on the clearance function of innate immune cells. Moreover, the products of the adaptive immune system will enhance the efficacy of innate immune responses. Thus, in vertebrates, innate and adaptive immunity act in a well-tuned concert to ascertain an effective antipathogen response.

## 1. INTRODUCTION

Multicellular organisms are threatened by infectious agents that might damage their complex cellular integrity, compete for essential nutrition, and might destroy their fine-tuned cellular specialization. Infectious agents (bacteria, viruses, and parasites) can be found almost everywhere, in primates the number of associated microorganisms outnumbers the number of somatic cell by a factor of at least 10. Thus, during phylogeny, a sophisticated defense system has evolved that prevents infection. This system comprises of multiple strategies of which some might have arisen independently from each other, whereas other mechanisms are interconnected and reflect a co-evolutionary development [1]. Although almost at the level of metazoan, antimicrobial peptides and phagocytosis have been observed, in later developmental stages further cellular and humoral mechanisms were added, cumulating in the development of an adaptive immune system in vertebrates.

Although various and distinct anti-infective effector strategies exist, they are based on primordial mechanisms: phagocytosis, cytotoxicity, and humoral defense. This complement of distinct mechanisms is termed innate immune system. Innate immunity has to sense infectious agents and to ensure a proper anti-infective response. For this, during phylogeny recognition, receptors were generated that sense infectious danger. These receptors are encoded within the germ-line and are expressed on innate immune cells or act as soluble humoral receptors [1]. The formerly used term “unspecific immunity" for the innate immune system is thus highly misleading. In vertebrates, innate immunity was further complemented by the adaptive immune response. To guarantee a further optimization of recognizable foreign structures, a receptor system was created that is generated in individuals by random somatic recombination of germ-line receptor DNA segments. Accordingly, the number of diverse receptors (receptor repertoire) generated by the action of the recombinase complex exceed by far the number of germ-line-encoded innate receptors. Thus, T and B cells generate during individual differentiation and selection an almost unlimited receptor repertoire. The adaptive immune system memorizes foreign structures on the level of individuals and, opposed to the classical innate immune systems, responds to secondary challenge with a faster and enhanced effector response. Fine specificity and memory as key features of adaptive immunity have attracted scientists for almost two decades. However, induction as well as subsequent effector functions of the adaptive immune system relies critically on efficient activation of the innate limb of the immune system [2,3]. Thus, without innate immune cells, an adaptive system would not function at all. This chapter will thus focus on the germ-line-encoded and recombinase-independent receptor repertoire of innate immunity and its interconnectivity with adaptive immunity.

## 2. RECOGNITION STRUCTURES OF INNATE IMMUNITY

### 2.1. The pathogen-associated molecular pattern concept

Infectious danger has to be sensed by the innate immune system. Obviously, recognition structures have to be present that are capable to discriminate between structures of infectious agents and self-structures of the host. When comparing microbe cells and eukaryotic host cells, several marked differences become apparent. While bacteria have developed complex cell walls, eukaryotic cells are lacking molecular structures like peptidoglycan (PG), lipopolysaccharides (LPSs), or certain lipoproteins. Accordingly, these entities could serve as an ideal target structure to recognize bacterial cells. The complement of such structures has been termed pathogen-associated molecular pattern (PAMP) [4]. PAMPs comprise of multiple molecular structures that are unique for viruses and prokaryotic cells and thus can be targeted by PAMP recognition structures (Table 1). Indeed, individual PAMPs are recognized by an array of

Table 1 Pathogen-associated molecular pattern (PAMP) and their receptors

*PAMPs*
Peptidoglycan (PG), muramyl dipeptide, lipoteichoic acid (LTA), lipopolysaccharides (LPS), bacterial CpG-DNA, flagellin, bacterial lipoproteins, bacterial carbohydrates, heat-shock proteins, dsRNA, glycolipids, porins

*Soluble PAMP receptors*
Mannose-binding lectin (MBL), C-reactive protein (CRP), peptidoglycan-binding protein (PGRP)

*Membrane-bound PAMP receptors*
Toll-like receptors (TLR), PGRP, scavenger receptors, mannose receptor

*Endosomal PAMP-receptors*
TLR 3, 7, 8, 9

*Cytosolic PAMP receptors*
NOD receptors
NALP receptors
RNA helicases (RIG-I, MDA5)

specific PAMP receptors (PAMPRs) of innate immunity. Even bacterial DNA can be recognized by innate immune cells and thus can be classified as PAMP [5]. In contrast to vertebrate DNA, bacterial DNA contains more frequent unmethylated CpG dinucleotides (CpG-DNA) that are recognized by PAMPR. Because the mixture of PAMPs can be characteristic for an individual pathogen, an array of receptors thus has the ability to discriminate between different pathogens, and therefore pathogen-adapted immune responses can be initiated.

### 2.2. PAMP recognition structures

Several recognition systems have evolved that act as soluble or membrane-bound receptor systems. These receptors are phylogenetically old and thus can be found also in non-vertebrates [6,7]. This fact has allowed to identify several PAMPRs by homology search. Recognition of PAMP is brought about by soluble, membrane-bound, endosome-restricted, and cytosolic PAMPRs.

Soluble PAMPRs mainly recognize carbohydrates specific for pathogens. Prototypic representatives are the mannose-binding lectin (MBL) [8,9], which is central during the lectin pathway of complement activation and the C-reactive protein (CRP) that recognizes, for example, the sugar moieties of bacterial capsules (Table 1). Soluble receptors for PG were found in mammals by homology cloning [10], which resemble the peptidoglycan recognition proteins (PGRPs) of drosophila [11]. Thus, PGRP seems to represent a further soluble detection system for PAMPs although additional membrane-bound forms exist [11].

As soluble PAMPRs broaden immune recognition in tissues, membrane-bound PAMPRs are expressed not only in specialized immune cells but also in different tissues including the central nervous system [12]. The most important membrane-bound receptor system for PAMPs is the Toll-like receptor (TLR) family [13,14]. TLRs were discerned by homology cloning using the drosophila Toll protein sequence. Toll is responsible for signaling during development of drosophila but also essential for recognition of PAMPs. Interestingly, in drosophila, Toll collaborates with PGRP, which was not found so far in mammals. Studies using knock-out mice suggest that opposing to Toll in drosophila, TLRs seem to have no role during embryonal differentiation. So far, 10 TLRs have been cloned in humans, which share a leucine-rich extracellular portion and an intracellular signal-transducing domain. This domain can also be found in the intracellular portion of the interleukin-1 (IL-1) receptor and thus has been termed TIR domain. The TIR domain is the adaptor region for intracellular signaling molecules like Myeloid differentiation primary response gene 88 (MyD88) and TIRAP/Mal and TRIF [15–18]. From this interaction, the intracellular signal transduction is started that leads to activation of the nuclear factor kappa B (NF-κB), mitogen-activated protein (MAP) kinases, and IRF-dependent signaling pathways [18]. The extracellular domains of TLRs are distinct and allow recognition of a wide variety of PAMPs (Fig. 1). It is remarkable that TLRs 1–6 are expressed on the cell surface while TLRs 3, 7–9, which are involved in recognition of oligonucleotides, are expressed in the endosomal compartment [19].

Although knock-out studies and transfection experiments clearly have demonstrated the involvement of certain TLR in recognition of distinct PAMPs, it is still probable that further adaptor/recognition proteins are involved in addition. Several studies have shown that at least dimerization of TLR is necessary for signaling to occur [20]. Even though TLRs share common intracellular signal pathways, their biological significance might be different. Some PAMPs seem to induce stronger and broader activation responses as others, for example, lipoteichoic acid (LTA) that interacts with TLR 2 induces the cytokine tumor necrosis factor-α (TNF-α) in comparable amounts to CpG-DNA; however, the latter is superb

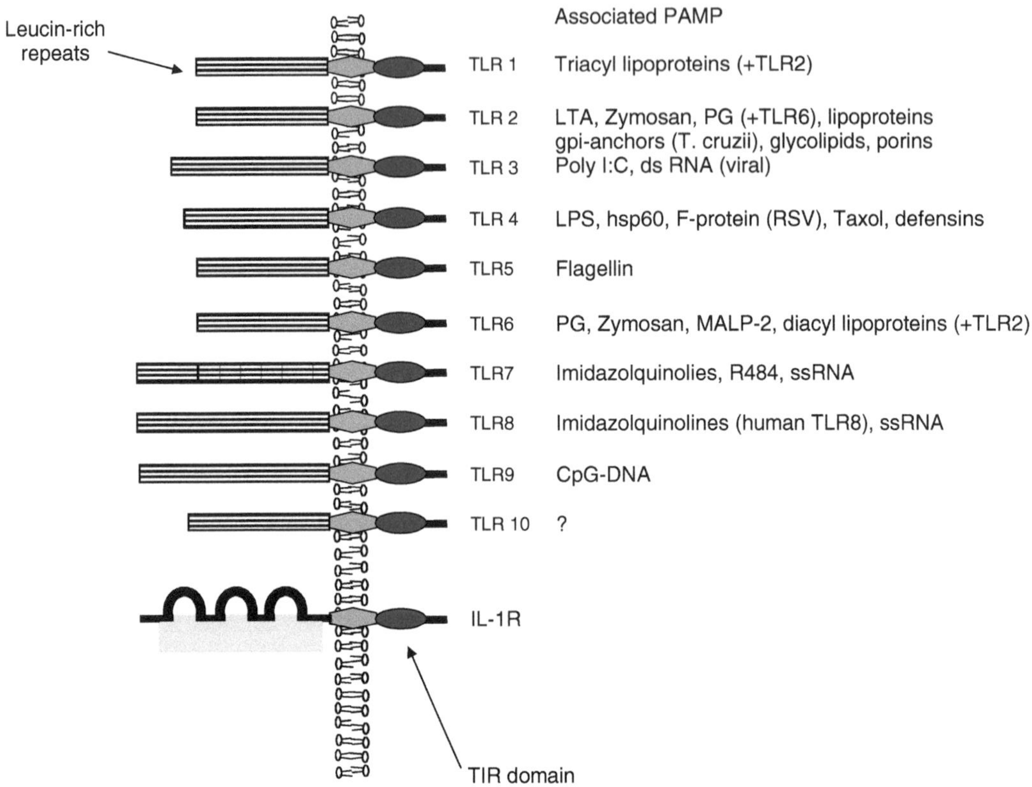


Figure 1. The TLR/IL-1R superfamily.

in inducing IL-12 [21]. Thus, different PAMPs might tune an innate immune response quite differentially.

Once a pathogen has entered the host-cell PAMP, recognition is mediated by cytosolic PAMPR. Viral RNA is recognized by a RNA helicase RIG-I and MDA-5 [22]. Cytosolic muramyl dipeptide is perceived via nucleotide-binding oligomerization domain 2 (NOD2) and its C-terminal leucine-rich repeats [23]. NOD proteins contain further a terminal caspase recruitment domain (CARD) that activates downstream signaling pathways [24]. Furthermore, there is indirect evidence that suggests a cytosolic recognition system for foreign CpG-DNA in addition [25,26].

The scavenger receptor system and the mannose receptor also qualify as PAMPRs on innate immune cells (Table 1) [27,28]. Both receptors recognize characteristic carbohydrate moieties and also signal infectious danger to the cell. Besides direct recognition of PAMP, indirect recognition takes place. During activation of the complement system (vide infra), cleavage products such as C3b can bind to C3b receptors on phagocytes thus enhancing phagocytic function. Moreover, even products of the adaptive immune system, antibodies, can bind to Fc receptors and thus enhance these functions too. This demonstrates the high complexity of interactions of innate and adaptive immunity. In conclusion, innate immune cells are armed with an array of specific receptors that recognize PAMP and thus are perfectly equipped to recognize infective danger.

## 3. EFFECTOR MECHANISMS OF INNATE IMMUNITY

### 3.1. Humoral components

In vertebrates, pathogens first encounter an unspecific physical/chemical barrier system. This comprises of mucus, fluid movement, cilia, and certain pH milieus. Interestingly, the efficacy of this system can be enhanced by products of the adaptive immune system, the immunoglobulin A (IgA) antibodies. After overcoming this barrier, pathogens are now confronted with innate immune cells. After sensing infection, the innate immune system subsequently launches an anti-infective defense program that consists of distinct humoral and cellular-effector mechanisms (Table 2).

#### *3.1.1. Defensins and complement*

Defensins are phylogenetically old effector molecules [29]. Upon contact with host cells, small basic peptides are released that directly insert into the membranes of pathogens leading to membrane disintegration and subsequent death of the pathogen. Defensins are primarily targeted to the cell membrane of prokaryotic cells and display a certain degree of specificity. Production is mainly achieved by epithelial cells whose TLRs are involved in induction of defensins [30]. However, a report showed that defensins by themselves were able to trigger TLR. This might indicate that some effector molecules of the innate immune system might be bifunctional: once they target the infectious agent on the other side, they lead via triggering of TLR to a broadened immune response [31].

In drosophila, proteolytic cascades are activated by invading pathogens that lead to activation of multiple proteins and effector functions. In vertebrates, the complement system seems to function analogically (Fig. 2). Key component of the complement system is the complement factor C3 [32]. Upon cleavage, C3 fragments can serve as opsonins to enhance phagocytosis, as anaphylatoxins that induce inflammation directly and indirectly, and as initiators of activation of the complement factor C5 whose cleavage products also function as anaphylatoxin. Thereafter, the components C6–C9 are activated, which ultimately form the membrane-attacking complex (MAC) that integrates into membranes and form stable pores. These effector mechanisms are directed against pathogens, yet complement activation might also result in damage of host cells and tissues. Complement activation therefore might induce undesired autoaggressive bystander effects that are well known in the course of certain infections. On the whole, this immunopathology is prevented by tight control of the activation steps of the complement system by multiple regulator proteins.

Activation of C3 is brought around by three different initiation pathways: the classical, the alternative, and the lectin pathways (Fig. 2). The latter pathway was recognized only recently although it represents the perhaps most important activation pathway. Accordingly, bacterial carbohydrates bind to the MBL, which in turn activates the adaptor proteins MASP1 and MASP2.

Table 2 Effector mechanisms/cells of the innate immune system

| Cells of the innate immune system | Humoral effector mechanisms |
|---|---|
| Phagocytes (macrophages, granulocytes) | Defensins |
| natural killer cell (NK cell) | Complement system |
| natural killer T-cell (NKT cells) | CRP |
| $\gamma\delta$ T cells | Pro-inflammatory cytokines |
| B1 B-lymphocytes, dendritic cells (DCs) | Interferons |

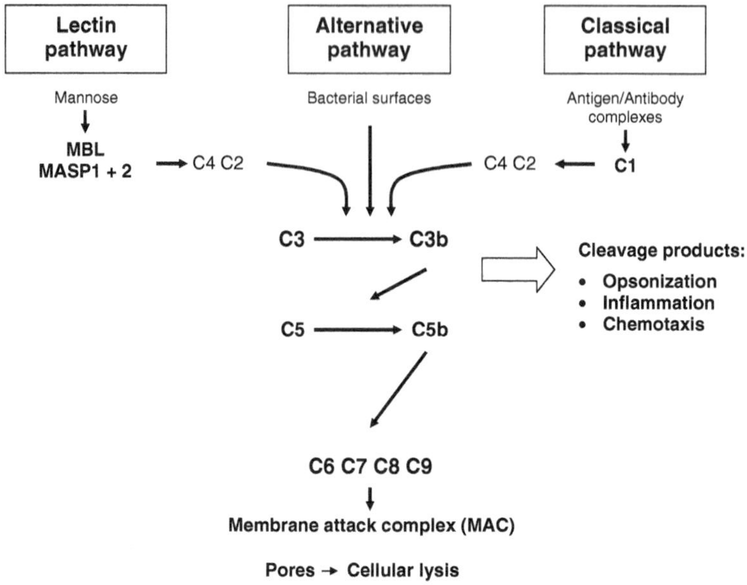


Figure 2. The complement system (simplified).

MASP proteins resemble in structure the C1 complex of the classical pathway and activate C4. The induced complex gains the activity of a C3 convertase, which activates the central cascade of the complement system. Recent knock-out experiments in mice suggest that this activation pathway (lectin pathway) is most important in combating an intraperitoneal infection [9,33,34]. Within the alternative pathway of complement activation, C3 is activated and cleaved on microbial surfaces and the C3 cascade is turned on. Whether minute amounts of C3b generated by the lectin pathway play an important role during this process is still under debate. Finally, the classical pathway bridges adaptive and innate immunity. Antibodies bound to antigen are capable of activating C1 which in turn initiates a C3 convertase comprising of cleavage products of C4 and C2. This convertase finally leads to activation of the central C3 cascade. Therefore, the complement system integrates multiple initiation signals to a common effector strategy that directly targets the invading pathogen and simultaneously signals the innate immune system.

### *3.1.2. Type 1 interferons and proinflammatory cytokines – endogenous PAMPs?*

Upon triggering of PAMPRs, innate effector cells like macrophages or dendritic cells (DCs) are activated. Direct effector functions are induced including cytokine release. Some of these cytokines are bifunctional. Effector cytokines trigger bactericidal effector functions of host cells, such as oxidative burst or induction of nitric oxide synthase (NO synthase). Other cytokines activate bystander cells using signal cascades that are nearly identical to TLR signal cascades, for example, TLR-triggered macrophages produce IL-1 that binds to IL-1 R and triggers via the TIR domain of IL-1 R almost identical intracellular responses [35]. Thus, cytokines trigger directly activation but also broaden the induced antipathogen response. Key players in this process are pro-inflammatory cytokines like IL-1, IL-6, and TNF-α. Type 1 interferons induce self-protecting defense mechanisms but also activate central regulatory cells of the immune system [36,37]. The numerous members of the family of chemokines will contribute to an inflammatory response by recruiting various cell types to the place of the infection.

#### *3.1.3. Effector cytokines*

Within this group, interferon gamma (IFN-γ) plays a pivotal role [38]. Produced by activated macrophages, natural killer (NK) cells, or T cells, IFN-γ controls critically the effectiveness of innate effector mechanisms. Phagocytosis, oxidative burst, and NO synthesis are activated by IFN-γ. Finally, several cytokines will also influence the pattern of responses of the adaptive immune system, for example, DC-derived IL-12 on the one hand will activate NK cells to produce IFN-γ [39], and on the other hand, IL-12 will bias T-cell response that is characterized by a T helper 1 phenotype.

### 3.2. Cellular components

Besides humoral effector mechanisms of innate immunity, cellular recognition and response pattern are developed. Innate cells can be characterized by distinct recognition systems for PAMPs. Opposed to adaptive immune cells, which express one clonally distributed receptor to recognize antigen, innate cells express multiple receptors (TLR, MannoseR) and detection systems (C3b receptor, Fc receptor). Furthermore, while only a subfraction of T and B cells is activated clonally by antigen, innate cells respond according to their distribution of PAMPRs. Therefore, the initial innate response is polyclonal in its nature and strong. Although innate effector cells can be further activated (mainly by IFN-γ) to enhance their effector response, it is unlikely that they memorize at the clonal level their encounter with PAMPs. Thus, secondary responses of innate immune cells show the same kinetics and effectiveness as primary responses. On the other hand, it has been shown that encounter with bacterial components early in life shapes the response pattern of the innate and adaptive immune systems and skews the reactivity toward a Type 1 helper cell (Th1)[40]. Accordingly, the innate immune system might possess "memory" yet at a non-clonal level.

#### *3.2.1. Phagocytes*

The major role of phagocytes is to take up pathogens and to destroy them subsequently. Clinical evidence from bone marrow transplantation suggests that granulocytes are the most important cellular population for clearance. Low numbers of granulocytes are correlated with an inability to cope with infections; increase of granulocyte numbers by supplementation of granulocyte colony-stimulating factor (G-CsF) will restore this capacity. However, granulocytes are short-lived and die soon. In contrast, macrophages are long-lived and not only digest the pathogen but also are involved in the regulatory circuits of the innate immune system. Macrophages express a broad spectrum of TLRs, scavenger, and mannose receptors on their surface. After encounter of pathogens, macrophages are activated and phagocytose the pathogen. This process is greatly enhanced by C3b bound to the pathogen's surface. Moreover, antibodies bound to pathogens will bind to Fc receptors on the macrophages and thus further enhance phagocytosis. This process is called opsonization. Opsonization is critical for all pathogens that synthesize a capsule, without opsonization the rate of phagocytosis is rather low [41]. Once taken up, the phagosomes fuse with lysosomes resulting in phagolysosomes. In this compartment, the final degradation of pathogens takes place. For this, a rather toxic milieu has to be created which is formed by reactive oxygen intermediates (ROI, oxidative burst) and NO [42,43]. Both enzyme complexes are triggered to higher efficacy by IFN-γ. Once degraded, material from the pathogen can be further processed and presented to T lymphocytes. Furthermore, macrophages produce pro-inflammatory cytokines, chemokines, and lymphokines, which further induce an inflammatory reaction.

### 3.2.2. *NK cells, γδ T cells, and B1 B cells*

Although NK cells seem not to express a PAMPR, they can be included in the realm of innate immunity because they respond immediately and show no immunological memory. NK cells make use of the fact that some intracellular pathogens (bacteria and viruses) down-regulate major histocompatibility complex (MHC) molecules from the surface of infected cells. Because MHC molecules on target cells inhibit NK cell activation [44], lack of MHC expression leads to unrestricted NK cell activation and subsequent triggering of target cell death. NK cells can be activated by DC-derived IL-12 and produce high amounts of IFN-γ which in turn activates macrophages. Thus, cellular activation of a broad variety of cell types is induced [45]. NKT cells are a heterogeneous group of immune cells that share features of T cells and NK cells. They express the NK cell-associated marker NK1.1 and recognize lipids and glycolipids presented via CDld molecules. Some NKT cells express an invariant TCR Valpha chain (iNKT cells) and respond with the production of effector cytokines such as IFN-γ, IL-4, and GM-CSF [46].

As NKT cells, γδ T cells seem to be in between the innate and the adaptive immune system. Whilst their receptors are generated through somatic recombination, a subgroup of γδ T cells recognizes conserved microbial pattern. This pattern is due to unique metabolic pathways of bacteria whose products are presented in context with MHC class 1b molecules [47]. This reflects a peculiarity because bacterial pattern is presented here by specialized molecules of the host [48]. Similar to γδ T cells, B1 B cells rearrange their germ-line DNA and generate a clonally distributed set of B-cell receptors. B1 B cells can be activated independent of T-cell help by antigens displaying multiple and repetitive structures such as phosphatidylcholine and carbohydrates. After activation, B1 cell produces antibodies of the IgM subclass. This subclass is conserved, class switching does not occur. B1 cells are responsible for the so-called natural antibodies whose levels increase after birth and remain for the whole lifetime. Natural antibodies recognize sugar moieties and thus also enhance the immune function of the innate immune system against pathogens [49,50].

### 3.2.3. *Dendritic cells*

DCs are essential in innate immunity as well as the most important population for induction of adaptive immune responses. DCs can be subdivided into sessile DCs which can be found in almost every tissue. There, they form long branches that contact many cells of the tissue. These DCs act as sentinels that control the infection status of the tissue. In the skin, the Langerhans cells are the prototype of a tissue-specific peripheral DC. Peripheral DCs are highly active in phagocytosis yet express low levels of co-stimulatory molecules and are poor stimulators of T cells. DC can be further found in the lymphatic tissues and in the circulation. According to surface markers and function DC can be further subdivided into monocyte-derived DC and plasmacytoid DC. These central DCs have ceased the phagocytic activity yet now are fully equipped to activate naïve T cells. On a per cell basis, DCs are by far the most efficient cell type to activate T lymphocytes [51,52].

Sessile phagocytic DCs are activated by recognition of PAMPs. They respond with lymphokines secretion (pro-inflammatory cytokines) and differentiation. During the latter process, DCs seize phagocytosis yet enhance their antigen presentation capacity and migrate to the central lymphatic tissues (lymph nodes) [53]. They present antigen to T cells and provide a strong co-stimulatory environment. This is accomplished by up regulation of MHC molecules (signal 1) and co-stimulatory molecules (signal 2) and by secretion of activating cytokines (signal 3). The activity of DC is not only controlled by PAMPs but also by cytokines directly acting on DC. Type 1 interferons activate DC while suppressive cytokines such as IL-10 and TGF-β inhibit their activity. Interestingly, it was shown that neuropeptides are also capable to influence the functional status of DC [54].

## 4. FIRST LEVEL OF AN ANTI-INFECTIVE RESPONSE: INFLAMMATION

During infection, these constituents of the innate immune system interact with each other in order to control the initial steps of an infection. Locally produced defensins complement activation and tissue phagocytes ensure initially a proper and immediate antipathogen response. During this initial step, further mediators and messengers are released and signal infectious danger to the innate immune system. Cytokines, perhaps defensins itself, chemokines, and cleavage products of proteolytic cascades (complement) recruit further phagocytes to the place of infection. For this, endothelial cells are activated to promote transcytosis of neutrophil granulocytes and leukocytes. Vasoactive substances released during initial infection induce vasodilatation and exudation from the dilated capillaries. Exudation and released mediators will activate the afferent limb of the peripheral neuronal system and thus induce pain.

Cytokines released during this first step of an infection play a dual role. As signal molecules, they act locally and activate, like an endogenous PAMP, further effector cells in the local tissue. Cytokines, however, also have systemic effects. IL-1 and TNF-$\alpha$ might enter the circulation and become effective on other sites of the infected host. One specific example for the long-range effects of cytokine action is the induction of fever. Specifically, IL-1, TNF, and Type 1 interferons are capable to reset the hypothalamic temperature circuits to a higher body temperature. This combination of events is termed inflammation and is a prerequisite to cope infection. The classical symptoms of inflammation *dolor* (pain), *calor* (enhanced temperature), *rubor* (vasodilatation), and *tumor* (exudation and accumulation of phagocytes) can be explained by these events.

It should be stressed that this profile of reactivity occurs within the first minutes to hours of infection. The response is immediate and assures in most cases elimination of the infectious danger. During initial contact with a pathogen, the adaptive immunity plays at this phase no essential role. Primary responses of the adaptive immune system require days to become efficient; thus control of many infections primarily relies on innate immunity. This picture changes notably when secondary immune responses are considered. Preformed products of the adaptive immune system (antibodies) will specifically enhance the efficacy of the innate response. After binding to antigen, antibodies might neutralize infectious agents, activate the complement system via the classical pathway, and might enhance phagocytosis by opsonization. T lymphocytes might contribute to this process by providing IFN-$\gamma$, which is capable to enhance the immune function of phagocytes. Thus, although innate immune responses lack by itself immunological clonal memory, effector molecules of the adaptive immune system might provide an indirect memory and thus enhance the effector function of innate immunity in secondary immune responses.

## 5. SECOND LEVEL OF AN ANTI-INFECTIVE RESPONSE: ACTIVATION OF THE ADAPTIVE IMMUNE SYSTEM

In adaptive immune responses, T and B lymphocytes are activated upon triggering their clonally distributed receptors by presented or, respectively, free antigen. In primary responses, only a small fraction of lymphocytes is activated and then induced to clonal proliferation, expansion, and subsequent differentiation to effector lymphocytes. This process requires several days; therefore a primary immune response of the adaptive system is delayed as compared to the immediate response of innate immune cells. For activation, lymphocytes need at least two signals, whereas antigen-induced triggering of the receptors represents signal 1. Signal 2 is provided by co-stimulatory signals that critically control activation of

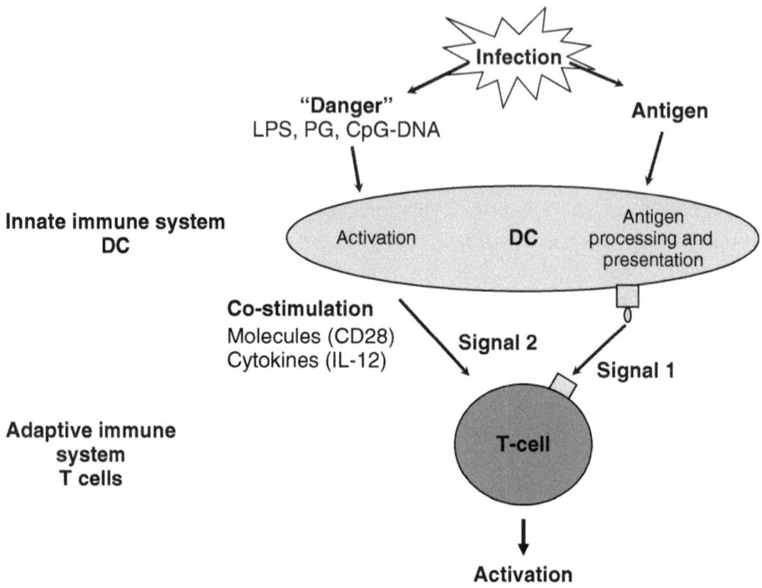


Figure 3. Steps in T-cell activation.

the immune system Fig. 3. The most potent cell to provide co-stimulatory signals is the DC. DCs thus bridge innate and adaptive immunity and are essentially involved in activation of adaptive immune cells.

Resting DCs are sessile in the tissues. Upon infection, they take up antigen and are activated by PAMPs. This activation leads to cytokine production and a change in the program of DC. They cease phagocytosis but enhance their antigen presentation capacity. Moreover, they express co-stimulatory molecules and start to produce co-stimulatory cytokines such as IL-12 and IL-18. At the same time, DCs leave the tissue and migrate to the local lymph nodes. Here, they provide both the necessary signals to the T lymphocytes: signal 1 (antigen) that is presented by their MHC molecules and signal 2 (co-stimulation) that is brought about by co-stimulatory molecules and co-stimulatory cytokines. DCs are thus professional antigen presenting cells that initiate a T-cell response (Fig. 3). Accordingly, activation of DC by PAMPs is crucial also for the induction of a response of the adaptive immune system. This concept has considerable implications for T-cell immunology as well as for vaccination approaches. Delivery of antigen alone is insufficient to induce a T-cell response, co-stimulatory signals have to be added in addition. PAMPs might serve as an inductor of such co-stimulatory signals (Fig. 3) [55]. Indeed, some PAMPs like CpG-DNA can serve as an adjuvant during vaccination and ensure proper activation of T-cell responses [56,57].

## 6. CONCLUDING REMARKS

The innate immune response represents the archetypical immune system. In non-vertebrates, innate immunity alone is capable to cope with infection. However, individual vertebrates are challenged with new set of infectious agents and furthermore, due to their relatively long life time, are challenged with certain pathogens repeatedly. Thus, an adaptive immune response, which can

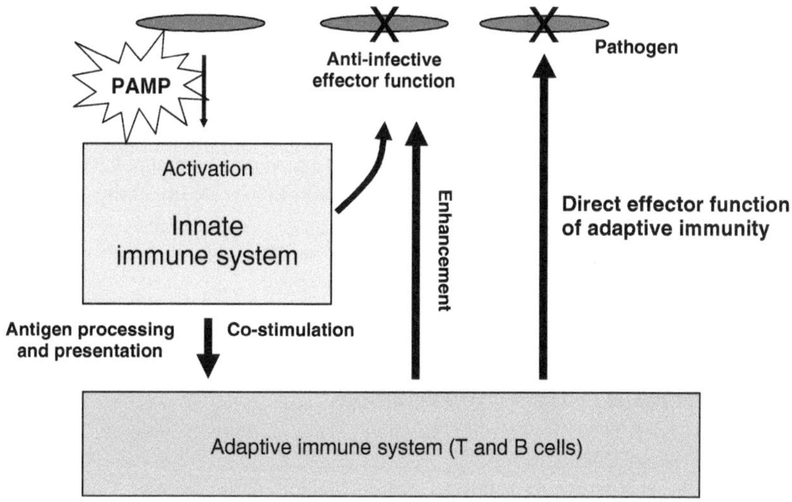


Figure 4. Interaction of innate and adaptive immunity.

memorize a pathogen, is of decisive importance. The last decade has witnessed a change in the perception of innate and adaptive immunity Fig. 4. It is now recognized that innate immunity is not only the first and immediate line of antipathogen defense but also a crucial prerequisite to activate the adaptive immune system. In turn, products of the adaptive immune system enhance the efficacy of the innate effector response. Finally, without the clearing capacity of the innate system, adaptive immunity alone could not function. Thus, both systems are intensely interconnected and represent in vertebrates an integrated immune system rather than two independent entities of immunity.

## REFERENCES

1. Medzhitov R, Janeway CA. Innate immunity: the virtues of a nonclonal system of recognition. Cell 1997;91:295–8.
2. Fearon DT, Locksley RM. The instructive role of innate immunity in the acquired immune response. Science 1996;272:50–4.
3. Medzhitov R, Janeway CA. Innate immunity: impact on the adaptive immune response. Curr Opin Immunol 1997;9:4–9.
4. Medzhitov R, Janeway CA, Jr. Innate immune recognition: mechanisms and pathways. Immunol Rev 2000;173:89–97.
5. Lipford GB, Heeg K, Wagner H. Bacterial DNA as immune cell activator. Trends Microbiol 1998;6:496–500.
6. Hoffmann JA, Kafatos FC, Janeway CA, Ezekowitz RA. Phylogenetic perspectives in innate immunity. Science 1999;284:1313–18.
7. Hoffmann JA, Reichhart JM. Drosophila innate immunity: an evolutionary perspective. Nat Immunol 2002;3:121–6.
8. Turner MW. Mannose-binding lectin: the pluripotent molecule of the innate immune system. Immunol Today 1996;17:532–40.
9. Jack DL, Klein NJ, Turner MW. Mannose-binding lectin: targeting the microbial world for complement attack and opsonophagocytosis. Immunol Rev 2002;180:86–99.

10. Liu C, Xz Z, Gupta D, Dziarski R. Peptidoglycan recognition proteins: a novel family of four human innate immunity pattern recognition molecules. J Biol Chem 2002;276:34686–94.
11. Werner T, Liu G, Kang D, Ekengren S, Steiner H, Hultmark D. A family of peptidoglycan recognition proteins in the fruit fly *Drosophila melanogaster*. Proc Natl Acad Sci USA 2000;97:13772–7.
12. Jack CS, Arbour N, Manusow J, Montgrain V, Blain M, McCrea E, Shapiro A, Antel JP. TLR signaling tailors innate immune responses in human microglia and astrocytes. J Immunol 2005;175:4320–30.
13. Kopp EB, Medzhitov R. The Toll-receptor family and control of innate immunity. Curr Opin Immunol 1999;11:13–18.
14. Medzhitov R, Janeway CA, Jr. The Toll receptor family and microbial recognition. Trends Microbiol 2000;8:452–6.
15. Kawai T, Adachi O, Ogawa T, Takeda K, Akira S. Unresponsiveness of MyD88-deficient mice to endotoxin. Immunity 1999;11:115–22.
16. Horng T, Barton GM, Flavell RA, Medzhitov R. The adaptor molecule TIRAP provides signalling specificity for Toll-like receptors. Nature 2002;420:329–33.
17. Akira S, Takeda K, Kaisho T. Toll-like receptors: critical proteins linking innate and acquired immunity. Nat Immunol 2001;2:675–80.
18. Takeda K, Akira S. TLR signaling pathways. Semin Immunol 2004;16:3–9.
19. Häcker H, Mischak H, Miethke T, Liptay S, Schmid R, Sparwasser T, Heeg K, Lipford GB, Wagner H. CpG-DNA specific activation of antigen presenting cells requires stress kinase activity and is proceeded by non-specific endocytosis and endosomal maturation. EMBO J 1998;17:6230–40.
20. Bell JK, Askins J, Hall PR, Davies DR, Segal DM. The dsRNA binding site of human Toll-like receptor 3. Proc Natl Acad Sci USA 2006;103:8792–7.
21. Dalpke A, Zimmermann S, Hartung T, Heeg K. Collaboration of lipoteichoic acid and CpG-DNA during activation of innate immune cells. Immunobiology 2002;206:392–407.
22. Pichlmair A, Schulz O, Tan CP, Naslund TI, Liljestrom P, Weber F, Reis e Sousa. RIG-I-mediated antiviral responses to single-stranded RNA bearing 5′-phosphates. Science 2006;314:997–1001.
23. Delbridge LM, O'Riordan MX. Innate recognition of intracellular bacteria. Curr Opin Immunol 2007;19:10–16.
24. Fritz JH, Ferrero RL, Philpott DJ, Girardin SE. Nod-like proteins in immunity, inflammation and disease. Nat Immunol 2006;7:1250–7.
25. Hochrein H, Schlatter B, O'Keeffe M, Wagner C, Schmitz F, Schiemann M, Bauer S, Suter M, Wagner H. Herpes simplex virus type-1 induces IFN-alpha production via Toll-like receptor 9-dependent and -independent pathways. Proc Natl Acad Sci USA 2004;101:11416–21.
26. Spies B, Hochrein H, Vabulas M, Huster K, Busch DH, Schmitz F, Heit A, Wagner H. Vaccination with plasmid DNA activates dendritic cells via Toll-like receptor 9 (TLR9) but functions in TLR9-deficient mice. J Immunol 2003;171:5908–12.
27. Pearson AM. Scavenger receptors in innate immunity. Curr Opin Immunol 1996;8:20–8.
28. Engering AJ, Cella M, Fluitsma D, Brockhaus M, Hoefsmit ECM, Lanzavecchia A, Pieters J. The mannose receptor functions as a high capacity and broad specificity antigen receptor in human dendritic cells. Eur J Immunol 1997;27:2417–25.
29. Zasloff M. Antimicrobial peptides of multicellular organisms. Nature 2002;415:389–95.
30. Hertz CJ, Wu Q, Porter EM, Zhang YJ, Weismuller KH, Godowski PJ, Ganz T, Randell SH, Modlin RL. Activation of Toll-like receptor 2 on human tracheobronchial epithelial cells induces the antimicrobial peptide human beta defensin-2. J Immunol 2003;171:6820–6.
31. Biragyn A, Ruffini PA, Leifer CA, Klyshnenkova E, Shakhov A, Chertov O, Shirakawa AK, Farber JM, Segal DM, Oppenheim JJ, Kwak LW. Toll-like recptor 4 dependent activation of dendritic cells by beta-defensin 2. Science 2002;298:1025–9.
32. Barrington R, Zhang M, Fischer M, Carroll MC. The role of complement in inflammation and adaptive immunity. Immunol Rev 2002;180:5–15.
33. Matsushita M, Fujita T. Ficolins and the lectin complement pathway. Immunol Rev 2002;180:78–85.
34. Stover CM, Thiel S, Lynch NJ, Schwaeble WJ. The rat and mouse homologues of MASP-2 and MAp19, components of the lectin activation pathway of complement. J Immunol 1999;163:6848–59.

35. O'Neill LA, Greene C. Signal transduction pathways activated by the IL-1 receptor family: ancient signaling machinery in mammals, insects, and plants. J Leukoc Biol 1998;63:650–7.
36. Kadowaki N, Antonenko S, Lau JY, Liu YJ. Natural interferon alpha/beta-producing cells link innate and adaptive immunity. J Exp Med 2000;192:219–26.
37. Biron CA. Interferons alpha and beta as immune regulators – a new look. Immunity 2001;14:661–4.
38. Billiau A. Interferon-gamma: Biology and role in pathogenesis. Adv Immunol 1996;62:61–130.
39. Okamura H, Kashiwamura S, Tsutsui H, Yoshimoto T, Nakanishi K. Regulation of interferon-gamma production by IL-12 and IL-18. Curr Opin Immunol 1998;10:259–64.
40. Braun-Fahrlander C, Riedler J, Herz U, Eder W, Waser M, Grize L, Maisch S, Carr D, Gerlach F, Bufe A, Lauener RP, Schierl R, Renz H, Nowak D, von Mutius E. Environmental exposure to endotoxin and its relation to asthma in school-age children. N Engl J Med 2002;347:869–77.
41. Allen LAH, Aderem A. Mechanisms of phagocytosis. Curr Opin Immunol 1996;8:36–40.
42. Yi AK, Tuetken R, Redford T, Waldschmidt M, Kirsch J, Krieg AM. CpG motifs in bacterial DNA activate leukocytes through the pH-dependent generation of reactive oxygen species. J Immunol 1998;160:4755–61.
43. Evans TJ, Cohen J. Mediators: nitric oxide and other toxic oxygen species. Curr Top Microbiol Immunol 1996;216:189–207.
44. Colonna M. Natural killer cell receptors specific for MHC class I molecules. Curr Opin Immunol 1996;8:101–7.
45. Lanier LL. Turning on natural killer cells. J Exp Med 2000;191:1259–62.
46. Bendelac A, Savage PB, Teyton L. The biology of NKT cells. Annu Rev Immunol 2007;25:297–336.
47. Bürk MR, Mori L, de Libero G. Human Vgamma9 – Vdelta2 cells are stimulated in a crossreactive fashion by a variety of phosphorylated metabolites. Eur J Immunol 1995;25:2052–8.
48. Bukowski JF, Morita CT, Brenner MB. Human gamma/delta T cells recognize alkylamines derived from microbes, edible plants, and tea: implications for innate immunity. Immunity 1999;11:57–65.
49. Fagarasan S, Watabe N, Honjo T. Generation, expansion, migration and activation of mouse B1 cells. Immunol Rev 2002;176:205–15.
50. Martin F, Kearney JF. B cell subsets and the mature preimmune repertoire. Marginal zone and B1 B cells as part of a "natural immune memory". Immunol Rev 2002;175:70–9.
51. Liu YJ, Kanzler H, Soumelis V, Gilliet M. Dendritic cell lineage, plasticity and cross-regulation. Nat Immunol 2001;2:585–9.
52. Pulendran B, Palucka K, Banchereau J. Sensing pathogens and tuning immune responses. Science 2001;293:253–6.
53. Ban E, Dupre L, Hermann E, Rohn W, Vendeville C, Quatannens B, Ricciardi-Castagnoli P, Capron A, Riveau G. CpG motifs induce Langerhans cell migration in vivo. Int Immunol 2000;12:737–45.
54. Dunzendorfer S, Kaser A, Meierhofer C, Tilg H, Wiedermann CJ. Cutting edge: peripheral neuropeptides attract immature and arrest mature blood-derived dendritic cells. J Immunol 2001;166:2167–72.
55. Bendelac A, Medzhitov R. Adjuvants of immunity: harnessing innate immunity to promote adaptive immunity. J Exp Med 2002;195:F19–23.
56. Weeratna RD, McCluskie MJ, Xu Y, Davis HL. CpG DNA induces stronger immune responses with less toxicity than other adjuvants. Vaccine 2000;18:1755–62.
57. Lipford GB, Bauer M, Blank C, Reiter R, Wagner H, Heeg K. CpG-containing synthetic oligonucleotides promote B and cytotoxic T cell responses to protein antigen: a new class of vaccine adjuvants. Eur J Immunol 1997;27:2340–4.

*The Hypothalamus–Pituitary–Adrenal Axis*
Edited by A. del Rey, G. P. Chrousos and H. O. Besedovsky



# *Specific Immune Response*

IVAN LEFKOVITS[1] and LOUIS DU PASQUIER[2]

[1]*Research Department of the University Clinics, Basel, Switzerland*
[2]*Institute of Zoology and Evolutionary Biology, University of Basel, Switzerland*

## ABSTRACT

This chapter describes the molecular and cellular components involved in the specific immune response. It starts with explaining the evolutionary aspects, especially the "common roots" of many structural elements that underlay both the nervous system and the immune system. After describing the immunoglobulin superfamily members and their duplication in the evolution, the work addresses the issue of the milieu of the immune response and the nature of the immunogens, especially the wide-ranging spectrum of epitopes. The work portrays the primary antibody response, involvement of B and T cells, antigen processing, antigen presentation, activation of T cells, formation of B cell–T cell conjugates, and the clonal expansion of B cells. Details of the antibody response (including the repertoires of antibody) are given, and special attention is paid to isotype switch and affinity maturation. Thereupon the secondary immune response and formation of memory is explained, and also the processes of induction and maintenance of tolerance, anergy, and apoptosis are given. Besides the T-dependent responses, also responses not requiring T cell help, specific cytotoxic responses as well as involvement of regulatory T cells (Tregs) are portrayed. This chapter does not address the innate immune response (nor the field of Toll-like receptor family) that is left to another chapter(s) of this volume.

**Keywords:** antibody, antigen, immunogen, epitope, paratope, idiotope, affinity, avidity, primary response, secondary response, memory, tolerance, apoptosis, lymphocytes, B cells, T cells, clonal selection, antibody repertoire, antigen processing, antigen presentation, receptors, T-cell receptor, vaccination, humoral response, cellular response

## ABBREVIATIONS

| | |
|---|---|
| APC | antigen-presenting cell |
| BcR | B-cell receptor |
| C | constant (region) |
| CD4 | cluster of differentiation 4 |
| CD8 | cluster of differentiation 8 |

| | |
|---|---|
| CD40 and CD 40L | cluster of differentiation 40 and 40 ligand |
| CDR | complementarity determining region |
| CRTAM | class I restricted T cell associated molecule |
| CTX | cortical thymocyte marker of xenopus |
| DSCAM | down syndrome cell adhesion molecule |
| H chain | heavy chain |
| IFN | interferon |
| IgSF | immunoglobulin superfamily |
| IL-2 | interleukin-2 |
| JAM | junctional adhesion molecule |
| KIR | killer cell immunoglobulin-like receptor |
| L chain | light chain |
| LRR | leucine-rich repeat |
| MHC | major histocompatibility complex |
| Necl2 | Nectin like 2 |
| RAG | recombination-activating-gene |
| SIRP | signal regulatory protein |
| Tc cell | T cytotoxic cell |
| TcR | T-cell receptor |
| TGF | transforming growth factor |
| Th cell | T helper cell |
| TM | transmembrane segment |
| TNF | tumor necrosis factor |
| Treg | T regulatory cell (also suppressor cell) |
| V | variable (region) |

## 1. INTRODUCTION

Although in the context of this volume the structural characteristics of the components of the nervous system and the immune system are of utmost importance, the specific immunity as an adaptive process cannot be understood without comprehending the broad evolutionary context. There is a structural relationship between the nervous and the immune systems, which has been underestimated throughout the history of vertebrate development [1]. The discoveries have been facilitated by the availability of the deciphered genomes from many different species of vertebrates and invertebrates. This allowed multiple comparisons across phyla for related members of several gene families, including the immunoglobulin superfamily (IgSF). The understanding of these relationships revealed how interwoven have been the histories of the nervous and immune systems during evolution.*

*To understand what will follow, one has to be reminded of some definitions (Web definitions):

Homologous (as applied to proteins): a sequence/structure whose resemblance to a previously characterized sequence is strong enough to suggest a common genetic ancestry.

Homologous (in molecular biology): used loosely to mean similar, regardless of genetic relationship.

Paralogous: homologous sequences (e.g., sequences that share a common evolutionary ancestor) that diverged by gene duplication, as opposed to orthologous, which diverged by speciation.

## 2. EVOLUTIONARY CONTEXT

### 2.1. The nervous and the immune systems use immunoglobulin superfamily (IgSF) receptors with similar structural characteristics

It has become possible to trace back the origin of many receptors of the surface of lymphocytes and/or nerve cells. In this chapter assigned to the "immune system" we start with the receptors of the immune system and explore their connection(s) to the structures of the nervous system.

### 2.2. Origin of immunoglobulin superfamily (IgSF) members participating in the immune system

The special structure of the antigen-specific receptors on B lymphocytes (BcR) and on T lymphocytes (TcR), so typical of the adaptive immune system, has suggested that the immediate ancestor of the BcR and TcR should be a molecule with the same structure, but whose variable (V) domains responsible for antigen binding would not be generated by somatic rearrangement. In other words, the ancestral V domain should be encoded by a complete gene. The structure should be associated with a constant region domain C1 or C2, a transmembrane segment (TM) and a cytoplasmic tail with a direct or indirect capacity of signaling (Cy). When searching for such members in invertebrates or vertebrates, several IgSF integral membrane protein candidates fulfilling these criteria were discovered. The closest relative is perhaps the signal regulatory protein (SIRP) (V-C1-C1-TM-Cy) found in the vertebrate genome, followed by the VC1-like core of nectins (V-C1-like-C2-TM-Cy) as well as VC2 molecules. Of those, only the nectins and V-C2n-TM-Cy of the CTX/JAM family are found in invertebrates. True C1 domains are absent in species that do not have an adaptive immune system. So it is difficult to assess whether SIRP is really the primitive form since it does not exist in a recognizable form in any invertebrate. Its C1 component is perhaps of recent origin and is "borrowed" from an antigen receptor gene by exon shuffling. With respect to the VC1-like and VC2 members they seem to belong to a large family of genes conserved across many phyla of the animal kingdom that were involved later in the lympho-myeloid cell interactions in vertebrates (e.g., CD80–86, CD96, CD166, CD90, JAM/CTX, CD47) and also in invertebrates. Species such as *Branchiostoma* (amphioxus), *Botryllus*, *Ciona* (sea squirts) (prochordates), *Drosophila* (insects) contain several genes of this category.

The transformation of one such ancestor into the vertebrate BcR and TcR implied a split inside a V domain between the strand F and G, introduction of a mobile element, and a "recombination-activating-gene" (RAG) capable of mediating the rearrangement. Ancestors of RAG have now been identified in sea urchin and in *Drosophila* among members of the transposon family.

### 2.3. Paralogs and homologs of IgSF members involved in the nervous system

The most interesting aspect of the members of this last category is that many of them are related either to immunity (CD96 human) or to the nervous system (CD96 homolog amalgam in *Drosophila*) or to both (CD166 in *Homo*) and that the homolog of one gene in one phylum can be involved in immunity whereas its ortholog is involved in the nervous system. This kind of swap of commitment is also true for the above-mentioned SIRPs that were first discovered in the brain and later at the interface between T cells and antigen-presenting cell (APC).

In insects a molecule known for its involvement in axon guidance, the homolog of the human down syndrome cell adhesion molecule (DSCAM), has its IgSF domains diversified by alternate splicing and the different forms are used in specific immune recognition of the insect. It is a case where analogous pressures for obtaining single-cell specificity have found analogous but not homologous solutions. The result of these evolutionary processes is a somatically created diversity with a very large potential, vastly exceeding the number of cells able to express all the forms. In immunology, one of the consequences of this enormous potential diversity is that too many receptors per surface unit of membrane would be expressed, with a risk of autoimmunity. These risks, in the adaptive immune system, have been solved in evolutionary terms by clonal selection: one cell expresses only one receptor and the autoreactive cells are deleted in the thymus during the ontogenic development. In the case of DSCAM the great peculiarity is that the very same molecule is used in both systems. However the diversity used by one system is not exactly the same as the diversity of the other – in insects the diversity of DSCAM is mandatory for a good build-up of nervous connections. However, this diversity is different in each insect species studied so far, as if what mattered was only the presence of *a* diversity. Therefore, it is more likely that the specific diversity of each insect species is driven by the environment surrounding the immune system. In support of this statement the human DSCAM, highly homologous of insect DSCAM, is not diversified and represents the default option since it is also found in a primitive flatworm. The adaptation to immunity might be specific to the arthropods [2].

The example of DSCAM was explored in more details because it represents a case of dual selection on the same gene product and illustrates the concept of interwoven evolutionary history between the nervous and immune systems. The way DSCAM is expressed and presented on circulating single cells (hemocytes) might have affected the selection of some recognition options not accessible on cells that are part of a tissue from the nervous system.

Both the immune and the nervous system have chosen their interacting molecules from a rather limited arsenal of cell adhesion molecules. An NK cell encountering a target has a brief adhesion phase followed by a separation phase, not so different from the defasciculation of neurons during branching. The interacting molecules that share the ability to promote or to suppress cell interactions have similar outlook for NK cells (CD96, CD155, CRTAM, Necl2) and for the nervous system (Beat/Fasciclin). All are related members of the IgSF with a very similar architecture (V-C IgSF domains at the distal end of the molecule). IgSF can promote cell interactions or adhesion but it can also prevent such interactions and this they do both in the immune and the nervous systems of many different species.

Which involvement came first then during evolution? The involvement of IgSF in immunity or in the nervous system? How did the distribution of functions take place during evolution?

It is well possible that the IgSF original pool of vertebrates was essentially involved in nervous system and later in the evolution it gave rise to the structures involved in immunity. Immunity could therefore have made use of other non-IgSF molecules of which many examples exist in invertebrate among LRR-containing receptors (LRR – leucine-rich repeat), scavenger receptors, peptidoglycan and beta 1–3 glucan receptors, or complement-like factors. A good example showing how such categories evolve their own adaptive systems is provided by the jawless fish by the LRR variable lymphocyte receptors of the lamprey that generate unicommitted lymphocytes by a mechanism of somatic rearrangement. The contingencies of evolution oriented the gnathostomes vertebrates into another direction making use of IgSF receptors for generating antigen-specific receptors [3].

## 2.4. The role of gene duplications in the evolution of immune systems

The history of the vertebrate immune system is likely to have been influenced by the two rounds of polyploidization that are thought to have occurred about 400–450 million years ago and that are responsible for the observation of the tetrads of paralogous genes [4].

The positioning of all the human homologs of IgSF domains V-C1-like or V-C2 domains discovered in invertebrates on a tetrad of chromosome segments (1q, 3q, 11q, 21q) suggests that a single region containing the homologous ancestors of those genes was indeed present in invertebrate before duplication. A similar finding has been made for the major histocompatibility complex (MHC) region whose paralogs involved in other functions are found on chromosomes 6p, 1q, 9q, 19p. In the case of the 1q, 3q, 11q, 21q tetrad the fate of the duplicate has been manifold. The above tetrad is unique in that most of the IgSF members involved in APC/lymphocyte interaction and in NK/target interaction are clustering in these regions. While some IgSF genes of this region got involved in immunity, other ones were selected by the nervous system and in some instances by both [5]. Homologous receptors can be involved in the immune system in one phylum and in the nervous system in a different phylum.

Polyploidization is a form of duplication that may occur without selective pressures on many of the genes that are concerned. It is known to be the result of meiotic abnormalities in many cases and differs from generation of multigene families by tandem duplication. These better reflect a specific adaptation of one gene family to a need for diversification, like the killer cell immunoglobulin-like receptor (KIR) genes, or the alternate exons of DSCAM, or the different V families of antigen receptors. On a short-term basis polyploidization may perhaps be the equivalent of a rapid tandem duplication creating four copies of a gene for which there might be a selective advantage, but for the adjacent genes it may not be the case. Therefore, following polyploidization the variation on four copies may not imply that the four genes will remain allotted to the exact same function. Therefore, there has been always an evolutionary chance of altering the commitment from nervous system for one paralog into immunity for another paralog (and vice versa). Many of the invertebrate genes of this family have not yet been proven to be involved in immunity and only speculations have been made as to how these IgSF have been recruited in the immune system and what were their relationships to the TCR and BcR. The reason for such recruitment for the immune recognition function could be related to their ability to bind virus, other pathogens, or to modified self. Homologous interactions between some members could have been at the origin of the recruitment of other members into the pairs of lymphocyte-interaction receptors.

One result of duplication is that vertebrate ancestors inherited a larger number of genes than deuterostome or protostome invertebrates. This might have been evolutionarily very useful, since dual recognition properties of a single molecule or receptor are exploited in invertebrates that have only one copy of the gene, whereas in vertebrates division of labor has been accomplished by attributing one function to one gene and another one to the duplicated member.

In summary whatever the causes, to the favor of duplication during the history of the vertebrate lineage, genes of the very same family have different utilizations in very different contexts but under analogous selection pressures, which resulted in analogies between nervous and immune systems.

## 2.5. Evolution of the gnathostome vertebrate immune system

The immune system of vertebrates, once equipped with TcR, BcR, APC, and polymorphic MHC class I and II system, did not evolve much further. Instead, only fine-tuning is seen comparing one class to the other. However, the variations seen are very informative with respect to what must have been the selection pressures affecting the evolution of the immune system. Diversity of repertoires seems to be of paramount importance because it is achieved in every class but with slightly different mechanisms: somatic RAG-mediated recombination involving combinatorial usage of gene segments, gene conversion, somatic mutation, usage of a large number of germ line–joined V (D) J segments. All these mechanisms concur to establish a potential repertoire, usually much larger than the number of cells that can express it at a given moment in one individual. Similarly, in invertebrates DSCAM repertoire is much larger than the number of hemocytes present in a single *Drosophila*. Somatic mutations detected in IgSF of mollusk immune effector molecules (FREP) occur in the absence of rearrangement. They could perhaps play a role in generating a vast repertoire of recognition structures in this invertebrate as they do in mammals. The usage of these repertoires will depend largely on the way the mutants or the variants are selected. With good capacities of selection a good affinity maturation of the antibody response follows as in warm-blooded vertebrates. In the absence of good selection the increase in affinity is poor. The architecture of the lymphoid organs seems to be responsible for this state and the presence or the absence of the germinal centers seems to be an important factor [6].

This chapter focuses on immune receptors *sensu stricto* and the evolution of MHC is not considered in detail. The reader is referred to special reviews on this subject [7]. For the MHC the number of loci of classical class I can vary from "a single one" in frogs and chicken to six in mammals like human. In a species like the chicken, limited in its diversity of presentation by the existence of a single locus, the diversity in peptide presentation is more under the control of the transport associated proteins (TAP).

The relationship of innate and adaptive immunity in vertebrates may depend upon the level of pressure to exploit diversity. In those species with limited maturation of their immune response, the adaptive response is under lesser selective pressure during the response, and the innate system is functioning as a prominent defense tool rather than a primer of the adaptive system. To be convinced of the efficiency of innate immunity in cold-blooded vertebrates suffices to observe the fate of a wound in a frog and in human. It is rare that a frog dies from infection following an amputation or a major skin injury. So the balances of innate and adaptive immunities are not the same across classes, although when looking at the conserved elements of the immune system, one could have anticipated a much higher level of similarity between sharks, frogs, and human immune responses. Some affinity maturation linked to somatic mutation has been found to occur in all vertebrate species including sharks and teleost fish and frogs. Still, shark maturation differs by orders of magnitude from that seen in mammals. In fact, the lymphoid organs of those species with low affinity maturation are simpler and lack germinal center traditionally considered as good breeders and selectors of mutants.

Another aspect of the evolution of the vertebrate immune system is a quantitative one: complement components are more numerous in fish than in mammals, some families of receptors disappear from one order of mammals to the other, the number of Ig subclasses can vary for heavy or light chain from species to species. Independent adaptations, evolutionary time of observation, and random chance, all contribute to the observed level of diversity and multiplicity, the selecting pressure of the environment being the driving force for achieving an efficient immune system.

All in all, the complexity of the vertebrate immune system is larger than that of invertebrates. This is not only due to the introduction of the RAG-mediated generation of TcR and BcR, but due to the simple increase in numbers of paralogs following duplication. In terms of categories of receptors the *Drosophila* or *Ciona* genetic regions contribute fewer genes than the four paralogous regions identified in human that in sum encode dozens of receptors involved in the immune system or in the nervous system. Many duplicated members have been conserved on the paralogous chromosome segments and even some have undergone *cis* duplication. This diversity of lymphomyeloid cell receptors in vertebrates implies the genesis of numerous networks of cell to cell interactions during evolution.

The large number of genes involved from the onset of vertebrate history due to massive duplication events has not only favored specialization within the immune system but also has offered more accurate specialization for the members to be involved in the nervous system.

In the same way the pressure to obtain diversity of immune recognition has been satisfied in many phyla with very different mechanisms: by alternate splicing of scavenger receptors or DSCAM, by somatic mutation in IgSF, as well as by rearrangement, gene conversion, and somatic mutation. Through this pressure a hierarchy of adaptiveness was achieved independently at many different levels of the animal kingdom, which in turn reflects the beneficial trend towards individualization of the immune responses.

## 3. BASIC CONSIDERATIONS ABOUT THE MILIEU OF THE IMMUNE RESPONSE AND THE NATURE OF THE IMMUNOGENS

### 3.1. The milieu in which the specific immune response takes place

Specific immune response can manifest itself everywhere in the organism where lymphocytes are localized or where they circulate, and where body fluids have access. The surveillance is accomplished through the interdigitated circulatory system of blood and lymphatic vessels, which connect the solid organs of the peripheral immune system.

There are several anatomical locations at which the specific cells of the immune system exert their surveillance activity. Especially relevant are blood, spleen, lymph nodes, bone marrow, thymus, tonsils, skin, respiratory tract, gastrointestinal, and genital tract. It is noteworthy that besides the 6 L of blood and several thousands of lymph nodes, in an adult human there is a mucosal surface of an area of about 400 $m^2$ (about a tennis court surface). Taken together, this is just the right environment for the strategic defense – both the specific and non-specific one.

### 3.2. The universe of immunogens and the spectrum of epitopes

Antigens, molecular structures to which specific immune responses can be elicited, might be composed of several antigenic determinants, epitopes. The universe of antigenic shapes to which the organism is – or might be – exposed is called the "antigenic spectrum." The ability of the organism to produce antibody to a certain antigen is giving it the property of being immunogenic, and indeed the same antigen that is capable of eliciting a response in one organism might be nonimmunogenic in another organism. Besides that, an immunogen in one species might be a non-immunogen (thus an inert structure) in another one.

Since we know that the immune system has – in the evolutionary survival game – developed into a defense system against invading pathogens, like viruses, bacteria, fungi, and parasites, we often drop the expression "immunogen" for a clinically more apt "pathogen." In these terms pathogen is a microorganism which is capable of proliferating or being in some other way harmful to the invaded organism.

Whether we refer to antigens, immunogens, or pathogens, and whether we have in mind viruses, bacteria, fungi, or parasites, the actual immune response develops against molecular structures of proteins, lipids, sugars, nucleic acid components, and so on.

### 3.3. Site of entry of the immunogens into the organism

The initial stimulus toward eliciting a specific response depends on the entry port of the immunogen. An immunogen enters the organism most commonly via skin, respiratory tract, or blood stream. Depending on the antigen, it penetrates the skin, might be trapped in the respiratory tract, bind to epithelial surfaces, or form antigen–antibody complexes in the blood.

Some microbial pathogens have exploited a niche that allows them to survive in face of an active immune response. If the invading organism is an obligate intracellular parasite (all viruses and various bacteria like *Rickettsia prowazekii*, *Chlamydia trachomatis*, *Chlamydia pneumoniae* are obligate intracellular parasites), it will attach to (and taken up by) the prospective host cell and find there the niche to hide.

*Mycobacterium tuberculosis*, *Mycobacterium leprae*, *Salmonella paratyphi*, *Listeria monocytogenes*, and *Legionella pneumophila* will find their habitat inside of mononuclear phagocytes and macrophages, while Shigellae and Yersiniae will transiently invade some cells and at some later stages continue to live in extracellular spaces or in the blood stream. All intracellular bacteria – whether facultative or obligate – abuse the inside of the host cells as a highly protective recess.

## 4. BASIC TENETS OF THE SPECIFIC IMMUNE RESPONSE

When Jerne postulated the natural selection theory of the antibody response [8] and Burnet adapted it to the – now generally accepted – clonal selection theory [9], the T-cell involvement in the immune response was not yet considered; thus the theory did concern only the antibody response. Today we know that the theory is valid both for the B- and T-cell responses.

Throughout this chapter we consider the following basic properties of the immune system:

- The organism as a whole possesses a large repertoire of precommitted precursor cells, and for a particular antigenic stimulus only a distinct subset of cells is capable of responding.
- The responding subset is a small fraction of the whole population of precursor cells, and upon antigenic stimulation it is amplified by a sequence of rapid mitotic divisions.
- Each cell of the responding subset expresses on its surface an antigen-specific receptor of one molecular species only (V region), and the progeny of each cell is expressing the same specificity as the ancestor cell (except rare mutations).
- The specific cell is either (a) a B cell, effector function of which is exerted by clonal progeny producing specific antibody molecules or (b) a T cell, effector function of which is exerted by clonal progeny through direct cell contact action.
- The conversion of a cell from uncommitted to the committed state is an antigen-independent process; uncommitted cells cannot be triggered by antigen.
- Information on the primary structure of the antigenic determinant (epitope) is not required at any stage of antibody synthesis, folding or secretion. Information on the primary structure of peptide fragments is utilized in the T-cell recognition process.

## 5. ANTIBODY AND MEMBRANE-BOUND RECEPTORS

### 5.1. The immunoglobulin domains of lymphocyte receptors

The immunoglobulin domain is central to many receptors involved in the immune system. A very large number of receptors important for the interaction of lymphocytes with other cells or tissues are made of these domains. At least 76 IgSF receptor families can be expressed on lymphomyeloid cells, besides the two families of antigen-specific receptor, the BcR and the TcR. These domains represent an evolutionary success across the animal kingdom, since they are used widely not only for immunity but also in cell in general and in differentiation of the nervous system in particular.

Individually, these domains adopt a shape that many other surface protein domain (calycins, fibronectin 3, cadherins, cytokine R domains) adopt for intrinsic structural constraint reasons, a form of a beta barrel. These domains are assembled in many different ways – they can be concatenated or assembled in multimers, and several regions of their external surface can be used for binding. Ig domain participates not only in the edification of antigen receptors but also in many co-receptors that are important for a harmonious functioning of the cells, migration, homing, etc.

The domains consist of a sandwich of two beta sheets, each one made up of three strands (ABE) on one side and three strands (CFG) on the other side, attached through a disulfide bridge between strands B and F. There can be variation in the distribution and in the nature of those strands according to the types of domains. The V (variable) domain is the most complex one and is made of ABCC′C″DEFG strands. The constant C2 domain is the simplest one with ABCEFG strands. The constant C1 type (ABCDEFG) is found only in gnathostome vertebrates with an adaptive immune system. The intermediary set I (with AA′BCDEFG strands) is an intermediary form between C and V domains. In the antigen receptor molecules, the loops between the strands B–C, C′–C″, and F–G contribute to the combining site of the molecule. The heavy (H) and light (L) chains of antibodies as well as the α, β, γ and δ chains of the TcR are made up of various numbers of Ig C domains, and the V domains are at the distal end of the polypeptide [10].

V domains that do not rearrange form a category of IgSF domains that can be found in many other receptors and they could represent relatives of the molecules, the genes of which gave rise to the BcR and TcR during the evolution toward the vertebrate adaptive immune system. A large number of IgSF cell surface markers using these non-rearranging V domains are involved in cell interactions between lymphocytes and APCs or between NK cells and their targets. They often form pairs of ligands and receptors (e.g., CD28 or CTLA4 with CD80 or CD86).

The major evolutionary innovation in the adaptive immune system is that, in the germ line configuration, the V domain gene is split at the junction between the F and G strands, and somatic rearrangement is necessary during lymphopoiesis to provide a complete functional variable receptor gene.

The molecular structure produced and secreted by B cells with the property of binding to antigen is called "antibody." On the cell surface of B and T lymphocytes there are displayed B-cell receptors (BcR) and T-cell receptors (TcR). BcR is of the same molecular quality as the secreted antibody, while the TcR has several molecular features different from antibody. As it has been defined above, one cell (lymphocyte) produces a single molecular species of antibody. The entire population of all antibody molecules is referred to as "immunoglobulins." Immunoglobulin molecules consist of several molecular classes (isotypes) and in some instances the function (not the specificity) is related to the given isotype of immunoglobulins; thus – as will be detailed later – for example, IgA is related to mucosal immunity, while IgE plays a role in allergy.

Antibody can be raised not only against antigens but also against other antibody molecules. Epitopes on the antibody molecules are called "allotopes" and "idiotopes." Allotope is an epitope on the constant portion of the antibody molecule or on that stretch of the variable region that remains invariant within the given species (or strain). Idiotope is an epitope on the variable portion of the antibody molecule, usually at (or near) the combining site. The set of all allotopes and the set of all idiotopes characteristic for the given molecular type is referred to as "allotypes" and "idiotypes." Modern term for the antigen-binding site of antibody is "paratope" while the structure to which the paratopes bind are called "epitopes." The terms "idotype," "epitope," and "paratope" were introduced by Jerne [11] in 1960.

## 5.2. Association of heavy and light chains

The antibody molecule is a "Y"-shaped structure composed of the unique combination of two identical heavy (H) and two identical light (L) polypeptide chains. H chains are polypeptide molecules of about 440 amino acid residues length, while L chains are made of about 220 amino acid residues. Each L chain is aligned with an H chain along the amino terminal end of the molecule. The two H chains are aligned along the carboxy terminal regions of the molecule. The L–H associations form the two arms of the "letter Y," while the H–H association defines the stem of the "letter Y." The structure of the molecule is held together by covalent bonds (S–S bridges) and non-covalent interactions.

There are five different types of heavy chains: $\mu$, $\delta$, $\gamma$, $\alpha$, $\varepsilon$ and the $\gamma$ isotype falls into four subtypes designated $\gamma1$, $\gamma2$, $\gamma3$, $\gamma4$ in humans and $\gamma1$, $\gamma2a$, $\gamma2b$, $\gamma3$ in mice. There are two types of light chains: $\kappa$ and $\lambda$. The Greek letter designation gives the name to the isotype of the immunoglobulin, thus $\mu$ heavy chain yields IgM; $\delta$ gives IgD; $\gamma$ gives IgG, and so on. Each heavy chain type forms antibody molecules either with $\kappa$ or $\lambda$ light chains. Both in mice and humans, the usage of light chains is skewed, thus in humans the ratio of $\kappa$ to $\lambda$ is 2:1, while in mice it is 20:1.

## 5.3. Variable and constant regions of H and L chains

The variable (V) regions of the L and H chains are confined to about 110 amino acid residues at the N terminal end of each of the polypeptide chains. The remaining amino acid residues (about 110 in L chains and 330 in H chains) are making up the constant (C) portion of the molecule. The amino acid variability of V regions is not distributed evenly throughout the length of these regions. There are three hypervariable regions corresponding to the interstrand loops of B–C, C–D and F–G of Ig domain, also referred to as "complementarity determining regions" (CDR) that are separated by intervening framework regions (FR). Each of the hypervariable regions is about 10 amino acid residue long (starting from residue 28, 49 and 92). The amino acid sequence of the V region of heavy chains differs from the sequence of the V region of the light chains. Antigen binding occurs through the "binding site" that is localized at the combined tertiary folding structure of the tip of the V region. The binding site of the antibody is referred to as "paratope" [11].

## 5.4. Isotypes of immunoglobulin

In the mouse and human species there are five isotypes of antibodies IgM, IgD, IgG, IgA, and IgE, each with a specialized function. The designation of the isotypes follows the heavy chain nomenclature ($\mu$, $\delta$, $\gamma$, $\alpha$, $\varepsilon$), since it is the genetic organization of the heavy chain gene segments that decides on the isotype form. (The IgG type is further subdivided into IgG1, IgG2, IgG3, IgG4 in human and IgG1, IgG2a, IgG2b, IgG3 in mice.) The change from the expression of one isotype of immunoglobulin to another one occurs in a process termed "immunoglobulin switch" (see Section 9.1.). Isotypes of IgM, IgD, IgG1, IgG2, IgG3, IgG4, IgA, IgE appear during a response in this order, and the switch events are irreversible.

IgM is the most prominent antibody of the primary immune response. In circulation it is present as a pentamer, thus possessing 10 epitope-binding sites. The five monomeric components are held together by a J chain. Occasionally, IgM can be found in serum as a hexamer. This conformation occurs without involvement of the J chain. IgM efficiently activates complement. IgD is a minor component and it probably plays a role in induction of memory. Its further biological role is unknown. IgG is the most abundant immunoglobulin in serum and the principal antibody class of a secondary immune response. IgG is the only isotype that is actively transported across the placenta. It is worthwhile to note that newborns possess a full repertoire of maternal IgG. IgA is present in the blood either as a monomer or alternatively as a dimer joined by J chain. In mucous milieu it is present always as a dimer. Due to the J chain, the molecule can be actively secreted across mucous membranes. This is further facilitated by attachment of a specialized secretory component (SC). IgA is present in tears, saliva, respiratory, gastrointestinal, and genitourinary systems (it is relatively resistant to enzymatic digestion). In colostrum its concentration is 50 times higher than in serum (providing passive immunity to a nursing neonate).

IgE, the least abundant immunoglobulin isotype in the serum, is very short-lived with a half-life of 2–3 days in humans. It exists as a monomer, and it does not bind the J chain. IgE is responsible for immediate-type hypersensitivity reaction including systemic anaphylaxis, and for defense against parasitic infestation. Considerably higher concentrations of IgE than in serum can be found in respiratory nasal secretions. IgE synthesis is of protective nature toward helminthic infections and is of unfavorable nature in inducing immediate hypersensitivity responses.

## 5.5. Specific membrane receptors

All antibody isotypes that are produced for secretion are also formed in a slightly altered form (containing an additional transmembrane domain, 25 amino acid residues long) as immunoglobulin receptors that are expressed on the membrane. Immunoglobulin receptor on the membrane of B cell is referred to as "B-cell receptor" (BcR). The molecule that is present on T cells and has the specific property of recognizing peptides in conjunction of MHC molecules is called "T-cell receptor" (TcR). Since receptors on B cells were identified decades earlier than the receptors on the T cells, the use of BcR starts to be established only now, and this is done in order to point out the symmetry to TcR.

B-cell receptors are formed of four chains, while the T-cell receptors consist of two chains. In terms of the binding sites, the B-cell receptors are bivalent, while the T-cell receptors are monovalent. All BcR molecules on a B cell and all TcR molecules on a T cell are of a single molecular species (clonal property). There are some hundred thousand BcRs on a B cell, while there are some ten thousand TcRs on a T cell.

### *5.5.1. BcR and its anchoring domain*

The paratope of the antibody is identical both on the molecule displayed on the membrane (BcR) and on the secreted molecule (antibody). The constant portion of the BcR molecule has an added transmembrane anchoring domain consisting of a 25 amino acid hydrophobic stretch and a short cytoplasmic tail (3 amino acid long for IgM and IgD, and somewhat longer for other isotypes). The cytoplasmic stretch is too short for being of use for any kind of intracellular signaling function.

The BcR molecule is associated with a transmembrane heterodimer designated Igα/Igβ with a cytoplasmic tail of about 50 amino acid residues, which function as signal transmitting elements.

### 5.5.2. *TcR and the signaling complex*

T-cell receptor consists of two polypeptide chains bound to one another by disulfide bridge. Most of the T cells express the TcRα,β chain combination while there is a smaller portion of cells that express the TcRγ,δ chain combination. The T-cell machinery for epitope recognition and for the transfer of signals to cell interior is of a much more complex nature than that of the B-cell system. T-cell receptor is expressed on the membrane in association of a molecular complex that can be defined as an assembly of six polypeptide chains on the outer surface of the membrane (TcRα, TcRβ, CD3γ, CD3δ, 2 × CD3ε) and two signaling molecules (ζ) directed toward the cytoplasm. Once the antigen binding occurs, it is the function of the TcR complex to transmit the proper signal to the interior. It was mentioned above that there is a small portion of TcRs that is made up of γ and δ chains. These cells are of distinct origin; in some species, for example in sheep, they do appear in higher proportions.

## 5.6. Repertoire of antibody, repertoire of receptors

A healthy individual is capable of synthesizing such a large number of different molecular species of antibody that any non-self epitope can bind to some paratopes. It is assumed that about one million different antibody molecules provide a "complete" coverage, that is, all specificities required for recognizing any three-dimensional form of antigenic determinants are available in the immune system of the individual. The ensemble of all specificities – the "antibody repertoire" – is in qualitative terms the same for secreted antibody and for the BcRs. Although – as mentioned above – one million specificities are satisfactory, it is the case that an individual is capable of producing millions of distinctly different antibody molecules generated somatically by RAG-mediated rearrangement of variable gene segments. During the life span of the individual the repertoire might change – some specificities are newly "added" to the system and some disappear. Furthermore, the molecular species produced in one individual differ from those produced in another individual. Thus it is useful to distinguish between the repertoire of the individual, repertoire of the species, repertoire of the entire life span. The estimates vary from $10^7$ to $10^{11}$ antibody specificities [12]. This is a small fraction of the possible combinations, which can arise from the immunoglobulin gene rearrangements – from which a maximal diversity of $10^{14}$ different antibody molecules is deduced.

The molecular machinery for the generation of TcR diversity might yield up to $10^{18}$ different TcRs. Upon removal – by negative selection – of specificities that would lead to self-reactivity, the theoretical repertoire could be still of the order of $10^{15}$ different TcRs [13]. In the human species with about $10^{11}$ mature T cells the upper limit of the repertoire is given by this cell number.

The repertoire of a healthy individual is considered to be complete, which means that it can recognize any conformational epitope via antibody or BcR, or any peptide epitope in conjunction with MHC via TcR. The repertoire of B cells might include at least a portion of B cells with antiself specificities (unresponsiveness enforced by anergy), while from the T-cell repertoire self-specific T cells are excluded [14].

## 5.7. Affinity of antigen–antibody binding

The strength of the total non-covalent interactions between a single antigen-binding site of an antibody (paratope) and a single antigenic determinant (epitope) is termed "affinity."

In another words, affinity describes how strong a paratope and an epitope bind to each other. The binding capacity of a paratope to an epitope can be precisely measured through establishing the equilibrium association constant for the paratope–epitope pair. In formal sense it is the reciprocal of the concentration of epitopes at which half of the paratopes will be occupied, and in simpler terms it measures how likely it is that epitope and paratope will be found in a complex. Low-affinity epitope–paratope complexes have $K$ values between $10^4$ and $10^5$ L/mol while high-affinity complexes have $K$ values as high as $10^{11}$ L/mol.

When complex antigens (repeating epitopes) are mixed with antibodies containing multiple paratopes, the interaction is much stronger. The strength of these interactions is called "avidity." High avidity interactions can lead to $K$ values of up to $10^{14}$ L/mol. We refer to avidity, for example, when polysaccharide molecules consisting of several epitopes bind to (up to 10) combining sites in the antibody molecule. In instances when interactions occur on a cell surface we consider avidity rather than affinity.

## 6. MHC AND THE CELLS EXPRESSING CLASS I AND CLASS II MOLECULES

Molecules of the MHC are glycoproteins with a specially adapted three-dimensional structure for binding short stretches of peptide epitopes. They are structures that play a crucial role in inter-cellular recognition and in discrimination between self and non-self. They are indispensable in the development of both humoral and cellular immune responses.

MHC molecules are falling into two classes: I and II. MHC class I molecules are present virtually on all nucleated cells of the organism, while MHC class II molecules are present only on APCs. As a rule (with some exceptions), peptides which bind to MHC class I molecules are those which are derived from polypeptides synthesized by the cell, while MHC class II molecules bind those peptides which have been internalized from the extracellular milieu.

### 6.1. Structure of the class I and class II molecules

MHC class I molecules are formed by a single heavy chain (α chain) that is non-covalently associated with β2-microglobulin molecule. The α chain is an integral membrane protein, while the β2 chain is a non-membrane bound molecule. MHC class II molecule consists of two polypeptide chains α and β, of approximately equal size, each having a transmembrane region. The transmembrane chains of both classes have a cytoplasmic tail. A common feature of both classes is the quaternary structure with a cleft, function of which is to bind peptidic structures.

The cleft of the class I molecules has a closed shape such that into the groove, only peptides not longer than 9 amino acid residues fit. The cleft of the class II molecules is open at both ends. Peptides longer than the actual cleft (13–18 amino acid residues) are readily bound to these molecules. Each MHC molecule can bind many different peptides. The MHC molecule is not peptide-specific.

### 6.2. Cells expressing class I and class II molecules

MHC class I molecules are expressed on almost all somatic cells. The density of MHC class I molecules varies between 10,000 and 500,000 molecules per cell. Fibroblasts, muscle cells, hepatocytes, and brain cells express very low levels of MHC class I molecules. A healthy cell of any kind will express self-peptides resulting from normal turnover of self-proteins.

MHC class II molecules are expressed only on cells specialized for antigen-presenting function. The primary APCs are macrophages, dendritic cells, and B cells. The level of MHC class II molecule expression depends on the cell's differentiation stage. At certain stages of differentiation less than 100,000 MHC class II molecules might be expressed on the cell, while upon antigenic stimulation the level might increase to 500,000 or even more.

## 7. CELLS OF THE IMMUNE SYSTEM

An adult man possesses some $10^{12}$ lymphocytes present in the blood, lymph nodes, spleen, skin, mucosal tissue, bone marrow, thymus, and other organs. The overall mass of all lymphocytes in the body is about 1 kg. The largest compartment of the lymphocytic population in terms of volume is the blood. An average adult has about 6 L of blood within his/her vasculature, 1 mL of blood containing some $10^{16}$ antibody molecules.

### 7.1. B cells

Precursors of antibody-forming cells are small non-dividing B lymphocytes resting in the $G_o$ phase of the mitotic cycle. They synthesize very little, if any, DNA, but they do synthesize immunoglobulin molecules that become located on the cell surface where they serve as receptor molecules for antigen recognition. Some 100,000 antibody molecules – B-cell receptors (BcR) – are distributed on the cell membrane of B lymphocytes. The lymphocyte membrane behaves as a two-dimensional fluid and the receptors and other surface molecules are free to move about in the plane of the membrane.

The BcRs possess the same molecular specificity as the antibody ultimately produced by the cell. The cell must not have "wrong" receptors, to prevent unwanted antibody production. The corollary to the above is that the cell not only possesses correct BcRs, but no other immunoglobulin is expressed on the cell surface. When a precursor of an antibody-forming cell meets an antigen for the first time, the cell is already committed to formation of antibody of a single specificity.

### 7.2. T cells

T cells are precommitted to a single specificity in the same manner as it has been described above for the B cells. There are, however, two major differences. *First*, the T cell expresses the antigen-specific molecule on the membrane of the T cells, but does not produce this molecule for the purpose of secretion. T cells do not secrete any antigen-specific products, while they do secret certain lymphokines which facilitate various short-range cellular interactions. *Second*, the T-cell receptor does not recognize the *shape* of the epitope, but rather the *primary structure* of a portion of the epitope in context of a cell surface molecule of the MHC.

T cells are considered to form two distinct sets, differing both in their functions and in their molecular markers. The T helper (Th) cells and T cytotoxic (Tc) cells express characteristic receptors of CD types. These two major types are defined by molecules known as CD4 and CD8, present on their membranes. CD4 and CD8 molecules play a decisive role in the differential recognition of MHC class I and class II molecules. Both CD4 and CD8 molecules upon interaction with the MHC molecule stabilize the interaction between TcR and antigen, thus increasing the effectiveness of the stimulus.

It is now an accepted view that both Th and Tc cells form two distinct types of cell populations:

| | | |
|---|---|---|
| Th | cell type 1 (Th1) | $CD4^+$ expresses IFN-γ (and others) |
| Th | cell type 2 (Th2) | $CD4^+$ secrets IL-4 and IL-5 (and others) |
| Tc | cell type 1 (Tc1) | $CD8^+$ expresses IFN-γ (and others) |
| Tc | cell type 2 (Tc2) | $CD8^+$ secrets IL-4 and IL-5 (and others) |

Besides these sets, in the Th cell types there is a further set, dubbed Th3, which is involved in mucosal immunity. In addition there are claims of further "stratification" of Th and Tc cells, with a likelihood that various phenotypic combinations are permissible, depending on the type of the immune response. Not surprisingly, Th0 and Tc0 types as silent precursors to the particular effector cell types are also described.

### *7.2.1. T cell subset expressing CD4 molecules*

CD4 is a single-chain polypeptide molecule expressed about in 20,000 copies on the membrane of the cell capable of recognizing MHC class II molecules on APCs. The co-aggregation of the TcR molecules and the CD4 molecules is a prerequisite for successful signal transfer. The actual binding of CD4 molecules to the MHC class II molecules initiates the stimulation events. $CD4^+$ T cells with their TcRs engaged in the recognition of polypeptidic epitopes bound to MHC class II molecules on the APC develop in most instances into Th cells.

Depending on the circumstances of stimulation $CD4^+$ T cells develop either into Th1 or Th2 cells. The cytokine profiles of Th1 and Th2 are mutually inhibitory; thus the milieu of Th1 response is not favorable to promote Th2 response on the same site – and vice versa. The mutual antagonism of the two processes is referred to as "polarized T-cell responses."

### *7.2.2. Characteristics of Th1 cells*

- Conduct cell-mediated immunity by activating macrophages or CD8 cytotoxic T cells
- Th1 response eradicates intracellular pathogens, including bacteria, protozoa, and fungi (e.g., microorganisms such as *Leishmania* and *Brucella*)
- Characterized by production of IL-2, IFN-γ, TNF-α.

### *7.2.3. Characteristics of Th2 cells*

- Provide help to B cells in antibody production
- Th2 response is in charge of controlling extracellular pathogens
- Characterized by production of IL-4, 5, 6, 10, and 13. Especially important is the pattern of IL-4, IL-5, and IL-10, which plays a major role in humoral immunity against extracellular and toxin-producing pathogens.

### *7.2.4. T-cell subset expressing CD8 molecules*

CD8 molecule is a heterodimer of an α and β chain associated by a disulfide bridge. The molecule is expressed in about 10,000 copies on the membrane of the cell made to recognize MHC class I molecules on a large variety of cells. The co-aggregation of the TcR molecules with the CD8 molecules is required for successful signal transfer from the peptide-loaded MHC class I expressing cell. $CD8^+$ T cells with their TcRs engaged in the recognition of polypeptide epitopes bound to MHC class I molecules on various cell types develop in most instances into Tc cells.

### 7.3. Antigen-presenting cells

Cells that express MHC class II molecules are predominantly B cells, macrophages, and dendritic cells, and they are termed APCs. All cells expressing MHC class II molecules are capable of presenting peptide antigens. Some cells are especially well suited for this purpose, and those are referred to as professional presenting cells. One such cell type is dendritic cells, and the special feature of high level of MHC class II molecules on their surface is predestining them for this function.

APCs take up protein antigens and partially digest them. Peptides are then bound to MHC class II molecules and transported to the surface of the APC, where they are presented to the T-cell receptors of the $CD4^+$ T cells. The presentation does not discriminate between self and non-self peptides. All those exogenous peptides, which occur in the acidic endosomal vacuoles, are loaded into the groove of the MHC class II molecules. Langerhans cells are large dendritic cells present in skin; they participate in cutaneous immune/allergic reactions. Langerhans cells may comprise the first antigen-processing cells that antigen encounters. In B lymphocytes endocytosed antigen–BcR complexes are delivered to the intracellular processing in a more efficient manner than antigens internalized by pinocytosis.

## 8. PRIMARY ANTIBODY RESPONSE

Specific antibody response is elicited upon cooperation of two types of cells – B cells and Th cells. The function of B cells is to produce antibody, while the function of Th cells is to ensure that the required controlling mechanisms are in operation.

For both types of cells involved in the response, that is, for B cells and for T cells the initial activation is a two-signal event. Signal 1 results from the binding of the epitope to the BcR. Signal 2 involves acceptance of cytokine signals from Th cells, or the signal induced by interaction of CD40 on B cells with CD40L on T cells.

Binding of divalent antibodies (directed against determinants of the surface immunoglobulin as well as binding of multivalent antigens) induces a dramatic redistribution of the membrane immunoglobulin molecules such that these first aggregate into patches and finally form a "cap" over one pole of the cell. Such events are a crucial step in triggering the lymphocyte to transform into blast cells, to divide, and to initiate the antibody response.

Three sets of molecules are responsible for specificity of immune responses by virtue of their capacity to bind foreign antigen: Ig $\leftrightarrow$ TcR $\leftrightarrow$ MHC. All are members of the immunoglobulin superfamily (IgSF) of adhesion molecules.

### 8.1. Antigen entry into the organism

The tenets of the clonal selection theory assert that antigen invading the organism encounters specific B cells and these are then selected to respond. Such assumptions are certainly correct, though the events leading to a specific response might be of a somewhat more complex nature depending on the port of antigen entry.

If antigen enters the organism through the blood stream, it will encounter specific antibody rather than specific B cells. In a milliliter of blood there are about $10^{16}$ immunoglobulin molecules and about $10^6$ B cells. In the mentioned volume of blood there maybe $10^{11}$ specific antibody molecules and say 3 to 5 specific B cells. Clearly, the initial event is that antigen forms antigen–antibody complexes. If the antigenic load is limited, there will be very little free unbound (residual) antigen in the blood.

The epitope–paratope complexes will further aggregate and the Fc portions will be recognized by cells displaying Fc receptors with a consequence of an efficient phagocytosis. High antigen load implies that besides the epitope–paratope complexes the superfluous free antigen will bind to the circulating specific cells. Due to high surface density of BcR molecules the binding of antigen molecules will result in a high avidity interaction. All the involved components – whether of cellular or molecular nature – will find an optimal playground in the spleen and in the mucosal tissues.

Pathogens that penetrate the tissues will end up in the draining lymph nodes. Soon after antigen is present in the lymph nodes, cytokine signals will alter the pattern of cellular trafficking through the lymph nodes. The cells will be allowed to enter the nodes, while the departure of cells from the efferent vessels of the lymph nodes will be "closed down" with a consequence of a massive recruitment of B and T cells from the blood. Thus, swelling of lymph nodes at the onset of the infection is not due to clonal amplification of lymphocytes but rather due to altered cell traffic yielding the mentioned accumulation of cells. The follow-up mitogenic and antigenic stimuli lead to clonal expansion of the relevant cell populations. Lymph nodes enlarge enormously (lymphoadenopathy) because of a many fold increase in the cellularity.

The probable reason why the primary immune response is in most instances channeled to lymph nodes – thousands of them in each individual – is that they have a special standing as quiet nurseries where the antigen-specific immune responses can develop in a most efficient manner. A primary immune response is unlikely to develop at sites such as infected respiratory surface, where mucous secretion, lack of proper anatomic niche contribute to the diffusion of helper cytokines and disruption of cell–cell contact. Antigens which enter the organism through upper respiratory tract or intestine will be trapped by mucous lymphoid tissue.

It should be noted that at various stages of the immune response different short-range signals modulate the involvement of cells combating the infection. At early stage, IL-12 and IL-18 are produced rapidly, and these molecules serve as the central signal for the Th1 development. Thereafter, cognate receptor interactions (e.g., CD40-CD154) provide means for balanced amplification and tight control of the response.

## 8.2. Antigen processing

Antigen processing involves two steps: (a) internalization of the antigen and (b) degradation of the macromolecules into peptide components. APCs are able to internalize antigens either by phagocytosis or endocytosis. All three types of APCs, that is macrophages, dendritic cells, and B cells, are capable of internalizing extracellular structures. Antigen–antibody complexes (and also some aggregated forms of protein molecules) can be phagocytosed by macrophages and dendritic cells, while antigen-specific B cells are readily available to bind antigen and to initiate a receptor-mediated endocytosis. B cells with irrelevant specificity will not participate in the process, since no antigen is bound to their receptors.

Inside the cell, antigens are rapidly hydrolyzed in both neutral and acidic endosomal vesicles. Processing involves endosomal and lysosomal proteases; often, fusion of endosomes with lysosomes is required. Antigen processing is a highly efficient process; within 30–60 min after the contact with antigen, the peptides are ready for presentation.

## 8.3. Antigen presentation

Newly synthesized MHC class II molecules associate with so-called invariant chains and then enter the endocytic compartment of the APC. The invariant chains are digested, leaving behind small peptidic fragments in the cleft of the MHC molecules. These fragments are then replaced by the peptides derived from the processed antigen. The resulting peptide–MHC molecular complex anchored in the vesicle membrane becomes finally displayed on the outer surface of the presenting cell. The rule is that the peptide–MHC class II complex is recognized only by $CD4^+$ T cells, while $CD8^+$ T cells remain unchallenged. It should be noted that "empty" MHC class II molecules are unstable and become degraded, and also free "unbound" peptides are easily degraded.

All three principal kinds of APCs (dendritic cells, macrophages, and B cells) fulfill the task of presenting peptides to $CD4^+$ T cells. B cells have an additional function: their peptide–MHC molecules are supposed to attract activated T cells that are expected to supply an activation signal which drives the B cells toward clonal responses. The presentation of antigen occurs in the T cell–rich zones of the spleen and lymph nodes.

## 8.4. Activation of TH cells

The epitope-specific $CD4^+$ T cells upon receiving signal from the APC will become activated and start to proliferate. This proliferation phase occurs in the T cell–rich zones of spleen and lymph nodes. According to the classification given above (Section 7.2.1) these helper cells are referred to as Th2 cells.

In spite of the high efficiency of B cells in their function of presenting cells, it is probable that for the first stimulus, macrophages and dendritic cells are required as APCs. About 100 specific peptide–MHC complexes are needed to trigger a T cell expressing CD4 molecules. (Incidentally, stimulation can occur also in T cells that do not express CD4 molecules, though in such an instance about 10,000 peptide–MHC complexes are necessary to perform the stimulation.) Activation of T cells starts within a few hours after the TcR complex transmits the signal to the cell interior, and proliferation of T cells yields in about 4 days clonally amplified "armed" Th cells.

## 8.5. B cell–T cell conjugates

The specific activation of B cells by "activated" or "armed" Th cells is initiated by antigen binding to the BcR molecules. Antigen cross-links the BcR molecules causing clustering of the signaling complexes. Cross-linked BcR molecules (facilitated by two additional molecules, Igα and Igβ that help to overcome the inadequacy of the BcR's cytoplasmatic tail for signaling) transmit the signal to cell interior. Once a Th cell recognizes the peptide-bearing MHC molecules on the B cell, the two cells interact to form a T–B conjugate. The helper signal generated by the physical contact between activated Th cells and B cells leads not only to directional (T–B) release of cytokine signals produced by Th cells but also to up-regulation of various important receptors and ligands on both cells. Armed Th cells will help only those B cells whose receptors bind such antigen that contains the stretch of the peptide they recognize. Upon such cognate interaction the B cells start the course of clonal expansion. First, the B cells become larger, additional BcRs are synthesized and displayed on the membrane so that B cells are able to bind more antigen molecules. Then, the cells start to divide and small foci of antibody-forming cells develop at the edges of T cell–rich zones of spleen or lymph node.

## 8.6. Clonal expansion of B cells and antibody response

The kinetics and other characteristics of the primary antibody response differ considerably depending on the nature of antigen, route of antigen administration, and the frequency of epitope-specific cells in the organism. Typically some $10^6$ B cells in man (or $10^3$ B cells in a mouse) are *selected* by antigen and are *activated* by armed T cells. Once B cell activation is completed and clonal proliferation is initiated, formation of germinal centers in the involved lymph nodes will be established within about 4 days. Eight or nine successive cell divisions yield at the peak of the response (attained about at day 5) to about 1000 progeny from a single B cell. Thus some $10^9$ B cells – $10^6$ clones – participate in the response of an adult human organism. A large proportion of these cells will generate terminally differentiated plasma cells. Since a plasma cell is assumed to synthesize and to secrete some 10,000 antibody molecules per second, at the peak of the response the entire organism will produce about $10^{18}$ antibody molecules (250 mg) per day. It should be noted that the peak of the primary antibody response is attained by around day 7–9. Antibody-producing plasma cells are found in the red pulp of the spleen and the medulla of lymph nodes. Antibody-secreting cells also migrate and reside in the bone marrow and a few weeks after immunization this may be the prominent site of antibody production.

# 9. ISOTYPE SWITCH, AFFINITY MATURATION

The primary immune response is based – in most instances – on secretion of antibody belonging to the pentameric IgM type. Members of the activated clones under certain conditions and in certain milieu might be induced to switch toward forming antibody of other isotypes. The "new" clones retain the genetic information for the variable part of molecule, while they will be equipped with proper information for the formation of the new isotype-defining constant parts.

During the events related to isotype switch, the clones undergo also somatic hypermutation, which together with the selection of well-performing B cell clones results in a profound increase of average affinity to the challenging antigen.

## 9.1. Isotype switch

In the process termed "immunoglobulin switch" change from the expression of one isotype of immunoglobulin to another one occurs. From the topography of the immunoglobulin gene segments the order in which any given isotype can be expressed is precisely determined. IgM, IgD, IgG1, IgG2, IgG3, IgG4, IgA, IgE appear in this order, and the switch events are irreversible.

Isotype switching requires Th cells and certain lymphokines, though in molecular terms is mediated by increased accessibility of specific switch regions of the immunoglobulin genes to a "switch recombinase." It has been found that Th1 cells promote switch to IgG2 and IgG3 isotype, while Th2 cells promote switch to IgG4 and IgE isotype. IL-4, IFN-γ and TGF-β are meant to be switch factors for IgE, IgG2a (in mice), and IgA, respectively. Certain milieus, for example the mucosal system, contain lymphoid follicles specialized for the development of cells, which ultimately favor IgA production. IgA-specific switch differentiation occurs in Peyer's patches.

### 9.2. Affinity maturation

The average affinity of antibody molecules produced during an antigenic challenge increases as much as 100–10,000 folds. It has been shown that affinity maturation is result of (a) somatic hypermutation and (b) selection of high-affinity B-cell clones during the course of the response.

Average affinity of antibodies resulting from the primary response is about $10^7$/mol, while for the secondary response this value is of the order of $10^{11}$/mol. Affinity maturation is a phenomenon highly dependent on Th cells, and interestingly it occurs only after immunization with protein antigens.

## 10. SECONDARY IMMUNE RESPONSES AND MEMORY

Upon completion of the primary response several components of the immune system are altered. The number of specific B cells and T cells is increased. The density of BcR, TcR, and MHC molecules on those cells is augmented. In the blood there is an elevated concentration of circulating antibody. The second encounter of the immune system with the antigen is of a different kind.

### 10.1. Fast kinetics

When antigen enters the organism for the second time, it might happen that antigen will be eliminated without setting the secondary immune response in motion, since in the circulation there is an elevated concentration of antigen-neutralizing antibody. If, however, the response is induced, it will occur with a faster kinetics due to availability of high-affinity clonal members, and also that other molecular components are better "tuned" to the next response, for example that cells display higher density of MHC class II molecules. The immune system responds with faster- and larger-amplitude responses. The response lasts longer. The secondary response is characterized by formation of other isotypes than IgM.

### 10.2. Memory

Memory is the ability of the immune system to respond more effectively to the repeated exposure to the immunogen than at first encounter. It reflects the presence of clonally amplified population of antigen-specific cells. It is assumed, but not proven, that for the maintenance of memory the persisting presence of antigen is required, and that repetitive re-stimulation of cells even in the absence of further encounter with the immunogen is required. If this is the case, antigen trapped in the follicular dendritic cells could serve as a source of re-stimulating signal or alternatively the repetitive re-stimulation could be achieved by the presence of cross-reactive antigens.

About a month after the primary immune response, memory B cells and memory T cells can be detected in the organism. Increased affinity for antigen (and increased levels of MHC class II expression) allows memory B cells to profit from the T cell help at lower doses of antigen. Memory cells are recruited from the clonal pool of proliferating cells. Plasma cells are terminal cells, and cannot be recruited to the memory pool. It is probable that also within the pool of T cells the memory portion is established before the cells perform their effector function. Plasma cells and effector T cells – upon completion of their task – are eliminated from the system.

## 10.3. Altered repertoire upon repeated antigenic exposure

Repeated, recurrent antigenic exposure might extensively enrich the ability to respond to the challenging antigen, but it might impoverish the overall repertoire of the organism. It is the case that moderate alterations of cellular balance allow for return to the original homeostatic state, while profound recurrent changes happen at the expense of other clones.

## 10.4. Vaccination

Vaccination is a procedure in which application of a "disarmed pathogen" induces in the individual the ability to respond to the pathogen. The resting population of precursor cells is recruited to a clonal expansion and antibody production by the vaccine. The individual is usually capable of responding to the antigen prior to encounter of the pathogen but the response might be so inefficient that the fast multiplying pathogen would win the strategic race. By vaccination the organism "learns" to respond to the pathogen prior to arrival of pathogen.

The incidence of diseases such as diphtheria, measles, mumps, pertusis, rubella, poliomyelitis, and tetanus has declined dramatically as vaccination has become more common. Vaccines are either in form of whole organisms or as purified macromolecules.

## 10.5. Tolerance, anergy, apoptosis

Specific tolerance develops when reactive clones of T cells or B cells are deleted or otherwise incapacitated. Th1 cells are more easily tolerized than other T cells. Apoptosis is a highly regulated process leading to cytoplasm shrinking, chromatin condensation, DNA fragmentation followed by cell death. There are definite signals leading to T-cell anergy, while other signaling events induce T-cell apoptosis. Exposure of lymphocytes to antigen in the absence of an activation signal leads to anergy.

Clearance of activated T cells after the end of infection is as important as the clonal expansion of the responding cells. Failure to clear activated T cells increases the risk of cross-reactivity with self-antigens and as a consequence an autoimmune reaction.

# 11. ANTIBODY RESPONSE NOT REQUIRING T CELL HELP

B cells can respond directly to an antigen if that antigen is able to cross-link the antibodies on the B cell surface. Cell walls of various microbes are polysaccharides with multiple identical epitopes, and such antigens are able to stimulate B cells "directly," inducing strong IgM responses. As a rule, switch to other isotypes does not occur. Besides polysaccharides, glycolipids and also DNA molecules function as T-cell independent antigens.

Whether thymus-independent antigens – upon binding to the BcRs of the specific B cells – are endocytosed or not is of no relevance to the response, since these antigens cannot be associated with MHC molecules, and thus are not recognized by Th cells.

# 12. RESPONSE TO SUPERANTIGENS

Certain polypeptide components of some pathogens are called "superantigens." Instead of binding in the combining site of the TcR, these antigens bind to the outside structure of the

TcR molecules, and succeed to stimulate a much larger portion of T cells than do conventional peptide antigens, up to 2%–10% of all T cells of the organism. Most commonly known superantigen is staphylococcal endotoxin. Besides bacterial superantigens, there exist also viral superantigens. Since the stimulation of T cells does not involve the cascade of events needed for establishing a lasting T-cell memory, the organism does not acquire memory to this type of antigens. Any new pathogenic encounter resembles a primary infection.

This stimulation induces a massive production of various cytokines with the consequence of causing systemic toxicity and immune response suppression. Excessive responses to superantigens are often harmful to the organism.

## 13. DELAYED TYPE OF HYPERSENSITIVITY

Delayed type of hypersensitivity (DTH) response is a transient and local tissue response to certain antigens, involving antigen-specific $CD4^+$ T cells and mononuclear phagocytes. $CD8^+$ T cells are not challenged by DTH antigens. At the onset, antigen is taken up by Langerhans cells in the epidermis and carried to the draining lymph nodes. Peptide fragments in conjunction with MHC class II molecules are displayed on the membrane of Langerhans cells, and here they are recognized by epitope-specific $CD4^+$ T cells. Activated T cells proliferate and then cross endothelial barriers. Once the T cells are activated by antigen, they secrete high amounts of various cytokines. Although the original APC is a Langerhans cell, the response is amplified and sustained by macrophages. The vast majority of T cells that appear at later stage are independent of antigen. Massive cellular infiltration develops between 40 and 70 h after the application of antigens.

If the antigen is of microbial origin (e.g., Listeria), DTH ensures through the enhanced microbicidal function of phagocytes (induced by T cell–derived cytokines) the eradication of the microbe. Thus, the DTH response is of protective nature for the organism but some kinds of epitopes might result in tissue injury without performing a protective function.

Mycobacterial protein, when injected into the skin of individuals who had been exposed earlier to infection with the tuberculosis bacillus, will elicit a DTH reaction. In more general terms, the test serves as a useful diagnostic indicator for specific immunity against many intracellular bacteria.

(Note that "immediate hypersensitivity" in contrast to the above-described DTH begins within minutes after antigen challenge. The reaction is initiated by IgE. Hay fever, asthma, urticaria, and chronic eczema are various forms of immediate sensitivity.)

## 14. CYTOTOXICITY

Antigen-specific cytotoxic response is supposed to eliminate cells infected by viruses, intracellular bacteria, or cells forming altered or undesired polypeptides. When viral infection or an infection with intracellular parasites achieves its primary goal and viruses or bacteria reach intracellular compartments, they are in a safe environment and cannot be eliminated anymore by antibody. Utilizing the property of the MHC class I system in presenting portions of proteins synthesized endogenously by the cell, a surveillance system has evolved in which $CD8^+$ T cells recognize those cells which became infected by viruses or parasites and which "advertise" the infected status in terms of their peptide–MHC molecules. All viruses upon usurping the cellular machinery of the infected cell do express viral protein in the cytosolic compartments of the cells they infect, and these cells are killed by cytotoxic lymphocytes. Intracellular bacteria in most instances "flag" their presence in a similar manner and are marked for destruction as well. Similarly, tumor cells often synthesize proteins that are not part of normal cells, and their peptide–MHC molecules are the "tags" for cytotoxic cell attack.

## 14.1. MHC class I loading

Each and every nucleated cell of an individual expresses on its surface MHC class I molecules. Each and every such cell could be a target of viral or bacterial attack, or it can loose its proper regulatory function and become a tumor cell. Both healthy and infected cells are capable of presenting samples of the endogenous intracellular proteins on their surface. Such peptides move from cytosol to endoplasmic reticulum, are loaded into MHC class I molecule, and are delivered to the surface of the cell. Specific $CD8^{+}$ T cells readily recognize MHC class I molecules loaded with epitope components of such "non-self origin."

Peptide fragments of protein antigens of viruses and of intracellular bacteria are loaded into newly synthesized class I molecules in the endoplasmic reticulum and presented on MHC class I molecules on the membrane of the infected cell. MHC class I molecules are distinctly specialized for presenting peptides derived from proteins synthesized on endogenous polysomes; thus any newly synthesized polypeptide – self and non-self – has a chance to be presented. Such a polypeptide, if entering the hollow cylindrical structure of the proteasome, yields peptides of exactly the size range needed for MHC class I binding.

Although it has been stated that intracellular infections lead to peptide presentation by MHC class I molecules, this needs some further clarification. Not all intracellular infections are eliminated via MHC class I pathway, the fine dividing line being whether or not the intracellular parasites remain in the vacuolar compartment or if they escape to the cytosol. If they remain in the vacuolar compartment, the peptide ends up in the groove of the MHC class II molecule, while the cytosolic location shifts the presentation via MHC class I.

## 14.2. Cytotoxic killing

Between day 2 and 3 after infection, proliferation of CD8 T cells specific for the pathogen begins. Antigen-specific CD8 T cells may divide as rapidly as every 6 h. Once induced, cytotoxic cells join the pool of recirculating cells and then migrate to the site of antigen load. Upon contact of the cytotoxic cell with the target cell, perforin monomers and proteolytic granules are released; and upon formation of pores in the target cells, a killing process is induced, and then accomplished by apoptotic signals. The killing is unidirectional; the CTL itself is not harmed and after transmission of the lethal hit it can detach and seek another target. CTL can retain granules in reserve or can resynthesize them, allowing to kill multiple target cells in a serial fashion.

The process of T cell-mediated clearance (e.g., of influenza virus) is extremely efficient. Although the titers of virus peak at day 7, one day later might be no longer possible to recover infectious virus from any place of the organism. Exposure to many viruses induces lifelong immunity and greatly increased frequency of antigen-specific CTL precursors. Memory CTLs are long-lived cells that can exist either in a dormant state or they are sporadically stimulated by cross-reactive environmental or self-antigens. (Note that a small subset of cytotoxic T cells expresses CD4 instead of CD8. $CD4^{+}$ T cells are especially suited to kill macrophages infected with mycobacteria. To do so, they utilize the Fas/FasL pathway that is an alternative for the lytic granules.)

## 15. REGULATORY T-CELL RESPONSES

Some antigens can stimulate cells of the immune system in such a manner that the resulting response is down regulated. Inhibition and suppression of aberrant or excessive immune responses provides an adequate control mechanism for physiological and pathological responses including autoimmunity, tumor immunity, microbial immunity, allergy, and feto-maternal tolerance. The homeostasis of processes so regulated is maintained by regulatory T cells (Tregs), once called Suppressor T cells. Numerous putative mechanisms of Treg-mediated suppression have been proposed. They include (a) direct cell killing, (b) perforin- and granzyme B-dependent killing, (c) IL-10-mediated suppression, (d) dendritic cell-induced suppression, and (e) IL-2 exhaustion by Tregs.

There exist various sets and subsets of Tregs, and it seems that the immune system needs a redundancy of such cells:

- $CD25^+CD4^+$ Tregs
- IL-10 secreting Tr1 cells
- Qa1-1 restricted $CD8^+$ T cells
- $CD28^+CD8^+$ T cells
- $CD112^+CD8^+$ T cells
- NK T cells
- and others.

The common theme seems to be that the transcription factor Foxp3 is a key control molecule for the development and function of Tregs.

The regulatory effect of the resulting responses includes (i) production of an excess of cytokines with inhibitory functions, (ii) absorption or capturing of growth factors or differentiation factors. Suppressor responses are often induced under immunization with high concentrations of protein antigens, or administration of chemically reactive ligands injected without adjuvants. Antigen-specific suppression might be a strategy of the immune system to disrupt the cytokine balance that determines whether a Th1 or a Th2 response takes place [15].

## 16. IMMUNE SYSTEM AS AN AUTONOMOUS SET OF LYMPHOCYTES AND AS A NETWORK

The dichotomy of the immune system is that it can be considered as a collection of independently operating lymphocytes but that it can be considered also as a network. Idiotypes are epitopes on immunoglobulin molecules that are potentially immunogenic. In a normal healthy individual most of the epitope-specific cells are present at a low frequency, and therefore the expressed idiotypes (both in antibody and in BcR molecules) are at very low abundance. Upon a specific immune response, the concentration of idiotypes raises to such a level that it becomes immunogenic, and antibody response against idiotypes is elicited. This anti-antibody response is named "anti-idiotypic response." The result of such a response is an increase in the new set of idiotypes ("second wave" idiotypes). These new epitopes elicit a response that can be considered as an anti-anti-idiotypic response, but more importantly it is a "mirror image" of the original epitope, that is, of the antigen that started the response. Several waves of anti-idiotypic responses as predicted by the "idiotypic network hypothesis" might include a profound perturbation of the cellular balance of the immune system [16].

## 17. CONCLUDING REMARKS

Although there exist many comprehensive review articles on the specific immune response, there are only scarce attempt to bridge in evolutionary terms the nervous system and the immune system. In the context of this volume it is of importance to learn not only how does the specific immune response function but also what are the "common roots" of the immune system and the nervous system. Since the "commonality" is in most instances recognizable only as a "final result" of evolutionary forces, the only suitable approach is to trace back the roots through evolutionary panorama. This is offered here in the first part of the chapter, with a special emphasis on the structural characteristics of the components of the nervous system and the immune system, retracing several gene families, including the immunoglobulin superfamily (IgSF) in vertebrates and invertebrates.

From these interwoven histories of the two complex systems and upon deciphering the main components of the adaptive immune system, it is only a logical extension to describe the mechanism of the specific immune response. The major portion of the chapter relates to the molecular and cellular components that play a role in specific epitopal recognition and paratopal response. Another chapter of this volume, complementary to this one, addresses the innate immune response, while the field of Toll-like receptor (TLR) family system is left to a special chapter of this volume.

## ACKNOWLEDGMENTS

Both authors are indebted to their "new," post-retirement positions: Ivan Lefkovits to the Research Department of the University Clinics of Basel and Louis Du Pasquier to the Institute of Zoology, University of Basel, for being integrated in a new environment of research activity. For both authors their new environment was created after spending over 30 years at the Basel Institute for Immunology, function of which has been terminated in 2001.

## DEDICATION

This chapter has been written in a synergistic manner by Ivan Lefkovits and Louis Du Pasquier. In the context of this and earlier writings both of them think of late Charley Steinberg – a friend, colleague, and mentor.

Ivan Lefkovits dedicates this work to the memory of Dr Ernö Novy, the oldest person on this planet who read and commented this and also earlier work. Dr Ernö Novy died at the age of 105; as a Medical Doctor he kept a profound interest in immunology. Almost a century ago, he had been a good friend to father of Ivan Lefkovits.

## REFERENCES

1. Boulanger LM, Huh GS, Shatz CJ. Neuronal plasticity and cellular immunity: shared molecular mechanisms. Curr Opin Neurobiol 2001 Oct;11(5):568–78.
2. Du Pasquier L. Insects diversify one molecule to serve two systems. Science 2005 Sep 16;309(5742):1826–7.
3. Pancer Z, Cooper MD. The evolution of adaptive immunity. Annu Rev Immunol 2006;24:497–518.

4. Kasahara M. What do the paralogous regions in the genome tell us about the origin of the adaptive immune system? Immunol Rev 1998 Dec;166:159–75.
5. Du Pasquier L, Zucchetti I, De Santis R. Immunoglobulin superfamily receptors in protochordates: before RAG time. Immunol Rev 2004 Apr;198:233–48.
6. Flajnik MF, Miller K, Du Pasquier L. Evolution of the Immune System in Fundamental Immunology 5th Edn. (ed. Paul, WE). Philadelphia: Lippincott Williams & Wilkins, 2003; pp. 519–70.
7. Flajnik MF, Kasahara M. Comparative genomics of the MHC: glimpses into the evolution of the adaptive immune system. Immunity 2001 Sep;15(3):351–62.
8. Jerne NK. The natural selection theory of antibody formation. Proc Nat Acad Sci (USA) 1955;41: 849–57.
9. Burnet FM. A modification of Jerne's theory of antibody production using the concept of clonal selection. Aust J Sci 1957;20:67–8.
10. Williams AF, Barclay AN. The immunoglobulin superfamily: domains for cell surface recognition. Annu Rev Immunol 1988;6:381–405.
11. Jerne NK. Immunological speculations. Annu Rev Microbiol 1960;14:341–58.
12. Berek C, Milstein C. The dynamic nature of antibody repertoire. Immunol Rev 1988;105:5–26.
13. Davis MM, Bjorkman PJ. A model for T cell receptor and MHC/peptide interaction. Adv Exp Med Biol 1989;254:13–16.
14. von Boehmer H. T cell lineage fate: instructed by receptor signals. Curr Biol 2000;10:642–5.
15. Sakaguchi S. Regulatory T cells: meden agan. Immunol Rev 2006;212:8–27.
16. Jerne NK. Towards a network theory of the immune system. Ann Immunol (Institut Pasteur) 1974;125c:373–89.

*The Hypothalamus–Pituitary–Adrenal Axis*
Edited by A. del Rey, G. P. Chrousos and H. O. Besedovsky



# *The Immune System of the Brain*

T. BUCKY JONES[1], KURT M. LUCIN[2–4], and PHILLIP G. POPOVICH[1–5]

[1]*Neuroscience Graduate Studies Program, The Ohio State University College of Medicine, Columbus, Ohio, USA*
[2]*Department of Molecular Virology, Immunology and Medical Genetics*
[3]*The Integrated Biomedical Science Graduate Program*
[4]*Center for Brain and Spinal Cord Repair*
[5]*Institute for Behavioral Medicine Research, The Ohio State University College of Medicine, Columbus, Ohio, USA*

## ABSTRACT

Historically, the central nervous system (CNS) was believed to be anatomically and physiologically isolated from the immune system. The existence of a book entitled "*Neuroimmune Biology*" is evidence that this concept of CNS "immune privilege" has evolved and neuroimmunology has emerged at the cutting edge of experimental and clinical neuroscience. Today, we know that cells and mediators of the immune system routinely access the intact brain and spinal cord. Furthermore, neurons and glia interact with and regulate inflammatory leukocytes under normal and pathological conditions. This chapter provides an overview of the historical concept of immune privilege and presents evidence that mandates a reappraisal of this dogma. Specifically, new data suggest that the CNS and the immune system actively regulate each other through discrete cellular and molecular mechanisms and when this control is disrupted, pathology ensues.

## 1. OVERVIEW

The emergence of neuroimmunology as a unique scientific discipline can be traced back to the 1970s when the National Institutes of Health created a Neuroimmunology Branch. Initially, the research focus of this branch was biased toward immune-mediated destruction of the central nervous system (CNS) myelin. As a result, neuroimmunology became inextricably linked with multiple sclerosis (MS) and its animal model, experimental autoimmune encephalomyelitis (EAE). Even today, the principles of neuroimmunology, as described by most textbooks and review articles, are discussed in the context of MS and EAE. However, the basic principles of neuroimmune interactions as defined by MS and EAE now transcend these diseases. Clinical trials designed to attenuate or boost parameters of immune function in Alzheimer's disease (AD), stroke, and spinal cord injury (SCI) are evidence of the impact of lessons learned from EAE and MS. Although models of immune-mediated myelin destruction will remain a cornerstone of this chapter, we will also discuss neuroimmune interactions in the context of trauma and neurodegenerative disease.

## 2. IMMUNE PRIVILEGE AND THE CNS

The concept of CNS immune privilege arose from early studies in transplantation. Medawar observed that heterologous tissues placed into the CNS were spared from immunological rejection [1]. In subsequent years, this apparent failure to develop an immune response within the CNS to heterotopic tissue grafts was explained by (1) sequestration of antigens behind the blood–brain barrier (BBB) and impairment of T-cell migration across this interface; (2) the absence of conventional lymphatic drainage from the CNS extracellular space into peripheral lymphoid tissues; and (3) limited parenchymal expression of major histocompatibility complex (MHC) or accessory molecules required for antigen presentation in the CNS [1,2]. Tight junctions between cerebrovascular endothelia provide a structural barrier inhibiting blood-to-brain transfer of leukocytes, antibodies, complement proteins, and select cytokines [3]. Early attempts to reveal antigen-presenting cells (APCs) in the CNS were met with limited success. Indeed, of all resident CNS cell populations, only astrocytes and microglia could be induced to express MHC class II molecules *in vitro* [4–6].

However, each of these categorical definitions of immune privilege has since been challenged. For example, it is now accepted that activated T lymphocytes routinely survey the intact CNS [7,8]. Furthermore, MHC class II molecules are found in normal rodent and human brain on parenchymal and perivascular microglia [9–12]. Although there is still limited anatomical evidence of a conventional lymphatic system in the CNS, elegant physiological and anatomical studies have defined specific routes for flow of cerebrospinal and extracellular fluid from the CNS parenchyma to the deep cervical lymph nodes [13–16]. Thus, the healthy CNS possesses the molecular and cellular requirements needed to support an immune response. Yet, under normal circumstances, this potential appears to be masked or actively suppressed.

## 3. CNS IMMUNE REGULATION: REDEFINING IMMUNE PRIVILEGE

Today, rather than consider the CNS as "immune privileged," it is perhaps better to view it as a site where a collection of regulatory networks govern interactions between resident CNS cells (glia and neurons) and inflammatory leukocytes [17–19]. Within the CNS, parenchymal microglia and astroglia are capable of controlling leukocyte function via the release of immunosuppressive cytokines like transforming growth factor β (TGFβ) and interleukin (IL)-10 [17,20–26]. *In vitro*, TGFβ inhibits leukocyte migration across CNS endothelia and can inhibit antigen-specific T-cell proliferation [27,28]. An immunomodulatory role for TGFβ is also suspected *in vivo*. Clinical data support this notion, showing that TGFβ is elevated in cerebrospinal fluid of MS patients in remission, but not those with active disease [29]. IL-10 may play a similar role in minimizing CNS-immune cross-talk. Transgenic mice expressing human IL-10 under control of the MHC II promoter show no evidence of inflammatory cell infiltration into the CNS after induction of EAE [30]. Astroglia and microglia also may induce T-cell anergy or apoptosis due to their normally low levels of expression of co-stimulatory molecules (e.g., B7-1) and MHC class II glycoproteins [20,24] or by facilitating Fas (CD95)/Fas ligand (CD95L or FasL) interactions. Fas is a member of the tumor necrosis factor (TNF) receptor gene family and contains a cytoplasmic death domain, which activates apoptotic-inducing proteins (e.g., caspases). FasL is constitutively expressed on astrocytes and microglia and can be upregulated by nerve injury [31,32].

Neurons also contribute to immunomodulation. Electrically active neurons may influence glial and lymphocyte functions via a variety of mechanisms including changes in ion concentrations, the release of soluble mediators, or direct ligand–receptor interactions. Constitutively expressed neuronal FasL [31] has been shown to contribute to killing of CNS autoreactive T cells. Indeed, blockade of neuronal FasL increases encephalitogenic T-cell survival [33]. Neurons can also downregulate the expression of glial surface antigens which are required for establishing communication with lymphocytes [34]. In the face of an inflammatory challenge, electrically active neurons can suppress interferon (IFN)γ-mediated induction of MHC on astrocytes and microglia [35]. However, when neuronal function is impaired (e.g., glutamate antagonists or tetrodotoxin), there is marked upregulation of MHC on microglia and astrocytes [35]. Thus, under normal conditions, low or undetectable levels of MHC expression on glia could be explained by tonic neuronal inhibition. Accordingly, the *de novo* increase in MHC expression in rodent and human CNS after traumatic injury or in chronic neurodegeneration could be explained by the loss of neuronal function.

Neuronal immunomodulation may also occur through the release of neurotransmitters or neuropeptides including serotonin, dopamine, substance P, calcitonin gene-related peptide, and norepinephrine [36–39]. For example, norepinephrine binding to adrenoreceptors on leukocytes can simultaneously inhibit the production of proinflammatory cytokines (i.e., IFNγ and TNF-α) while stimulating the production of TGFβ and IL-10 [40]. Classical neurotransmitters (e.g., glutamate), neuropeptides (e.g., vasoactive intestinal peptide), and neurotrophins (e.g., nerve growth factor, neurotrophin-3 [NT-3]) can regulate glial expression of MHC II molecules [41–43].

Direct neuronal–glial or neuronal–leukocyte interactions are also involved in immune modulation. CD200, a recently defined membrane glycoprotein (expressed on neurons) appears to be involved in regulating microglia. In the CNS, the cognate receptor for CD200 (i.e., CD200R) is restricted to microglia. In mice genetically deficient in CD200, microglial reactivity is enhanced under normal and pathological conditions (i.e., EAE and facial nerve axotomy) [44,45]. Interestingly, CD200 shares structural homology with molecules involved in the regulation of T cells and is genetically linked to B7 co-stimulatory molecules [46,47].

In AD, stroke, or brain/SCI, the process of neurodegeneration disrupts these CNS immunomodulatory mechanisms. Regardless of the triggering event, glial activation is rapid and ubiquitous. Accompanied with this activation are phenotypic alterations (e.g., increased expression of various cell surface proteins) and increased production of proinflammatory cytokines and chemokines; subsequently propagating the recruitment of hematogenous leukocytes. It is the cross-talk between neurons, glia, and leukocytes that then ultimately influences the reparative and destructive potential of neuroinflammation.

## 4. RESIDENT CNS IMMUNE EFFECTOR CELLS

### 4.1. Astrocytes

Astrocytes comprise 40%–50% of all glia [48]. Given their prevalence in the CNS, their responsiveness to all forms of CNS pathology, and their ability to modulate immune cell function, astrocytes are central to any discussion of CNS–immune interactions. Once activated, astrocytes undergo a characteristic pattern of proliferation and cellular hypertrophy that is readily visible using antibodies against glial fibrillary acidic protein (GFAP). This response, referred to as "reactive astrogliosis," appears to be a physiological attempt to preserve or repair the integrity of the injured CNS. Activated astrocytes metabolize extracellular neurotransmitters, produce extracellular matrix (ECM) molecules, and provide neurotrophic support to damaged neurons [49]. Elegant studies by Bush et al. demonstrate the pivotal role astrocytes play in modulating these post-injury sequelae. Specifically, ablation of reactive astrocytes adjacent to a cerebral stab wound resulted in pronounced leukocyte infiltration, impaired BBB function, and exacerbated neuronal degeneration [50]. When this same approach was used to limit astrogliosis in a model of SCI, demyelination and inflammation were exacerbated and locomotor recovery was impaired [51].

Some of the earliest evidence that astrocytes might participate as immune effector cells was provided by Fontana et al. who showed that astrocytes augment mitogen-driven lymphocyte proliferation and IL-2 production *in vitro* [52]. This effect was mediated by an IL-1-like factor [52]. It is now well established that astrocytes produce a plethora of growth factors, cytokines, chemokines, and adhesion molecules with immunomodulatory functions [53,54]. Although most of these factors are not expressed constitutively, they are induced in activated astrocytes via exposure to proinflammatory stimuli (e.g., IFNγ or lipopolysaccharide [LPS]) [55].

*In vivo*, cytokines and chemokines produced by astrocytes are elevated in autoimmune demyelination (EAE and MS), stroke/ischemic CNS injury, neurodegeneration (i.e., AD), and nerve trauma [53,56,57]. In each case, the initiation of astrocyte cytokine production appears to be temporally and spatially linked to the production of microglial-derived mediators, suggesting that glial cross-talk is in part mediated by cytokines ([58,59]; also see below). Astrocytes have been implicated in inducing microglial activation through the release of colony-stimulating factors (i.e., M-CSF and GM-CSF) [60,61], lymphotoxin-alpha [62], and ATP [63]. Additionally, astrocytes may control synthesis of ECM molecules by microglia [64]. Microglia reciprocate by producing cytokines and growth factors (e.g., IL-1 and TGFβ) that trigger astroglial proliferation and production of ECM molecules [58,65,66].

The landmark studies of Fontana et al. demonstrated that, *in vitro*, astrocytes could present antigen to T cells in an MHC-restricted fashion [67]. In the mid 1980s, additional *in vitro* studies confirmed that astrocytes express MHC II molecules, either constitutively or following induction by IFNγ [68,69]. However, many still question the antigen-presenting capacity of astrocytes since most studies have failed to demonstrate robust or prolonged expression of either MHC II or co-stimulatory molecules (i.e., B7-1/B7-2) on astrocytes *in vivo* [68,70–72].

Benveniste and colleagues have shown that the class II transactivator (CIITA), a protein essential for regulating MHC class II expression, is active in astrocytes and is responsive to IFNγ [73,74]. However, a variety of cytokines (e.g., TGFβ), neurotransmitters/neuropeptides (e.g., neuroepinephrine), and tissue substrates (e.g., gangliosides) that are abundant *in vivo* can suppress CIITA expression. These normal constituents of the CNS may explain why consistent expression of MHC class II on astrocytes is only found in cell/tissue culture models (reviewed in Ref. [54]).

Whether astrocytes express the co-stimulatory molecules, B7-1 or B7-2, is also controversial [75–78]. Expression of these molecules is critical for effective antigen presentation. Indeed, MHC-restricted antigen presentation (signal one) along with B7 ligation of CD28 (signal two) results in T-cell activation, clonal expansion, and initiation or propagation of an antigen-specific immune response. However, without "signal two" delivered by co-stimulatory molecules, T cells become anergic or undergo apoptosis [79]. *In vitro*, presentation of antigen by astrocytes results in suppression of T-cell proliferation and cytokine production or the induction of lymphocyte apoptosis [21,80,81]. Under "*in vivo*-like" conditions, astrocytes are equally ineffective in supporting T-cell proliferation [82]. As described above, this deficiency in astroglial signaling may represent a physiological mechanism for controlling the expansion of potentially deleterious lymphocytes within the CNS. With such functional diversity, astrocytes are likely to participate in all phases of the inflammatory response – initiation and propagation of immune responses by the production of proinflammatory cytokines and chemokines, then resolution of immune responses by inefficient presentation of antigen, and the production of immunosuppressive cytokines (e.g., TGFβ).

## 4.2. Microglia

Of the cells in the CNS that react to trauma, infection, or ischemia, microglia are arguably the most responsive [57,83]. Microglia constitute as much as 20% of all glia, but once activated, they encompass more surface area than astrocytes [83,84]. Microglia are derived from bone marrow precursors and infiltrate the CNS during embryonic development [85]. Originally thought to be functionally quiescent, microglia are now known to constantly survey the extracellular environment [63,86]. The large surface area of resting microglia is believed to facilitate pinocytosis and surveillance of the extracellular milieu (e.g., shifts in ions or extracellular metabolites) [87]. Following injury or infection, microglia proliferate, retract their elaborate cellular processes, and increase the expression of cell surface molecules including MHC [88]. Although this morphological and phenotypic transformation implies a heightened functional status, the biological impact that these cells have on the surrounding milieu is ill-defined. Nevertheless, the increased expression of co-stimulatory molecules (e.g., B7-1), MHC class I and II molecules, complement receptors (CR)-3, and cytokine mRNA (e.g., TGFβ and TNF) suggests that activated microglia are poised for action. Further activation renders microglia phagocytic at which point they are referred to as "brain macrophages" and are phenotypically and morphologically indistinguishable from macrophages that infiltrate the injured or diseased CNS from the circulation (see below). Microglial-derived brain macrophages are phagocytic and can secrete cytokines, growth factors, oxygen- and nitrogen-free radicals, neurotransmitters, and proteolytic enzymes [65,83,89]. It is through the release of these mediators that microglia influence the differentiation and survival of neurons, astrocytes, and oligodendrocytes. Recent data show that microglia express p75, trkA, trkB, and trkC receptors and respond to NT-3 and brain-derived neurotrophic factor (BDNF) [90,91] suggesting that microglia comprise a resident neurotrophin network. Although the majority of these secretory products have been associated with brain macrophage activation *in vitro*, immunohistochemical detection of inducible nitric oxide synthase, matrix metalloproteinases, cytokines, quinolinic acid, and neurotrophic factors in tissue sections suggests a similar functional competency *in vivo* [92–95].

Because activated microglia predominate in pathological CNS tissue, there is a tendency to assume a cause–effect relationship. In fact, microglia-derived cytokines (i.e., IL-1, TNF-α), chemokines, quinolinic acid, and reactive oxygen and nitrogen intermediates (i.e., nitric oxide) have been implicated in the progression of neuropathology associated with MS, AD, stroke/ischemia, and trauma (brain and SCI) [56,96–98]. However, profound microglial activation does occur without notable injury to neurons or glia. Intraparenchymal or intrathecal injection of LPS, TNF-α, or IFNγ induces microglial activation and upregulation of cytokine mRNA but without notable damage to axons, myelin, or neurons [99–102]. After facial nerve axotomy, activated microglia are temporally and spatially associated with regenerating facial motor neurons [57,103]. Although microglial-derived TNF may contribute to oligodendrocyte cell death in MS, cerebral ischemia, and SCI [104–106], neuronal degeneration is enhanced following brain injury in TNF receptor-deficient mice – presumably due to the absence of microglial-derived TNF in promoting neuronal survival [107,108]. Interestingly, cuprizone-mediated demyelination in TNF receptor-deficient mice is associated with delayed remyelination coupled with a reduced pool of proliferating oligodendrocyte precursors (OPCs) [109]. Microglia may be pivotal in triggering proliferation, migration, and differentiation in the OPC pool [110]. Microglia also produce TGFβ, which can inhibit macrophage activation and the production of damaging reactive oxygen species and cytokines [111,112]. Microglia produce and respond to BDNF and NT-3, both of which can enhance neuronal survival and stimulate remyelination through proliferation of oligodendrocyte progenitor cells [90,91,113,114]. Whether microglia become neurotoxic or neurotrophic effector cells is likely determined by the microenvironment in which they are activated. Future research is needed to define the cellular or molecular cues that trigger the diverse functions of microglia.

In addition to their ability to influence survival and repair through the release of soluble mediators, microglia can modulate cell survival through cell–cell interactions (see above discussion regarding FasL and CD200R). Also, subsets of microglia constitutively express MHC class II molecules [57], and there is evidence that they express co-stimulatory molecules during inflammation [115,116]. Thus, microglia appear to be endowed with the molecular machinery necessary for initiating antigen-specific T-cell responses [82,117–119]. However, similar to astrocytes, parenchymal microglia do not seem to be effective APCs *in vivo*. Indeed, microglia stimulate cytokine production and differentiation in T cells but do not promote T-cell proliferation [22,24,120]. There is also evidence that microglia induce T-cell apoptosis [24]. This may trigger an immunoregulatory feedback cycle whereby phagocytosis of apoptotic T cells by activated microglia downregulates subsequent activation of microglia and reduces the secretion of chemokines involved in recruitment of encephalitogenic T cells [121,122].

### 4.3. Perivascular microglia/perivascular cells

Since astrocytes and microglia downregulate adaptive (i.e., lymphocyte-mediated) immune responses in the CNS, how are myelin-reactive lymphocytes propagated in MS and EAE? Hickey and Kimura demonstrated that this function is most likely carried out by perivascular cells located at the blood–CNS interface [12]. These cells constitutively express co-stimulatory and MHC class II molecules and are phagocytic – necessary prerequisites for antigen presentation [72,123]. Moreover, they appear to be a dynamic cell population that continuously repopulates the perivascular space [124]. Because of their location outside the basal lamina of the vasculature, perivascular cells are ideally positioned for the initial confrontation with activated T cells as they traverse the endothelia.

## 5. CYTOKINES AND CHEMOKINES: SOLUBLE EFFECTORS OF CNS IMMUNITY

Cytokines and chemokines are a diverse group of polypeptides typically associated with immune cell function and chemotaxis. However, within the CNS, virtually all cells have been shown to produce and respond to a range of chemokines and cytokines. Current understanding of cytokine function in the CNS is based largely on *in vitro* studies, transgenic or knockout animals, blockade of endogenous cytokine production, or exogenous administration of cytokines into the periphery or directly into the CNS parenchyma. Collectively, these studies have improved our understanding of the functional potential of these mediators within the CNS. What remains unclear is how cytokine production by resident or recruited cells contributes to processes of neurodegeneration, neuroprotection, and neural repair. Similarly, while chemokine regulation in the CNS has been a focus in modulating the pathogenesis of MS, viral encephalopathies, neurodegeneration, and trauma, chemokines may also be involved in neural development and physiological maintenance of the CNS. It is beyond the scope of this chapter to describe the diversity of cytokine/chemokine cascades involved in CNS physiology and pathology. However, such a task has been accomplished in a number of excellent reviews [125–127]. Thus, only an overview of cytokine/chemokine involvement in neuroimmune reactions will be provided here, with an emphasis on conflicting data in the autoimmune and neurodegeneration literature.

### 5.1. Cytokines

A large number of studies implicate cytokines in the pathogenesis of various neurological disorders [98,125,128–130]. Here, our discussion will focus on the divergent functions of two extensively studied cytokines (IL-1 and TNF-α) in the context of MS/EAE, acute neurodegeneration (e.g., AD, stroke, ischemia), brain trauma, and SCI.

The increased expression of mRNA encoding IL-1 and TNF-α (along with IL-6) occurs within minutes to hours following injury/infection and typically precedes biochemical and morphological evidence of neuronal injury/death and infiltration of circulating leukocytes [131]. This rapid induction of cytokine expression may represent a physiological mechanism of intercellular signaling between glia and neurons. However, depending on the magnitude of expression, time of induction, and context of the disease in question, cytokine cascades can precipitate pathology. The most convincing evidence of this potential is found when CNS-specific cytokine production is controlled through glial (i.e., GFAP and MBP) and neuronal (i.e., neurofilament) promoters (reviewed in Ref. [132]). Generally, these models reveal that cytokine overexpression in the normal CNS elicits pathology with functional deficits. Various *in vitro* and *in vivo* models have confirmed the potential involvement of cytokine-mediated pathology in the CNS.

Elevated IL-1 and TNF-α are observed and suspected to enhance pathology in ischemic brain injury, EAE, and traumatic brain injury [133–136]. Blockade of endogenous IL-1 in these disorders, via infusion of the naturally occurring IL-1 receptor antagonist (IL-1ra), reduces CNS pathology [137–139]. Additionally, IL-1 is overexpressed in Alzheimer's brain and is directly related to the formation of plaques and dystrophic neurites [140]. TNF-α is known to cause myelin and oligodendrocyte damage *in vitro* and *in vivo* [141–144]. Similar to IL-1, inhibition of endogenous TNF-α with antibodies, antisense oligonucleotides, or soluble TNF-α receptors attenuates the pathology and deficits associated with stroke, ischemia, EAE, and traumatic brain injury [145–149].

Recent data suggest that TNF-α and IL-1 may also be essential for repair of the pathological CNS. In EAE, TNF-α-deficient mice show exacerbated pathology and neurological impairment with higher incidence of mortality compared with wild-type mice [108]. These effects are reversed by exogenous administration of TNF-α. Similarly, impaired functional recovery and exacerbated neuron loss have been observed in models of SCI, traumatic brain injury, and cerebral ischemia using mice deficient in TNF-α receptors [107,147,150]. Similarly, the proliferative capacity of oligodendrocyte progenitors and remyelination of chemically demyelinated lesions are impaired in mice lacking IL-1, TNF-α, or TNF receptors [109,151].

### 5.2. Chemokines

Chemokines are small (7–10 kDA) chemoattractant peptides, typically associated with recruitment of leukocytes to inflammatory foci [152]. Accordingly, regulation of chemokine ligand-receptor networks in the CNS is being studied as a therapeutic target in MS/EAE, stroke, AD, and SCI [129,130,153]. In EAE, chemokine expression parallels or follows leukocyte recruitment [154]. These findings suggest that amplification of chemokine pathways – presumably the result of an acute increase in *glial-derived cytokines*, enhanced cross-talk between resident CNS cells, and the recruitment of inflammatory cells – might propagate pathology and neurological impairment. For example, IFNγ-inducible protein of 10 kDa (IP-10) and monocyte chemoattractant protein-1 (MCP-1) are upregulated by IFNγ and TNF-α and have been localized to astrocytes in EAE lesions [154,155]. In MS, IP-10, "regulated upon activation, normal T-cell expressed and secreted" (RANTES), MCP-1, and macrophage inflammatory protein-1α (MIP-1α) have been localized to astrocytes and microglia in demyelinating plaques [156,157]. IP-10 and MCP-1 are pivotal in eliciting T-cell and monocyte migration from the circulation into the CNS parenchyma. These leukocytes are central to the propagation of clinical and histological signs of EAE (see discussion below). MCP-1 has also been localized to astrocytes after mechanical brain injury [158]. MCP-1, IP-10, and MIP-1α are elevated in the traumatically injured spinal cord and select inhibition of these chemokines has been associated with neuroprotection [159–163]. Similarly, broad spectrum inhibition of chemokine upregulation after focal ischemia is neuroprotective [164]. AD pathology may be propagated by chemokine signaling in microglia. In response to β-amyloid, microglia produce MCP-1 which can induce the migration of astrocytes and microglia to regions of plaque deposition thereby enhancing gliosis [130,165]. Thus, chemokines are integral in targeting glia and leukocytes to regions of pathology; however, the onset of injurious effector functions in these newly recruited cells may be determined by other signals. Indeed, transgenic expression of the chemokines MCP-1 or growth-related oncogene-alpha (GRO-α) causes florid intraparenchymal inflammation but without acute neuropathology or functional impairment [166,167]. Another chemokine, fractalkine (CX3CL1), may limit the effector potential of inflammatory cells and glia in the CNS. In the intact brain and spinal cord, only microglia express the receptor for CX3CL1 (i.e., CX3CR1), and mice deficient in CX3CR1 exhibit enhanced activation of microglia and a corresponding increase in neuropathology in models of neurodegenerative disease and neuroinflammation [168]. During excitotoxic brain damage, fractalkine is cleaved from neurons – possibly contributing to enhanced pathology [169].

## 6. HEMATOGENOUS LEUKOCYTES AND CNS INFLAMMATION

Although astrocytes and microglia are among the first cells to respond to disruption of the CNS, recruitment of immune cells from the circulation is typical of most forms of progressive neurological disease. The onset of bleeding and subsequent clotting, together with reactive astrogliosis and microglial activation, causes the production of a chemoattractant gradient that elicits leukocyte recruitment. Leukocyte entry into the CNS is tightly regulated and is largely dictated by the regulated expression of specific chemokines (reviewed in Refs [130,170]) and adhesion molecules (reviewed in Ref. [171]). In general, glial activation precedes the infiltration of neutrophils, monocytes, and lymphocytes.

GRO-α and cytokine-induced neutrophil chemoattractant-(CINC-1), neutrophil chemoattractants, are increased early during inflammation. In models of SCI and cerebral ischemia, upregulation of these chemokines corresponds with the acute neutrophil influx [159,160,162,172,173]. Attenuation of neutrophil function results in improved anatomical and functional recovery implicating neutrophils in the acute secondary pathology of ischemic and traumatic CNS injury [174–176]. However, in EAE, neutrophils may suppress local immune responses and disease severity [177,178].

The infiltration and differentiation of blood monocytes into macrophages within the pathological CNS can exacerbate neurotoxicity and demyelination. Thus, it is not surprising that limiting or depleting hematogenous (i.e., blood-derived) macrophages reduces tissue injury and accelerates neurological recovery in stroke, SCI, and EAE [179–184]. In fact, limiting the involvement of circulating monocytes has been shown to be beneficial in a number of models of neurologic disease. However, macrophages are essential for efficient wound repair, and there is evidence that macrophages promote axon regeneration and remyelination after traumatic CNS injury [185–188]. It is possible that this functional dichotomy is the result of a heterogenous macrophage response consisting of macrophages derived from resident microglia and those originating from infiltrating monocytes [182,189,190].

Just as glia, monocytes, and neutrophils have been shown to have "protective" effects in the CNS, T lymphocytes have been shown to be important for repair and trophic support after axotomy or crush of the facial and optic nerves [191–195]. However, these studies are controversial and independent replication studies have failed to find a neuroprotective effect of autoreactive T cells [196].

## 7. SUMMARY

It is now recognized that the immune and nervous systems are inextricably linked. Indeed, interleukins and neurotrophins, named for their roles in the immune and nervous systems, respectively, are known to have similar functions in both systems and are in fact not restricted in their expression or responsiveness. In pathological conditions, astrocytes and microglia perform many of the functions previously associated only with immune cells. Moreover, hematogenous leukocytes interact with parenchymal cells of the CNS just as they do within the confines of peripheral lymphoid tissues. While this cross-talk appears important in the maintenance of normal CNS and immune function, in the face of a challenge to either system, it becomes increasingly difficult to determine where therapeutic interventions should be targeted. The remaining chapters in this volume emphasize bidirectional effects and the complexity of interactions of the nervous and immune systems at various levels of the neuraxis. The final section in which clinical implications of these interactions are discussed provides a glimpse of the challenges we are faced with when establishing appropriate therapeutic interventions designed to target specific aspects of these systems.

## ACKNOWLEDGMENTS

The authors would like to thank Dr Dana McTigue for her contributions in final editing of the manuscript. This work is supported in part by grants from NINDS (NS37846 and NS047175).

## REFERENCES

1. Medawar PB. Immunity to homologous grafted skin. III. The fate of skin homografts transplanted to the brain, to subcutaneous tissue, and to the anterior chamber of the eye. British J Pathol 1948;29: 58–69.
2. Reese TS, Karnovsky MJ. Fine structural localization of a blood-brain barrier to exogenous peroxidase. J Cell Biol 1967;34:207–17.
3. Risau W, Wolburg H. Development of the blood-brain barrier. Trends Neurosci 1990;13(5):174–8.
4. Hart DN, Fabre JW. Demonstration and characterization of Ia-positive dendritic cells in the interstitial connective tissues of rat heart and other tissues, but not brain. J Exp Med 8-1-1981;154(2):347–61.
5. Lassmann H, Zimprich F, Vass K, Hickey WF. Microglial cells are a component of the perivascular glia. J Neurosci Res 1991;28:236–43.
6. Streit WJ, Graeber MB, Kreutzberg GW. Expression of Ia antigen on perivascular and microglial cells after sublethal and lethal motor neuron injury. Exp Neurol 1989;105:115–26.

7. Wekerle H, Linington C, Lassmann H, Meyermann R. Cellular immune reactivity within the CNS. Trends Neurosci 1986;9:271–7.
8. Hickey WF, Hsu BL, Kimura H. T-lymphocyte entry into the central nervous system. J Neurosci Res 1991;28:254–60.
9. Graeber MB, Streit WJ, Büringer D, Sparks DL, Kreutzberg GW. Ultrastructural location of major histocompatibility complex (MHC) class II positive perivascular cells in histologically normal human brain. J Neuropathol Exp Neurol 1992;51(3):303–11.
10. Sedgwick JD, Schwender S, Gregersen R, Dorries R, ter Meulen V. Resident macrophages (ramified microglia) of the adult brown Norway rat central nervous system are constitutively major histocompatibility complex class II positive. J Exp Med 1993;177(4):1145–52.
11. Jander S, Stoll G. Strain-specific expression of microglial keratan sulfate proteoglycans in the normal rat central nervous system: inverse correlation with constitutive expression of major histocompatibility complex class II antigens. Glia 1996;18(3):255–60.
12. Hickey WF, Kimura H. Perivascular microglial cells of the CNS are bone marrow-derived and present antigen in vivo. Science 1988;239:290–2.
13. Bradbury MW, Cserr HF, Westrop RJ. Drainage of cerebral interstitial fluid into deep cervical lymph of the rabbit. Am J Physiol 1981;240:F329–36.
14. Hochwald GM, Van Driel A, Robinson ME, Thorbecke GJ. Immune response in draining lymph nodes and spleen after intraventricular injection of antigen. Int J Neurosci 1988;39(3–4):299–306.
15. Bradbury MWB, Cole DF. The role of the lymphatic system in drainage of cerebrospinal fluid and aqueous humor. J Physiol London 1980;299:353–65.
16. Kida S, Pantazis A, Weller RO. CSF drains directly from the subarachnoid space into nasal lymphatics in the rat. Anatomy, histology and immunological significance. Neuropathol Appl Neurobiol 1993;19:480–8.
17. Streilein JW. Immune privilege as the result of local tissue barriers and immunosuppressive microenvironments. Curr Opin Immunol 1993;5(3):428–32.
18. Streilein JW. Unraveling immune privilege. Science 1995;270:1158–9.
19. Carson MJ, Sutcliffe JG. Balancing function vs. self defense: the CNS as an active regulator of immune responses. J Neurosci Res 1-1-1999;55(1):1–8.
20. Sun D, Coleclough C, Whitaker JN. Nonactivated astrocytes downregulate T cell receptor expression and reduce antigen-specific proliferation and cytokine production of myelin basic protein (MBP)-reactive T cells. J Neuroimmunol 1997;78(1–2):69–78.
21. Gold R, Schmied M, Tontsch U, Hartung HP, Wekerle H, Toyka KV et al. Antigen presentation by astrocytes primes rat T lymphocytes for apoptotic cell death. A model for T-cell apoptosis in vivo. Brain 1996;119(Pt 2):651–9.
22. Carson MJ, Sutcliffe JG, Campbell IL. Microglia stimulate naive T-cell differentiation without stimulating T-cell proliferation. J Neurosci Res 1-1-1999;55(1):127–34.
23. Weber GF, Abromson-Leeman S, Cantor H. A signaling pathway coupled to T cell receptor ligation by MMTV superantigen leading to transient activation and programmed cell death. Immunity 1995;2(4):363–72.
24. Ford AL, Foulcher E, Lemckert FA, Sedgwick JD. Microglia induce CD4 T lymphocyte final effector function and death. J Exp Med 11-1-1996;184(5):1737–45.
25. Wilbanks GA, Streilein JW. Fluids from immune privileged sites endow macrophages with the capacity to induce antigen-specific immune deviation via a mechanism involving transforming growth factor-beta. Eur J Immunol 1992;22:1031–6.
26. Xiao BG, Bai XF, Zhang GX, Hojeberg B, Link H. Shift from anti- to proinflammatory cytokine profiles in microglia through LPS- or IFN-gamma-mediated pathways. Neuroreport 8-12-1996;7(12):1893–98.
27. Fabry Z, Topham DJ, Fee D, Herlein J, Carlino JA, Hart MN et al. TGF-beta 2 decreases migration of lymphocytes in vitro and homing of cells into the central nervous system in vivo. J Immunol 1995;155(1):325–32.
28. Taylor AW, Streilein JW. Inhibition of antigen-stimulated effector T cells by human cerebrospinal fluid. Neuroimmunomodulation 1996;3(2–3):112–18.

29. Carrieri PB, Provitera V, Bruno R, Perrella M, Tartaglia G, Busto A et al. Possible role of transforming growth factor-beta in relapsing-remitting multiple sclerosis. Neurol Res 1997;19(6):599–600.
30. Cua DJ, Groux H, Hinton DR, Stohlman SA, Coffman RL. Transgenic interleukin 10 prevents induction of experimental autoimmune encephalomyelitis. J Exp Med 3-15-1999;189(6):1005–10.
31. Bechmann I, Mor G, Nilsen J, Eliza M, Nitsch R, Naftolin F. FasL (CD95L, Apo1L) is expressed in the normal rat and human brain: evidence for the existence of an immunological brain barrier. Glia 1999;27(1):62–74.
32. Frigerio S, Silei V, Ciusani E, Massa G, Lauro GM, Salmaggi A. Modulation of fas-ligand (Fas-L) on human microglial cells: an in vitro study. J Neuroimmunol 2000;105(2):109–14.
33. Flugel A, Schwaiger FW, Neumann H, Medana I, Willem M, Wekerle H et al. Neuronal FasL induces cell death of encephalitogenic T lymphocytes. Brain Pathol 2000;10(3):353–64.
34. Neumann H. Control of glial immune function by neurons. Glia 2001;36(2):191–9.
35. Neumann H, Boucraut J, Hahnel C, Misgeld T, Wekerle H. Neuronal control of MHC Class II inducibility in rat astrocytes and microglia. Eur J Neurosci 1996;8:2582–90.
36. Regoli D, Boudon A, Fauchere JL. Receptors and antagonists for substance P and related peptides. Pharmacol Rev 1994;46(4):551–99.
37. McGillis JP, Humphreys S, Reid S. Characterization of functional calcitonin gene-related peptide receptors on rat lymphocytes. J Immunol 1991;147(10):3482–9.
38. Santambrogio L, Lipartiti M, Bruni A, Dal Toso R. Dopamine receptors on human T- and B-lymphocytes. J Neuroimmunol 1993;45(1–2):113–19.
39. Aune TM, McGrath KM, Sarr T, Bombara MP, Kelley KA. Expression of 5HT1a receptors on activated human T cells. Regulation of cyclic AMP levels and T cell proliferation by 5-hydroxy-tryptamine. J Immunol 1993;151(3):1175–83.
40. Elenkov IJ, Wilder RL, Chrousos GP, Vizi ES. The sympathetic nerve–an integrative interface between two supersystems: the brain and the immune system. Pharmacol Rev 2000;52(4):595–38.
41. Lee SC, Collins M, Vanguri P, Shin ML. Glutamate differentially inhibits the expression of Class II MHC antigens on astrocytes and microglia. J Immunol 1992;148(11):3391–7.
42. Frohman EM, Vayuvegula B, Gupta S, van den NS. Norepinephrine inhibits gamma-interferon-induced major histocompatibility class II (Ia) antigen expression on cultured astrocytes via beta-2-adrenergic signal transduction mechanisms. Proc Natl Acad Sci USA 1988;85(4):1292–6.
43. Neumann H, Wekerle H. Neuronal control of the immune response in the central nervous system: linking brain immunity to neurodegeneration. J Neuropathol Exp Neurol 1998;57(1):1–9.
44. Hoek RM, Ruuls SR, Murphy CA, Wright GJ, Goddard R, Zurawski SM et al. Down-regulation of the macrophage lineage through interaction with OX2 (CD200). Science 2000;290(5497):1768–71.
45. Deckert M, Sedgwick JD, Fischer E, Schluter D. Regulation of microglial cell responses in murine Toxoplasma encephalitis by CD200/CD200 receptor interaction. Acta Neuropathol (Berl) 2006;111(6):548–58.
46. Davis SJ, Ikemizu S, Wild MK, van der Merwe PA. CD2 and the nature of protein interactions mediating cell-cell recognition. Immunol Rev 1998;163:217–36.
47. Borriello F, Tizard R, Rue E, Reeves R. Characterization and localization of Mox2, the gene encoding the murine homolog of the rat MRC OX-2 membrane glycoprotein. Mamm Genome 1998;9(2):114–18.
48. Aldskogius H, Kozlova EN. Central neuron-glial and glial-glial interactions following axon injury. Prog Neurobiol 1998;55(1):1–26.
49. Darlington CL. Astrocytes as targets for neuroprotective drugs. Curr Opin Investig Drugs 2005;6(7):700–3.
50. Bush TG, Puvanachandra N, Horner CH, Polito A, Ostenfeld T, Svendsen CN et al. Leukocyte infiltration, neuronal degeneration, and neurite outgrowth after ablation of scar-forming, reactive astrocytes in adult transgenic mice. Neuron 1999;23(2):297–308.
51. Okada S, Nakamura M, Katoh H, Miyao T, Shimazaki T, Ishii K et al. Conditional ablation of Stat3 or Socs3 discloses a dual role for reactive astrocytes after spinal cord injury. Nat Med 2006;12(7):829–34.

52. Fontana A, Kristensen F, Dubs R, Gemsa D, Weber E. Production of prostaglandin E and an interleukin-1 like factor by cultured astrocytes and C6 glioma cells. J Immunol 1982;129(6): 2413–19.
53. Eddleston M, Mucke L. Molecular profile of reactive astrocytes – implications for their role in neurologic disease. Neuroscience 1993;54(1):15–36.
54. Dong Y, Benveniste EN. Immune function of astrocytes. Glia 2001;36(2):180–90.
55. Chung IY, Benveniste EN. Tumor necrosis factor-alpha production by astrocytes. Induction by lipopolysaccharide, IFN-gamma, and IL-1 beta. J Immunol 1990;144:2999–3007.
56. Raivich G, Bohatschek M, Kloss CUA, Werner A, Jones LL, Kreutzberg GW. Neuroglial activation repertoire in the injured brain: graded response, molecular mechanisms and cues to physiological function. Brain Res Rev 1999;30:77–105.
57. Kreutzberg GW. Microglia: A sensor for pathological events in the CNS. Trends in Neurosciences 1996;19(8):312–18.
58. Giulian D, Lachman LB. Interleukin-1 stimulation of astroglial proliferation after brain injury. Science 1985;228:497–9.
59. Giulian D, Baker TJ. Peptides released by ameboid microglia regulate astroglial proliferation. J Cell Biol 1985;101:2411–15.
60. Giulian D, Ingeman JE. Colony-stimulating factors as promoters of ameboid microglia. J Neurosci 1988;8:4707–17.
61. Sawada M, Suzumura A, Yamamoto HAUM. Activation and proliferation of the isolated microglia by colony stimulating factor-1 and possible involvement of protein kinase C. Brain Res 1990;509:119–24.
62. Plant SR, Arnett HA, Ting JP. Astroglial-derived lymphotoxin-alpha exacerbates inflammation and demyelination, but not remyelination. GLIA 1-1-2005;49(1):1–14.
63. Davalos D, Grutzendler J, Yang G, Kim JV, Zuo Y, Jung S et al. ATP mediates rapid microglial response to local brain injury in vivo. Nat Neurosci 2005;8(6):752–8.
64. Rabchevsky AG, Streit WJ. Grafting of cultured microglial cells into the lesioned spinal cord of adult rats enhances neurite outgrowth. J Neurosci Res 1997;47(1):34–48.
65. Giulian D, Baker TJ, Shih LN, Lachman LB. Interleukin 1 of the central nervous system is produced by ameboid microglia. J Exp Med 1986;164:594–604.
66. Smith GM, Hale JH. Macrophage/Microglia regulation of astrocytic tenascin: synergistic action of transforming growth factor-beta and basic fibroblast growth factor. J Neurosci 12-15-1997;17(24): 9624–33.
67. Fontana A, Fierz W, Wekerle H. Astrocytes present myelin basic protein to encephalitogenic T-cell lines. Nature 1984;307:273–6.
68. Wong GH, Bartlett PF, Clark-Lewis I, Battye F, Schrader JW. Inducible expression of H-2 and Ia antigens on brain cells. Nature 8-23-1984;310(5979):688–91.
69. DuBois JH, Hammond-Tooke GD, Cuzner ML. Expression of major histocompatibility complex antigens in neonate rat primary mixed glial cultures. J Neuroimmunol 1985;9(6):363–77.
70. Horwitz MS, Evans CF, Klier FG, Oldstone MB. Detailed in vivo analysis of interferon-gamma induced major histocompatibility complex expression in the central nervous system: astrocytes fail to express major histocompatibility complex class I and II molecules. Lab Invest 1999;79(2):235–42.
71. Zeinstra E, Wilczak N, Streefland C, De Keyser J. Astrocytes in chronic active multiple sclerosis plaques express MHC class II molecules. Neuroreport 1-17-2000;11(1):89–91.
72. Bo L, Mork S, Kong PA, Nyland H, Pardo CA, Trapp BD. Detection of MHC class II-antigens on macrophages and microglia, but not on astrocytes and endothelia in active multiple sclerosis lesions. J Neuroimmunol 1994;51(2):135–46.
73. Panek RB, Lee YJ, Itoh-Lindstrom Y, Ting JP, Benveniste EN. Characterization of astrocyte nuclear proteins involved in IFN-gamma- and TNF-alpha-mediated class II MHC gene expression. J Immunol 11-15-1994;153(10):4555–64.
74. Dong Y, Rohn WM, Benveniste EN. IFN-gamma regulation of the type IV class II transactivator promoter in astrocytes. J Immunol 4-15-1999;162(8):4731–9.

75. Nikcevich KM, Gordon KB, Tan L, Hurst SD, Kroepfl JF, Gardinier M et al. IFN-gamma-activated primary murine astrocytes express B7 costimulatory molecules and prime naive antigen-specific T cells. J Immunol 1-15-1997;158(2):614–21.
76. Satoh J, Lee YB, Kim SU. T-cell costimulatory molecules B7-1 (CD80) and B7-2 (CD86) are expressed in human microglia but not in astrocytes in culture. Brain Res 12-15-1995;704(1):92–6.
77. Soos JM, Ashley TA, Morrow J, Patarroyo JC, Szente BE, Zamvil SS. Differential expression of B7 co-stimulatory molecules by astrocytes correlates with T cell activation and cytokine production. Int Immunol 1999;11(7):1169–79.
78. Cross AH, Ku G. Astrocytes and central nervous system endothelial cells do not express B7-1 (CD80) or B7-2 (CD86) immunoreactivity during experimental autoimmune encephalomyelitis. J Neuroimmunol 10-2-2000;110(1–2):76–82.
79. Mueller DL, Jenkins MK, Schwartz RH. Clonal expansion versus functional clonal inactivation: a costimulatory signalling pathway determines the outcome of T cell antigen receptor occupancy. Annu Rev Immunol 1989;7:445–80.
80. Sedgwick JD, Mossner R, Schwender S, ter Meulen V. Major histocompatibility complex-expressing nonhematopoietic astroglial cells prime only CD8+ T lymphocytes: astroglial cells as perpetuators but not initiators of CD4+ T cell responses in the central nervous system. J Exp Med 1991;173(5):1235–46.
81. Weber F, MeinL E, Aloisi F, Nevinny-Stickel C, Albert E, Wekerle H et al. Human astrocytes are only partially competent antigen presenting cells. Possible implications for lesion development in multiple sclerosis. Brain 1994;117(Pt 1):59–69.
82. Matsumoto Y, Ohmori K, Fujiwara M. Immune regulation by brain cells in the central nervous system: microglia but not astrocytes present myelin basic protein to encephalitogenic T cells under in vivo-mimicking conditions. Immunology 1992;76(2):209–16.
83. Banati RB, Graeber MB. Surveillance, intervention and cytotoxicity: is there a protective role of microglia. Dev Neurosci 1994;16(3–4):114–27.
84. Lawson LJ, Perry VH, Dri P, Gordon S. Heterogeneity in the distribution and morphology of microglia in the normal adult mouse brain. Neuroscience 1990;39(1):151–70.
85. Ling EA, Wong WC. The origin and nature of ramified and amoeboid microglia: a historical review and current concepts. Glia 1993;7:9–18.
86. Nimmerjahn A, Kirchhoff F, Helmchen F. Resting microglial cells are highly dynamic surveillants of brain parenchyma in vivo. Science 2005;308(5726):1314–18.
87. Ranson PA, Thomas WE. Pinocytosis as a select marker of ramified microglia in vivo and in vitro. J Histochem Cytochem 1991;39:853–8.
88. Streit WJ, Walter SA, Pennell NA. Reactive microgliosis. Prog Neurobiol 1999;57(6):563–81.
89. Gehrmann J, Banati RB, Kreutzberg GW. Microglia in the immune surveillance of the brain: human microglia constitutively express HLA-DR molecules. J Neuroimmunol 1993;48(2):189–98.
90. Elkabes S, DiCicco-Bloom EM, Black IB. Brain microglia/macrophages express neurotrophins that selectively regulate microglial proliferation and function. J Neurosci 1996;16(8):2508–21.
91. Nakajima K, Kikuchi Y, Ikoma E, Honda S, Ishikawa M, Liu Y et al. Neurotrophins regulate the function of cultured microglia. Glia 1998;24:272–89.
92. Maeda A, Sobel RA. Matrix metalloproteinases in the normal human central nervous system, microglial nodules, and multiple sclerosis lesions. J Neuropathol Exp Neurol 1996;55(3):300–9.
93. Moffett JR, Els T, Espey MG, Walter SA, Streit WJ, Namboodiri MA. Quinolinate immunoreactivity in experimental rat brain tumors is present in macrophages but not in astrocytes. Exp Neurol 1997;144(2):287–301.
94. Vezzani A, Conti M, De Luigi A, Ravizza T, Moneta D, Marchesi F et al. Interleukin-1beta immunoreactivity and microglia are enhanced in the rat hippocampus by focal kainate application: functional evidence for enhancement of electrographic seizures. J Neurosci 1999;19(12):5054–65.
95. Possel H, Noack H, Putzke J, Wolf G, Sies H. Selective upregulation of inducible nitric oxide synthase (iNOS) by lipopolysaccharide (LPS) and cytokines in microglia: in vitro and in vivo studies. Glia 2000;32(1):51–9.
96. Banati RB, Gehrmann J, Schubert P, Kreutzberg GW. Cytotoxicity of microglia. Glia 1993;7:111–18.

97. Heyes MP. The kynurenine pathway and neurologic disease. Therapeutic strategies. Adv Exp Med Biol 1996;398:125–9.
98. Benveniste EN. Cytokine actions in the central nervous system. Cytokine Growth Factor Rev 1998;9(3–4):259–75.
99. Vass K, Lassmann H. Intrathecal application of interferon gamma. Progressive appearance of MHC antigens within the rat nervous system. Am J Pathol 1990;137(4):789–800.
100. Andersson PB, Perry VH, Gordon S. Intracerebral injection of proinflammatory cytokines or leukocyte chemotaxins induces minimal myelomonocytic cell recruitment to the parenchyma of the central nervous system. J Exp Med 1992;176(1):255–9.
101. Kim WG, Mohney RP, Wilson B, Jeohn GH, Liu B, Hong JS. Regional difference in susceptibility to lipopolysaccharide-induced neurotoxicity in the rat brain: role of microglia. J Neurosci 2000;20(16):6309–16.
102. Stern EL, Quan N, Proescholdt MG, Herkenham M. Spatiotemporal induction patterns of cytokine and related immune signal molecule mRNAs in response to intrastriatal injection of lipopolysaccharide [In Process Citation]. J Neuroimmunol 2000;109(2):245–60.
103. Streit WJ, Kreutzberg GW. Response of endogenous glial cells to motor neuron degeneration induced by toxic ricin. J Comp Neurol 1988;268(2):248–63.
104. Shuman SL, Bresnahan JC, Beattie MS. Apoptosis of microglia and oligodendrocytes after spinal cord contusion in rats. J Neurosci Res 1997;50(5):798–808.
105. Zajicek JP, Wing M, Scolding NJ, Compston DA. Interactions between oligodendrocytes and microglia. A major role for complement and tumour necrosis factor in oligodendrocyte adherence and killing. Brain 1992;115(Pt 6):1611–31.
106. Gregersen R, Lambertsen K, Finsen B. Microglia and macrophages are the major source of tumor necrosis factor in permanent middle cerebral artery occlusion in mice. J Cereb Blood Flow Metab 2000;20(1):53–65.
107. Bruce AJ, Boling W, Kindy MS, Peschon J, Kraemer PJ, Carpenter MK et al. Altered neuronal and microglial responses to excitotoxic and ischemic brain injury in mice lacking TNF receptors. Nat Med 1996;2(7):788–94.
108. Liu J, Marino MW, Wong G, Grail D, Dunn A, Bettadapura J et al. TNF is a potent anti-inflammatory cytokine in autoimmune-mediated demyelination. Nat Med 1998;4(1):78–83.
109. Arnett HA, Mason J, Marino M, Suzuki K, Matsushima GK, Ting JP. TNF alpha promotes proliferation of oligodendrocyte progenitors and remyelination. Nat Neurosci 2001;4(11):1116–22.
110. Li WW, Setzu A, Zhao C, Franklin RJ. Minocycline-mediated inhibition of microglia activation impairs oligodendrocyte progenitor cell responses and remyelination in a non-immune model of demyelination. J Neuroimmunol 2005;158(1–2):58–66.
111. Lodge PA, Sriram S. Regulation of microglial activation by TGF-b, IL-10, and CSF-1. J Leukoc Biol 1996;60:502–8.
112. Tan J, Town T, Saxe M, Paris D, Wu Y, Mullan M. Ligation of microglial CD40 results in p44/42 mitogen-activated protein kinase-dependent TNF-alpha production that is opposed by TGF–beta 1 and IL-10. J Immunol 1999;163(12):6614–21.
113. Novikova LN, Novikov LN, Kellerth JO. Survival effects of BDNF and NT-3 on axotomized rubrospinal neurons depend on the temporal pattern of neurotrophin administration. Eur J Neurosci 2000;12(2):776–80.
114. McTigue DM, Horner PJ, Stokes BT, Gage FH. Neurotrophin-3 and brain-derived neurotrophic factor induce oligodendrocyte proliferation and myelination of regenerating axons in the contused adult rat spinal cord. J Neurosci 1998;18(14):5354–65.
115. De Simone R, Giampaolo A, Giometto B, Gallo P, Levi G, Peschle C et al. The costimulatory molecule B7 is expressed on human microglia in culture and in multiple sclerosis acute lesions. J Neuropathol Exp Neurol 1995;54(2):175–87.
116. Issazadeh S, Navikas V, Schaub M, Sayegh M, Khoury S. Kinetics of expression of costimulatory molecules and their ligands in murine relapsing experimental autoimmune encephalomyelitis in vivo. J Immunol 8-1-1998;161(3):1104–12.

117. Frei K, Siepl C, Groscurth P, Bodmer S, Schwerdel C, Fontana A. Antigen presentation and tumor cytotoxicity by interferon-gamma-treated microglial cells. Eur J Immunol 1987;17:1271–8.
118. Aloisi F, Ria F, Penna G, Adorini L. Microglia are more efficient than astrocytes in antigen processing and in Th1 but not Th2 cell activation. J Immunol 1998;160(10):4671–80.
119. Cash E, Rott O. Microglial cells qualify as the stimulators of unprimed CD4+ and CD8+ T lymphocytes in the central nervous system. Clin Exp Immunol 1994;98(2):313–18.
120. Sedgwick JD, Ford AL, Foulcher E, Airriess R. Central nervous system microglial cell activation and proliferation follows direct interaction with tissue-infiltrating T cell blasts. J Immunol 1998;160:5320–30.
121. Magnus T, Chan A, Grauer O, Toyka KV, Gold R. Microglial phagocytosis of apoptotic inflammatory T cells leads to down-regulation of microglial immune activation. J Immunol 2001;167(9):5004–10.
122. Chan A, Hummel V, Weilbach FX, Kieseier BC, Gold R. Phagocytosis of apoptotic inflammatory cells downregulates microglial chemoattractive function and migration of encephalitogenic T cells. J Neurosci Res 2006;84(6):1217–24.
123. Williams K, Ulvestad E, Antel JP. B7/BB-1 antigen expression on adult human microglia studied in vitro and in situ. Eur J Immunol 1994;24(12):3031–7.
124. Bechmann I, Kwidzinski E, Kovac AD, Simburger E, Horvath T, Gimsa U et al. Turnover of rat brain perivascular cells. Exp Neurol 2001;168(2):242–9.
125. Allan SM, Allan SM, Rothwell NJ. Cytokines and acute neurodegeneration. Nat Rev Neurosci 2001;2(10):734–44.
126. Minami M, Katayama T, Satoh M. Brain cytokines and chemokines: roles in ischemic injury and pain. J Pharmacol Sci 2006;100(5):461–70.
127. Biber K, de Jong EK, van Weering HR, Boddeke HW. Chemokines and their receptors in central nervous system disease. Curr Drug Targets 2006;7(1):29–46.
128. Barone FC, Feuerstein GZ. Inflammatory mediators and stroke: new opportunities for novel therapeutics. J Cereb Blood Flow Metab 1999;19(8):819–34.
129. Ransohoff RM, Tani M. Do chemokines mediate leukocyte recruitment in post-traumatic CNS inflammation? Trends Neurosci 1998;21(4):154–9.
130. Asensio VC, Campbell IL. Chemokines in the CNS: plurifunctional mediators in diverse states. Trends Neurosci 1999;22(11):504–12.
131. Streit WJ, Semple-Rowland SL, Hurley SD, Miller RC, Popovich PG, Stokes BT. Cytokine mRNA profiles in contused spinal cord and axotomized facial nucleus suggest a beneficial role for inflammation and gliosis. Exp Neurol 1998;152(1):74–87.
132. Owens T, Wekerle H, Antel J. Genetic models for CNS inflammation. Nat Med 2001;7(2):161–6.
133. Wang X, Yue TL, Barone FC, White RF, Gagnon RC, Feuerstein GZ. Concomitant cortical expression of TNF-alpha and IL-1 beta mRNAs follows early response gene expression in transient focal ischemia. Mol Chem Neuropathol 1994;23(2–3):103–14.
134. Laman JD, Van Meurs M, Schellekens MM, de Boer M, Melchers B, Massacesi L et al. Expression of accessory molecules and cytokines in acute EAE in marmoset monkeys (Callithrix jacchus). J Neuroimmunol 6-1-1998;86(1):30–45.
135. Kamm K, Vanderkolk W, Lawrence C, Jonker M, Davis AT. The effect of traumatic brain injury upon the concentration and expression of interleukin-1beta and interleukin-10 in the rat. J Trauma 2006;60(1):152–7.
136. Vitarbo EA, Chatzipanteli K, Kinoshita K, Truettner JS, Alonso OF, Dietrich WD. Tumor necrosis factor alpha expression and protein levels after fluid percussion injury in rats: the effect of injury severity and brain temperature. Neurosurgery 2004;55(2):416–24.
137. Relton JK, Rothwell NJ. Interleukin-1 receptor antagonist inhibits ischaemic and excitotoxic neuronal damage in the rat. Brain Res Bull 1992;29(2):243–6.
138. Toulmond S, Rothwell NJ. Interleukin-1 receptor antagonist inhibits neuronal damage caused by fluid percussion injury in the rat. Brain Res 2-13-1995;671(2):261–6.
139. Martin D, Near SL. Protective effect of the interleukin-1 receptor antagonist (IL-1ra) on experimental allergic encephalomyelitis in rats. J Neuroimmunol 1995;61(2):241–5.

140. Mrak RE, Griffin WS. Interleukin-1, neuroinflammation, and Alzheimer's disease. Neurobiol Aging 2001;22(6):903–8.
141. Selmaj KW, Raine CS. Tumor necrosis factor mediates myelin and oligodendrocyte damage in vitro. Ann Neurol 1988;23:339–46.
142. Bitsch A, Kuhlmann T, Da Costa C, Bunkowski S, Polak T, Bruck W. Tumour necrosis factor alpha mRNA expression in early multiple sclerosis lesions: correlation with demyelinating activity and oligodendrocyte pathology. Glia 2000;29(4):366–75.
143. Selmaj K, Raine CS, Cannella B, Brosnan CF. Identification of lymphotoxin and tumor necrosis factor in multiple sclerosis lesions. J Clin Invest 1991;87:949–54.
144. Hofman FM, Hinton DR, Johnson K, Merrill JE. Tumor necrosis factor identified in multiple sclerosis brain. J Exp Med 1989;170(2):607–12.
145. Nawashiro H, Martin D, Hallenbeck JM. Inhibition of tumor necrosis factor and amelioration of brain infarction in mice. J Cereb Blood Flow Metab 1997;17(2):229–32.
146. Mayne M, Ni W, Yan HJ, Xue M, Johnston JB, Del Bigio MR et al. Antisense oligodeoxynucleotide inhibition of tumor necrosis factor-alpha expression is neuroprotective after intracerebral hemorrhage. Stroke 2001;32(1):240–8.
147. Scherbel U, Raghupathi R, Nakamura M, Saatman KE, Trojanowski JQ, Neugebauer E et al. Differential acute and chronic responses of tumor necrosis factor-deficient mice to experimental brain injury. Proc Natl Acad Sci USA 1999;96(15):8721–6.
148. Korner H, Lemckert FA, Chaudhri G, Etteldorf S, Sedgwick JD. Tumor necrosis factor blockade in actively induced experimental autoimmune encephalomyelitis prevents clinical disease despite activated T cell infiltration to the central nervous system. Eur J Immunol 1997;27(8): 1973–81.
149. Selmaj KW, Raine CS. Experimental autoimmune encephalomyelitis: immunotherapy with anti-tumor necrosis factor antibodies and soluble tumor necrosis factor receptors. Neurology 1995; 45(6 Suppl 6):S44–9.
150. Kim GM, Xu J, Xu J, Song SK, Yan P, Ku G et al. Tumor necrosis factor receptor deletion reduces nuclear factor-kappaB activation, cellular inhibitor of apoptosis protein 2 expression, and functional recovery after traumatic spinal cord injury. J Neurosci 2001;21(17):6617–25.
151. Mason JL, Suzuki K, Chaplin DD, Matsushima GK. Interleukin-1beta promotes repair of the CNS. J Neurosci 2001;21(18):7046–52.
152. Luster AD. Chemokines-chemotactic cytokines that mediate inflammation. N Engl J Med 1998;338(7):436–45.
153. Ransohoff RM. Chemokines in neurological disease models: correlation between chemokine expression patterns and inflammatory pathology. J Leukoc Biol 1997;62(5):645–52.
154. Glabinski AR, Tani M, Tuohy VK, Tuthill RJ, Ransohoff RM. Central nervous system chemokine mRNA accumulation follows initial leukocyte entry at the onset of acute murine experimental autoimmune encephalomyelitis. Brain Behav Immun 1995;9(4):315–30.
155. Ransohoff RM, Hamilton TA, Tani M, Stoler MH, Shick HE, Major JA et al. Astrocyte expression of mRNA encoding cytokines IP-10 and JE/MCP- 1 in experimental autoimmune encephalomyelitis. FASEB J 1993;7(6):592–600.
156. Ransohoff RM. Mechanisms of inflammation in MS tissue: adhesion molecules and chemokines. J Neuroimmunol 1999;98(1):57–68.
157. Simpson JE, Newcombe J, Cuzner ML, Woodroofe MN. Expression of monocyte chemoattractant protein-1 and other beta-chemokines by resident glia and inflammatory cells in multiple sclerosis lesions. J Neuroimmunol 1998;84(2):238–49.
158. Glabinski AR, Balasingam V, Tani M, Kunkel SL, Strieter RM, Yong VW et al. Chemokine monocyte chemoattractant protein-1 is expressed by astrocytes after mechanical injury to the brain. J Immunol 1996;156(11):4363–8.
159. Lee YL, Shih K, Bao P, Ghirnikar RS, Eng LF. Cytokine chemokine expression in contused rat spinal cord. Neurochem Int 2000;36(4–5):417–25.

160. McTigue DM, Tani M, Krivacic K, Chernosky A, Kelner GS, Maciejewski D et al. Selective chemokine mRNA accumulation in the rat spinal cord after contusion injury. J Neurosci Res 1998;53(3):368–76.
161. Ghirnikar RS, Lee YL, Eng LF. Chemokine antagonist infusion promotes axonal sparing after spinal cord contusion injury in rat. J Neurosci Res 2001;64(6):582–9.
162. Bartholdi D, Schwab ME. Expression of pro-inflammatory cytokine and chemokine mRNA upon experimental spinal cord injury in mouse: an in situ hybridization study. Eur J Neurosci 1997;9(7):1422–38.
163. Glaser J, Gonzalez R, Perreau VM, Cotman CW, Keirstead HS. Neutralization of the chemokine CXCL10 enhances tissue sparing and angiogenesis following spinal cord injury. J Neurosci Res 9-1-2004;77(5):701–8.
164. Beech JS, Reckless J, Mosedale DE, Grainger DJ, Williams SC, Menon DK. Neuroprotection in ischemia-reperfusion injury: an anti-inflammatory approach using a novel broad-spectrum chemokine inhibitor. J Cereb Blood Flow Metab 2001;21(6):683–9.
165. Ishizuka K, Kimura T, Igata-yi R, Katsuragi S, Takamatsu J, Miyakawa T. Identification of monocyte chemoattractant protein-1 in senile plaques and reactive microglia of Alzheimer's disease. Psychiatry Clin Neurosci 1997;51(3):135–8.
166. Fuentes ME, Durham SK, Swerdel MR, Lewin AC, Barton DS, Megill JR et al. Controlled recruitment of monocytes and macrophages to specific organs through transgenic expression of monocyte chemoattractant protein-1. J Immunol 1995;155(12):5769–76.
167. Tani M, Fuentes ME, Peterson JW, Trapp BD, Durham SK, Loy JK et al. Neutrophil infiltration, glial reaction, and neurological disease in transgenic mice expressing the chemokine N51/KC in oligodendrocytes. J Clin Invest 1996;98(2):529–39.
168. Cardona AE, Pioro EP, Sasse ME, Kostenko V, Cardona SM, Dijkstra IM et al. Control of microglial neurotoxicity by the fractalkine receptor. Nat Neurosci 2006;9(7):917–24.
169. Chapman GA, Moores K, Harrison D, Campbell CA, Stewart BR, Strijbos PJ. Fractalkine Cleavage from Neuronal Membranes Represents an Acute Event in the Inflammatory Response to Excitotoxic Brain Damage. J Neurosci (Online) 8-1-2000;20(15):RC87.
170. Ubogu EE, Cossoy MB, Ransohoff RM. The expression and function of chemokines involved in CNS inflammation. Trends Pharmacol Sci 2006;27(1):48–55.
171. Archelos JJ, Previtali SC, Hartung HP. The role of integrins in immune-mediated diseases of the nervous system. Trends Neurosci 1999;22(1):30–8.
172. Popovich PG, Wei P, Stokes BT. Cellular inflammatory response after spinal cord injury in Sprague-Dawley and Lewis rats. J Comp Neurol 1997;377(3):443–64.
173. Liu T, Young PR, McDonnell PC, White RF, Barone FC, Feuerstein GZ. Cytokine-induced neutrophil chemoattractant mRNA expressed in cerebral ischemia. Neurosci Lett 1993;164(1–2):125–8.
174. Tonai T, Shiba K, Taketani Y, Ohmoto Y, Murata K, Muraguchi M et al. A neutrophil elastase inhibitor (ONO-5046) reduces neurologic damage after spinal cord injury in rats. J Neurochem 2001;78(5):1064–72.
175. Taoka Y, Okajima K. Spinal cord injury in the rat. Prog Neurobiol 1998;56:341–58.
176. Connolly ES, Jr., Winfree CJ, Springer TA, Naka Y, Liao H et al. Cerebral protection in homozygous null ICAM-1 mice after middle cerebral artery occlusion. Role of neutrophil adhesion in the pathogenesis of stroke. J Clin Invest 1996;97(1):209–16.
177. Zehntner SP, Brickman C, Bourbonniere L, Remington L, Caruso M, Owens T. Neutrophils that infiltrate the central nervous system regulate T cell responses. J Immunol 2005;174(8):5124–31.
178. McColl SR, Staykova MA, Wozniak A, Fordham S, Bruce J, Willenborg DO. Treatment with anti-granulocyte antibodies inhibits the effector phase of experimental autoimmune encephalomyelitis. J Immunol 1998;161(11):6421–6.
179. Liu XZ, Xu XM, Hu R, Du C, Zhang SX, McDonald JW et al. Neuronal and glial apoptosis after traumatic spinal cord injury. J Neurosci 1997;17(14):5395–06.
180. Blight AR. Effects of silica on the outcome from experimental spinal cord injury: implication of macrophages in secondary tissue damage. Neuroscience 1994;60(1):263–73.

181. Giulian D, Robertson C. Inhibition of mononuclear phagocytes reduces ischemic injury in the spinal cord. Ann Neurol 1990;27:33–42.
182. Popovich PG, Guan Z, Wei P, Huitinga I, van Rooijen N, Stokes BT. Depletion of hematogenous macrophages promotes partial hind-limb recovery and neuroanatomical repair after experimental spinal cord injury. Exp Neurol 1999;158:351–65.
183. Schroeter M, Jander S, Huitinga I, Witte OW, Stoll G. Phagocytic response in photochemically induced infarction of rat cerebral cortex. The role of resident microglia. Stroke 1997;28(2):382–6.
184 Huitinga I, van Rooijen N, de Groot CJ, Uitdehaag BM, Dijkstra CD. Suppression of experimental allergic encephalomyelitis in Lewis rats after elimination of macrophages. J Exp Med 1990;172:1025–33.
185. Lazarov-Spiegler O, Rapalino O, Agranov G, Schwartz M. Restricted inflammatory response in the CNS: a key impediment to axonal regeneration. Mol Med Today 1998;4:337–42.
186. Batchelor PE, Liberatore GT, Wong JYF, Porritt MJ, Frerichs F, Donnan GA et al. Activated macrophages and microglia induce dopaminergic sprouting in the injured striatum and express brain-derived neurotrophic factor and glial cell line-derived neurotrophic factor. J Neurosci 1999;19(5):1708–16.
187. Schwartz M, Lazarov-Spiegler O, Rapalino O, Agranov I, Velan G, Hadani M. Potential repair of rat spinal cord injuries using stimulated homologous macrophages. Neurosurgery 1999;44(5):1041–6.
188. Kotter MR, Setzu A, Sim FJ, van Rooijen N, Franklin RJ. Macrophage depletion impairs oligodendrocyte remyelination following lysolecithin-induced demyelination. Glia 2001;35(3):204–12.
189. Popovich PG, Hickey WF. Bone marrow chimeric rats reveal the unique distribution of resident and recruited macrophages in the contused rat spinal cord. J Neuropathol Exp Neurol 2001;60(7):676–85.
190. Schroeter M, Jander S, Huitinga I, Stoll G. CD8+ phagocytes in focal ischemia of the rat brain: predominant origin from hematogenous macrophages and targeting to areas of pannecrosis. Acta Neuropathol (Berl) 2001;101(5):440–8.
191. Serpe CJ, Kohm AP, Huppenbauer CB, Sanders VM, Jones KJ. Exacerbation of facial motoneuron loss after facial nerve transection in severe conbined immunodefiecient (scid) mice. J Neurosci 1999;19(RC7):1–5.
192. Serpe CJ, Sanders VM, Jones KJ. Kinetics of facial motoneuron loss following facial nerve transection in severe combined immunodeficient mice. J Neurosci Res 2000;62(2):273–8.
193. Moalem G, Leibowitz-Amit R, Yoles E, Mor F, Cohen IR, Schwartz M. Autoimmune T-cells protect neurons from secondary degeneration after central nervous system axotomy. Nature Med 1999;5(1):49–55.
194. Moalem G, Yoles E, Leibowitz-Amit R, Muller-Gilor S, Mor F, Cohen IR et al. Autoimmune T cells retard the loss of function in injured rat optic nerves. J Neuroimmunol 2000;106(1–2):189–97.
195. Moalem G, Gdalyahu A, Leibowitz-Amit R, Yoles E, Muller-Gilor S, Shani Y et al. Production of neurotrophins by activated T-cells: implications for neuroprotective autoimmunity. J Autoimmun 2000;15:331–45.
196. Jones TB, Ankeny DP, Guan Z, McGaughy V, Fisher LC, Basso DM et al. Passive or active immunization with myelin basic protein impairs neurological function and exacerbates neuropathology after spinal cord injury in rats. J Neurosci 2004;24:3752–76.

# II. EFFECT OF THE HPA AXIS AND THE SNS ON THE IMMUNE SYSTEM

*The Hypothalamus–Pituitary–Adrenal Axis*
Edited by A. del Rey, G. P. Chrousos and H. O. Besedovsky



# *Glucocorticoid Effects on Immune Responses*

PAUL M. GUYRE[1], MARK P. YEAGER[2], and ALLAN MUNCK[3]

[1]*Department of Physiology, Dartmouth Medical School, Lebanon*
[2]*Dartmouth Medical School, Lebanon*
[3]*Emeritus Department of Physiology, Dartmouth Medical School, Lebanon*

## ABSTRACT

Recent studies have brought additional clarity to the mechanisms by which glucocorticoid (GC) steroid hormones exert both suppressive and permissive effects on immune responses. Widely used for their immunosuppressive and anti-inflammatory actions, stress-elevated and therapeutic levels of GCs suppress defense mechanisms that would otherwise overshoot and damage the organism. Basal levels, on the other hand, are essential for permissive effects that prime defense mechanisms for more efficient action against a challenge. Both suppressive and permissive actions on immunity and inflammation have been documented in intact organisms. Suppressive actions are often due to inhibition by GCs of the production of cytokines and other mediators, whereas permissive effects include both induction of cytokine receptors and enhancement of activity of certain cytokines. GCs interact with both innate and adaptive immune responses. This is exemplified by immune–adrenal crosstalk wherein exposure to endotoxin stimulates the hypothalamic–pituitary–adrenal (HPA) axis and GC production, largely via endotoxin-induced cytokines. These stress-induced levels of GCs are, in turn, essential for survival in the face of endotoxic shock. GCs influence adaptive immunity in many ways. They inhibit antigen-specific T-lymphocyte activation, slow or redirect the maturation of macrophages and dendritic cells, favor Th2 over Th1 responses, and enhance the development as well as actions of specialized T-regulatory cells. GCs also induce T-cell apoptosis selectively on different T-cell populations, and can protect thymocytes from apoptosis induced by activation of the T-cell receptor. Moreover, experiments with rats that are highly susceptible to autoimmune disorders demonstrate the importance of a normal stress-induced increase in GC levels for the physiological control of autoimmunity. At the molecular level, at least three mechanisms contribute to GC effects on immunity and inflammation. Transactivation and transrepression are the "classical" mechanisms whereby ligand-activated glucocorticoid receptors (GRs) bind to DNA sequences called GC response elements and either activate or repress transcription of the targeted gene. Transactivation of genes encoding inhibitory proteins and transrepression of inflammatory genes have been described. However, the majority of anti-inflammatory effects are due to so-called cross-talk in which GC-liganded GRs interact with transcription factor proteins such as nuclear factor-κB (NF-κB) and activator protein-1(AP-1), interfering with their ability to activate transcription of target genes.

## 1. INTRODUCTION

GCs made headlines in 1949 with the announcement that in high doses they produced dramatic anti-inflammatory effects [1]. Almost overnight these hitherto obscure hormones became miracle drugs. Clinicians and their patients celebrated the news. Physiologists were bewildered. They had believed that the high levels of glucocorticoids (GCs) induced by stress, which were known to protect against stress, did so by enhancing – not suppressing – defense mechanisms like inflammation. Consequently for decades, despite a penetrating but almost universally overlooked 1951 review by Marius Tausk [2], physiologists stigmatized anti-inflammatory and immunosuppressive effects as high-dose pharmacological artifacts [3]. The major clinical applications of GCs were thus left without apparent mechanisms based on physiology, a remarkable situation unique among hormones.

In 1984 we proposed that stress-induced levels of GCs in reality benefit the organism by suppressing stress-activated defense mechanisms, preventing them from overshooting and causing damage [3]. That idea, applied to cardiovascular, neural, immune, and other physiological defense systems, has been supported experimentally and won wide acceptance [4]. Anti-inflammatory and immunosuppressive actions are seen in that context as prototypical physiological effects that limit tissue injury and prevent development of autoimmunity during an immune response. A similar view of immunosuppressive actions had been advanced earlier by Besedovsky and Sorkin [5] and was already explicit in Tausk's review [2].

Although stress-induced levels of GCs are usually suppressive, basal levels can be enhancing or "permissive." Permissive actions of GCs, originally described by Ingle, are essential for priming many defense systems for rapid action against a challenge. They are almost certainly required for the immune response. Schematically, we can visualize the physiological actions of GCs on the activity of a hypothetical defense mechanism as illustrated in Fig. 1 [4,6,7]. As basal

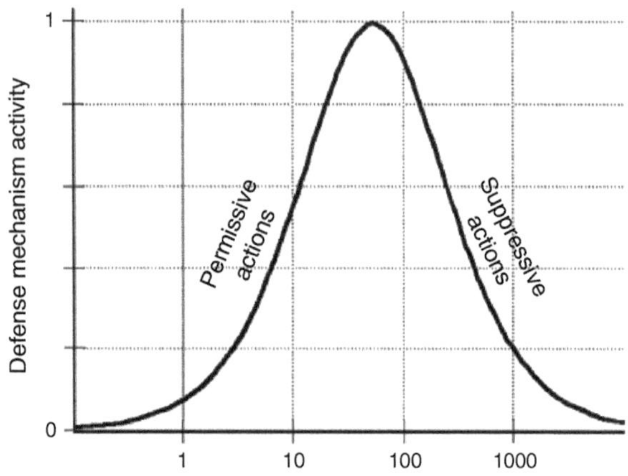


Figure 1. Regulation by GCs of defense mechanisms through permissive and suppressive actions. The bell-shaped curve, derived from a mathematical model, depicts how cortisol regulates the activity (arbitrary scale) of a defense mechanism composed of a cytokine, its receptor, and the cytokine-receptor complex responsible for activity [6]. Cortisol is assumed to permissively enhance activity by inducing cytokine receptors, and to suppress activity by lowering cytokine levels. $Kd = 30\,\text{nM}$ for binding of cortisol to GC receptors is used in the calculations. Some permissive actions of GCs may be mediated through mineralocorticoid receptors, for which cortisol and corticosterone have much higher affinities. The bell-shaped curve in that case is shifted to the left by roughly a factor of 10 in cortisol concentration [4]. In humans, basal free cortisol levels vary from almost immeasurable in the evening up to about 50 nM in the morning. Stress-induced levels can exceed 1000 nM.

levels of cortisol (or corticosterone, the other natural GC) rise in the course of normal diurnal variation to a maximum at around the start of daily activity (morning for humans, evening for mice and rats), permissive actions raise activity to a peak. If cortisol levels increase further, suppressive actions dominate, generating a bell-shaped curve. In the mathematical model underlying Fig. 1 [6], activity is proportional to binding of a cytokine or other mediator to its receptor. At lower concentrations, GCs act permissively to induce cytokine receptors and thereby increase cytokine signaling. At higher concentrations, cytokine effects are suppressed by GC inhibition of cytokine production. There is much evidence for both these kinds of actions, but undoubtedly many other molecular mechanisms for permissive and suppressive actions come into play. Within this overall framework, we now update and broaden our earlier contribution to Neuroimmune Biology. We touch on a wide range of the effects of GCs on immune reactions that are addressed in more detail by other authors both in this volume and elsewhere [8–12].

## 2. GC EFFECTS ON CYTOKINES AND OTHER IMMUNE AND INFLAMMATORY MEDIATORS

The most general signaling pathways through which GCs influence immune and inflammatory responses, summarized in Table 1, are those of cytokines and other mediators, and their receptors. GCs suppress levels of mediators that promote those responses by inhibiting their synthesis, release, or efficacy.

Although initial observations on GC suppression of mediators were made with cell cultures, there is ample evidence that these responses occur in intact organisms [4], and it has been noted that production of interferon-γ (IFN-γ), tumor necrosis factor-α (TNF-α), interleukin (IL)-1, IL-6, and IL-12 in whole blood from normal human subjects reaches peak values during the

Table 1 Cytokines and other mediators suppressed by glucocorticoids[a]

| Cytokines | Chemokines | Inflammatory agents | Hormones neurotransmitters | Cell adhesion molecules |
|---|---|---|---|---|
| IL-1b | IL-8 | Eicosanoids | CRHb | ICAM-1 |
| IL-2b | MCP-1 | Bradykinin | VPb | LFA-1 |
| IL-3 | RANTES | 5-HTb | Oxytocin | E-Selectin |
| IL-4b(+−) | MIP-1α | Histamine | ACTH | |
| IL-5 | CINC/gro | Collagenase | β-endorphin | |
| IL-6b | SCF | Elastase | Substance P | |
| IL-8 | LIX | MMP-9 | LH | |
| IL-10(+−) | | Plasminogen activator | Insulinb | |
| IL-11 | | | Norepinephrineb | |
| IL-12 | | | Epinephrine(+−) | |
| IL-16 | | | Nitric oxide | |
| IFN-γ[b] | | | | |
| TNF-α | | | | |
| GM-CSFb | | | | |
| TGF-β(+−) | | | | |

(+−) GCs suppress mediator in some cases and enhance in others.
[a] Reviewed or reported in Refs [4,13–24].
[b] GCs induce receptors for these mediators.

night and early morning, when plasma cortisol levels are lowest [25]. This is also the time at which it has been shown, experimentally, that animals are most susceptible to inflammatory effects of administered cytokines [26]. Decreased production of cytokines and inflammatory agents goes a long way toward explaining how GCs suppress immune and inflammatory reactions, while effects on chemokines and cell adhesion molecules may account for how GCs affect cell traffic. Activities of some mediators like the cytokines IL-4 and IL-10 can be suppressed or enhanced by GCs, depending on the stage of cell activation and whether a high concentration of GC is present before the cell stimulus or after the stimulus [27,28].

As also indicated in Table 1, GCs raise levels of several mediator receptors. Besides the cases in Table 1, which lists only mediators suppressed by GCs, GCs have been shown to induce the chemokine receptor CXCR4 [29] and receptors for IL-7 [30]. Upregulation of hormone, cytokine, and growth factor receptors by GCs may underlie their permissive activation of several physiological systems. GC upregulation of receptors for GM-CSF (granulocyte-macrophage colony-stimulating factor) has been proposed to account for synergism of GCs with GM-CSF to increase MHC class II expression [31]. Beneficial actions of GCs in culture media may also stem from such effects [32]. Through their differing effects on cytokines and their receptors, GCs have been proposed not only to affect the magnitude or direction of a cytokine-induced response, but also to increase the rate at which the response proceeds [13].

At the cellular level, not only are cytokine actions regulated by GCs, but GC actions may be regulated by cytokines. For example, TNF-α increases the transcription-regulating effects of GCs in several cell types [33], whereas IL-2 decreases it, possibly by direct interaction of the signal transducer and activator of transcription STAT5 with the GR [34]. TNF-α furthermore can increase the concentration of cortisol or corticosterone in certain cells by increasing levels of 11β-hydroxysteroid dehydrogenase type 1 or decreasing levels of 11β-hydroxysteroid dehydrogenase type 2 [35]. The type 1 enzyme that converts inactive GCs to active ones, in turn, has been shown to increase macrophage phagocytosis of apoptotic neutrophils and thus plays an essential role in the resolution of inflammation [36,37]. Macrophage migration inhibitory factor (MIF) is another cytokine with important effects on GC actions. Induced by GCs, MIF is a proinflammatory mediator with the ability to induce various immunomodulatory responses and override GC-driven immunosuppression [38–41]. MIF potentiates cytokine production that is induced by LPS (bacterial lipopolysaccharide or endotoxin) [42], and may block or directly counter-regulate GC inhibition of other cytokines. For example, using macrophages from MIF-/- mice, Aeberli and co-workers demonstrated that MIF impairs GC inhibition of LPS-induced TNF release. Interestingly, MIF deficiency was associated with increased expression of mitogen-activated protein kinase (MAPK) phosphatase (MKP)-1, concomitant with decreased phosphorylation of p38 MAPK. However, there appeared to be no effect of MIF on nuclear factor-κB (NF-κB) [43,44]. As noted below, MKP-1 now appears to be a key anti-inflammatory protein that plays a central role in the cross-talk between GRs and the Toll-like receptors (TLR) of the innate immune system [45,46].

In the acute phase response, the general response to immune and inflammatory reactions that is set off by injury or infection, GCs appear to play both permissive and suppressive roles. They increase the sensitivity to mediators like IL-1 and TNF-α that are released into the circulation, and stimulate synthesis in the liver of acute phase proteins such as serum amyloid A, C-reactive protein, and components of complement [47,48]. This effect may be due, at least in part, to upregulation of receptors for cytokines that directly stimulate the acute phase reaction [49]. At the same time, GCs control the overall response by suppressing mediator production [47].

## 3. GC EFFECTS ON INNATE IMMUNITY

The innate immune system uses a limited number of germline-encoded and thus invariant receptors (pattern recognition receptors [PRRs]) to detect and respond to pathogen-associated molecular patterns (PAMPs). PAMPs are unique to microbes, are shared by both pathogens and non-pathogens, and are identical among microbes of a given class [50]. PAMPs signal through PRRs to rapidly stimulate protective responses by leukocytes, endothelial, epithelial, and stromal cells. These responses result in activation of the complement and coagulation pathways, opsonization of the microbe, cytokine and chemokine production, and upregulation of adhesion molecules and receptors – all leading to enhanced leukocyte migration, phagocytosis, and killing of the infectious agent [51]. Perhaps the best-known example of a PAMP is LPS which binds to TLR4, a PRR that plays a crucial role in immune adrenal cross-talk (see chapters in this volume by Heeg, Moynagh, and Dhabar as well as Refs [45,52,53]).

The critical role of GCs in regulating the innate response to PAMPs has been demonstrated in both human and animal systems. A GC surge is a crucial component of the physiological response to endotoxin, and endotoxin-induced lethality is nearly certain in its absence [54]. Mice engineered to have overexpressed GC receptors (GRs) show enhanced resistance to endotoxic shock, with significantly lower production of the proinflammatory cytokine IL-6 [55]. Conversely, macrophage-specific GR knockout results in greater LPS-induced mortality and cytokine production, as well as markedly diminished inhibition of inflammatory genes by the synthetic GC dexamethasone (Dex) [56]. GC regulation of the anti-inflammatory cytokine IL-10 can also be an important variable in the innate response to LPS since neutralization of IL-10 considerably enhances lethal responses to LPS in mice [57,58], whereas IL-10 administration is protective [59,60]. GCs have been reported to upregulate constitutive IL-10 production [61]. Moreover, an endogenous GC surge during human cardiac surgery is required for optimal IL-10 production in that setting since, in the absence of a surge, IL-10 production is minimal [62]. In both human cardiac surgery and human experimental endotoxemia, high anti-inflammatory levels of GCs not only inhibit TNF-$\alpha$, IL-6, and IL-8, but also enhance IL-10 production in a highly dose-dependent manner [63]. Thus GCs not only inhibit, but also enhance production of cytokines, depending upon the context.

There is compelling evidence that the innate immune response plays an instructive role in adaptive immunity [64]. For example, environmental exposure to LPS, a prototypical PAMP, seems to play a crucial role in the development of allergies. Higher endotoxin levels in dust samples of children's mattresses were associated with lower cytokine production by leukocytes as well as lower occurrence of hay fever, atopic asthma, and atopic sensitization [65]. How might this occur? It appears that the innate immune activation program – that is, the array of cytokines, chemokines, adhesion molecules, and so on that are activated by PAMPs in a given individual – dramatically influences whether the adaptive response is predominantly cellular or humoral, T helper 1 (Th1) or T helper 2 (Th2), activating or tolerogenic. The genetics of the individual, the molecular nature and concentrations of the PAMPs, and the timing and duration of exposure certainly influence the magnitude as well as the types of innate pathways that are activated. In addition, we would suggest that an incompletely understood interaction between PAMPs and the hypothalamic–pituitary–adrenal (HPA) axis is likely to play an important role. Shanks and co-workers showed that exposure of neonatal rats to LPS had long-lasting effects on the HPA axis [66]. Adults that had been exposed to LPS as neonates had earlier peak as well as higher mean corticosterone levels in their basal, unstimulated state than did non-LPS-exposed littermates. LPS-induced splenocyte proliferation following restraint stress, as well as adjuvant arthritis, was significantly lower in the neonatal LPS exposure group. These results demonstrate that LPS (and presumably other PAMPs) can exert long-lasting effects on the timing and levels of GCs that are produced following a subsequent stressful episode, and thus can have a marked influence on immunity and immunopathology.

Gene profiling is beginning to reveal previously unknown enhancing and suppressive effects of both LPS [67] and GCs [68], and suggests that GCs induce higher expression of a majority of innate immunity genes while inhibiting expression of a majority of genes required for activation of the adaptive immune response [68]. Gene expression profiling also has the potential to identify molecules that dictate sensitivity or resistance to GCs. Donn et al. administered a dexamethasone suppression test to 100 healthy volunteers, and then selected the 10% most GC sensitive and 10% most GC resistant subjects for further analysis. Following lymphocyte activation with phytohemagglutinin (PHA) and treatment with varying doses of dexamethasone, 24 genes were identified to be differentially activated in GC "resistant" versus "sensitive" subjects [69]. Other interactions between GCs and immune mediators are also likely to be important with respect to innate immune function. For example, TNF-$\alpha$ and IL-10 have been shown respectively to decrease and increase monocyte sensitivity to Dex [70]. Also, GCs can induce human monocyte apoptosis, an effect that is prevented by IL-1$\beta$ [71] and probably by other cytokines that are induced by PAMPs. It thus seems certain that GC modulation of cytokines and cytokine receptors play a critical role in preventing a lethal response to LPS and other PAMPs. In addition, there is evidence that, through similar mechanisms, GCs help to shape the adaptive immune response to maximize protection and minimize tissue-damaging inflammation.

## 4. GC EFFECTS ON ADAPTIVE IMMUNITY/ANTIGEN PRESENTATION

GCs were shown in the early 1980s to profoundly inhibit antigen-specific T-cell activation by mouse macrophages and human monocytes. Although inhibition of MHC class II antigen expression appeared to account for this effect in the case of mouse macrophages [72], this was not true for human monocytes in which HLA DR expression was actually increased [73]. It was subsequently found that GCs inhibit activation-dependent expression of the costimulatory molecule CD80 [74] as well as the production of stimulatory cytokines such as IL-2 and IFN-$\gamma$ (Table 1). Interestingly, when cell-specific gene knockout studies were done to assess how GCs suppress inflammation in contact hypersensitivity, inactivation of GR in T cells and keratinocytes had no effect. Rather, GR ablation in macrophages and neutrophils abolished the anti-inflammatory effect of GC treatment [75].

Other recent studies have focused on dendritic cells (DCs) which are now recognized as the most potent antigen-presenting cells [76]. Just as with monocytes, GCs have scant direct effects on the expression of MHC class II or costimulatory molecules on mature DCs [77,78], but markedly inhibit DC differentiation and maturation. For example, when Dex was present during differentiation of monocytes to DCs, these "Dex-DCs" expressed low levels of CD1a (a DC marker) and high levels of CD14 and CD16 (macrophage markers). While Dex-DCs had higher expression of the mannose receptor and higher endocytic activity, they manifested impaired TNF-$\alpha$- and CD40 ligand-induced maturation [79]. Dex-DCs were also unresponsive to induction of maturation by CD40 ligand as well as LPS, failing in both cases to upregulate CD80, CD83, CD86, or MHC class II [80]. In contrast, the development of Langerhan's cells from $CD34^+$ hematopoietic precursors was unaffected by Dex or prednisolone [81].

de Jong et al. [82] and Rea and co-workers [83] have suggested that GCs redirect, rather than simply block, DC maturation. Rea et al. noted that while Dex-DCs failed to acquire high levels of costimulatory, adhesion, and MHC molecules, antigen uptake was unaffected. And while IL-12 production by Dex-DCs was blocked, IL-10 secretion was greatly enhanced. Moreover, interaction of these Dex-DCs with $CD4^+$ Th1 T cells rendered the T cells unresponsive to further antigen-specific restimulation. de Jong et al. reported that GC-treated DCs, due to impaired production of IL-12 p70, stimulated uncommitted T cells to become Th2 biased cells [82]. Such effects may underlie the finding that transient GC treatment amplifies the Th2 response in a murine model of asthma [84].

Lutz and Schuler summarized the DC signals that are proposed to induce tolerance versus immunity [76]. Immature tissue-resident DCs capable of inducing T-cell anergy have most of the characteristics just described of DCs in which maturation has been held in check by GCs. These include low expression of MHC class II, CD80 and CD86, and low production of IL-12, IL-10, IL-6, and TNF-α. Fully mature migratory DCs capable of activating protective immunity have high expression of MHC and costimulatory molecules as well as high production of cytokines, particularly IL-12. Semi-mature migratory DCs that are capable of activating $IL\text{-}10^+$ regulatory T cells (Treg) also have high expression of the appropriate surface molecules but lack production of IL-12, IL-6, and TNF-α [76]. More recently, it was shown that human DCs and macrophages treated with GCs or IL-10 upregulate expression of the GC-induced leucine zipper (GILZ) [85,86]. Antigen presentation by GILZ-expressing DC generated CD25(high)FOXP3(+)CTLA-4/CD152(+) and IL-10-producing Treg cells that inhibited the response of $CD4^+$ and $CD8^+$ T lymphocytes. This inhibition was specific to the antigen presented [87]. GILZ is also expressed on human airway epithelial cells where its upregulation by dexamethasone appears to play a role in local suppression of cytokine production and T-cell activation [88].

Another important GC-induced regulatory molecule expressed on Treg cells is the GC-induced tumor necrosis factor receptor, GITR. GITR was first described as a molecule present on dexamethasone-treated T-cell hybridomas that were capable of inhibiting T-cell receptor-induced apoptosis. Its natural ligand, GITRL, is a TNF superfamily member expressed on both mature and immature DCs. Engagement of GITR by GITRL can reverse Treg-mediated suppression, leading to costimulation of effector T cells (reviewed in Refs [89,90]). It was therefore surprising that coordinate induction of GITR on $CD4^+$ T cells and GITRL on plasmacytoid DCs was responsible, at least in part, for dexamethasone-induced protection against allergic bronchopulmonary aspergillosis. Grohmann et al. determined that reverse signaling of plasmacytoid DCs through GITRL led to activation of IDO (indoleamine 2,3-dioxygenase), a tryptophan degradation pathway that plays a key role in peripheral tolerance and is best known for its importance in protecting the fetus from maternal immunity [91]. Thus, the influence of GCs on the phenotype and function of DCs, and presumably of other antigen-presenting cells, is likely to play a very significant role in the type, intensity, and duration of immune responses that are elicited.

## 5. GC EFFECTS ON T-CELL ACTIVATION

In 1951, Billingham et al. demonstrated that cortisone prolonged the survival of allogeneic skin grafts in rabbits [92]. This led to half a century of studies demonstrating the suppressive effects of GCs on T-lymphocyte growth and activation. These actions include GC-induced thymocyte apoptosis and the inhibition of a wide variety of activation-induced gene products [4,93]. In this section, we will focus on the physiological significance of GC effects on thymocyte selection and apoptosis, production of regulatory Treg cells, and Th2-biased responses.

Thymocyte killing by GCs is a dramatic phenomenon that was discovered in stressed, intact organisms decades before the immunological significance of the thymus in T-cell development was recognized. GC-induced thymocyte apoptosis occurs even in subsets of thymocytes with low GR expression [94]. Nonetheless, GR numbers are physiologically relevant since thymocytes from transgenic mice and rats that have increased GR expression are significantly more sensitive to GC-induced apoptosis [55]. Moreover, cell-specific targeting of thymocytes or peripheral T cells for higher or lower GR expression demonstrates that increased GC sensitivity causes marked reductions in both thymocyte and T-cell numbers [95].

Despite these and many other observations, the physiological significance of GC-induced thymocyte apoptosis is still obscure. A central question is whether GCs are necessary for T-cell development and selection in the thymus [93,96]. Substantial evidence has been cited in favor of this proposition, including the fact that thymic epithelial cells produce GCs, and that although T-cell receptor (TCR) and GR activation independently induce thymocyte apoptosis, together they promote T-cell survival and GCs antagonize TCR-mediated signals for cell death [93,97]. TCR signaling inhibits GC-induced apoptosis differently depending on the stage of thymocyte development [98]. GCs may attenuate TCR signaling by affecting early phosphorylation events induced after TCR ligation [99]. Two important signaling pathways involve IL-7, a key cytokine for T-cell development, and GILZ that inhibits the transcription factors activator protein 1 (AP-1) and NF-κB [100]. GCs increase IL-7Rα expression in both naïve and activated CD4+ T cells, and it has been proposed that increased sensitivity to IL-7 accounts, in part, for IL-7-mediated rescue from GC-induced apoptosis [30].

Results that cast doubt on a significant role for GCs in the thymus come particularly from experiments with mice lacking functional GRs altogether. Thymocytes taken from 15- or 18-day embryos of such mice and cultured for up to 12 days display normal T-cell development, including positive and negative selection, when compared to thymocytes from wild-type mice [101]. The physiological function of GCs in the thymus thus remains uncertain.

The concept that immunity to infection is controlled by distinct Th1 and Th2 subpopulations of T lymphocytes is well established, as is the fact that GCs tend to favor Th2 and inhibit Th1 responses [102]. A related but controversial concept involved GC effects on so-called T-suppressor (Ts) cells. As introduced above, a regulatory T-cell subset that can restrain immune responses from overshooting and thus limit damage to host tissues has been well documented and is now widely accepted [103]. Treg cells can be generated by stimulation of naive CD4+ T cells in the presence of IL-10, or IL-4 plus IL-10, resulting in a cell population with high production of IL-10. However, these Treg cells also secrete low amounts of the cytokines IL-5 and IFN-γ which can have proinflammatory actions [104,105]. One report demonstrates that in addition to the already noted positive effect of GCs on IL-10 production, which would seem to favor the development of Tregs, GCs can be used to generate a unique population of Treg cells. Culture of naïve human or murine T cells in the presence of 10 nM Dex plus 40 nM vitamin D3 gives rise to a population of Treg cells that produce only IL-10 and retain strong proliferative capacity [106]. DCs or other antigen-presenting cells were not required to generate these cells, as immobilized CD3 plus soluble anti-CD28 were a sufficient activation stimulus. Development of Tregs was enhanced by neutralization of the Th1 and Th2 cytokines IL-4, IL-12, and IFN-γ, but this was not essential. Regulatory function of these cells was demonstrated by their ability to prevent central nervous system damage in an animal model of experimental allergic encephalomyelitis. In another study, Dex treatment of mice was shown to favor survival of Tregs by a mechanism that involved selective protection by IL-2 of these high CD25 (IL-2 receptor)-positive cells [107]. It seems likely that Treg cells play an important role in the previously reported ability of GCs to promote a Th2 cytokine response [102,108].

## 6. GC EFFECTS ON INFLAMMATION

High concentrations of GCs result in broad suppression of immune and inflammatory reactions (reviewed elsewhere in this volume by Franchimont and Chrousos and by Webster and Sternberg, as well as in Refs [11,12,109]). There is little doubt that the protective effects of administered GCs result largely from overwhelming inhibition of the mediators listed in Table 1. As noted earlier, there is also substantial evidence for more discriminating regulation of immune and inflammatory mediators by GCs at lower concentrations [16,110]. Both permissive and suppressive actions appear to be important for normal homeostatic control of inflammation and suppression of autoimmunity.

A review of the work of Sternberg and others sheds substantial light on the importance of a normal stress-induced surge in GC secretion [110]. Lewis rats are exceptionally susceptible to development of a wide range of autoimmune/inflammatory diseases in response to antigenic stimuli. Fisher rats are largely histocompatible to Lewis rats, but resistant to illnesses after exposure to the same antigens. These strains show related differences in their HPA response to antigen. The inflammation-susceptible Lewis rat exhibits low corticosterone production while the inflammation-resistant Fisher rat has an excessive HPA response compared to outbred rats. When Fisher rats were adrenalectomized or treated with the GR antagonist RU486, their inflammatory response to streptococcal cell walls or *Salmonella typhimurium* was exaggerated, resulting in high mortality. Conversely, when Lewis rats were treated with low-dose Dex or transplanted intracerebroventricularly with Fisher hypothalamic tissue, carageenan-induced inflammation and arthritis were significantly attenuated. These studies demonstrate that HPA feedback and modulation of GC levels within the normal range can play a critical role in the resolution of inflammation.

Although most reports indicate that GCs suppress cytokine production, enhancement has also been observed. Several studies in humans show that high-dose GCs decrease blood IL-6 but act to increase blood interleukin-10 during cardiac surgery with cardiopulmonary bypass [62,111,112]. Suppression of GC synthesis by etomidate anesthesia, which eliminates the normal stress-induced rise in plasma cortisol levels, leads to markedly lower production of IL-10 during surgery [62]. Thus, the stress-induced GC surge during cardiac surgery appears to be important for optimal production of anti-inflammatory cytokines since neither surgery alone in the absence of GC nor GC alone will induce substantial elevations of IL-10 [62,113]. An earlier example of permissive actions of GCs was reported by Barber et al. [28]. They showed, in accord with the anti-inflammatory actions of GCs, that TNF-α and IL-6 responses to intravenous (i.v.) endotoxin were suppressed by cortisol administered just before (within 6 h) or during experimental endotoxemia in humans. They gave cortisol (as hemisuccinate) in 6-h i.v. infusions that raised plasma cortisol levels to the micromolar range, corresponding to levels that were approximately twice those that are normally observed during the stress of endotoxemia or major surgery in humans. However, when cortisol was given 12, 36, 72, or 144 h before the endotoxin, there was markedly enhanced secretion of TNF-α and IL-6. In mice, low permissive GCs enhance both nitric oxide and cytokine production, while higher levels are inhibitory [114]. Inactivation of the GR by siRNAs abrogated both the stimulatory and suppressive GC actions. Furthermore, GC conditioning during the differentiation of myeloid progenitors into macrophages resulted in their enhanced LPS responsiveness, demonstrated by an overexpression of the inflammatory cytokines TNF-α, IL-6, and IL-12 [115]. These studies demonstrate that permissive actions of GCs on the production of proinflammatory cytokines can be induced even by high GC concentrations, and that the timing in relation to the stressor is a critical variable in the physiologic outcome.

Time is an important variable that can be easily overlooked during *in vitro* testing of GC actions. An important relationship between GCs and time is implicit in the diurnal variations of GCs that is seen in animals, in the concept of permissive actions of GCs (since the GCs must be present before a stimulus to have an effect), and in the very nature of the *in vivo* inflammatory response to injury which is sequentially orchestrated and time-dependent. It is readily seen in systemic models that inflammation is not an isolated event but an evolving process. Following *in vivo* administration of endotoxin to humans, for example, TNF is rapidly released (within 60 min) into the blood where it promotes a variety of processes including a surge of cortisol release that does not peak until 4 h after LPS administration, thus allowing TNF to exert its potent proinflammatory events early in the process. As noted, however, the subsequent response of an organism to the systemic stimulus of endotoxin can be altered depending on recent GC exposure of the individual. A transient (6-h) elevation of total plasma cortisol (to 70–80 μg/dL) from 1 to 6 days before an acute systemic inflammatory stimulus (2 ng/kg i.v. endotoxin) actually increases the proinflammatory plasma response (TNF, IL-6) without altering the anti-inflammatory IL-10 response while minimally affecting the response of soluble TNF receptor [27,28]. In contrast, administration of GCs coincident with the same systemic inflammatory stimulus substantially reduces proinflammatory mediators and increases IL-10 [62,63]. The anti-inflammatory effects of IL-10, in turn, are also time-dependent with suppression of inflammatory cytokine release observed when a high-systemic concentration of IL-10 is present before, but not 1 h after LPS exposure [116]. When administered 1 h after LPS exposure, IL-10 is associated with substantially enhanced release of IFN-γ and IFN-γ-inducible chemokines [117].

At least, some of the anti-inflammatory effects of GCs in the acute setting are not only dose-dependent, but exhibit a maximal effect at usual "stress"-induced plasma concentrations of cortisol (30–40 μg/dL) with no further suppression of inflammation at higher doses [62,118]. Finally, even when GCs exhibit inhibitory systemic responses to injury, there is evidence that they may coincidentally enhance local responses via cellular effects. GCs substantially delay in vitro apoptosis of neutrophils in a dose-dependent and GR-sensitive manner [119,120]. This finding has a clinical correlate in the delayed apoptosis of polymorphonuclear leukocytes (PMNs) that is observed during systemic sepsis or following major trauma [121,122] and may be a component of the "two-hit" hypothesis in which an increasing inflammatory response to sequential stimuli leads to increased morbidity following the second event [123]. As leukocytes traffic out of the central circulation in response to a local inflammatory stimulus, their subsequent activity may also be affected by prior exposure to GCs. Pre-treatment of monocytes, for example, with GCs leads to a persistent cell response in which the treated cells show enhanced phagocytosis of apoptotic neutrophils, an action that could contribute to resolution of inflammation in vivo [124]. In this way, the time span of inflammation is shortened by GCs as proposed for T-cell activation [13].

One avenue by which GCs may suppress inflammation is via their influence on alternatively activated hematopoietic cells. We already noted that GCs play a key role in the generation of regulatory T cells and a subset of DCs that produce anti-inflammatory cytokines. "Alternatively activated macrophages," a term introduced by Gordon and colleagues [125], represent another cell type that may actively participate in anti-inflammatory processes, tolerance induction, and healing processes. Transcriptional profiling of M1 versus M2 human monocyte-derived macrophages has revealed new molecules and patterns of gene expression [126]. Alternatively activated macrophages may be induced with IL-4 and GCs as well as with IL-10, IL-13, and TGFβ [127,128]. They preferentially express receptors associated with innate immunity, such as macrophage mannose receptor, β-glucan receptor, scavenger receptor type 1, and CD163 [129].

How each of these cell types responds to GCs and inflammatory stimuli such as PAMPs is just beginning to be clarified by gene profiling. Some of the earliest findings are consistent with previously reported information regarding GCs and the mediators they affect. For example, GC-induced increases in peripheral blood mononuclear cells (PBMCs) of mRNA for three mediators – IL-10, TGFβ, and IL-1R antagonist – are consistent with their known anti-inflammatory effects [68]. Also, while TLR4, the important innate immune receptor for LPS, was increased by Dex in resting PBMCs, it was decreased in activated cells. Thus, GCs may enhance signaling via TLR4 in resting cells but serve to dampen an otherwise damage-provoking signal once cells are activated [115]. Furthermore, although expression of many molecules associated with adaptive immunity was markedly suppressed by 100 nM Dex, expression of others, including TCRαβ, HLA-DR, and CD74, appeared to be permissively enhanced by lower GC concentrations. These examples show that the cell activation profile is influenced by GCs through effects on a wide array of molecules that contribute to enhancement as well as to suppression of immunity and inflammation.

## 7. MOLECULAR MECHANISMS OF GC ACTIONS ON IMMUNE FUNCTIONS

The molecular mechanisms of GC action have been discussed in several recent reviews [4,11,93,96,130,131]. Relying on those sources for earliest references, here we give a brief account of GRs and mechanisms of suppressive effects of GCs on immune and inflammatory responses, in particular on activities of cytokines [132].

Evolution has linked GCs to cytokines by numerous inhibitory paths. GCs lower cytokine activity by suppressing cytokine gene transcription (as with IL-1, IL-2, IL-3, IL-8), by decreasing cytokine secretion (IL-1), by destabilizing cytokine mRNAs through AU sequences in the 3′-untranslated regions (IL-1, TNF, GM-CSF), and by inducing decoy receptors that bind and inactivate the cytokine (IL-1). Increased levels of cytokine receptors and receptor subunits induced by GCs (IL-2Rα, IL-4Rα, IL-6Rα, IFN-γR, GM-CSFRα, CSF-1R, TNF-R) are associated with increasing levels of receptor mRNAs.

After penetrating cell membranes freely, GCs initiate most known actions in their target cells by binding to and activating cytoplasmic receptors that then translocate to the nucleus. In the nucleus the GC-receptor complexes regulate transcription of target genes. Transactivation and transrepression are the "classical" mechanisms whereby GCs either activate or repress transcription of the targeted genes. GC–GR complexes can also bind directly to transcription factors such as NF-κB and AP-1 to inhibit their activation of inflammatory gene transcription. Rapid effects by what seem to be non-genomic pathways have also been reported [133,134]. Whether such pathways have significant roles in the immune system remains an open question [135,136] and we will not discuss them further.

There are at least two GC receptors: classical GRs through which most known GC actions are exerted; and mineralocorticoid receptors (MRs), originally identified by their high affinity for aldosterone. Among GRs there are various isoforms, the differential expression of which may be important with respect to GC sensitivity or resistance to stress [137] as well as diseases such as rheumatoid arthritis [138] and asthma [139]. Notable is GRβ which arises via alternative splicing of the gene that encodes the normal GRα [130]. GRβ lacks hormone-binding capacity but can form both heterodimers with GRα as well as homodimers. GRα is thought to be an antagonist to GRβ [140]. In support of this concept, Li and co-workers showed that human monocytes expressed higher levels of GRβ than did T cells, and were less responsive than T cells to Dex-induced suppression of IL-6 production in response to phorbol ester [141]. Moreover, silencing of GRβ in monocytes resulted in enhanced steroid-induced GRα translocation to the nucleus and transrepression of cytokine gene transcription [141]. Human neutrophils also have much higher ratios of GRβ to GRα than lymphocytes, and are resistant to the apoptotic activity of GCs. This resistance may be due in part to antagonism by GRβ. IL-8 further raises levels of neutrophil GRβ, possibly enhancing resistance to GCs [142]. Below, we describe results with transgenic mice carrying other mutant GRs.

Cortisol and corticosterone, the natural GCs, have affinities for MRs similar to that of aldosterone, and much higher than their affinities for GRs. Thus, GCs have the potential to influence inflammation through actions on the MR as well as GR [143]. Since circulating levels of GCs exceed many-fold the levels of aldosterone and can saturate MRs, GCs would prevent aldosterone from binding to MRs except for the presence in mineralocorticoid target cells of the enzyme 11β-hydroxysteroid dehydrogenase type 2. This enzyme rapidly oxidizes cortisol and corticosterone to their respective 11-keto forms, which bind only weakly to MRs and GRs. Not all cells have MRs that are "protected" in this way from GCs. Most unprotected MRs identified to date are in the neural system where both inflammatory and protective actions of GCs are well established [11].

Mechanisms by which GCs regulate activation or repression of genes vary widely. Some target-gene promoters contain short palindromic sequences of nucleotides called glucocorticoid response elements (GREs) to which GC–GR homodimers bind and recruit either co-activator or co-repressor proteins. Transcription of the associated gene is thereby stimulated or inhibited. With respect to the immune system, transactivation is particularly important for induction of anti-inflammatory proteins such as annexin 1, MKP-1, and IκBα. ~MKP-1 is a pivotal regulator of inflammation and its knockout substantially sensitizes mice to endotoxic shock induced by LPS, with enhanced production of IL-6 and TNF-α [46]. Dex induces MKP-1 in human T cells and monocytes [141], in human endothelial cells [144], and in wild-type but not GR-deficient mouse macrophages [56]. Dex-induced MKP-1 is associated with decreased phosphorylation of p38 MAP kinase in the absence of an effect on NF-κB in a number of systems. For example, 1–100 nM Dex dose-dependently inhibited TNF-α-induced E-selectin on human endothelial cells without affecting NF-κB [144]. In MKP-1-deficient cells, Dex did not inhibit E-selectin expression. Annexin 1, a GC-induced protein known chiefly for its inhibition of phospholipase A2 [130,145], was recently shown to be required for upregulation of MKP-1 and inhibition of IL-6 expression in mice [146]. Transrepression of genes containing negative GREs in their promoters can also be important [130,147,148]. And as noted earlier, steric occlusion of an NF-κB responsive element in the Fas ligand promoter has been reported as an additional mechanism by which GC–GR complexes inhibit activation of that gene [149].

Most GC actions within the immune system, on the other hand, do not involve direct binding of activated GRs to GREs or even to DNA, but operate via so-called transcriptional "cross-talk" [17,23,96,147,148,150–152]. Here GRs, probably in monomer form, bind directly to transcription factors that in turn regulate transcription through DNA-binding sites. Among these factors are NF-κB, AP-1 proteins cJun and cFos, the cAMP response element binding protein (CREB), the CREB binding protein (CBP), and the signal transducer and activator of transcription (STAT5). Other possible targets for GR cross-talk are general transcription factors that compose the RNA polymerase II transcription complex, and co-activators that link the basic transcriptional machinery to nuclear receptors or to other DNA-binding proteins [153–155]. GRs usually interfere with a factor but may synergize with it. For example, upon interaction with proteins at the AP-1 site, GRs suppress if bound to cJun–cFos heterodimers, but stimulate if bound to cJun–cJun homodimers. Modulation by GRs of factor activity may be reciprocated by modulation by the factor of GR activity [151].

NF-κB mediates immune and inflammatory signals from several cytokines as well as from antigens. It regulates TNF-α, IL-1β, IL-2, IL-6, GM-CSF, M-CSF (monocyte colony-stimulating factor), G-CSF (granulocyte colony-stimulating factor), IL-8 and other chemokines, nitric oxide synthase, COX-2, ICAM-1, the IL-2Rα subunit, the TCRβ subunit, the serum amyloid A protein, and matrix metalloproteinase 9 (MMP-9) [24,156]. A member of the Rel/NF-κB family, NF-κB, is a cytoplasmic protein found in most cells. In inactive form it is bound to the inhibitor protein IκB. Activated NF-κB is a dimer composed of the proteins p65 (also known as relA) and p50. When IκB is phosphorylated, it releases NF-κB, which then enters the nucleus and activates transcription of target genes by binding to NF-κB sites.

GC inhibition of NF-κB is believed to be a key step in immunosuppressive and anti-inflammatory actions. Several mechanisms for inhibition have been proposed. Among these are direct protein–protein interactions between GRs and NF-κB, induction of transcription of IκBβ, "squelching" or sequestration of some limiting factor (like CBP) required for NF-κB activity, and interference with phosphorylation of the RNA polymerase II carboxy-terminal domain [17,23,131,151,157–159]. Different mechanisms operate in different cell types. GRs and NF-κB display mutual transcriptional antagonism or "cross-repression" in which CBP [160] and the catalytic subunit of protein kinase A [161] are important.

Cross-talk between GRs and AP-1 was revealed over a decade ago in studies on the mechanism of GC suppression of basal and phorbol ester-activated transcription of the gene for collagenase, a major inflammatory mediator. Protein–protein interactions between GRs and c-jun were implicated. Since then, other genes found to be similarly controlled include those for IL-2, IFN-γ, and GM-CSF. IL-2 gene expression is mediated by AP-1 in synergy with the nuclear factor of activated T cells (NFAT). Suppression by GRs reduces the synergy, and may in part require GC-induced transcription of the GILZ gene [100]. GCs repress the IFN-γ gene via interaction of GRs with complexes of AP-1, CREB, and ATF (activating transcription factor) [162]. Repression of the GM-CSF gene has been proposed to involve inhibition of enhancer activity from composite NF-AT/AP-1 elements by binding of GRs to those elements [18].

Certain transgenic mice have revealed strikingly the importance of cross-talk mechanisms for immunosuppressive and anti-inflammatory effects of GCs. Known as GRdim mice, they carry mutated GRs that are unable to dimerize and bind to GREs but can still engage in cross-talk. Results with such mice show that cross-talk is responsible for almost all immunosuppressive and anti-inflammatory GC actions. GC administration to GRdim mice is as effective as GC administration to wild-type mice in suppressing PMA-induced skin inflammation and the resulting rise in serum IL-6, as well as in suppressing the rise in serum TNF-α following LPS. When added to macrophages from GRdim mice, GCs inhibit PMA-induced increases in IL-1β, IL-6, TNF-α, and cyclooxygenase-2 (Cox-2) mRNA, and when added to activated thymocytes, they suppress IL-2 and IFN-γ mRNA. They also repress PMA-activated collagenase-3 gene transcription in immortalized GRdim fibroblasts [163–165]. Most of these responses, as already noted, are mediated by NF-κB. Since GC-treated cells from GRdim mice show no increased synthesis of IκBα mRNA or protein, induction of IκBα is clearly not essential for most NF-κB-mediated immunosuppressive actions of GCs.

One prominent immunosuppressive effect of GCs that is defective in GRdim mice is thymocyte and lymphocyte apoptosis. Since GCs can inhibit activation-induced T-cell apoptosis by downregulating the expression of Fas-ligand, an NF-κB-inducible gene, Novac and co-workers examined the interaction between GR and NF-κB within the Fas-ligand promoter [149]. They found that GR dimers bind to a negative GRE that overlaps with a known NF-κB binding site. Thus, Fas-ligand transcription appears to be suppressed by GR-mediated steric occlusion of NF-κB binding to DNA [149].

## REFERENCES


1. Hench PS, Kendall EC, Slocumb CH, Polley HF. The effect of a hormone of the adrenal cortex (17-hydroxy-11-dehydrocorticosterone: compound E) and of pituitary adrenocorticotropic hormone on rheumatoid arthritis. Mayo Clin Proc 1949;24:181–97.
2. Tausk M. Hat die Nebenniere tatsächlich eine Verteidigungsfunktion? Das Hormon (Organon, Holland) 1951;3:1–24.
3. Munck A, Guyre PM, Holbrook NJ. Physiological functions of glucocorticoids in stress and their relation to pharmacological actions. Endocr Rev 1984;5:25–44.
4. Sapolsky RM, Romero LM, Munck A. How do glucocorticoids influence stress-responses? Integrating permissive, suppressive, stimulatory, and preparative actions. Endocr Rev 2000;21:55–89.
5. Besedovsky H, Sorkin E. Network of immune-neuroendocrine interactions. Clin Exp Immunol 1977;27:1–12.
6. Munck A, Náray-Fejes-Tóth A. The ups and downs of glucocorticoid physiology. Permissive and suppressive effects revisited. Mol Cell Endocrinol 1992;90:C1–4.
7. Munck A, Náray-Fejes-Tóth A. Glucocorticoids and stress: permissive and suppressive actions. Ann N Y Acad Sci 1994;746:115–30.
8. Charmandari E, Tsigos C, Chrousos G. Endocrinology of the stress response. Annu Rev Physiol 2005;67:259–84.
9. Franchimont D. Overview of the actions of glucocorticoids on the immune response: a good model to characterize new pathways of immunosuppression for new treatment strategies. Ann N Y Acad Sci 2004;1024:124–37.
10. Herold MJ, McPherson KG, Reichardt HM. Glucocorticoids in T cell apoptosis and function. Cell Mol Life Sci 2006;63:60–72.
11. Sorrells SF, Sapolsky RM. An inflammatory review of glucocorticoid actions in the CNS. Brain Behav Immun 2007;21:259–72.
12. Yeager MP, Guyre PM, Munck AU. Glucocorticoid regulation of the inflammatory response to injury. Acta Anaesthesiol Scand 2004;48:799–813.
13. Wiegers GJ, Reul JMHM. Induction of cytokine receptors by glucocorticoids: functional and pathological significance. Trends Pharmacol Sci 1998;19:317–21.
14. Wang J, Zhu Z, Nolfo R, Elias JA. Dexamethasone regulation of lung epithelial cell and fibroblast interleukin-11 production. Am J Physiol 1999;276:L175–85.
15. Yoshikawa H, Nakajima Y, Tasaka K. Glucocorticoid suppresses autocrine survival of mast cells by inhibiting IL-4 production and ICAM-1 expression. J Immunol 1999;162:6162–70.
16. Sapolsky RM. Glucocorticoids and hippocampal atrophy in neuropsychiatric disorders. Arch Gen Psychiatry 2000;57:925–35.
17. Nissen RM, Yamamoto KR. The glucocorticoid receptor inhibits NFkB by interfering with serine-2 phosphorylation of the RNA polymerase II carboxy-terminal domain. Genes Dev 2000;14:2314–29.
18. Smith PJ, Cousins DJ, Jee YK, Staynov DZ, Lee TH, Lavender P. Suppression of granulocyte-macrophage colony-stimulating factor expression by glucocorticoids involves inhibition of enhancer function by the glucocorticoid receptor binding to composite NF-AT/activator protein-1 elements. J Immunol 2001;167:2502–10.
19. Ammit AJ, Lazaar AL, Irani C et al. Tumor necrosis factor-α-induced secretion of RANTES and interleukin-6 from human airway smooth muscle cells: modulation by glucocorticoids and β-agonists. Am J Respir Cell Mol Biol 2002;26:465–74.
20. Pitzalis C, Pipitone N, Perretti M. Regulation of leukocyte-endothelial interactions by glucocorticoids. Ann N Y Acad Sci 2002;966:108–18.
21. Umland SP, Schleimer RP, Johnston SL. Review of the molecular and cellular mechanisms of action of glucocorticoids for use in asthma. Pulm Pharmacol Ther 2002;15:35–50.
22. Da Silva CA, Kassel O, Mathieu E, Massard G, Gasser B, Frossard N. Inhibition by glucocorticoids of the interleukin-1β-enhanced expression of the mast cell growth factor SCF. Br J Pharmacol 2002;135:1634–40.

23. Almawi WY, Melemedjian OK. Molecular mechanisms of glucocorticoid antiproliferative effects: antagonism of transcription factor activity by glucocorticoid receptor. J Leukoc Biol 2002;71:9–15.
24. Eberhardt W, Schulze M, Engels C, Klasmeier E, Pfeilschifter J. Glucocorticoid-mediated suppression of cytokine-induced matrix metalloproteinase-9 expression in rat mesangial cells: involvement of nuclear factor-κB and Ets transcription factors. Mol Endocrinol 2002;16:1752–66.
25. Petrovsky N, Harrison LC. The chronobiology of human cytokine production. Int Rev Immunol 1998;16:635–49.
26. Hrushesky WJ, Langevin T, Kim YJ, Wood PA. Circadian dynamics of tumor necrosis factor alpha (cachectin) lethality. J Exp Med 1994;180:1059–65.
27. Barber AE, Coyle SM, Fischer E et al. Influence of hypercortisolemia on soluble tumor necrosis factor receptor II and interleukin-1 receptor antagonist responses to endotoxin in human beings. Surgery 1995;118:406–10; discussion 410–1.
28. Barber AE, Coyle SM, Marano MA et al. Glucocorticoid therapy alters hormonal and cytokine responses to endotoxin in man. J Immunol 1993;150:1999–2006.
29. Caulfield J, Fernandez M, Snetkov V, Lee T, Hawrylowicz C. CXCR4 expression on monocytes is up-regulated by dexamethasone and is modulated by autologous CD3+ T cells. Immunology 2002;105:155–62.
30. Franchimont D, Galon J, Vacchio MS et al. Positive effects of glucocorticoids on T cell function by up-regulation of IL-7 receptor α. J Immunol 2002;168:2212–18.
31. Hawrylowicz CM, Guida L, Paleolog E. Dexamethasone up-regulates granulocyte-macrophage colony-stimulating factor receptor expression on human monocytes. Immunology 1994;83:274–80.
32. Baker JB, Barsh GS, Carney DH, Cunningham DD. Dexamethasone modulates binding and action of epidermal growth factor in serum-free culture. Proc Natl Acad Sci USA 1978;75:1882–6.
33. Arzt E, Kovalovsky D, Igaz LM et al. Functional cross-talk among cytokines, T-cell receptor, and glucocorticoid receptor transcriptional activity and action. Ann N Y Acad Sci 2000;917:672–7.
34. Biola A, Lefebvre P, Perrin-Wolff M, Sturm M, Bertoglio J, Pallardy M. Interleukin-2 inhibits glucocorticoid receptor transcriptional activity through a mechanism involving STAT5 (signal transducer and activator of transcription 5) but not AP-1. Mol Endocrinol 2001;15:1062–76.
35. Heiniger CD, Rochat MK, Frey FJ, Frey BM. TNF-α enhances intracellular glucocorticoid availability. FEBS Lett 2001;507:351–6.
36. Chapman KE, Coutinho A, Gray M, Gilmour JS, Savill JS, Seckl JR. Local amplification of glucocorticoids by 11beta-hydroxysteroid dehydrogenase type 1 and its role in the inflammatory response. Ann N Y Acad Sci 2006;1088:265–73.
37. Chapman KE, Gilmour JS, Coutinho AE, Savill JS, Seckl JR. 11Beta-hydroxysteroid dehydrogenase type 1 – a role in inflammation? Mol Cell Endocrinol 2006;248:3–8.
38. Calandra T, Bernhagen J, Metz CN et al. MIF as a glucocorticoid-induced modulator of cytokine production. Nature 1995;377:68–71.
39. Lolis E. Glucocorticoid counter regulation: macrophage migration inhibitory factor as a target for drug discovery. Curr Opin Pharmacol 2001;1:662–8.
40. Flaster H, Bernhagen J, Calandra T, Bucala R. The macrophage migration inhibitory factor-glucocorticoid dyad: regulation of inflammation and immunity. Mol Endocrinol 2007;21:1267–80.
41. Morand EF, Leech M, Bernhagen J. MIF: a new cytokine link between rheumatoid arthritis and atherosclerosis. Nat Rev Drug Discov 2006;5:399–410.
42. Kudrin A, Scott M, Martin S et al. Human macrophage migration inhibitory factor: a proven immunomodulatory cytokine? J Biol Chem 2006;281:29641–51.
43. Aeberli D, Leech M, Morand EF. Macrophage migration inhibitory factor and glucocorticoid sensitivity. Rheumatology (Oxford) 2006;45:937–43.
44. Aeberli D, Yang Y, Mansell A, Santos L, Leech M, Morand EF. Endogenous macrophage migration inhibitory factor modulates glucocorticoid sensitivity in macrophages via effects on MAP kinase phosphatase-1 and p38 MAP kinase. FEBS Lett 2006;580:974–81.
45. Bornstein SR, Ziegler CG, Krug AW et al. The role of Toll-like receptors in the immune-adrenal crosstalk. Ann N Y Acad Sci 2006;1088:307–18.

46. Zhao Q, Wang X, Nelin LD et al. MAP kinase phosphatase 1 controls innate immune responses and suppresses endotoxic shock. J Exp Med 2006;203:131–40.
47. Baumann H, Gauldie J. The acute phase response. Immunol Today 1994;15:74–80.
48. Thorn CF, Whitehead AS. Differential glucocorticoid enhancement of the cytokine-driven transcriptional activation of the human acute phase serum amyloid A genes, SAA1 and SAA2. J Immunol 2002;169:399–406.
49. Fischer CP, Bode BP, Takahashi K, Tanabe KK, Souba WW. Glucocorticoid-dependent induction of interleukin-6 receptor expression in human hepatocytes facilitates interleukin-6 stimulation of amino acid transport. Ann Surg 1996;223:610–18; discussion 618–19.
50. Medzhitov R, Janeway CA Jr. Decoding the patterns of self and nonself by the innate immune system. Science 2002;296:298–300.
51. Janeway CA Jr, Medzhitov R. Innate immune recognition. Annu Rev Immunol 2002;20:197–216.
52. Zacharowski K, Zacharowski PA, Koch A et al. Toll-like receptor 4 plays a crucial role in the immune-adrenal response to systemic inflammatory response syndrome. Proc Natl Acad Sci USA 2006;103:6392–7.
53. Ogawa S, Lozach J, Benner C et al. Molecular determinants of crosstalk between nuclear receptors and Toll-like receptors. Cell 2005;122:707–21.
54. Kapcala LP, Chautard T, Eskay RL. The protective role of the hypothalamic-pituitary-adrenal axis against lethality produced by immune, infectious, and inflammatory stress. Ann N Y Acad Sci 1995;771:419–37.
55. Reichardt HM, Umland T, Bauer A, Kretz O, Schutz G. Mice with an increased glucocorticoid receptor gene dosage show enhanced resistance to stress and endotoxic shock. Mol Cell Biol 2000;20:9009–17.
56. Bhattacharyya S, Brown DE, Brewer JA, Vogt SK, Muglia LJ. Macrophage glucocorticoid receptors regulate Toll-like receptor-4-mediated inflammatory responses by selective inhibition of p38 MAP kinase. Blood 2007;109:4313–19.
57. Marchant A, Bruyns C, Vandenabeele P et al. Interleukin-10 controls interferon-γ and tumor necrosis factor production during experimental endotoxemia. Eur J Immunol 1994;24:1167–71.
58. Gerard C, Bruyns C, Marchant A et al. Interleukin 10 reduces the release of tumor necrosis factor and prevents lethality in experimental endotoxemia. J Exp Med 1993;177:547–50.
59. Howard M, Muchamuel T, Andrade S, Menon S. Interleukin 10 protects mice from lethal endotoxemia. J Exp Med 1993;177:1205–8.
60. Standiford TJ, Strieter RM, Lukacs NW, Kunkel SL. Neutralization of IL-10 increases lethality in endotoxemia. Cooperative effects of macrophage inflammatory protein-2 and tumor necrosis factor. J Immunol 1995;155:2222–9.
61. Mozo L, Suarez A, Gutierrez C. Glucocorticoids up-regulate constitutive interleukin-10 production by human monocytes. Clin Exp Allergy 2004;34:406–12.
62. Fillinger MP, Rassias AJ, Guyre PM et al. Glucocorticoid effects on the inflammatory and clinical responses to cardiac surgery. J Cardiothorac Vasc Anesth 2002;16:163–9.
63. van der Poll T, Barber AE, Coyle SM, Lowry SF. Hypercortisolemia increases plasma interleukin-10 concentrations during human endotoxemia – a clinical research center study. J Clin Endocrinol Metab 1996;81:3604–6.
64. Fearon DT, Locksley RM. The instructive role of innate immunity in the acquired immune response. Science 1996;272:50–3.
65. Braun-Fahrlander C, Riedler J, Herz U et al. Environmental exposure to endotoxin and its relation to asthma in school-age children. N Engl J Med 2002;347:869–77.
66. Shanks N, Windle RJ, Perks PA et al. Early-life exposure to endotoxin alters hypothalamic-pituitary-adrenal function and predisposition to inflammation. Proc Natl Acad Sci USA 2000;97:5645–50.
67. Nau GJ, Richmond JF, Schlesinger A, Jennings EG, Lander ES, Young RA. Human macrophage activation programs induced by bacterial pathogens. Proc Natl Acad Sci USA 2002;99:1503–8.
68. Galon J, Franchimont D, Hiroi N et al. Gene profiling reveals unknown enhancing and suppressive actions of glucocorticoids on immune cells. FASEB J 2002;16:61–71.

69. Donn R, Berry A, Stevens A et al. Use of gene expression profiling to identify a novel glucocorticoid sensitivity determining gene, BMPRII. FASEB J 2007;21:402–14.
70. Franchimont D, Martens H, Hagelstein M-T et al. Tumor necrosis factor α decreases, and interleukin-10 increases, the sensitivity of human monocytes to dexamethasone: potential regulation of the glucocorticoid receptor. J Clin Endocrinol Metab 1999;84:2834–9.
71. Schmidt M, Pauels H-G, Lügering N, Lügering A, Domschke W, Kucharzik T. Glucocorticoids induce apoptosis in human monocytes: potential role of IL-1β. J Immunol 1999;163:3484–90.
72. Snyder DS, Unanue ER. Corticosteroids inhibit murine macrophage Ia expression and interleukin 1 production. J Immunol 1982;129:1803–5.
73. Gerrard TL, Cupps TR, Jurgensen CH, Fauci AS. Hydrocortisone-mediated inhibition of monocyte antigen presentation: dissociation of inhibitory effect and expression of DR antigens. Cell Immunol 1984;85:330–9.
74. Girndt M, Sester U, Kaul H, Hunger F, Kohler H. Glucocorticoids inhibit activation-dependent expression of costimulatory molecule B7–1 in human monocytes. Transplantation 1998;66:370–5.
75. Tuckermann JP, Kleiman A, Moriggl R et al. Macrophages and neutrophils are the targets for immune suppression by glucocorticoids in contact allergy. J Clin Invest 2007;117:1381–90.
76. Lutz M, Schuler G. Immature, semi-mature and fully mature dendritic cells: which signals induce tolerance or immunity? Trends Immunol 2002;23:445.
77. van den Heuvel MM, van Beek NM, Broug-Holub E et al. Glucocorticoids modulate the development of dendritic cells from blood precursors. Clin Exp Immunol 1999;115:577–83.
78. Vanderheyde N, Verhasselt V, Goldman M, Willems F. Inhibition of human dendritic cell functions by methylprednisolone. Transplantation 1999;67:1342–7.
79. Piemonti L, Monti P, Allavena P et al. Glucocorticoids affect human dendritic cell differentiation and maturation. J Immunol 1999;162:6473–81.
80. Woltman AM, de Fijter JW, Kamerling SW, Paul LC, Daha MR, van Kooten C. The effect of calcineurin inhibitors and corticosteroids on the differentiation of human dendritic cells. Eur J Immunol 2000;30:1807–12.
81. Woltman AM, Massacrier C, de Fijter JW, Caux C, van Kooten C. Corticosteroids prevent generation of CD34+-derived dermal dendritic cells but do not inhibit Langerhans cell development. J Immunol 2002;168:6181–8.
82. de Jong EC, Vieira PL, Kalinski P, Kapsenberg ML. Corticosteroids inhibit the production of inflammatory mediators in immature monocyte-derived DC and induce the development of tolerogenic DC3. J Leukoc Biol 1999;66:201–4.
83. Rea D, van Kooten C, van Meijgaarden KE, Ottenhoff TH, Melief CJ, Offringa R. Glucocorticoids transform CD40-triggering of dendritic cells into an alternative activation pathway resulting in antigen-presenting cells that secrete IL-10. Blood 2000;95:3162–7.
84. Wiley RE, Cwiartka M, Alvarez D et al. Transient corticosteroid treatment permanently amplifies the Th2 response in a murine model of asthma. J Immunol 2004;172:4995–5005.
85. Cohen N, Mouly E, Hamdi H et al. GILZ expression in human dendritic cells redirects their maturation and prevents antigen-specific T lymphocyte response. Blood 2006;107:2037–44.
86. Berrebi D, Bruscoli S, Cohen N et al. Synthesis of glucocorticoid-induced leucine zipper (GILZ) by macrophages: an anti-inflammatory and immunosuppressive mechanism shared by glucocorticoids and IL-10. Blood 2003;101:729–38.
87. Hamdi H, Godot V, Maillot MC et al. Induction of antigen-specific regulatory T lymphocytes by human dendritic cells expressing the glucocorticoid-induced leucine zipper. Blood 2007;110:211–19.
88. Eddleston J, Herschbach J, Wagelie-Steffen AL, Christiansen SC, Zuraw BL. The anti-inflammatory effect of glucocorticoids is mediated by glucocorticoid-induced leucine zipper in epithelial cells. J Allergy Clin Immunol 2007;119:115–22.
89. Nocentini G, Riccardi C. GITR, a multifaceted regulator of immunity belonging to the tumor necrosis factor receptor superfamily. Eur J Immunol 2005;35:1016–22.
90. Shevach EM, Stephens GL. The GITR-GITRL interaction: co-stimulation or contrasuppression of regulatory activity? Nat Rev Immunol 2006;6:613–18.

91. Grohmann U, Volpi C, Fallarino F et al. Reverse signaling through GITR ligand enables dexamethasone to activate IDO in allergy. Nat Med 2007;13:579–86.
92. Billingham RE, Krohn PL, Medawar PB. Effect of cortisone on survival of skin homografts in rabbits. Br Med J 1951;1:1157–63.
93. Ashwell JD, Lu FW, Vacchio MS. Glucocorticoids in T cell development and function. Annu Rev Immunol 2000;18:309–45.
94. Wiegers GJ, Knoflach M, Bock G et al. CD4(+)CD8(+)TCR(low) thymocytes express low levels of glucocorticoid receptors while being sensitive to glucocorticoid-induced apoptosis. Eur J Immunol 2001;31:2293–301.
95. Pazirandeh A, Xue Y, Prestegaard T, Jondal M, Okret S. Effects of altered glucocorticoid sensitivity in the T cell lineage on thymocyte and T cell homeostasis. FASEB J 2002;16:727–9.
96. Winoto A, Littman DR. Nuclear hormone receptors in T lymphocytes. Cell 2002;109(Suppl):S57–66.
97. Zacharchuk CM, Mercep M, Chakraborti PK, Simons SS Jr, Ashwell JD. Programmed T lymphocyte death. Cell activation- and steroid-induced pathways are mutually antagonistic. J Immunol 1990;145:4037–45.
98. Erlacher M, Knoflach M, Stec IE, Bock G, Wick G, Wiegers GJ. TCR signaling inhibits glucocorticoid-induced apoptosis in murine thymocytes depending on the stage of development. Eur J Immunol 2005;35:3287–96.
99. Van Laethem F, Baus E, Smyth LA et al. Glucocorticoids attenuate T cell receptor signaling. J Exp Med 2001;193:803–14.
100. Mittelstadt PR, Ashwell JD. Inhibition of AP-1 by the glucocorticoid-inducible protein GILZ. J Biol Chem 2001;276:29603–10.
101. Purton JF, Boyd RL, Cole TJ, Godfrey DI. Intrathymic T cell development and selection proceeds normally in the absence of glucocorticoid receptor signaling. Immunity 2000;13:179–86.
102. Ramírez F, Fowell DJ, Puclavec M, Simmonds S, Mason D. Glucocorticoids promote a Th2 cytokine response by CD4+ T cells in vitro. J Immunol 1996;156:2406–12.
103. Shevach EM. CD4+ CD25+ suppressor T cells: more questions than answers. Nat Rev Immunol 2002;2:389–400.
104. McGuirk P, Mills K. Pathogen-specific regulatory T cells provoke a shift in the Th1/Th2 paradigm in immunity to infectious diseases. Trends Immunol 2002;23:450.
105. Zelenika D, Adams E, Humm S et al. Regulatory T cells overexpress a subset of Th2 gene transcripts. J Immunol 2002;168:1069–79.
106. Barrat FJ, Cua DJ, Boonstra A et al. *In vitro* generation of interleukin 10-producing regulatory CD4(+) T cells is induced by immunosuppressive drugs and inhibited by T helper type 1 (Th1)- and Th2-inducing cytokines. J Exp Med 2002;195:603–16.
107. Chen X, Murakami T, Oppenheim JJ, Howard OM. Differential response of murine CD4+CD25+ and CD4+CD25− T cells to dexamethasone-induced cell death. Eur J Immunol 2004;34:859–69.
108. Agarwal SK, Marshall GD Jr. Dexamethasone promotes type 2 cytokine production primarily through inhibition of type 1 cytokines. J Interferon Cytokine Res 2001;21:147–55.
109. Morand EF. Effects of glucocorticoids on inflammation and arthritis. Curr Opin Rheumatol 2007;19:302–7.
110. Webster JI, Tonelli L, Sternberg EM. Neuroendocrine regulation of immunity. Annu Rev Immunol 2002;20:125–63.
111. Tabardel Y, Duchateau J, Schmartz D et al. Corticosteroids increase blood interleukin-10 levels during cardiopulmonary bypass in men. Surgery 1996;119:76–80.
112. Tassani P, Richter JA, Barankay A et al. Does high-dose methylprednisolone in aprotinin-treated patients attenuate the systemic inflammatory response during coronary artery bypass grafting procedures? J Cardiothorac Vasc Anesth 1999;13:165–72.
113. Dandona P, Aljada A, Garg R, Mohanty P. Increase in plasma interleukin-10 following hydrocortisone injection. J Clin Endocrinol Metab 1999;84:1141–4.
114. Lim HY, Muller N, Herold MJ, van den Brandt J, Reichardt HM. Glucocorticoids exert opposing effects on macrophage function dependent on their concentration. Immunology 2007;122:47–53.

115. Zhang TY, Daynes RA. Glucocorticoid conditioning of myeloid progenitors enhances TLR4 signaling via negative regulation of the phosphatidylinositol 3-kinase-Akt pathway. J Immunol 2007;178:2517–26.
116. Pajkrt D, Camoglio L, Tiel-van Buul MC et al. Attenuation of proinflammatory response by recombinant human IL-10 in human endotoxemia: effect of timing of recombinant human IL-10 administration. J Immunol 1997;158:3971–7.
117. Lauw FN, Pajkrt D, Hack CE, Kurimoto M, van Deventer SJ, van der Poll T. Proinflammatory effects of IL-10 during human endotoxemia. J Immunol 2000;165:2783–9.
118. de Kruif MD, Lemaire LC, Giebelen IA et al. Prednisolone dose-dependently influences inflammation and coagulation during human endotoxemia. J Immunol 2007;178:1845–51.
119. Cox G. Glucocorticoid treatment inhibits apoptosis in human neutrophils. Separation of survival and activation outcomes. J Immunol 1995;154:4719–25.
120. Meagher LC, Cousin JM, Seckl JR, Haslett C. Opposing effects of glucocorticoids on the rate of apoptosis in neutrophilic and eosinophilic granulocytes. J Immunol 1996;156:4422–8.
121. Ertel W, Keel M, Infanger M, Ungethum U, Steckholzer U, Trentz O. Circulating mediators in serum of injured patients with septic complications inhibit neutrophil apoptosis through up-regulation of protein-tyrosine phosphorylation. J Trauma 1998;44:767–75; discussion 775–6.
122. Ogura H, Tanaka H, Koh T et al. Priming, second-hit priming, and apoptosis in leukocytes from trauma patients. J Trauma 1999;46:774–81; discussion 781–3.
123. Murphy TJ, Paterson HM, Kriynovich S et al. Linking the "two-hit" response following injury to enhanced TLR4 reactivity. J Leukoc Biol 2005;77:16–23.
124. Heasman SJ, Giles KM, Ward C, Rossi AG, Haslett C, Dransfield I. Glucocorticoid-mediated regulation of granulocyte apoptosis and macrophage phagocytosis of apoptotic cells: implications for the resolution of inflammation. J Endocrinol 2003;178:29–36.
125. McWilliam AS, Tree P, Gordon S. Interleukin 4 regulates induction of sialoadhesin, the macrophage sialic acid-specific receptor. Proc Natl Acad Sci USA 1992;89:10522–6.
126. Martinez FO, Gordon S, Locati M, Mantovani A. Transcriptional profiling of the human monocyte-to-macrophage differentiation and polarization: new molecules and patterns of gene expression. J Immunol 2006;177:7303–11.
127. Goerdt S, Orfanos CE. Other functions, other genes: alternative activation of antigen-presenting cells. Immunity 1999;10:137–42.
128. Schebesch C, Kodelja V, Muller C et al. Alternatively activated macrophages actively inhibit proliferation of peripheral blood lymphocytes and CD4+ T cells in vitro. Immunology 1997;92:478–86.
129. Gratchev A, Guillot P, Hakiy N et al. Alternatively activated macrophages differentially express fibronectin and its splice variants and the extracellular matrix protein betaIG-H3. Scand J Immunol 2001;53:386–92.
130. Rhen T, Cidlowski JA. Antiinflammatory action of glucocorticoids – new mechanisms for old drugs. N Engl J Med 2005;353:1711–23.
131. Eggert M, Schulz M, Neeck G. Molecular mechanisms of glucocorticoid action in rheumatic autoimmune diseases. J Steroid Biochem Mol Biol 2001;77:185–91.
132. Liberman AC, Druker J, Perone MJ, Arzt E. Glucocorticoids in the regulation of transcription factors that control cytokine synthesis. Cytokine Growth Factor Rev 2007;18:45–56.
133. Wehling H. Specific, nongenomic actions of steroid hormones. Annu Rev Physiol 1997;59:365–93.
134. Hafezi-Moghadam A, Simoncini T, Yang E et al. Acute cardiovascular protective effects of corticosteroids are mediated by non-transcriptional activation of endothelial nitric oxide synthase. Nat Med 2002;8:473–9.
135. Watson CS, Gametchu B. Membrane estrogen and glucocorticoid receptors – implications for hormonal control of immune function and autoimmunity. Int Immunopharmacol 2001;1:1049–63.
136. Buttgereit F, Scheffold A. Rapid glucocorticoid effects on immune cells. Steroids 2002;67:529–34.
137. DeRijk R, de Kloet ER. Corticosteroid receptor genetic polymorphisms and stress responsivity. Endocrine 2005;28:263–70.

138. Kozaci DL, Chernajovsky Y, Chikanza IC. The differential expression of corticosteroid receptor isoforms in corticosteroid-resistant and -sensitive patients with rheumatoid arthritis. Rheumatology (Oxford) 2007;46:579–85.
139. Goleva E, Li LB, Eves PT, Strand MJ, Martin RJ, Leung DY. Increased glucocorticoid receptor beta alters steroid response in glucocorticoid-insensitive asthma. Am J Respir Crit Care Med 2006;173:607–16.
140. Yudt MR, Cidlowski JA. The glucocorticoid receptor: coding a diversity of proteins and responses through a single gene. Mol Endocrinol 2002;16:1719–26.
141. Li LB, Leung DY, Hall CF, Goleva E. Divergent expression and function of glucocorticoid receptor beta in human monocytes and T cells. J Leukoc Biol 2006;79:818–27.
142. Strickland I, Kisich K, Hauk PJ et al. High constitutive glucocorticoid receptor β in human neutrophils enables them to reduce their spontaneous rate of cell death in response to corticosteroids. J Exp Med 2001;193:585–94.
143. Funder JW. Aldosterone, mineralocorticoid receptors and vascular inflammation. Mol Cell Endocrinol 2004;217:263–9.
144. Furst R, Schroeder T, Eilken HM et al. MAPK phosphatase-1 represents a novel anti-inflammatory target of glucocorticoids in the human endothelium. FASEB J 2007;21:74–80.
145. Goulding NJ. The molecular complexity of glucocorticoid actions in inflammation – a four-ring circus. Curr Opin Pharmacol 2004;4:629–36.
146. Yang YH, Toh ML, Clyne CD et al. Annexin 1 negatively regulates IL-6 expression via effects on p38 MAPK and MAPK phosphatase-1. J Immunol 2006;177:8148–53.
147. Miner JN, Diamond MI, Yamamoto KR. Joints in the regulatory lattice: composite regulation by steroid receptor-AP1 complexes. Cell Growth Differ 1991;2:525–30.
148. Beato M, Herrlich P, Schütz G. Steroid hormone receptors: many actors in search of a plot. Cell 1995;83:851–7.
149. Novac N, Baus D, Dostert A, Heinzel T. Competition between glucocorticoid receptor and NFkappaB for control of the human FasL promoter. FASEB J 2006;20:1074–81.
150. Göttlicher M, Heck S, Herrlich P. Transcriptional cross-talk, the second mode of steroid receptor action. J Mol Med 1998;76:480–9.
151. Refojo D, Liberman AC, Holsboer F, Arzt E. Transcription factor-mediated molecular mechanisms involved in the functional cross-talk between cytokines and glucocorticoids. Immunol Cell Biol 2001;79:385–94.
152. Herrlich P. Cross-talk between glucocorticoid receptor and AP-1. Oncogene 2001;20:2465–75.
153. Jenkins BD, Pullen CB, Darimont BD. Novel glucocorticoid receptor coactivator effector mechanisms. Trends Endocrinol Metab 2001;12:122–6.
154. Kino T, Slobodskaya O, Pavlakis GN, Chrousos GP. Nuclear receptor coactivator p160 proteins enhance the HIV-1 long terminal repeat promoter by bridging promoter-bound factors and the Tat-P-TEFb complex. J Biol Chem 2002;277:2396–405.
155. Rudiger JJ, Roth M, Bihl MP et al. Interaction of C/EBPα and the glucocorticoid receptor in vivo and in nontransformed human cells. FASEB J 2002;16:177–84.
156. Jensen LE, Whitehead AS. Regulation of serum amyloid A protein expression during the acute-phase response. Biochem J 1998;334:489–503.
157. Sheppard K-A, Phelps KM, Williams AJ et al. Nuclear integration of glucocorticoid receptor and nuclear factor-κB signaling by CREB-binding protein and steroid receptor coactivator-1. J Biol Chem 1998;273:29291–4.
158. De Bosscher K, Vanden Berghe W, Vermeulen L, Plaisance S, Boone E, Haegeman G. Glucocorticoids repress NF-κB-driven genes by disturbing the interaction of p65 with the basal transcription machinery, irrespective of coactivator levels in the cell. Proc Natl Acad Sci USA 2000;97:3919–24.
159. Deroo BJ, Archer TK. Glucocorticoid receptor activation of the IκBα promoter within chromatin. Mol Biol cell 2001;12:3365–74.
160. McKay LI, Cidlowski JA. CBP (CREB binding protein) integrates NF-kappaB (nuclear factor-kappaB) and glucocorticoid receptor physical interactions and antagonism. Mol Endocrinol 2000;14:1222–34.

161. Doucas V, Shi Y, Miyamoto S, West A, Verma I, Evans RM. Cytoplasmic catalytic subunit of protein kinase A mediates cross- repression by NF-κB and the glucocorticoid receptor. Proc Natl Acad Sci USA 2000;97:11893–8.
162. Cipitelli M, Sica A, Viggiano V et al. Negative transcriptional regulation of the interferon-γ promoter by glucocortcoids and dominant negative mutants of c-Jun. J Biol Chem 1995;270:12548–56.
163. Reichardt HM, Kaestner KH, Tuckermann J et al. DNA binding of the glucocorticoid receptor is not essential for survival. Cell 1998;93:531–41.
164. Tuckermann JP, Reichardt HM, Arribas R, Richter KH, Schutz G, Angel P. The DNA binding-independent function of the glucocorticoid receptor mediates repression of AP-1-dependent genes in skin. J Cell Biol 1999;147:1365–70.
165. Reichardt HM, Tuckermann JP, Gottlicher M et al. Repression of inflammatory responses in the absence of DNA binding by the glucocorticoid receptor. EMBO J 2001;20:7168–73.

*The Hypothalamus–Pituitary–Adrenal Axis*
Edited by A. del Rey, G. P. Chrousos and H. O. Besedovsky


# *Effects of Glucocorticoids on the Developing Thymus*

ROSA SACEDÓN[1], ALBERTO VARAS[1], EVA JIMÉNEZ[1], CARMEN HERNÁNDEZ-LÓPEZ[2], JUAN J. MUÑOZ[2], ANGELES VICENTE[1], and AGUSTÍN G. ZAPATA[3]

[1]*Department of Cell Biology, Faculty of Medicine, Complutense University, Madrid, Spain*
[2]*Centre for Microscopy and Cytometry, Complutense University, Madrid, Spain*
[3]*Department of Cell Biology, Faculty of Biology, Complutense University, Madrid, Spain*

## ABSTRACT

Involvement of glucocorticoids (GCs) in the homeostasis of immune system has been extensively studied. However, numerous questions remain unresolved, including the role played by these hormones in the physiology of thymus gland, a central, key organ involved in T-cell maturation. In this study, we review the available information on the relationships between GCs and thymus gland with special emphasis on glucocorticoid receptor (GR) expression in the organ, the presumptive endogenous production of GCs in the thymus gland, and the role played by GCs in the development of thymocytes, thymic epithelial cells, and thymic dendritic cells (DCs). Our review concludes (1) the lack of correlation between GR expression on thymocyte subsets and sensitivity to GCs; (2) however, GCs could affect the maturation of lymphocyte progenitors which colonize the thymic primordium during early ontogeny; (3) furthermore, GCs affect also the development of main thymic stromal components, including epithelium and DCs; (4) the involvement of GCs in the determination of intrathymic T-cell repertoire is an important matter of discussion although recent data support a lack of effects of GCs on those processes.

## 1. INTRODUCTION

It has been repeatedly recognized that the neuroendocrine system and the immune system are interrelated by a bidirectional network. In this way, both systems maintain homeostasis. Today, it is impossible to understand the physiology of immune system without considering the homeostatic control carried out by the neuroendocrine system on the immune responses. In fact, several years ago, a negative feedback loop between both systems was proposed. Inflammatory cytokines such as interleukin-1 (IL-1), tumor necrosis factor alpha (TNF-α), and IL-6, which are generated during the immune responses and activate the hypothalamic–pituitary–adrenal (HPA) axis, cause an increased glucocorticoid (GC) release, which can suppress T-cell activation [1–4]

Thus, central to these neuroendocrine–immune relationships are the thymus and the HPA axis. The thymus is the first lymphoid organ involved in the production of T lymphocytes throughout a complex process, which begins with the homing of lymphoid precursors into the organ and ends with the emigration of simple positive (SP) (both $CD4^{+}CD8^{-}$ and $CD4^{-}CD8^{+}$)

thymocytes to the periphery. During this process of maturation, developing T cells undergo positive and negative selection to ensure that mature T cells that reach the peripheral lymphoid organs are capable of responding to foreign but not self-antigens presented by self-MHC-encoded molecules.

On the other hand, most thymocytes are extremely sensitive to GCs [5] and strongly express glucocorticoid receptors (GRs) [6], providing a cellular basis for the known effects of adrenalectomy [7] and GC administration on the thymus cellularity [8–10]. In addition, evidence supports an intrathymic production of GCs [11–13] and the thymus tissue expresses some mediators of the HPA axis such as corticotropin-releasing hormone (CRH) and adrenocorticotropic hormone (ACTH) [14,15], which could constitute a local HPA axis which some authors [12] have claimed to be important for controlling the thymus physiology.

These and other recent findings supporting a possible role for GCs in intrathymic T-cell development will be reviewed in this chapter. Involvement of GCs in T-cell precursor differentiation, their role in the processes of intrathymic T-cell selection, and the true relevance of endogenously produced GCs for the physiology of thymus gland will be some topics to be discussed. In addition, a few available data on GCs and thymic stromal cells, including dendritic cells (DCs), will be reviewed.

## 2. GR EXPRESSION IN THE THYMUS GLAND AND THYMOCYTE RESPONSES TO GCs

There are two types of GRs [6,16,17]: Type 1, mineralocorticoid receptor (MR), the distribution of which is restricted to nervous system and tissues concerned with $Na^+$ uptake [18]; and Type 2, GR, broadly expressed in most cell types. Furthermore, there are multiple N-terminal isoforms of GRs generated by translational mechanism from GR gene [19]. Remarkably, these isoforms seem to regulate different sets of genes, suggesting that they are not functionally equivalent [20].

Classical studies that identified the thymus gland and the thymocytes as the main targets of immune system for GCs indirectly suggested the occurrence of GR in the organ. Studies on GR expression in the thymus have confirmed that existence but provided, however, controversial results on the GR-containing thymocyte subpopulations partially due to the distinct protocols used for GR evaluation.

Ligand-binding assays have the disadvantage of requiring the removal of the endogenous ligand by adrenalectomy, a procedure that in itself modulates GR expression [21]. In addition, this protocol does not allow determining whether GR expression is homogeneously distributed within each thymocyte subpopulation. Studies by immunostaining on thymic cryosections, cytospin preparations, or, more frequently, for flow cytometry analysis are limited by the variable accessibility of reagents to the cytoplasm in different thymocyte subpopulations. Brewer and colleagues [22] generated knock-in mice in which a chimeric green fluorescent protein (GFP)–GR fusion protein expressed is in place of the endogenous GR alleles. In this case, the fluorescence intensity is an intrinsic property of the GFP–GR protein, eliminating the limitations created by the intracellular staining.

Studies by flow cytometry demonstrated the highest GR expression in DN ($CD4^-CD8^-$) thymocytes, whereas the lowest one was found in the DP ($CD4^+CD8^+$) $TcR\alpha\beta^{lo}$ cells, remarkably the most sensitive thymocyte subpopulation to GC-induced apoptosis [23]. The GR expression increased in parallel to that of TcRαβ within the DP cell compartment and, accordingly, the DP $TcR\alpha\beta^{hi}$ cells strongly expressed GR. Berki et al. [24] confirmed that DP thymocytes showed the lowest GR expression as compared with DN ($CD4^{-}CD8^{-}$) and mature SP cells. Mature $CD4^{+}CD8^{-}$ cells expressed lower GR levels than the $CD4^{-}CD8^{+}$ thymocytes. Studies on GFP–GR-knock-in mice have partially confirmed and extended those results [22]. In this experimental model, relatively low levels of GR protein were found in the mature $CD4^{+}CD8^{-}$ SP cells and DP ($CD4^{+}CD8^{+}$) thymocytes, whereas immature $CD4^{-}CD8^{+}TcR\beta^{lo}HSA^{hi}$ cells and, principally, DN ($CD4^{-}CD8^{-}$) cell subpopulations heavily expressed GR. Within this last cell subset, $CD25^{+}$DN cells express high levels of GR. Apparently, the GR expression increased concomitant with CD44 expression and peaked at the $CD25^{+}CD44^{-}$ stage of development [22].

These authors also analyzed the GR expression throughout thymus ontogeny, concluding that not only GR expression differs importantly in the fetal and adult thymus but also it gradually matures over a 2- to 3-week period after birth. In 15.5-day-old fetal mouse, thymus GRs are expressed at lower levels than those seen in $CD25^{+}$DN-cell population in the adult. Besides, an uniform GR expression occurs within the DN subset with minor differences between DN, immature $CD4^{-}CD8^{+}$, and DP thymocytes [22]. Newborn thymocytes also expressed intermediate levels of GR protein in both DN and CD8 cell compartment, but GR expression was already down-regulated in DP cells to level similar to those found in adult thymus. By postnatal day 7, the mean fluorescence intensity of $CD25^{+}$DN thymocytes approached that of adult cells, a condition reached from day 14 onward. Recently, other studies have confirmed the appearance of GR gene expression in the mouse thymus around 15 embryonic days as well as of 11β-hydroxysteroid dehydrogenase type 1 (11β-HSD1), which amplifies GC levels locally by regulating the interconversion of active GCs in their inactive metabolites [25]. Previously, we found GR expression in a 13-day-old rat's $CD45^{+}$ fetal liver cells, a population of cell progenitors that contains thymic precursors [26]. In addition, Kitraki et al. [27] had demonstrated by in situ hybridization presence of GR mRNA in rat fetal liver of 12 days of gestation when the hematopoietic activity of the organs begins. Also, early intrathymic cell precursors express GR in rat fetal thymus [26,27], and Ranelletti et al. [28] demonstrated that human intrathymic precursors ($CD3^{-}CD1^{-}$ cells) contained high numbers of GR sites per cell.

A first conclusion raises from these results [22,24]: although previous evidence suggested that the expressed amount of GR determines the level and nature of the response to GCs, these new data show that GR expression (or the relative GR protein levels) does not correlate with the sensitivity to GCs. Thus, exogenous GC administration induced important apoptosis in immature $CD4^{-}CD8^{+}$ cells, which express relatively high levels of GR, and DP thymocytes with low GR expression but not DN thymocytes, which also exhibit a high expression of the receptor [22]. Accordingly, apart from the DP thymocytes classically reported as the most steroid-sensitive thymic cells, GC sensitivity begins in the thymus before cells reach the DP cell compartment [22].

On the contrary, these results also suggest that GR affinity rather than GR expression, or factors others than GR, control the GC-induced death of thymocytes. Brewer and colleagues [22] proposed that one of those factors could be SRG3, a mouse homologous of yeast SW13 and human BAF 155 initially isolated as a gene expressed highly in the thymus but at a low level in periphery [29]. SRG3 is a component of mouse SWI/SNF complex, a chromatin-remodeling complex required for the regulation of transcriptional activity associated with development, cell differentiation, and proliferation [30,31]. SRG3 binds to the GR complex and seems to modulate GC-induced apoptosis because the expression of antisense RNA to SRG3 in a thymoma cell line decreases the cell death mediated by GCs [29,32], whereas its overexpression in peripheral T cells leads to the opposite effect [29,32]. Furthermore, Choi and co-workers [33] have demonstrated that the expression level of SRG3 determines GC sensitivity of the thymocytes so that lowering of SRG3 after positive selection confers GC resistance on developing thymocytes. This same group recently reported a role for Notch 1, a molecule known to be involved in T-cell selection, in this process. Apparently, Notch signaling confers developing thymocytes GC resistance by regulating SRG3 expression through Deltex, a Notch 1 ligand, blocking the recruitment of p300 to the E2A/HEB transcription factor [34]. On the contrary, preliminary studies by Brewer et al. [22] have not detected differences in SRG3 expression between DN and DP thymocytes. Indeed, other factors, for instance CD28 expression, seem to confer protection to GCs [35]. SP thymocytes from CD28 knock-out mice exhibit increased sensitivity to GC administration [20] and after becoming sensitive to GCs in culture restore the resistance to GC treatment reinforcing CD28 signaling without TcR engagement [35].

On the contrary, the use of alternative promoters of the GR gene has been shown to correlate with the sensitivity of lymphocytes to apoptosis [36,37]. In mice, the GR gene is transcribed from five different promoters named 1A, 1B, 1C, 1D, and 1E. Nearly, all cells express GR from promoters 1A–1E, but the activity of the 1A promoter appears to be restricted to thymus or lymphocyte cell lines. Furthermore, the relative level of expression of the 1A promoter to the 1B–1E promoters within a lymphocyte population directly correlates with its susceptibility to GCs. Thus, the highly GC-sensitive DP ($CD4^+CD8^+$) thymocytes have the highest levels of GR 1A promoter activity, whereas both GC-resistant mature thymocytes and peripheral T cells exhibit the lowest values [37].

## 3. ENDOGENOUS PRODUCTION OF GCs IN THE THYMUS GLAND

GC hormones are synthesized in the adrenal gland from the precursor cholesterol through sequential conversions performed by members of the cytochrome P450 superfamily of oxidases and 3β-hydroxysteroid dehydrogenase. Recent studies have demonstrated that the thymus also possesses the entire cohort of enzymes and cofactors required for GC production. However, whether the thymus is a source of functionally significant levels of GCs remains to be conclusively determined as well as the conditions in which thymus produces endogenous GCs and the thymic cell components concerned with their production.

In a pioneer study, Ashwell and colleagues [11] showed, as measured by RIA, that murine thymus mitochondria could synthesize pregnenolone from cholesterol, and fetal organ thymus cultures (FTOCs) produced 11-deoxycorticosterone. However, these authors only indirectly detected P450c11β activity within the thymus, an important enzyme responsible for the final step in GC synthesis that enables the hormone for binding to the GR. Other studies have confirmed that thymic epithelial cells express steroidogenic cytochrome P450 hydroxylases and can activate in a paracrine way a GR-dependent reported gene in co-cultured cells or induce apoptosis of co-incubated thymocytes suggesting that produced GCs, although in the range of fentomoles, are biologically active [12]. Furthermore, this activity was blocked by RU486, a GR antagonist, which competes with GC for binding site in the GR.

In situ immunohistochemical studies have identified both cortical and medullary epithelial cells as the steroidogenic cells within the thymus [13]. However, these authors have emphasized that an intact thymus architecture is necessary for GC production because 11β-hydroxylase activity could not be detected in irradiated thymus nor in a thymic epithelial cell line. On this regard, other studies have pointed out that thymic epithelial cell lines can produce GC only after contacting with thymocytes [38]. Thus, Jenkinson and colleagues [39] easily detected P450 scc mRNA expression in whole thymus but not in purified $MHCII^+$ thymic epithelial cells obtained from thymus depleted of lymphoid cells concluding that thymic epithelial cells are only induced to produce steroids after interaction with lymphoid cells. However, thymic epithelial cells still do not express this enzyme after 4 days of reaggregation culture with DP cells although they were able to support positive selection of thymocytes. On the contrary, the expression of cytochrome P450 sec transcripts was up-regulated in the positively selected DP $CD69^+$ thymocytes purified from the cultures [39]. Authors suggest the possibility of inducing endogenous steroid production in thymocytes rather than in thymic epithelial cells after TcR ligation although it was not measured and results need further confirmation. In supporting a role for thymocytes in determining intrathymus GC levels, medullary thymocytes express 11β-HSD1, the enzymatic machinery required for regenerating active GCs from inactive 11-ketometabolites [40].

Recently, Jondal and colleagues have generated a conditioned transgenic mouse [41] that specifically expressed the rat GR transgene in mouse T cells inducible by tetracycline under control of a human T-cell-specific CD2 promoter. In adrenalectomized transgenic animals, which do not produce systemic GCs, there is a dramatic increase in thymocyte death by apoptosis after doxycycline induction of the GR transgene that must be provoked by endogenously produced GC. Similar results were in vitro obtained by using organ cultures established from thymus lobes isolated from transgenic animals. Moreover, in vivo and in vitro addition of the GR antagonism RU 486 partially blocked the increased apoptosis. Despite these results, many questions remain to be answered on the functional significance of endogenous production of GCs in the thymus. Some authors have claimed that the intrathymic de novo steroidogenesis could be insufficient to exert any effect on thymocyte biology [42]. In fact, thymic content of cortisone is very low on the order of 1/100000th of the content of the adrenal gland [43]. On the contrary, the thymocyte number is strongly reduced in transgenic rats overexpressing a modified GR with increased ligand affinity, but adrenalectomy restores cellularity within 3 weeks after surgery [20].

## 4. GCs AND T-CELL DEVELOPMENT

Maturation of lymphoid precursors and selection of TcR repertoire are the two main processes that occur within the thymus. Numerous results have supported a role for GCs in both processes although data are controversial and remain to be fully understood.

GC involvement in early stages of T-cell maturation has been claimed by some authors, but most results need further confirmation and are difficult for comparing because they have been obtained in very different experimental models.

Mice with either targeted disruption of the GR gene [44] or expressing a GR unable to bind DNA [45] did not show evident changes in the thymus. Godfrey and colleagues [46] confirmed these results, concluding that GCs are not involved in the maturation of T-cell progenitors. Also, Brewer et al. [22] have reported, by analyzing fetal T-cell maturation in mice deficient in the GR, that direct GR signaling is not essential for normal thymocyte development. These authors found at embryonic day 16.5 no differences in thymocyte numbers and subset percentages between GR-deficient and GR-intact embryos. More recently, this same group confirmed its own results on unaffected T-cell development in GR-deficient mice and observed a high mortality after anti-CD3-induced polyclonal activation of GR-deficient T cells [47]. Neither, Finotto and colleagues [48] found altered thymic cellularity and T-cell subset distribution in a mouse containing an exon 3-deleted GR gene. GR-deficient mice that carry the point mutation A458T ($GR^{dim}$ mice) exhibited thymocyte numbers and subset proportions unaltered. Furthermore, thymocytes were refractory to GC-induced apoptosis, but the inhibition of cytokine expression was normal in both thymocytes and splenic T cells [45].

On the contrary, other results support a role for GCs in T-cell development. Mice that express a targeted GR antisense transgene, resulting in impaired GR expression in thymocytes, show blockade of the normal progression of DN thymocytes to the DP cell compartment claiming that GCs could be required to antagonize apoptosis signals delivered via pre-TcR [49]. In addition, recent results reveal that, at least in both naïve and activated $CD4^+$ T cells, GCs can up-regulate IL7R$\alpha$ expression [50] which, together with IL7, is required for the normal progression of DN cells to DP cell stage.

We studied other transgenic mice that presented a significant decrease of GC-binding capacity [51]. These mice exhibit a partial blockade of T-cell maturation and decreased proportion of apoptotic cells during fetal period but not in adult life. Seventeen-day-old transgenic fetal thymuses contained higher percentages of both DN TcR$\alpha\beta^-$ and $CD4^+CD8^-$ TcR$\alpha\beta^-$ cells than those of normal, control littermates and decreased proportions of DP TcR$\alpha\beta^{-/lo}$ thymocytes.

In addition, analysis of the distinct DN cells subsets in a 15-day-old Tg mice demonstrated a significant reduction of the percentage of most differentiated subpopulation $CD44^-$ $CD25^-$, together with a slight increase of $CD44^+$ $CD25^+$ DN cells. Confirming this blockade of T-cell maturation, 2 days later, the numbers of $CD44^-$ $CD25^-$ DN thymocytes remained unchanged and the increase of the proportion of $CD44^+$ $CD25^+$ DN cells was higher than that observed at day 15 of gestation.

We additionally used other experimental approach to analyze the relevance of GCs for the T-cell maturation during thymus ontogeny. This experimental model ensures, by adrenalectomy of rat pregnant mothers, the absence of circulating GCs during the earliest stages of development of their progeny (Adx fetuses) [26]. In this condition, the Adx fetal thymuses exhibit decreased numbers and accelerated maturation of thymocytes that is normalized after recovering of GC levels by the implantation of osmotic minipumps that release GCs. On day 15 of fetal life, the Adx thymuses already present immature $CD4^-$ $CD8^+$ TcR$\alpha\beta^{-/lo}$ cells as well as some DP thymocytes, which appear for the first time in control rat thymuses on fetal day 18. One day later, most Adx thymocytes have reached the DP stage and there is an important proportion of mature SP TcR$\alpha\beta$ cells, whereas in control rat thymuses they appear on days 20–21 of fetal life.

The reduced thymic cell content found in Adx fetuses, although it could be reflecting the homing in the thymic primordium of a low number of cell progenitors, and/or the migration of mature cells to the periphery, correlates well with the observed increased proportions of apoptotic cells. We assume that these increased levels of apoptosis of the Adx thymuses could be related to changes in the above-mentioned GC intrathymic production. Remarkably, a semiquantitative PCR analysis demonstrated RNA transcripts specific for Cyp 11$\beta$-1, a key enzyme in the GC synthesis pathway, in 16-day-old Adx thymuses. In contrast, the expression of this enzyme in control thymuses was below detection levels for the used PCR conditions, indirectly suggesting increased or accelerated production of GCs in the Adx thymuses [52].

Other authors [53] have commented these results on the basis that in adrenalectomized mothers both ACTH and CRH levels are elevated by the lack of negative feedback by GCs and that elevated ACTH and/or other HPA products could induce increased local thymic GC production. However, maternal ACTH does not seem to cross the placenta [54], and the fetal levels of circulating ACTH are similar in the progeny of adrenalectomized and control rats [55]. In fact, our study demonstrates a role for GC in the maturation of fetal liver lymphoid progenitors [26]. Firstly, ultrastructural analysis confirmed the more precocious arrival of lymphoid progenitors to the Adx thymic primordium than to control one.

On the contrary, both numbers and phenotype of these 13-day-old fetal liver lymphoid progenitors are similar in Adx and control fetuses but alymphoid FTOCs reconstituted for 12 days with fetal liver Adx precursors contained a significantly higher proportion of mature TcR$\alpha\beta^{hi}$ cells than those receiving control ones. This higher proportion of recovered mature thymocytes was not originated by expansion but by accelerated maturation of Adx fetal liver cell precursors that are affected by the lack of GC before reaching the thymus primordium [26].

In the last years, Jondal and colleagues have provided additional results for supporting the role of GCs in T-cell maturation and the above-mentioned relevance of endogenous intrathymic GC production in the process. In this case, the studies were performed on transgenic mice containing a rat GR transgene under the control of either the proximal lck promoter [56,57] or the human CD2 promoter [41]. In these models, there was a specific twofold increase of GR levels and, accordingly, increased sensitivity of thymocytes and T cells to corticosterone. Remarkably, overexpression of GR specifically in the T-cell lineage resulted in a reduction of the number of thymocytes that affected most subpopulations, largely to DP ($CD4^+CD8^+$) cells, except to CD8 SP thymocytes in young mice [56], but had the opposite effects in aged mice that delayed the onset of gradual disappearance of thymocytes [57]. In addition, adrenalectomized transgenic mice unable to produce systemic GCs exhibit important increased numbers of apoptotic cells, presumably induced by intrathymic secreted GCs [41].

During the process of T lymphopoiesis, a major function of the thymus is to carefully check developing thymocytes to eliminate those capable to recognize and react to the self-Ag/MHC-encoded protein complex. As known, T lymphocytes, unlike immunoglobulins, only recognize peptide fragments presented by MHC-encoded proteins. For carrying out this discrimination, DP thymocytes that weakly express TcR$\alpha\beta$ undergo antigen-specific selection. The outcome for DP thymocytes bearing TcRs unable to recognize self-peptide/MHC complex is death by neglect. In contrast, DP TcR$\alpha\beta$ cells that express receptors with moderate affinity/activity for self-Ag/MHC are rescued of apoptosis (positively selected), they survive and differentiate to SP ($CD4^+CD8^-$ or $CD4^-CD8^+$) thymocytes. On the contrary, thymocytes expressing TcRs with high affinity/avidity for self-Ag/MHC also undergo apoptosis, negative selection, a key process for establishing immune tolerance to self.

Several authors have claimed a major role for GCs in those selective processes formulating the so-called mutual antagonism hypothesis [4,12,53,58]. This hypothesis proposes that GR signaling antagonizes TcR signaling converting strong negative signals into a weaker positive selection signal. Likewise, when GR signaling is inhibited in association with TcR ligation, cells that might have been positively selected are in fact deleted.

Thus, in neglected DP TcR$\alpha\beta$ thymocytes that do not recognize self-Ag/MHC, there is a greater ratio of GR to TcR stimulation and the cell will die by GC-mediated apoptosis. If thymocytes strongly recognize self-Ag/MHC, with a resultant higher ratio of TcR to GR signaling, the cells will die by negative selection through activated-mediated apoptosis. In an intermediate condition, in which there is a weaker recognition of self-Ag/MHC, a balance exists between the GR and the TcR activation, and the thymocytes are positively selected. Other authors point out that really positive selection involves the rescue from apoptosis by TcR-dependent signaling, which would block the effects of GR signaling [59].

Evidence supporting the mutual antagonism between GR signaling and TcR signaling has been principally provided by Ashwell's and Jondal's groups although there are some differences in their respective proposals and, sometimes, reported results are contradictory [53,56,58–60]. Their main results are summarized as follows:

1. GCs prevent apoptosis produced by TcR ligation in T-cell hybridomas and thymocytes [61,62]. However, incubation of a thymic epithelial cell line, which shows certain P450 steroid hydroxylase activity, with RU486 protects co-cultured DP thymocytes from apoptosis [38].
2. Stimulation of thymocytes via the TcR/CD3 complex or GR triggers apoptosis. However, simultaneous stimulation of both TcR and GR results in protection of thymocytes [61].
3. Addition of metyrapone (an inhibitor of GC synthesis) or RU486, which blocks GR signaling, to FTOCs causes apoptosis of Ag-specific TcR transgenic DP thymocytes that normally are positively selected. However, there is no effect if the MHC haplotype is unable for presenting the selecting peptide, indicating that the process is dependent on TcR signaling [11,63]. However, Jondal and colleagues found that the GC antagonism RU486 inhibits apoptosis induced by a peptide in the context of MHC class II but not that induced by another peptide in a MHC class I context [64].
4. Apoptosis due to inhibition of GC biosynthesis in FTOCs is proportional to the number of expressed MHC-encoded molecules [65].
5. Mice that express a GR antisense transgene under control of lck proximal promoter in thymocytes but not in peripheral T lymphocytes have a small thymus with increased numbers of apoptotic thymocytes and DP cells unusually sensitive to TcR-induced cell death [49]. By varying the level of available TCR ligands and consequently the probability of TCR occupancy, DP thymocyte numbers are partially restored in these mice [66]. The number of T lymphocytes positively selected, as evaluated by the use of certain Vβ regions, is reduced in these mice. In fact, they have an altered TcR repertoire, being unable to respond to some complex antigens, that is pigeon cytochrome c, but not to others [67]. However, Jondal et al. [56,59] using the same $p56^{lck}$/GR antisense construct found an increased thymus size and thymocyte numbers, without a plausible explanation for these controversial results. It is important, however, to remark that these authors emphasize that lck proximal promoter is equally active in thymocytes and resting and activated peripheral T cells in contrast to the assumption by King et al. [49] and other authors [68] that the promoter is predominantly active in immature thymocytes and relatively inactive in peripheral T cells.
6. CD5, expression on DP cells, an indicator of TcR signaling, is MHC-dependently up-regulated when the GC signaling is in vivo or in vitro diminished. In agreement to these results, Berki et al. [24] reported that combined a low dose of anti-CD3 mAb and dexamethasone, treatment resulted in an increased maturation of thymocytes followed by the increase of positively selected $CD69^+$ DP cells.
7. Jondal and colleagues analyzed the GC sensitivity in thymocytes from DO 11.10 mice expressing a transgenic selecting TcR [69]. These mice, which express a transgenic TcR specific for $I\text{-}A^d$-restricted OVA 323–339 peptide, show enlarged thymus with increased cellularity and reduced apoptosis. Authors conclude that positive selection renders DP cells comparatively more resistant to apoptosis by GC in association with reduced levels of GR.
8. Increased expression of CBP/p300, a factor necessary for both the functioning of transcriptional factors essential for T-cell activation and the transcriptional activation of GR [70], inhibits the thymocyte apoptosis prevented by the TcR–GR antagonism [71]. In these transgenic mice, there is an important decrease in the number of thymocytes and increased apoptosis in the DP cell compartment, supporting the survival role for GCs pointed out by others authors [58]. However, in this model, the positive selection was minimally affected.

Thus, the selection of TcR, that recognize pigeon cytochrome c 81–104 peptide presented by I-$E^K$, was not altered in the p300 transgenic mice. Likewise, T-cell responses to hen egg lysozyme 81–96, ovalbumin 323–339, and repressor $\lambda$73–88 were not reduced by the presence of the p300 transgene. Authors conclude that only the selection of a very few thymocyte clones could be GC dependent [71]. In this respect, it is important to remark that TcR signaling reverse GC-induced apoptosis seems to be dependent on the developmental stage of thymocytes and could largely affect a minor DP $TcR^{lo}$ cell subpopulation [72].

9. Neonatal dexamethasone treatment induces long-lasting changes in the peripheral T-cell repertoire preceded by changes in intrathymic corticosterone production and in the CD4/CD8 thymocyte ratio [73].
10. The mutual antagonism hypothesis also could explain certain characteristics of autoimmune syndromes. For instance, elevated GC levels could compensate for the high-affinity/avidity signaling of autoreactive clones and rescue those thymocytes from apoptosis resulting in an increase of autoreactive T cells [13]. In contrast, decreased GC signaling in thymocytes alters the T-cell repertoire causing diminished autoimmunity in MRL/MP-$fas^{Lpr}$ mice [74].

It is evident from all these results that there is considerable controversy on the role played by GCs in the development and selection of thymocytes. It is particularly remarkable that different groups reach different results using the same experimental models. For instance, the analysis of transgenic mice bearing an antisense construct of the GR gene [49,51,75]. On the contrary, the high sensitivity of thymocytes to GCs and the endogenous production of GCs in the thymus strongly support a role for these molecules in T-cell maturation. In this respect, a more detailed analysis of the used experimental model, especially of the deficient mice, has demonstrated that most animals could retain certain GR-mediated functional capacities. For example, cells from deficient mice that contain an exon 2-deleted GR gene [46,76] expressed a truncated but ligand-responsive C-terminal fragment that retains the capacity to regulate gene expression [77]. In these mice, Mittelstadt and Ashwell [78] demonstrated that dexamethasone treatment induced similar changes in GR-deficient and normal fetal thymocytes. On the contrary, in the KO mice generated by Brewer and colleagues [47], also by deletion of exon 2, the removal of GR seems to be total, but western blot analysis showed a non-specific band at a position corresponding to a presumptive C-terminal fragment. Unfortunately, authors did not provide data on ligand binding, PCR, or microchip assays. Finally, cells from GR dim mice carrying a point mutation in the GR gene express GR monomers with undefined effects, and peripheral T lymphocytes from these mice are sensitive to GC treatment [45].

Jondal also emphasizes that the different results raised from distinct studies using GR signaling blocking with the antagonism RU486 could be due to the different doses used: 20 μM in the study by Jondal and colleagues [64] and 0.1 μM in that by the Ashwell's group [11]. In addition, RU486 blocks not only GR signaling but also that of the progesterone receptor, another hormone that could also influence thymocyte behavior [79]. On this respect, it is important to note that GC inhibitors used in many of these studies, such as metyrapone, increase the steroid precursors, including progesterone, the substrate prior to deoxycorticosterone [13] that could affect the in vitro T-cell differentiation.

On the contrary, there is a general criticism on the experiments performed on FTOCs. Recent data on GR expression in FTOCs show that, after 7 days of culture, FTOCs express intermediate GR levels in DN and immature $CD8^+$ thymocytes, decreasing to the basal levels in both DP and SP cells. Accordingly, it seems difficult to compare results obtained in FTOCs with those raising from adult mice differently expressing GR [22].

In fact, the only conclusion that it can be reached from all these results is that many new studies remain to be done to conclusively determine the role of GCs in T-cell development.

The molecular basis of the relationships between GR signaling and TcR signaling is also a matter of discussion. Although, as mentioned above, how and when endogenous GCs specifically affect thymocyte development remains unresolved, we know that they affect key molecules of TcR signaling, including Zap 70, the linker for activation of T-cells (LAT), AP-1, nuclear factor (NF)-κB, and others [58,80]. Apparently, GCs block an early step of TcR signaling resulting in impaired $Ca^{2+}$ flux, interfere with phosphorylation of CD3-associated substrates, including the $\zeta$ chain, and affect the submembrane compartmentalization of several key molecules for starting TcR signaling, such as p56 lck and p59 fyn [80]. Accordingly, these authors proposed that GCs could affect thymocyte development by modifying the early signals of TcR pathway, converting an agonist signal (resulting in negative selection) into a positive selecting partial agonist signal.

On the contrary, GCs could inhibit apoptosis caused by TcR signaling by, at least, two different ways. Activation-mediated apoptosis involves up-regulation of the Fas ligand, and GCs prevent this up-regulation probably throughout the so-called GC-induced leucine zipper (GILZ) gene. Also, GC could interfere with the process by activation of GC-induced TNFR family-related (GITR) gene, which can inhibit CD3-mediated but not Fas-mediated apoptosis. In this respect, two recent reports are important to mention. Zhan and colleagues [81] have demonstrated that TCR-mediated activation promotes GITR up-regulation in T cells and resistance to GC-induced death. In addition, anti-apoptotic effects of GCs on TcR-dependent cell death are mediated by direct binding of the GR to two recently identified GC-responsive DNA elements in the CD95L promoter [82]. On the contrary, Jamieson and Yamamoto [83] have described that TcR-mediated activation of the MEK/ERK cascade led to a reduced capacity of GR to mediate apoptosis.

Other results [71] recognize CBP/p300 as a critical element for regulating GR-mediated thymocyte death and GR–TcR antagonism in thymocytes. As described above, many transcriptional factors function throughout binding to CBP/p300. Some of these factors are essential for the GR signaling, whereas others, including cAMB response element-binding protein (CREB), AP-1, NF-kB, and nuclear factor of activated T-cells (NFAT), are key for T-cell activation [70]. On this respect, it has been proposed that the mutual inhibition between AP-1 and GR [84] may be mediated by competition for CBP/p300 [85]. For recent results on the molecular mechanisms governing GC-induced apoptosis, readers must review the works of Wang et al. [86] and Herold et al. [20].

## 5. GCs AND THYMIC STROMAL CELL COMPONENTS

Although thymic epithelial cells express ten times more GR than thymocytes [87], there are many few data on the effects of GC on thymic stromal cell components. We analyzed this issue in the thymus of the two above-mentioned experimental models: the progeny of adrenalectomized pregnant rats [88] and a transgenic mouse that has GR with decreased capability to bind GCs [51]. Previously, a few studies had demonstrated that GCs induce expression of certain keratins in the thymic epithelium [89,90] and regulate the pattern of secretion of epithelial cytokines [91] as well as that of extracellular matrix components, such as laminin and fibronectin [90,92,93].

As demonstrated for thymocytes, combined ultrastructural and immunohistochemical analyses demonstrated an accelerated maturation of the thymic epithelial meshwork in Adx fetal rats [88]. Specific markers for adult cortical epithelial cells are expressed in the thymus of 16-day-old Adx fetal rats, but they are not evident until 19 in control ones. In addition, although isolated cytokeratin positive cells appear on day 16 in the thymus of both groups of rats, a true epithelial cytoreticulum is appreciated in 17-day-old Adx fetal thymus but not in that of control fetuses. Likewise, early expression of MHC molecules is evident in both Adx and control thymus but, whereas class II positive cells constituted a compact reticulum in an incipient thymic medulla of 16-day-old control rats, in Adx thymuses the pattern is reticular and similar to that found in control thymuses on day 19. At that stage, the class II expression exhibits an adult pattern in Adx but not in control thymuses.

This accelerated maturation of the epithelial meshwork was confirmed by electron microscopy. In 13-day-old control fetuses, the thymic primordium is formed by a homogeneous network of primitive epithelial cells but in Adx ones consists of more differentiated epithelial cells provided with well-organized desmosomes and bundles of cytoplasmic microfilaments.

Differences were not so evident in the pattern of appearance and development of extracellular matrix components between Adx and control thymus. Fibronectin expression was similar in the two groups of fetuses, and the evolution of laminin showed after day 18 a slight delay in Adx thymuses compared with that in the pattern of control ones.

Finally, macrophages were detected by acid phosphatase reactivity and appearance of ED1-positive or R-MC42-positive cells. First, acid phosphatase positive cells appeared on day 17 in control fetal thymuses and 1 day earlier in Adx ones. Furthermore, although $ED1^+$ or $R\text{-}MC42^+$ cells appeared at the same stage in both groups of fetuses (days 19–20 of gestation), the number of positive cells was higher in the Adx thymuses.

Immunohistochemical analysis of the thymuses of control and transgenic mice, which show decreased GR signaling, using mAbs specific to different stromal elements demonstrated important alterations in the later ones. Although the pattern of development of thymic epithelial network during ontogeny and the levels of cytokeratin expression were similar in control and Tg mice, in these later ones, the cortical epithelial cytoreticulum was slightly reduced and exhibited an importantly decreased expression of MHC class II molecules. Previous evidence had pointed out down-regulated expression of MHC molecules after GC administration [94] although other reports show the GC capacity to stimulate MHC-molecule expression [95].

From the earliest stages of development, wt thymuses strongly expressed laminin associated to the basement membrane of thymic capsule and trabeculae, in the intralobular perivascular spaces and as a discontinuous network throughout the organ. The pattern was similar in Tg thymus except for the occurrence of blood vessels devoid of laminin. In addition, they exhibited an important vasodilatation in both laminin-positive and laminin-negative blood vessels. Also, the thymic epithelial network of Tg adult mice was profoundly altered. It contains large areas devoid of cytokeratin, laminin, and MHC class II molecules throughout both cortex and medulla that seem to be principally occupied by immature DP thymocytes. The class II expression on thymic reticulum was reduced, mainly under the connective tissue capsule, and enlarged blood vessels occurred in the cortico-medullary border.

The changes in both fetal and adult blood vessels in Tg thymuses could be associated with modifications in their permeability, which could explain the known property of GC to redistribute lymphocytes [96]. Nevertheless, decreased index of sympathetic activity has been described in these transgenic mice [97] and reduced sympathetic outflow might well participate in some of the observed effects.

On the contrary, reduced numbers of F4/80-positive macrophages occurred in both fetal and adult Tg thymus although their pattern of distribution throughout the organ was similar to that of control mice.

All these results suggest, therefore, the requirement of GC for a normal development of the thymic stromal cell components.

As mentioned above for the thymic epithelial cells, very little is known about the possible effects of GCs on other thymic components, apart from thymocytes, although they are very relevant for thymic function. This is the case of DCs known to be largely involved in intrathymic negative selection [98].

Numerous studies during recent years have analyzed the effects of GCs on distinct parameters of peripheral but not thymic DCs [99–104]. We demonstrated that overnight exposure of thymic DCs to physiological doses of dexamethasone impaired their allostimulatory capacity [105]. Into the scope of this review, however, only a few data support a role for GCs in DC development. Dexamethasone prevents the terminal maturation of an epidermal-derived DC line [106], but this synthetic steroid, as well as prednisolone, does not affect Langerhans cell development and function although they block the generation of dermal/interstitial type DC [101]. In addition, neonatal rats diminish the DC network of respiratory tree after GC inhalation [107].

We demonstrated that rat thymic DCs express GR and respond specifically to GCs [105]. In addition, the development of thymic DCs was profoundly altered in absence of GCs [108]. Their numbers were extremely high in the earliest fetal stages of Adx thymus, and their maturation and turnover were strikingly accelerated. Thus, mature DCs appeared in the Adx thymus on day 15 of fetal life, representing 25% of the total thymic DCs, whereas in control fetuses they did not occur until day 17 accounting for less than 9% of total DCs. According to these results, the life cycle of thymic DCs in Adx fetuses occurs in parallel to that of T lymphocytes. Accelerated maturation of both cell types indirectly supports the existence of a common precursor for both cell types, as suggested by other authors [109], the biology of which could be altered in absence of GCs.

Although we had demonstrated that IL7 treatment of rat FTOC resulted in a massive production of DCs [110] similar to that observed in the thymus of 15- to 16-day-old Adx fetal rats, it is difficult to assume a role for IL7 in this model if, as mentioned above, GCs are necessary for inducing IL7R$\alpha$ expression [50]. Other factors involved in thymic DC differentiation, such as FLT3L, GM-CSF, or other currently unknown, could be, therefore, affected by the lack of GCs.

## 6. CONCLUSIONS AND FURTHER DIRECTIONS

Taking together all these results, it is difficult to reach convincing conclusions on the mechanisms, which govern the unquestionable sensitivity of thymocytes to GCs.

Firstly, GR expression in different thymocytes subsets does not correlate with their sensitivity to GCs. Most studies recognize that DN ($CD4^-CD8^-$) thymocytes highly express GR, whereas the lowest expression corresponds to DP ($CD4^+CD8^+$) cells, the most sensitive thymocyte subpopulation to GCs. Accordingly, it appears necessary to conclude that GR affinity and the participation of other factors, rather than GR expression, may account for the GC sensitivity of thymus tissue.

There is also controversy on the physiological significance of the endogenous production of GCs by thymic epithelial cells. Although some authors emphasize their involvement in the intrathymic positive and negative selection of developing thymocytes [12,58], others claimed that the effects of GCs on T-cell differentiation were really mediated through changes in the levels of extrathymic circulating GCs [20,22].

On the contrary, the involvement of GCs in both T-cell maturation and thymocyte selection is controversial, particularly because different researchers working with identical experimental models reach opposite conclusions. Thus, mice deficient in GR do not show defects in T-cell maturation [22,44–46], but other mice suffering a total or partial impairment of GR signaling show a blockade of T-cell differentiation at the DN ($CD4^-CD8^-$) stage [49,51]. Furthermore, the absence of GCs during the earliest stages of development affects the maturation of fetal liver T-cell progenitors, resulting in accelerated maturation of developing thymocytes [26]. We conclude, despite the difficulties to comparatively analyze results obtained from so different experimental models, that GCs could largely affect primitive lymphoid precursors that home into the thymus early in ontogeny.

The involvement of GCs in the determination of T-cell repertoire within the thymus is also an important matter of discussion. Nevertheless, all these studies have been carried out in genetically manipulated mice, the physiology of which is profoundly altered.

The mutual antagonism hypothesis was proposed for explaining the different behavior of developing DP thymocytes during intrathymic selection in the presence or absence of GCs. The hypothesis assumes that GR signaling antagonizes TcR signaling modifying the strength of the signals, which determine the death or the rescue of developing thymocytes. However, some authors claim that T-cell selection results from the balance or imbalance between TcR signals and GR signals [58,60], whereas others suggest that indeed TcR signaling blocks the effects of GR signaling [59]. Moreover, while some results support increased sensitivity to apoptosis in absence of GCs [63,65], other suggest that this thymocyte sensitivity decreases after inhibition of GR signaling [12].

Despite some methodological criticism, more recent data seem to be very conclusive in supporting the lack of effects of GCs on those processes [22,46]. We have no convincing explanation for these contradictory results. Perhaps, a better knowledge of the molecules involved in GR signaling and TcR signaling, and on their relationships throughout common targets, could help to understand the true nature of intrathymic processes governed by GCs and so, to clarify this issue.

On the contrary, mice defective in GR signaling, previously used for analyzing the role of GCs in T-cell maturation and selection, could be an excellent tool to study the effects of GCs on thymic stromal cell components, a topic that remains largely unknown. On this respect, our results demonstrated a remarkable correlation between the accelerated development of thymocytes and that of thymic stromal cell components, including epithelial cells and DCs. Furthermore, other results support that partial blockade of GR signaling causes important changes in the cytoarchitecture of thymic epithelial network, vascularization, and expression of extracellular matrix components [51].

## ACKNOWLEDGMENT

This work was supported by grants BMC2003-01901, BFU2004-03132, and BFU2006-00651/BMC.

## REFERENCES

1. Besedovsky H, del Rey A, Sorkin E, Dinarello CA. Immunoregulatory feedback between interleukin-1 and glucocorticoid hormones. Science 1986;233:652–4.
2. Wick G, Hu Y, Schwarz S, Kroemer G. Immunoendocrine communication via the hypothalamo-pituitary-adrenal axis in autoimmune diseases. Endocr Rev 1993;14:539–63.

3. Chrousos GP. The hypothalamic-pituitary-adrenal axis and immune-mediated inflammation. N Engl J Med 1995;332:1351–62.
4. Webster JI, Tonelli L, Sternberg EM. Neuroendocrine regulation of immunity. Annu Rev Immunol 2002;20:125–63.
5. Scollay R, Shortman K. Thymocyte subpopulations: an experimental review, including flow cytometric cross-correlations between the major murine thymocyte markers. Thymus 1983;5: 245–95.
6. Miller AH, Spencer RL, Stein M, McEwen BS. Adrenal steroid receptor binding in spleen and thymus after stress or dexamethasone. Am J Physiol 1990;259:E405–12.
7. Akana SF, Cascio CS, Shinsako J, Dallman MF. Corticosterone: narrow range required for normal body and thymus weight and ACTH. Am J Physiol 1985;249:R527–32.
8. Gruber J, Sgonc R, Hu YH, Beug H, Wick G. Thymocyte apoptosis induced by elevated endogenous corticosterone levels. Eur J Immunol 1994;24:1115–21.
9. Compton MM, Caron LA, Cidlowski JA. Glucocorticoid action on the immune system. J Steroid Biochem 1987;27:201–8.
10. Screpanti I, Morrone S, Meco D, Santoni A, Gulino A, Paolini R, Crisanti A, Mathieson BJ, Frati L. Steroid sensitivity of thymocyte subpopulations during intrathymic differentiation. Effects of 17 beta-estradiol and dexamethasone on subsets expressing T cell antigen receptor or IL-2 receptor. J Immunol 1989;142:3378–83.
11. Vacchio MS, Papadopoulos V, Ashwell JD. Steroid production in the thymus: implications for thymocyte selection. J Exp Med 1994;179:1835–46.
12. Pazirandeh A, Xue Y, Rafter I, Sjovall J, Jondal M, Okret S. Paracrine glucocorticoid activity produced by mouse thymic epithelial cells. FASEB J 1999;13:893–901.
13. Lechner O, Wiegers GJ, Oliveira-Dos-Santos AJ, Dietrich H, Recheis H, Waterman M, Boyd R, Wick G. Glucocorticoid production in the murine thymus. Eur J Immunol 2000;30: 337–46.
14. Batanero E, de Leeuw FE, Jansen GH, van Wichen DF, Huber J, Schuurman HJ. The neural and neuro-endocrine component of the human thymus. II. Hormone immunoreactivity. Brain Behav Immun 1992;6:249–64.
15. Aird F, Clevenger CV, Prystowsky MB, Redei E. Corticotropin-releasing factor mRNA in rat thymus and spleen. Proc Natl Acad Sci USA 1993;90:7104–8.
16. Reul JM, de Kloet ER. Two receptor systems for corticosterone in rat brain: microdistribution and differential occupation. Endocrinology 1985;117:2505–11.
17. De Kloet ER, Ratka A, Reul JM, Sutanto W, Van Eekelen JA. Corticosteroid receptor types in brain: regulation and putative function. Ann NY Acad Sci 1987;512:351–61.
18. Arriza JL, Weinberger C, Cerelli G, Glaser TM, Handelin BL, Housman DE, Evans RM. Cloning of human mineralocorticoid receptor complementary DNA: structural and functional kinship with the glucocorticoid receptor. Science 1987;237:268–75.
19. Lu NZ, Cidlowski JA. Translational regulatory mechanisms generate N-terminal glucocorticoid receptor isoforms with unique transcriptional target genes. Mol Cell 2005;18:331–42.
20. Herold MJ, McPherson KG, Reichardt HM. Glucocorticoids in T cell apoptosis and function. Cell Mol Life Sci 2006;63:60–72.
21. Homo F, Picard F, Durant S, Gagne D, Simon J, Dardenne M, Duval D. Glucocorticoid receptors and their functions in lymphocytes. J Steroid Biochem 1980;12:433–43.
22. Brewer JA, Sleckman BP, Swat W, Muglia LJ. Green fluorescent protein-glucocorticoid receptor knockin mice reveal dynamic receptor modulation during thymocyte development. J Immunol 2002;169:1309–18.
23. Wiegers GJ, Knoflach M, Bock G, Niederegger H, Dietrich H, Falus A, Boyd R, Wick G. CD4(+)CD8(+)TCR(low) thymocytes express low levels of glucocorticoid receptors while being sensitive to glucocorticoid-induced apoptosis. Eur J Immunol 2001;31: 2293–301.
24. Berki T, Palinkas L, Boldizsar F, Nemeth P. Glucocorticoid (GC) sensitivity and GC receptor expression differ in thymocyte subpopulations. Int Immunol 2002;14:463–9.

25. Speirs HJ, Seckl JR, Brown RW. Ontogeny of glucocorticoid receptor and 11beta-hydroxysteroid dehydrogenase type-1 gene expression identifies potential critical periods of glucocorticoid susceptibility during development. J Endocrinol 2004;181:105–16.
26. Sacedon R, Vicente A, Varas A, Jimenez E, Munoz JJ, Zapata AG. Early maturation of T-cell progenitors in the absence of glucocorticoids. Blood 1999;94:2819–26.
27. Kitraki E, Kittas C, Stylianopoulou F. Glucocorticoid receptor gene expression during rat embryogenesis. An in situ hybridization study. Differentiation 1997;62:21–31.
28. Ranelletti FO, Maggiano N, Aiello FB, Carbone A, Larocca LM, Musiani P, Piantelli M. Glucocorticoid receptors and corticosensitivity of human thymocytes at discrete stages of intrathymic differentiation. J Immunol 1987;138:440–5.
29. Jeon SH, Kang MG, Kim YH, Jin YH, Lee C, Chung HY, Kwon H, Park SD, Seong RH. A new mouse gene, SRG3, related to the SWI3 of Saccharomyces cerevisiae, is required for apoptosis induced by glucocorticoids in a thymoma cell line. J Exp Med 1997;185:1827–36.
30. Bultman S, Gebuhr T, Yee D, La Mantia C, Nicholson J, Gilliam A, Randazzo F, Metzger D, Chambon P, Crabtree G, Magnuson T. A Brg1 null mutation in the mouse reveals functional differences among mammalian SWI/SNF complexes. Mol Cell 2000;6:1287–95.
31. Sawa H, Kouike H, Okano H. Components of the SWI/SNF complex are required for asymmetric cell division in C. elegans. Mol Cell 2000;6:617–24.
32. Han S, Choi H, Ko MG, Choi YI, Sohn DH, Kim JK, Shin D, Chung H, Lee HW, Kim JB, Park SD, Seong RH. Peripheral T cells become sensitive to glucocorticoid- and stress-induced apoptosis in transgenic mice overexpressing SRG3. J Immunol 2001;167:805–10.
33. Choi YI, Jeon SH, Jang J, Han S, Kim JK, Chung H, Lee HW, Chung HY, Park SD, Seong RH. Notch1 confers a resistance to glucocorticoid-induced apoptosis on developing thymocytes by down-regulating SRG3 expression. Proc Natl Acad Sci USA 2001;98:10267–72.
34. Jang J, Choi YI, Choi J, Lee KY, Chung H, Jeon SH, Seong RH. Notch1 confers thymocytes a resistance to GC-induced apoptosis through Deltex1 by blocking the recruitment of p300 to the SRG3 promoter. Cell Death Differ 2006;13:1495–505.
35. van den Brandt J, Wang D, Reichardt HM. Resistance of single-positive thymocytes to glucocorticoid-induced apoptosis is mediated by CD28 signaling. Mol Endocrinol 2004;18:687–95.
36. Diba F, Watson CS, Gametchu B. 5′UTR sequences of the glucocorticoid receptor 1A transcript encode a peptide associated with translational regulation of the glucocorticoid receptor. J Cell Biochem 2001;81:149–61.
37. Purton JF, Monk JA, Liddicoat DR, Kyparissoudis K, Sakkal S, Richardson SJ, Godfrey DI, Cole TJ. Expression of the glucocorticoid receptor from the 1A promoter correlates with T lymphocyte sensitivity to glucocorticoid-induced cell death. J Immunol 2004;173:3816–24.
38. Zilberman Y, Yefenof E, Oron E, Dorogin A, Guy R. T cell receptor-independent apoptosis of thymocyte clones induced by a thymic epithelial cell line is mediated by steroids. Cell Immunol 1996;170:78–84.
39. Jenkinson EJ, Parnell S, Shuttleworth J, Owen JJ, Anderson G. Specialized ability of thymic epithelial cells to mediate positive selection does not require expression of the steroidogenic enzyme p450scc. J Immunol 1999;163:5781–5.
40. Nuotio-Antar AM, Hasty AH, Kovacs WJ. Quantitation and cellular localization of 11beta-HSD1 expression in murine thymus. J Steroid Biochem Mol Biol 2006;99:93–9.
41. Pazirandeh A, Jondal M, Okret S. Conditional expression of a glucocorticoid receptor transgene in thymocytes reveals a role for thymic-derived glucocorticoids in thymopoiesis in vivo. Endocrinology 2005;146:2501–7.
42. Pruett SB, Padgett EL. Thymus-derived glucocorticoids are insufficient for normal thymus homeostasis in the adult mouse. BMC Immunol 2004;5:24.
43. Nakanishi T, Nishi Y, Sato EF, Ishii M, Hamada T, Inoue M. Thermal injury induces thymocyte apoptosis in the rat. J Trauma 1998;44:143–8.

44. Cole TJ, Blendy JA, Monaghan AP, Krieglstein K, Schmid W, Aguzzi A, Fantuzzi G, Hummler E, Unsicker K, Schutz G. Targeted disruption of the glucocorticoid receptor gene blocks adrenergic chromaffin cell development and severely retards lung maturation. Genes Dev 1995;9: 1608–21.
45. Reichardt HM, Kaestner KH, Tuckermann J, Kretz O, Wessely O, Bock R, Gass P, Schmid W, Herrlich P, Angel P, Schutz G. DNA binding of the glucocorticoid receptor is not essential for survival [see comments]. Cell 1998;93:531–41.
46. Purton JF, Boyd RL, Cole TJ, Godfrey DI. Intrathymic T cell development and selection proceeds normally in the absence of glucocorticoid receptor signaling. Immunity 2000;13:179–86.
47. Brewer JA, Khor B, Vogt SK, Muglia LM, Fujiwara H, Haegele KE, Sleckman BP, Muglia LJ. T-cell glucocorticoid receptor is required to suppress COX-2-mediated lethal immune activation. Nat Med 2003;9:1318–22.
48. Finotto S, Krieglstein K, Schober A, Deimling F, Lindner K, Bruhl B, Beier K, Metz J, Garcia-Arraras JE, Roig-Lopez JL, Monaghan P, Schmid W, Cole TJ, Kellendonk C, Tronche F, Schutz G, Unsicker K. Analysis of mice carrying targeted mutations of the glucocorticoid receptor gene argues against an essential role of glucocorticoid signalling for generating adrenal chromaffin cells. Development 1999;126:2935–44.
49. King LB, Vacchio MS, Dixon K, Hunziker R, Margulies DH, Ashwell JD. A targeted glucocorticoid receptor antisense transgene increases thymocyte apoptosis and alters thymocyte development. Immunity 1995;3:647–56.
50. Franchimont D, Galon J, Vacchio MS, Fan S, Visconti R, Frucht DM, Geenen V, Chrousos GP, Ashwell JD, O'Shea JJ. Positive effects of glucocorticoids on T cell function by up-regulation of IL-7 receptor alpha. J Immunol 2002;168:2212–18.
51. Sacedon R, Vicente A, Varas A, Morale MC, Barden N, Marchetti B, Zapata AG. Partial blockade of T-cell differentiation during ontogeny and marked alterations of the thymic microenvironment in transgenic mice with impaired glucocorticoid receptor function. J Neuroimmunol 1999;98:157–67.
52. Sacedon R, Vicente A, Varas A, Jimenez E, Munoz JJ, Zapata AG. Role of glucocorticoids in early T-cell differentiation. Ann NY Acad Sci 2000;917:732–40.
53. Vacchio MS, Ashwell JD. Glucocorticoids and thymocyte development. Semin Immunol 2000;12:475–85.
54. Dupouy JP, Chatelain A, Allaume P. Absence of transplacental passage of ACTH in the rat: direct experimental proof. Biol Neonate 1980;37:96–102.
55. Chatelain A, Dupouy JP, Allaume P. Fetal-maternal adrenocorticotropin and corticosterone relationships in the rat: effects of maternal adrenalectomy. Endocrinology 1980;106:1297–1303.
56. Pazirandeh A, Xue Y, Prestegaard T, Jondal M, Okret S. Effects of altered glucocorticoid sensitivity in the T cell lineage on thymocyte and T cell homeostasis. Faseb J 2002;16:727–9.
57. Pazirandeh A, Jondal M, Okret S. Glucocorticoids delay age-associated thymic involution through directly affecting the thymocytes. Endocrinology 2004;145:2392–401.
58. Ashwell JD, Lu FW, Vacchio MS. Glucocorticoids in T cell development and function. Annu Rev Immunol 2000;18:309–45.
59. Jondal M, Pazirandeh A, Okret S. A role for glucocorticoids in the thymus. Trends Immunol 2001;22:185–6.
60. Ashwell JD, Vacchio MS, Galon J. Do glucocorticoids participate in thymocyte development. Immunol Today 2000;21:644–6.
61. Zacharchuk CM, Mercep M, Chakraborti PK, Simons SS, Jr, Ashwell JD. Programmed T lymphocyte death. Cell activation- and steroid-induced pathways are mutually antagonistic. J Immunol 1990;145:4037–45.
62. Iwata M, Hanaoka S, Sato K. Rescue of thymocytes and T cell hybridomas from glucocorticoid-induced apoptosis by stimulation via the T cell receptor/CD3 complex: a possible in vitro model for positive selection of the T cell repertoire. Eur J Immunol 1991;21:643–8.
63. Vacchio MS, Ashwell JD. Thymus-derived glucocorticoids regulate antigen-specific positive selection. J Exp Med 1997;185:2033–8.

64. Xue Y, Murdjeva M, Okret S, McConkey D, Kiuossis D, Jondal M. Inhibition of I-Ad-, but not Db-restricted peptide-induced thymic apoptosis by glucocorticoid receptor antagonist RU486 in T cell receptor transgenic mice. Eur J Immunol 1996;26:428–34.
65. Vacchio MS, Lee JY, Ashwell JD. Thymus-derived glucocorticoids set the thresholds for thymocyte selection by inhibiting TCR-mediated thymocyte activation. J Immunol 1999;163:1327–33.
66. Stephens GL, Ashwell JD, Ignatowicz L. Mutually antagonistic signals regulate selection of the T cell repertoire. Int Immunol 2003;15:623–32.
67. Lu FW, Yasutomo K, Goodman GB, McHeyzer-Williams LJ, McHeyzer-Williams MG, Germain RN, Ashwell JD. Thymocyte resistance to glucocorticoids leads to antigen-specific unresponsiveness due to "holes" in the T cell repertoire. Immunity 2000;12:183–92.
68. Wildin RS, Garvin AM, Pawar S, Lewis DB, Abraham KM, Forbush KA, Ziegler SF, Allen JM, Perlmutter RM. Developmental regulation of lck gene expression in T lymphocytes. J Exp Med 1991;173:383–93.
69. Pazirandeh A, Xue Y, Okret S, Jondal M. Glucocorticoid resistance in thymocytes from mice expressing a T cell receptor transgene. Biochem Biophys Res Commun 2000;276:189–96.
70. Chakravarti D, LaMorte VJ, Nelson MC, Nakajima T, Schulman IG, Juguilon H, Montminy M, Evans RM. Role of CBP/P300 in nuclear receptor signalling. Nature 1996;383:99–103.
71. Yu CT, Feng MH, Shih HM, Lai MZ. Increased p300 expression inhibits glucocorticoid receptor-T-cell receptor antagonism but does not affect thymocyte positive selection. Mol Cell Biol 2002;22:4556–66.
72. Erlacher M, Knoflach M, Stec IE, Bock G, Wick G, Wiegers GJ. TCR signaling inhibits glucocorticoid-induced apoptosis in murine thymocytes depending on the stage of development. Eur J Immunol 2005;35:3287–96.
73. Bakker JM, Kavelaars A, Kamphuis PJ, Zijlstra J, van Bel F, Heijnen CJ. Neonatal dexamethasone treatment induces long-lasting changes in T-cell receptor vbeta repertoire in rats. J Neuroimmunol 2001;112:47–54.
74. Tolosa E, King LB, Ashwell JD. Thymocyte glucocorticoid resistance alters positive selection and inhibits autoimmunity and lymphoproliferative disease in MRL-lpr/lpr mice. Immunity 1998;8: 67–76.
75. Morale MC, Batticane N, Gallo F, Barden N, Marchetti B. Disruption of hypothalamic-pituitary-adrenocortical system in transgenic mice expressing type II glucocorticoid receptor antisense ribonucleic acid permanently impairs T cell function: effects on T cell trafficking and T cell responsiveness during postnatal development. Endocrinology 1995;136:3949–60.
76. Godfrey DI, Purton JF, Boyd RL, Cole TJ. Stress-free T-cell development: glucocorticoids are not obligatory. Immunol Today 2000;21:606–11.
77. Cole TJ, Myles K, Purton JF, Brereton PS, Solomon NM, Godfrey DI, Funder JW. GRKO mice express an aberrant dexamethasone-binding glucocorticoid receptor, but are profoundly glucocorticoid resistant. Mol Cell Endocrinol 2001;173:193–202.
78. Mittelstadt PR, Ashwell JD. Disruption of glucocorticoid receptor exon 2 yields a ligand-responsive C-terminal fragment that regulates gene expression. Mol Endocrinol 2003;17:1534–42.
79. Tibbetts TA, DeMayo F, Rich S, Conneely OM, O'Malley BW. Progesterone receptors in the thymus are required for thymic involution during pregnancy and for normal fertility. Proc Natl Acad Sci USA 1999;96:12021–6.
80. Van Laethem F, Baus E, Smyth LA, Andris F, Bex F, Urbain J, Kioussis D, Leo O. Glucocorticoids attenuate T cell receptor signaling. J Exp Med 2001;193:803–14.
81. Zhan Y, Funda DP, Every AL, Fundova P, Purton JF, Liddicoat DR, Cole TJ, Godfrey DI, Brady JL, Mannering SI, Harrison LC, Lew AM. TCR-mediated activation promotes GITR upregulation in T cells and resistance to glucocorticoid-induced death. Int Immunol 2004;16:1315–21.
82. Baumann S, Dostert A, Novac N, Bauer A, Schmid W, Fas SC, Krueger A, Heinzel T, Kirchhoff S, Schutz G, Krammer PH. Glucocorticoids inhibit activation-induced cell death (AICD) via direct DNA-dependent repression of the CD95 ligand gene by a glucocorticoid receptor dimer. Blood 2005;106:617–25.

83. Jamieson CA, Yamamoto KR. Crosstalk pathway for inhibition of glucocorticoid-induced apoptosis by T cell receptor signaling. Proc Natl Acad Sci USA 2000;97:7319–24.
84. Yang-Yen HF, Chambard JC, Sun YL, Smeal T, Schmidt TJ, Drouin J, Karin M. Transcriptional interference between c-Jun and the glucocorticoid receptor: mutual inhibition of DNA binding due to direct protein-protein interaction. Cell 1990;62:1205–15.
85. Kamei Y, Xu L, Heinzel T, Torchia J, Kurokawa R, Gloss B, Lin SC, Heyman RA, Rose DW, Glass CK, Rosenfeld MG. A CBP integrator complex mediates transcriptional activation and AP-1 inhibition by nuclear receptors. Cell 1996;85:403–14.
86. Wang D, Muller N, McPherson KG, Reichardt HM. Glucocorticoids engage different signal transduction pathways to induce apoptosis in thymocytes and mature T cells. J Immunol 2006; 176:1695–702.
87. Dardenne M, Itoh T, Homo-Delarche F. Presence of glucocorticoid receptors in cultured thymic epithelial cells. Cell Immunol 1986;100:112–18.
88. Sacedon R, Varas A, Jimenez E, Munoz JJ, Vicente A, Zapata AG. Accelerated maturation of the thymic stroma in the progeny of adrenalectomized pregnant rats. Neuroimmunomodulation 1999;6:23–30.
89. Savino W, Cirne-Lima EO, Soares JF, Leite-de-Moraes Mdo C, Ono IP, Dardenne M. Hydrocortisone increases the numbers of KL1+ cells, a discrete thymic epithelial cell subset characterized by high molecular weight cytokeratin expression. Endocrinology 1988;123:2557–64.
90. De Souza LR, Savino W. Modulation of cytokeratin expression in the hamster thymus: evidence for a plasticity of the thymic epithelium. Dev Immunol 1993;3:137–46.
91. Shu S, Naylor P, Touraine JL, Hadden JW. IL-1, ICAM-1, LFA-3, and hydrocortisone differentially regulate cytokine secretion by human fetal thymic epithelial cells. Thymus 1996;24:89–99.
92. Lannes-Vieira J, Dardenne M, Savino W. Extracellular matrix components of the mouse thymus microenvironment: ontogenetic studies and modulation by glucocorticoid hormones. J Histochem Cytochem 1991;39:1539–46.
93. Lannes-Vieira J, Cirne-Lima EO, Savino W. Hydrocortisone modulates cytokeratin and extracellular matrix expression by the thymic microenvironment. Ann NY Acad Sci 1992;650:81–4.
94. Fertsch D, Schoenberg DR, Germain RN, Tou JY, Vogel SN. Induction of macrophage Ia antigen expression by rIFN-gamma and down- regulation by IFN-alpha/beta and dexamethasone are mediated by changes in steady-state levels of Ia mRNA. J Immunol 1987;139:244–9.
95. Manyak CL, Tse H, Fischer P, Coker L, Sigal NH, Koo GC. Regulation of class II MHC molecules on human endothelial cells. Effects of IFN and dexamethasone. J Immunol 1988;140:3817–21.
96. Dhabhar FS, Miller AH, McEwen BS, Spencer RL. Stress-induced changes in blood leukocyte distribution. Role of adrenal steroid hormones. J Immunol 1996;157:1638–44.
97. Karanth S, Linthorst AC, Stalla GK, Barden N, Holsboer F, Reul JM. Hypothalamic-pituitary-adrenocortical axis changes in a transgenic mouse with impaired glucocorticoid receptor function. Endocrinology 1997;138:3476–85.
98. Sprent J, Webb SR. Intrathymic and extrathymic clonal deletion of T cells. Curr Opin Immunol 1995;7:196–205.
99. Moser M, De Smedt T, Sornasse T, Tielemans F, Chentoufi AA, Muraille E, Van Mechelen M, Urbain J, Leo O. Glucocorticoids down-regulate dendritic cell function in vitro and in vivo. Eur J Immunol 1995;25:2818–24.
100. Rea D, van Kooten C, van Meijgaarden KE, Ottenhoff TH, Melief CJ, Offringa R. Glucocorticoids transform CD40-triggering of dendritic cells into an alternative activation pathway resulting in antigen-presenting cells that secrete IL-10. Blood 2000;95:3162–7.
101. Woltman AM, Massacrier C, de Fijter JW, Caux C, van Kooten C. Corticosteroids prevent generation of CD34+-derived dermal dendritic cells but do not inhibit Langerhans cell development. J Immunol 2002;168:6181–8.
102. Woltman AM, de Fijter JW, Kamerling SW, Paul LC, Daha MR, van Kooten C. The effect of calcineurin inhibitors and corticosteroids on the differentiation of human dendritic cells. Eur J Immunol 2000;30:1807–12.

103. Holt PG, Thomas JA. Steroids inhibit uptake and/or processing but not presentation of antigen by airway dendritic cells. Immunology 1997;91:145–50.
104. Pan J, Ju D, Wang Q, Zhang M, Xia D, Zhang L, Yu H, Cao X. Dexamethasone inhibits the antigen presentation of dendritic cells in MHC class II pathway. Immunol Lett 2001;76:153–61.
105. Sacedon R, Vicente A, Varas A, Jimenez E, Munoz JJ, Zapata AG. Glucocorticoid-mediated regulation of thymic dendritic cell function. Int Immunol 1999;11:1217–24.
106. Kitajima T, Ariizumi K, Bergstresser PR, Takashima A. A novel mechanism of glucocorticoid-induced immune suppression: the inhibition of T cell-mediated terminal maturation of a murine dendritic cell line. J Clin Invest 1996;98:142–7.
107. Nelson DJ, McWilliam AS, Haining S, Holt PG. Modulation of airway intraepithelial dendritic cells following exposure to steroids. Am J Respir Crit Care Med 1995;151:475–81.
108. Sacedon R, Vicente A, Varas A, Jimenez E, Zapata AG. Early differentiation of thymic dendritic cells in the absence of glucocorticoids. J Neuroimmunol 1999;94:103–8.
109. Ardavin C. Thymic dendritic cells [see comments]. Immunol Today 1997;18:350–61.
110. Varas A, Vicente A, Sacedon R, Zapata AG. Interleukin-7 influences the development of thymic dendritic cells. Blood 1998;92:93–100.

*The Hypothalamus–Pituitary–Adrenal Axis*
Edited by A. del Rey, G. P. Chrousos and H. O. Besedovsky



# *Effects of Catecholamines on the Immune Response*

ILIA J. ELENKOV

*Italian National Research Council, Institute of Neurobiology and Molecular Medicine, Rome, Italy*

## ABSTRACT

Primary and secondary lymphoid organs and blood vessels receive extensive sympathetic/noradrenergic innervation. Immune cells express adrenoreceptors (ARs), and through stimulation of these receptors, locally released norepinephrine (NE), or circulating catecholamines (CAs), including epinephrine, released from the adrenal medulla, affect lymphocyte traffic, circulation and proliferation, and modulate cytokine and antibody production, and the functional activity of different lymphoid cells. Recent evidence indicates that NE and epinephrine, through stimulation of the $\beta_2$-adrenoreceptor–cAMP–PKA pathway, inhibit the production of type 1/pro-inflammatory cytokines, such as interleukin-12 (IL-12), tumor necrosis factor-$\alpha$ (TNF-$\alpha$), and interferon-$\gamma$ (IFN-$\gamma$) by antigen-presenting cells (APCs) and T helper (Th) 1 cells, whereas they stimulate the production of type 2/anti-inflammatory cytokines such as IL-10 and transforming growth factor-$\beta$ (TGF-$\beta$). Through this mechanism, systemically, endogenous CAs may cause a selective suppression of Th1 responses and cellular immunity and a Th2 shift toward dominance of humoral immunity. On the other hand, in certain local responses and under certain conditions, CAs may actually boost regional immune responses, through induction of IL-1, TNF-$\alpha$, and primarily IL-8 production. Thus, the activation of the sympathetic nervous system (SNS) during an immune response might be aimed to localize the inflammatory response, through induction of neutrophil accumulation and stimulation of more specific humoral immune responses, while systemically it may suppress Th1 responses and thus protects the organism from the detrimental effects of pro-inflammatory cytokines and other products of activated macrophages.

## 1. INTRODUCTION

The brain and the immune system are the two major *adaptive* systems of the body. The immune system is regulated by a variety of factors "from within" – regulatory T-cell subsets, cytokines, chemokines, complement, antibodies, and so on – and by factors "from without" (a term used by Medawear in 1973): different hormones, neurotransmitters, or neuropeptides present in the microenvironment of immunocompetent cells. During an immune response, the

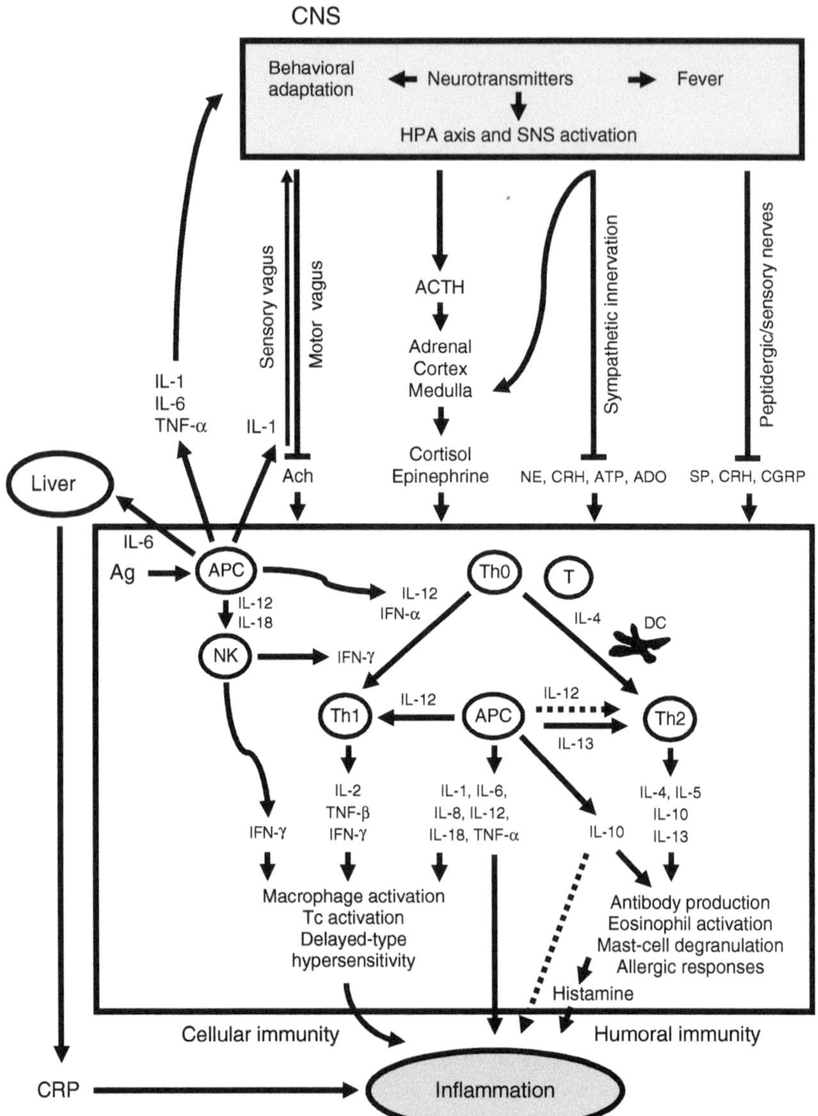


Figure 1. A simplified scheme of the bidirectional communication between the brain and the immune system; role of cytokines in the regulation of cellular and humoral immunity; role of neuroendocrine and immune adaptive responses in inflammation. Lymphoid organs, and particularly their parenchyma, similar to smooth muscles of the vasculature, receive predominantly sympathetic/noradrenergic and peptidergic/sensory innervation; the heart and the gastrointestinal tract receive both sympathetic and parasympathetic (cholinergic) innervation. Cellular immunity provides protection against intracellular bacteria, protozoa, fungi, and several viruses, whereas humoral immunity provides protection against multicellular parasites, extracellular bacteria, some viruses, soluble toxins, and allergens (see text). Ach, acetylcholine; ACTH, adrenocorticotropic hormone; ADO, adenosine; Ag, antigen; APC, antigen-presenting cell; CGRP, calcitonin gene-related peptide; CNS, central nervous system; CRP, C-reactive protein; DC, dendritic cell; HPA, hypothalamic–pituitary–adrenal axis; IFN, interferon; IL, interleukin, NE, norepinephrine; NK, natural killer; SNS, sympathetic nervous system; SP, substance P; Th, T helper cell; Tc, T-cytotoxic cell; TNF, tumor necrosis factor. From Ref. [4].

brain and the immune system "talk to each other," and this process is essential for maintaining *homeostasis* [1–3].

The central nervous system affects the immune system through the neuroendocrine humoral outflow via the pituitary and through direct neuronal influences via the sympathetic, parasympathetic (cholinergic), and peptidergic/sensory innervation of peripheral tissues. Thus, circulating hormones or locally released neurotransmitters and neuropeptides regulate major immune functions such as antigen presentation, secretion of cytokines and antibodies, selection of T helper (Th) 1 or Th2 responses, lymphocyte activity, proliferation, and traffic. Alternatively, certain cytokines, such as interleukin-1 (IL-1), IL-6, and tumor necrosis factor-α (TNF-α), released during an immune response activate the central components of the stress system, alter neurotransmitter networks activity, and induce fever, sleepiness, fatigue, loss of appetite, and decreased libido. In addition, they activate the hepatic synthesis of acute-phase proteins – changes referred to as "sickness behavior" and "acute-phase response" (Fig. 1).

The systemic/adrenomedullary sympathetic nervous system (SNS) and the hypothalamic–pituitary–adrenal (HPA) axis, which represents the peripheral limbs of the stress system, play a major role in immunoregulation [2,3]. This overview focuses on the SNS, its end-products, catecholamines (CAs), and their role in the regulation of the immune response. The SNS, a major component of the autonomic nervous system, innervates all lymphoid organs, adrenoreceptors (ARs) are expressed on different lymphoid cells, and the immune system is *tuned* by locally released norepinephrine (NE) or circulating epinephrine (adrenaline) secreted by the adrenal medulla. Thus, the SNS provides a major integrative and regulatory pathway between the brain and the immune system (Fig. 1).

CAs, similar to glucocorticoids, have been often regarded as immunosuppressive. Recently, however, there has been accumulating evidence that both CAs and glucocorticoids, under physiologic conditions or at levels that can be achieved during stress, influence the immune response in a less *monochromatic way*. This new understanding is discussed below, which helps explain some well-known, but often contradictory, effects of the neuroendocrine or stress system on the immunity and on the onset and course of common human diseases, such as infections, autoimmune/inflammatory, allergic, and neoplastic diseases (see also Refs [3,5–8]).

## 2. BRIEF HISTORICAL PERSPECTIVES

Evidence that lymphoid organs are innervated dates back to the end of last century when nerves, independent of blood vessels, were found to enter lymph nodes [9]. In 1898, Otto von Fürth isolated the first ever hormone from tissue and called this partly purified product "suprarenin." Three years later, Takamine and Aldrich independently isolated the responsible component in crystalline form. The substance was named adrenaline by Takamine, and Aldrich found the correct formula ($C_9H_{13}NO_3$). In 1904, Loeper and Crouzon described for the first time a pronounced leukocytosis after subcutaneous injection of epinephrine in humans. In 1919, Ishigami linked stress to immune system alterations showing decreased phagocytic activity of leukocytes during the periods of stress in subjects with chronic tuberculosis. In 1929, Pines and Majman and in 1935, Hammar used silver staining to demonstrate that the thymus gland is innervated. In the 1930s, Hans

Selye described involution of the thymus in animals exposed to stressors and developed the concept of stress response, whereas Walter Cannon called this response "fight or flight" reaction and linked the adaptive response to stress with CAs secretion and actions (for more details see Refs [3,10]).

In 1946, von Euler isolated NE from a lymphoid organ, the spleen, and later provided evidence that NE is the major neurotransmitter released from sympathetic nerves [11]. In the 1950s, Dougherty and Frank noticed about a 400% increase in circulating lymphocytes within 10 min after subcutaneous injection of epinephrine of what they called "stress lymphocytes" [12]. These cells had the morphology of large granular lymphocytes, or natural killer (NK) cells, whose function and characteristics were described in the late 1970s. In the 1970s and 1980s, Hugo Besedovsky and coworkers provided first evidence that classic hormones and newly described cytokines are involved in functionally relevant *cross-talk* between the brain and the immune system. They have shown that an immune response induces an increase of plasma corticosteroid levels, alters the activity of hypothalamic noradrenergic neurons, and drops the content of NE in the spleen [13–16]. Also in the 1970s, the first hormone/neurotransmitter receptor on lymphocytes was described functionally, when it was reported that adrenergic agents modulate lymphocyte proliferation [17]. In the 1970s and 1980s, the first comprehensive morphological studies provided evidence that both primary and secondary lymphoid organs are innervated by sympathetic/noradrenergic nerve fibers [18–20].

## 3. SYMPATHETIC INNERVATION OF LYMPHOID ORGANS

Lymphoid organs, similar to blood vessels, receive predominantly sympathetic innervation. Sympathetic/noradrenergic and sympathetic/neuropeptide Y (NPY) postganglionic nerve fibers innervate both the smooth muscle of the vasculature and the parenchyma of specific compartments of primary and secondary lymphoid organs. Noradrenergic innervation of lymphoid tissue appears to be regional and specific; generally, zones of T cells, macrophages, and plasma cells are richly innervated, whereas nodular and follicular zones of developing or maturing B cells are poorly innervated (for more details, see Refs [3,20–22]). In general, it appears that in lymphoid organs, NE is released *non-synaptically*, that is, from varicose axon terminals, which do not make synaptic contacts – NE released from perivascular or connective tissue septa plexuses of nerve terminals diffuses away through surrounding adventitia or collagenous fibrils, in a paracrine fashion. Thus, ARs on immune cells are targets of remote control, and thus, NE may play a modulatory role in signal transmission at the sympathetic–immune interface [3,22]. Similar *non-synaptic* transmission operates in the blood vessel wall (between varicose nerve terminals and smooth muscle cells) of these organs, where NE and NPY are involved in regulation of blood flow and lymphocyte traffic.

In addition to the autonomic/sympathetic innervation, all lymphoid organs also receive sensory peptidergic innervation that is confined mostly to the parenchyma [23]. Similar to other organs, the tachykinins/CGRP fibers most likely have sensory origins. Peptidergic nerves also appear to be sparse in pure B-cell regions. Neuro-mast cell contacts are seen in all lymphoid organs, with the exception of the spleen – noradrenergic nerve fibers are closely associated with mast cells in both perivascular and parenchymal zones, and peptidergic nerve fibers are in close proximity to mast cells, T cells, and macrophages. These anatomical organizations suggest NE – peptidergic mediators – histamine interactions in the blood vessels and lymphoid organs that may play an important role in immunoregulation [3,24].

## 4. SYMPATHETIC CONTROL OF LYMPHOCYTE TRAFFIC AND CIRCULATION

Two phases are recognized after CAs administration in humans: a quick (<30 min) mobilization of lymphocytes, followed by an increase of granulocytes with relative lymphopenia (maximal response at 2–4 h) [10]. CAs predominantly affect NK cells and granulocytes circulation, whereas T- and B-cell numbers remain relatively unaffected. Similarly, acute psychological stress or physical exercise induces a transient increase in lymphocyte numbers, in particular NK-cell numbers. The role of increased CAs levels in this phenomenon has also been documented [10,25]. By contrast, a reduction of NK-cell number is observed after 7 days of treatment with terbutaline, a $\beta_2$-adrenergic receptor ($\beta_2$-AR) selective agonist, changes identical to that seen in congestive heart failure patients [26]. Thus, in the short term, CAs acutely mobilize NK cells from depots, whereas in the long term, chronically, CAs decrease the number of lymphocytes and particularly of NK cells in the peripheral blood. This is substantiated by the observation that in humans, the percentage of NK cells in peripheral blood in normal subjects is negatively correlated to plasma epinephrine levels [27].

## 5. EFFECTS OF CAs ON LYMPHOCYTE PROLIFERATION

CAs or β-AR agonists inhibit the T-cell proliferation induced by mitogens or anti-CD3 monoclonal antibody through the CD3/TCR complex [17,28–30]. This is usually accompanied by an increase of cAMP in lymphocytes, and the amount of cAMP produced by T cells stimulated with isoproterenol, a β-AR agonist, is proportional to the degree of inhibition of the proliferation [30,31]. The proliferative response of $CD8^+$ T cells is inhibited to a greater extent than that of $CD4^+$ T cells, presumably because $CD8^+$ T cells have higher number of β-ARs [30]. The elevation of cAMP also inhibits IL-2 secretion by T cells [30], thus suggesting that the inhibition of T-cell proliferation by CAs might be due, at least in part, to the inhibition of the production of IL-2, a cytokine that is an important co-stimulatory molecule in T-cell proliferation.

Voltage-dependent potassium ($K^+$) channels (i.e., the opening rate of the channel increases with membrane depolarization) represent the predominant ion channels in lymphocytes. NE or the β-AR agonist isoproterenol decreases the peak current amplitude and increases the rate of inactivation of the $K^+$ currents recorded from human $CD8^+$ peripheral lymphocytes or rat thymocytes [22,32]. Because $K^+$ channels are involved in the processes of lymphocyte activation and proliferation, the above-mentioned observations suggest that CAs might be involved in the regulation of these processes also through modulation of $K^+$-channel gating [22].

## 6. THE TH1/TH2 PARADIGM AND PRO- AND ANTI-INFLAMMATORY CYTOKINES

Homeostasis within the immune system is largely dependent on cytokines, the chemical messengers between immune cells, which play crucial roles in mediating inflammatory and immune responses. These diverse groups of proteins may be regarded as hormones of the immune system. Cytokines act in an autocrine, paracrine, or endocrine fashion to control the proliferation, differentiation, and activity of immune cells. For instance, Th1 cells primarily secrete interferon-γ (IFN-γ), IL-2, and TNF-β (TNF-β), which promote cellular immunity, whereas Th2 cells secrete a different set of cytokines, primarily IL-4, IL-10, and IL-13, which promote humoral immunity [33–35] (Fig. 1).

Naïve $CD4^+$ (antigen-inexperienced) Th0 cells are bipotential and serve as precursors of Th1 and Th2 cells. IL-12, produced by antigen-presenting cells (APCs), such as monocytes/macrophages and dendritic cells (DCs), is the major inducer of Th1 differentiation and, hence, cellular immunity. IL-12 also synergizes with IL-18 to induce the production of IFN-γ by NK cells. Thus, IL-12 in concert with IL-18, IFN-α, and IFN-γ promotes the differentiation of Th0 cells toward the Th1 phenotype. IL-1, IL-12, TNF-α, and IFN-γ also stimulate the functional activity of T-cytotoxic cells (Tcs), NK cells, and activated macrophages, which are the major components of cellular immunity. The type 1 cytokines IL-12, TNF-α, and IFN-γ also stimulate the synthesis of nitric oxide and other inflammatory mediators that drive chronic, delayed-type inflammatory responses. Because of their synergistic roles in stimulating inflammation, IL-12, TNF-α, and IFN-γ are considered the major pro-inflammatory cytokines [33–35].

Th1 and Th2 responses are mutually inhibitory. Thus, IL-12 and IFN-γ inhibit Th2, whereas IL-4 and IL-10 inhibit Th1 cell activities. IL-4 and IL-10 promote humoral immunity by stimulating the growth and activation of mast cells and eosinophils (Eos), the differentiation of B cells into antibody-secreting B cells, and the B-cell immunoglobulin switching to IgE. Importantly, these cytokines also inhibit macrophage activation, T-cell proliferation, and the production of pro-inflammatory cytokines [33–35]. Therefore, the Th2 cytokines IL-4 and IL-10 are the major anti-inflammatory cytokines.

## 7. EFFECTS OF CAs ON CYTOKINE PRODUCTION AND THE TH1/TH2 BALANCE

### 7.1. Systemic effects

CAs suppress type1/pro-inflammatory cytokine production, Th1 activities, and cellular immunity but boost type2/anti-inflammatory cytokine production, Th2, and humoral responses at the level of APCs and Th1 cells, or through a direct effect on the cellular components of both cellular and humoral immunity (Fig. 1). Both NE and epinephrine through stimulation of $\beta_2$-ARs potently inhibit the production by monocytes and DCs of the main inducer of Th1 responses, IL-12 [36–38]. Epinephrine appears to be a strong inhibitor of IL-12 production, exhibiting an $EC_{50}$ of $10^{-9}$ M [36]. Because IL-12 is extremely potent in enhancing IFN-γ and inhibiting IL-4 synthesis by Th1 and Th2 cells, respectively, the inhibition of IL-12 production may represent one of the major mechanisms by which CAs affect the Th1/Th2 balance. Thus, in conjunction with their ability to suppress IL-12 production, $\beta_2$-AR agonists inhibit the development of Th1-type cells while promoting Th2-cell differentiation [37]. NE, epinephrine, and $\beta_2$-AR agonists inhibit the production of TNF-α by monocytes, microglial cells, and astrocytes [39–41]. CAs also suppress the production of IL-1, an effect that is mostly indirect via inhibition of TNF-α, and potentiation of IL-10 production [42,43].

While suppressing type 1 cytokine production, CAs appear to up-regulate the production of type 2 cytokines by APCs. Thus, the production of IL-10, one of the most potent anti-inflammatory cytokines, induced by LPS in human monocytes or mouse peritoneal macrophages, is potentiated by NE, epinephrine, and $\beta_2$-AR agonists, an effect that is $\beta_2$-AR-mediated and cAMP-PKA-dependent [36,44]. Similarly, the production of IL-6, a cytokine that exerts both pro- and anti-inflammatory effects but possesses mostly Th2-type activities (previously known as BCDF, B-cell differentiation factor), is also up-regulated by CAs [45,46].

It appears that $\beta_2$-ARs are expressed on Th1 cells, but not on Th2 cells [47]. This may provide an additional mechanistic basis for the differential effect of CAs on Th1/Th2 functions. In both murine and human systems, $\beta_2$-AR agonists inhibit IFN-$\gamma$ production by Th1 cells, but do not affect IL-4 production by Th2 cells [47,48]. Furthermore, cAMP levels increase in Th1 cells following terbutaline exposure, but not in Th2 cells [47].

In vivo, increasing sympathetic outflow and endogenous production of CAs in mice by selective $\alpha_2$-AR antagonists, or application of exogenous CAs, or $\beta$-AR agonists results in inhibition of LPS-induced TNF-$\alpha$ and IL-12 production [49,50]. CAs also appear to exert tonic inhibition on the production of pro-inflammatory cytokines in vivo. Thus, application of propranolol, a $\beta$-AR antagonist which blocks their inhibitory effect on cytokine-producing cells, results in substantial increases of LPS-induced secretion of TNF-$\alpha$ and IL-12 in mice [49,50]. In IL-10 deficient C57BL/6 IL-10 (−/−) mice, plasma levels of IL-12 are about 70-fold higher than in their counterparts, suggesting tonic inhibitory effect of IL-10 on IL-12 production. Injection of isoproterenol, while augmenting the IL-10 response in C57BL/6 IL-10 (+/+) mice, inhibits IL-12 production in both C57BL/6 IL-10 (+/+) and C57BL/6 IL-10 (−/−) mice [50]. Thus, the inhibition of IL-12 production appears to be independent of the increased release of IL-10. In humans, the administration of the $\beta_2$-AR agonist salbutamol results in inhibition of IL-12 production ex vivo [37], whereas acute brain trauma that is followed by massive release of CAs triggers secretion of substantial amounts of systemic IL-10 [51].

### 7.2. Local versus systemic effects

The above-described Th2-inducing effects of CAs may not pertain to certain conditions or local responses in specific compartments of the body. Thus, hemorrhage, a condition associated with elevations of systemic CAs concentrations, increases the expression of TNF-$\alpha$ and IL-1 by lung mononuclear cells via stimulation of $\alpha$-ARs [52]. Exposure of rats to mild inescapable electrical footshock stress also results in increased IL-1$\beta$ and TNF-$\alpha$ production by alveolar macrophages [53]. The up-regulation of pro-inflammatory cytokine production, in this case, appears to be dependent on intact sympathetic innervation and $\beta$-ARs. However, the effect is most likely indirect, because, in vitro, a direct modulatory effect of CAs on LPS-induced IL-1$\beta$ by alveolar macrophages was not demonstrated. It can be envisaged that the stress-induced changes in alveolar macrophage activity result from $\beta$-ARs-mediated alveolar type 2 epithelial cell activation, leading to release of surfactant and/or other factors [53]. Furthermore, CAs (through $\beta$2/$\beta$3-ARs) up-regulate IL-6 production by human adipocytes [54]. IL-6 is the major inducer of C-reactive protein (CRP) production by the liver and CAs enhance this induction [55]. Interestingly, chronic $\beta$-AR stimulation induces myocardial, but not systemic, production of TNF-$\alpha$, IL-1$\beta$, and IL-6 [56]. Thus, while systemically CAs suppress Th1 and cellular immune responses, locally, they may actually boost certain cellular immune responses in a transitory fashion. Further studies are needed to address this question (Fig. 2, see also below effects of CAs on IL-8 production and macrophage functions).

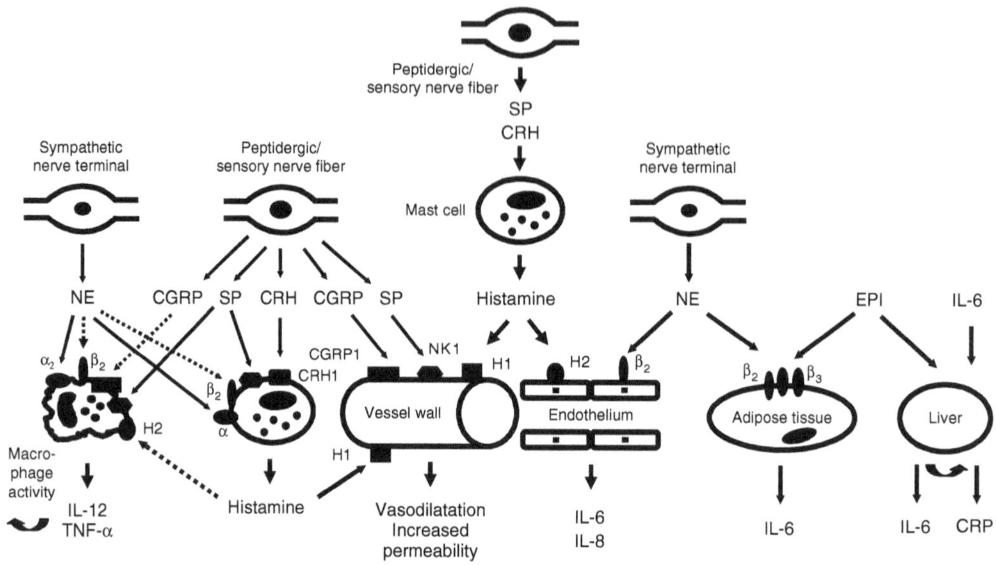


Figure 2. Simplified scheme of the complex interactions between CAs, neuropeptides, and CRH/SP–mast cell–histamine axis and their pro- and anti-inflammatory effects in certain local responses. Solid lines represent stimulation, whereas dashed lines represent inhibition. CGRP, calcitonin gene-related peptide; CRH, corticotropin-releasing hormone (peripheral); EPI, epinephrine; IL, interleukin; NE, norepinephrine; SP, substance P; TNF, tumor necrosis factor. From Ref. [4].

## 8. EFFECT OF CAs ON CHEMOKINE PRODUCTION

The recruitment of T cells, macrophages, and polymorphonuclear cells to an inflammatory site is greatly enhanced by the action of chemotactic cytokines termed *chemokines*, a large family of secreted 8- to 10-kDa proteins. In general, chemokines of the CXC subfamily or α-chemokines, such as IL-8, are specific in recruiting neutrophils and to varying extents, lymphocytes, whereas chemokines from the CC subfamily or β-chemokines, such as monocyte chemotactic proteins (MCP-1, MCP-2, and MCP-3) and macrophage inflammatory proteins (MIP-1α and MIP-1β), are chemotactic for monocytes and variably for NK cells, basophils, and Eos. Thus, the system of chemokines might serve to focus the immune defences around the invading microorganisms.

CAs potentiate the production of IL-8 by monocytes and epithelial cells of the lung [57,58], thus probably promoting recruitment of polymorphonuclear leukocytes in local inflammation. Interestingly, epinephrine promotes IL-8 production by human leukocytes via an indirect effect on platelets. Thus, IL-8 levels in samples containing platelets and stimulated with LPS and epinephrine are significantly higher than in control samples containing no platelets [59]. In fact, as shown by Kaplanski et al. [60], activated platelets are able to induce endothelial secretion of IL-8. The CC-chemokine MIP-1α is produced by a number of cells including neutrophils, activated lymphocytes, and fibroblasts. Recently, Hasko et al. [61] demonstrated that exogenous and endogenous CAs inhibit the production of MIP-1α via a β-AR-mediated mechanism. Thus, CAs are probably important endogenous regulators of the chemokine system; however, the interactions CAs – chemokine production – local inflammation remain poorly understood.

## 9. EFFECTS OF CAs ON THE CELLULAR COMPONENTS OF IMMUNITY

### 9.1. NK-cell activity

Application of epinephrine and isoproterenol in vitro elevates cAMP about 2.5-fold and induces an inhibition of NK-cell activity [62,63]. This effect is blocked by propranolol and mimicked by terbutaline, a β2-AR agonist, indicating the involvement of β2-ARs in this process [62,63].

Administration of the β2-AR agonist in vivo or stimulation of the splenic nerve in rats results in suppression of NK activity, an effect that is blocked by nadolol, a peripherally acting β-AR antagonist [64,65]. Central administration of corticotropin-releasing hormone (CRH), which is known to increase the sympathetic autonomic outflow, is accompanied by decreased NK activity in the periphery. This effect is independent of the adrenocortical activation, because chlorisondamine, a ganglionic blocker of the peripheral sympathetic neurotransmission, or propranolol, an β-ARs antagonist, prevents the CRH-induced inhibition of NK activity [66–68]. In patients with heart failure, a disease characterized by chronically high levels of plasma NE, these levels correlate with anergy in the cytotoxicity of circulating NK cells and with their response to the stimulation with IL-2 and IFN-γ [69].

Moreover, several lines of evidence suggest that stress, which is accompanied by increased levels of peripheral CAs, also inhibits NK cell activity, an effect that is mediated mainly by the CRH–SNS axis [67]. Thus, in animals, the application of anti-CRH antibodies completely blocks the inhibitory effect of footshock stress on NK activity [67]. It appears that NK cells are the most "sensitive" cells to the suppressive effect of stress, and not surprisingly, NK-cell activity has become a bona fide index of stress-induced suppression of cellular immunity, employed in many studies (for review, see Ref. [67]). The potent suppressive effect of CAs on NK-cell activity might be due to the fact that NK cells probably possess the highest number of $\beta_2$-ARs among lymphoid cells. Apart from a direct effect, CAs may suppress NK activity indirectly, through their selective suppression of Th1-type cytokines and particularly through the above-discussed potent inhibition of the production of IL-12 and IFN-γ, cytokines essential for NK activity.

### 9.2. Macrophage activity

After activation and/or migration to a particular organ, monocytes differentiate into macrophages. The effects of CAs on macrophage functions appear to be complex and subject of some controversies. Thus, epinephrine and NE block the capacity of IFN-γ to activate murine macrophages to a tumoricidal state as measured by the lysis of 125I-UdR-labeled melanoma target cells or to a cytotoxic state capable of selectively killing herpes simplex virus-infected cells, and treatment with epinephrine decreases the ability of human macrophages to kill *Mycobacterium avium* [70–72]. This is in agreement with several studies demonstrating that in both mouse and rat peritoneal macrophages, or macrophage cell lines, CAs inhibit TNF-α and IL-1 that potentiate macrophage activity, but they increase the production of IL-10, a cytokine that has potent suppressive effects on macrophage activity [42,73,74].

In contrast, both NE and epinephrine were reported to stimulate resident peritoneal macrophages from BALB/c mice to suppress the growth of *M. avium*, an effect mediated by $\alpha_2$-ARs [75]. This is in accord with the study of Spengler et al. [76] showing that NE, via stimulation of $\alpha_2$-ARs, can augment LPS-stimulated production of TNF-α from mouse peritoneal macrophages.

The apparent discrepancy between stimulatory and inhibitory activities of CAs may be attributed to the state of activation of macrophage populations: antigen challenge and activation of macrophages may result in an increase in β receptors and suppression of the response. It is highly likely, however, that there is a transient stage of differentiation, when monocytes (during maturation to macrophages) lose their β-AR responsiveness (see in Ref. [77]). Thus, naïve cells may preferentially express α-ARs, which will result in stimulation of macrophage activity. In fact, the $\alpha_2$-ARs-mediated stimulatory effect of NE on TNF-α production was observed in peritoneal macrophages elicited from specific pathogen-free mice [76], a condition that may reflect naïve, antigen-inexperienced macrophages. Furthermore, recent studies suggest that the β-AR responsiveness of peritoneal macrophages is a dynamic process [78,79]. Thus, minimal responsiveness to isoproterenol, a β-AR agonist, was observed with resident macrophages; maximum isoproterenol-induced inhibition of TNF-α production was observed with complete Freund's adjuvant-elicited macrophages and significantly less with macrophages from streptococcal cell wall-injected rats [78].

As already mentioned, anatomically, a close spatial relationship between sympathetic and peptidergic nerve fibers on one hand, and macrophages and mast cells on the other, is frequently observed [3]. Neuro-macrophage and neuro-mast cell connections are not restricted to lymphoid organs and tissues but are also regularly encountered in virtually all somatic and visceral tissues. Substance P (SP) and peripheral CRH, which are released from sensory peptidergic neurons, are two of the most potent mast cell secretagogues [80–83]. Furthermore, recent evidence indicates that SP up-regulates both TNF-α and IL-12 production by human and murine monocytes and macrophages [84–86]. This adds further complexity to the local effects of stress hormones, in conjunction with other neurotransmitters and/or mediators (Fig. 2; for more details, see Refs [3,87]).

Thus, the suppressive versus stimulatory effects of CAs on macrophage activities might be related to several factors, such as availability of type 1/pro-inflammatory cytokines; presence or absence of antigen; the nature of the antigen; presence in the microenvironment of pro-inflammatory mediators, such as SP, peripheral CRH, and histamine, released from the sensory neurons and mast cells; and the state of activation or differentiation of macrophages. All the above factors may influence the β-AR responsiveness and the expression of α-ARs.

### 9.3. T-cytotoxic activity

Relatively few observations are available on the effect of CAs on Tc ($CD8^+$) cell cytotoxicity. Epinephrine, NE, and isoproterenol suppress the in vitro generation of anti-MOPC-315 tumor cytotoxicity by mouse splenic Tc lymphocytes [88]. Increasing cAMP with either a cAMP analog or the β-ARs agonist metaproterenol significantly inhibits the in vitro development of memory Tc activity in mice using an anti-influenza cytotoxic Tc assay [89]. In contrast, CAs or β-AR agonists, when added at the initiation of a 5-day-sensitization phase, increased the generation of Tc-mediated cytotoxicity using the mixed lymphocyte culture method in Balb/c mice [90]. Thus, CAs may exert enhancing effects on the initiation of Tc responses, in contrast to inhibition of the effector cell function.

## 10. EFFECT OF CAs ON ANTIBODY PRODUCTION (HUMORAL IMMUNITY)

Binding of antigen to B cells induces an activation and subsequent proliferation and differentiation of these cells into antibody-secreting plasma cells. However, to proceed through these steps, the B cells require "help," and $CD4^+$ Th are the cells that provide this help through cell contact- and cytokine-mediated signals. When B cells and Th cells are exposed to Th cell-dependent antigens, NE, through stimulation of $\beta_2$ receptors, exerts an enhancing effect on B-cell antibody (Ab) production [47,91]. IFN-γ-producing Th1 cells induce B cells to produce IgG2a (in humans, IgG1), whereas IL-4-producing Th2 cells induce B cells to produce IgE and IgG1 (in humans, IgG4) [34]. Thus, in the study of Sanders et al. [47], the inhibition of IFN-γ production by Th1 cells induced by the $\beta_2$-AR agonist terbutaline was associated with subsequent suppressed IgG2a production by mouse B cells, whereas the lack of effect on IL-4 was associated with no changes in IgG1 production by B cells.

Moreover, Th cells not only activate B cells during cell-to-cell interaction, but Th2 cells also provide the cytokines necessary for B-cell growth. Thus, the β-AR agonists, salbutamol and fenoterol, potentiate IL-4-induced IgE production by human PBMC while they inhibit IFN-γ production by these cells [92]. Furthermore, salbutamol induces an increase of the ex vivo release of IL-4, IL-6, and IL-10 [93]. This might have resulted from the reversal of the restraining inputs of type 1 cytokines on Th2 cells and by a direct potentiation of the production of IL-6 and IL-10 by APC. Taken together, the enhancement of Ab production further supports the hypothesis discussed above that CAs selectively inhibit Th1 functions and mediate a Th2 shift that potentiate humoral immunity.

## 11. CONCLUSIONS AND CLINICAL IMPLICATIONS

CA–immune interactions are undoubtedly complex. Recent studies suggest that endogenous CAs modulate the function of primary lymphoid organs, such as the bone marrow and the thymus [22,94–96]. However, the role of the sympathetic innervation in regulation of hematopoiesis and thymocyte development remains poorly understood. In addition, there is almost complete lack of knowledge of how CAs might affect mucosal immunity. Overall, there is substantial evidence indicating that local or systemic CAs are involved in *fine tuning* of immune responses. The effects of CAs are quick, within minutes. This modulation might be ideally designed for quick adjustment of lymphoid cell traffic, cytokine/chemokine production, or cellular responsiveness and activity.

Although interest in the Th2 response was initially directed at its protective role in helminthic infections and its pathogenic role in allergy, this response may have important regulatory functions in countering the tissue-damaging effects of macrophages and Th1 cells [34]. Some pro-inflammatory cytokines, and particularly IL-1, via its central effects, stimulate the SNS output [15,97]. Thus, an excessive immune response and the subsequent activation of the SNS, and the release of CAs in the periphery, may trigger a mechanism that inhibits, systemically, Th1 functions and pro-inflammatory cytokine production but potentiates Th2 and anti-inflammatory responses [3,6,36]. This appears to be complemented by the locally released other sympathetic neuromediators such as ATP and adenosine while NPY may amplify these effects. This mechanism may protect the organism from systemic "overshooting" with type 1/pro-inflammatory cytokines and other products of activated macrophages with tissue-damaging potential.

On the contrary, in certain local responses, and under certain conditions, CAs may actually boost regional innate immune responses in a transitory fashion, through induction of IL-8, IL-6, IL-1, and TNF-α production and through short-term increase of NK cell, monocyte, and neutrophils numbers. This might be aimed to localize the inflammatory response, via stimulation of neutrophils accumulation and stimulation of macrophage activity. Although a complete discussion of this problem is beyond the scope of this chapter, an abnormal sympathetic–immune interface or activity of the SNS may play a role in the pathogenesis of infections, autoimmune and atopic/allergic reactions, and tumor growth.

Acute or chronic stress and subsequent CA-induced Th2 shift might specifically increase the susceptibility of the individual to intracellular infections, the defence against which is primarily through Th1-regulated cellular immunity – for example, mycobacterial, *Helicobacter pylori*, HIV, or common cold viral infections. The stress-induced Th2 shift may also trigger herpes simplex viral reactivation [6,36,98–102]. Additionally, NE directly accelerates HIV-1 replication and may affect anti-viral therapy while some of the instability in some inflammatory responses, such as leprosy, might be secondary to the damage of sensory C- and sympathetic nerve fibers and dysregulation of inflammation [8,103,104]. Major injury (serious traumatic injury and major burns or major surgical procedures) often triggers a "sympathetic storm" – the subsequent massive release of CAs via an induction of a Th2 shift may contribute to the severe immunosuppression and the severe infectious complications observed in these conditions [6,51,105–107]. Importantly, a recent study also demonstrates that the activation of the SNS may represent the major immunosuppressive mechanism leading to a high incidence of infections after stroke [108].

A hypoactive SNS system may facilitate or sustain the Th1 shift, observed in autoimmune diseases, such as RA or multiple sclerosis (MS). An additional factor might be the preponderance of about 10:1 for primary sensory, SP-positive fibers as compared with sympathetic fibers in synovial tissues of RA patients (Fig. 3). Alternatively, SNS hyperactivity may intensify the Th2 shift and induce or facilitate flares of systemic lupus erythematosus [3,6,87,109–112].

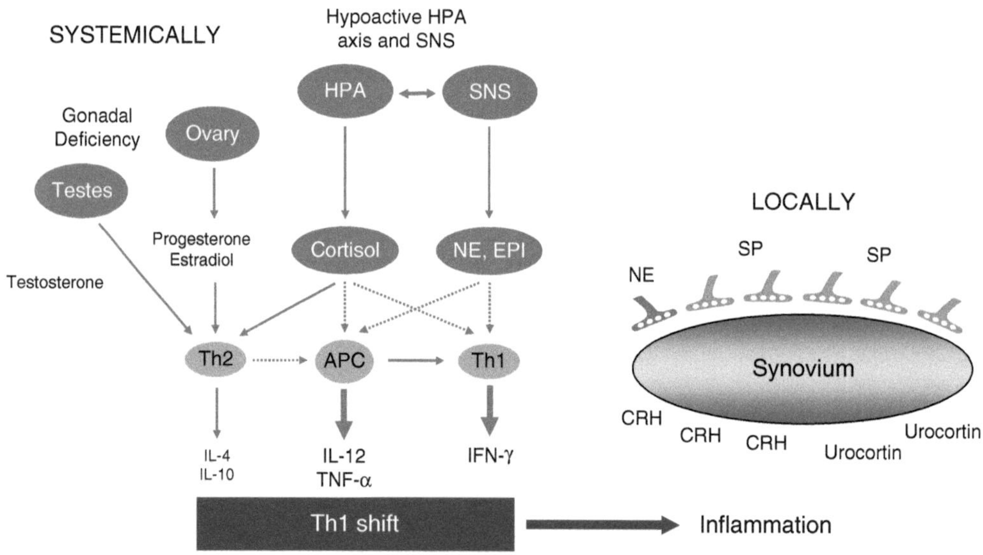


Figure 3. Role of systemic and local neuroendocrine factors in the pathogenesis of rheumatoid arthritis. The hypoactive stress system results in less inhibition of the Th1 responses by cortisol and CAs, systemically. In aging men and postmenopausal women, the gonadal deficiency (i.e., less stimulation of the Th2 responses) will further intensify the Th1 shift. These neuroendocrine abnormalities may sustain and facilitate the Th1 shift observed in RA and further promote the local inflammation. Locally, the preponderance of primary sensory, SP-positive fibers as compared with sympathetic fibers and the over-expression of CRH and urocortin result in a dominance of the autocrine and paracrine pro-inflammatory factors in the synovium of RA patients. Solid lines represent stimulation, whereas dashed lines represent inhibition. APC, antigen-presenting cell; CRH, corticotropin-releasing hormone (peripheral); EPI, epinephrine; HPA, hypothalamic–pituitary–adrenal axis; IL, interleukin; NE, norepinephrine; SNS, sympathetic nervous system; SP, substance P; Th, T helper lymphocyte; TNF, tumor necrosis factor. From Ref. [112].

Allergic reactions of type 1 hypersensitivity (atopy), such as asthma, eczema, hay fever, urticaria and food allergy, are characterized by dominant Th2 responses, overproduction of histamine, and a shift to IgE production. The effects of stress on atopic reactions are complex, at multiple levels, and can be in either direction. CAs acting at the level of APCs and lymphocytes may induce a Th2 shift and thus, facilitate or sustain atopic reactions; however, this can be antagonized by their effects on the mast cell [6,7,113]. In addition, the activation of CRH/SP–mast cell–histamine axis may also contribute to inflammation in allergic reactions.

Low levels of IL-12 and local overproduction of IL-10 and transforming growth factor-β (TGF-β) have been associated with tumor growth [114,115]. These data suggest that CAs and adenosine-induced inhibition of IL-12 and potentiation of IL-10 and TGF-β production and subsequent suppression of cellular immunity may contribute to the increased growth of certain tumors [116–119]. Clearly, these hypotheses require further investigation, but the answers should provide critical insights into mechanisms underlying a variety of common human diseases.

## REFERENCES


1. Besedovsky HO, del Rey AE, Sorkin E. What do the immune system and the brain know about each other? Immunol Today 1983;4:342–6.
2. Chrousos GP. The hypothalamic-pituitary-adrenal axis and immune-mediated inflammation. N Engl J Med 1995;332(20):1351–62.
3. Elenkov IJ, Wilder RL, Chrousos GP, Vizi ES. The sympathetic nerve-an integrative interface between two supersystems: the brain and the immune system. Pharmacol Rev 2000;52(4): 595–638.
4. Elenkov IJ, Iezzoni DJ, Daly A, Harris AJ, Chrousos GP. Cytokine dysregulation, inflammation and well-being. Neuroimmunomodulation 2005;12:255–69.
5. Rook GA, Hernandez-Pando R, Lightman SL. Hormones, peripherally activated prohormones and regulation of the Th1/Th2 balance. Immunol Today 1994;15(7):301–3.
6. Elenkov IJ, Chrousos GP. Stress hormones, Th1/Th2 patterns, pro/anti-inflammatory cytokines and susceptibility to disease. Trends Endocrinol Metab 1999;10(9):359–68.
7. Marshall GD, Jr, Agarwal SK. Stress, immune regulation, and immunity: applications for asthma. Allergy Asthma Proc 2000;21(4):241–6.
8. Rook GA, Lightman SL, Heijnen CJ. Can nerve damage disrupt neuroendocrine immune homeostasis? Leprosy as a case in point. Trends Immunol 2002;23(1):18–22.
9. Tonkoff W. Zur Kenntnis der Nerven der Lymphdrüsen. Anat Anz 1899;16:456–9.
10. Benschop RJ, Rodriguez-Feuerhahn M, Schedlowski M. Catecholamine-induced leukocytosis: early observations, current research, and future directions. Brain Behav Immun 1996;10(2):77–91.
11. von Euler US. The presence of a substance with sympathin E properties in spleen extracts. Acta Physiol Scand 1946;11:168–73.
12. Dougherty TF, Frank JA. The quantitative and qualitative responses of blood lymphocytes to stress stimuli. J Lab Clin Med 1953;42:530–7.
13. Besedovsky H, Sorkin E, Keller M, Muller J. Changes in blood hormone levels during the immune response. Proc Soc Exp Biol Med 1975;150(2):466–70.
14. Besedovsky H, Del Rey A, Sorkin E, Dinarello CA. Immunoregulatory feedback between interleukin-1 and glucocorticoid hormones. Science 1986;233(4764):652–4.
15. Besedovsky H, Del Rey A, Sorkin E, Da Prada M, Burri R, Honegger C. The immune response evokes changes in brain noradrenergic neurons. Science 1983;221(4610):564–6.
16. Besedovsky HO, del RA, Sorkin E, Da PM, Keller HH. Immunoregulation mediated by the sympathetic nervous system. Cell Immunol 1979;48(2):346–55.

17. Hadden JW, Hadden EM, Middleton E Jr. Lymphocyte blast transformation I. Demonstration of adrenergic receptors in human peripheral lymphocytes. Cell Immunol 1970;1:583–95.
18. Reilly FD, McCuskey RS, Meineke HA. Studies of the hemopoietic microenvironment. VIII. Andrenergic and cholinergic innervation of the murine spleen. Anat Rec 1976;185:109–17.
19. Williams JM, Felten DL. Sympathetic innervation of murine thymus and spleen: a comparative histofluorescence study. Anat Rec 1981;199:531–42.
20. Felten DL, Felten SY, Carlson SL, Olschowka JA, Livnat S. Noradrenergic and peptidergic innervation of lymphoid tissue. J Immunol 1985;135(2 Suppl):755s–65s.
21. Williams JM, Felten DL. Sympathetic innervation of murine thymus and spleen: a comparative histofluorescence study. Anat Rec 1981;199(4):531–542.
22. Vizi ES, Orso E, Osipenko ON, Hasko G, Elenkov IJ. Neurochemical, electrophysiological and immunocytochemical evidence for a noradrenergic link between the sympathetic nervous system and thymocytes. Neuroscience 1995;68(4):1263–76.
23. Weihe E, Nohr D, Michel S, Muller S, Zentel HJ, Fink T et al. Molecular anatomy of the neuro-immune connection. Int J Neurosci 1991;59:1–23.
24. Weihe E, Muller S, Fink T, Zentel HJ. Tachykinins, calcitonin gene-related peptide and neuropeptide Y in nerves of the mammalian thymus: interactions with mast cells in autonomic and sensory neuroimmunomodulation? Neurosci Lett 1989;100:77–82.
25. Schedlowski M, Hosch W, Oberbeck R, Benschop RJ, Jacobs R, Raab HR et al. Catecholamines modulate human NK cell circulation and function via spleen-independent beta 2-adrenergic mechanisms. J Immunol 1996;156(1):93–9.
26. Maisel AS, Michel MC. Beta-adrenoceptor control of immune function in congestive heart failure. Br J Clin Pharmacol 1990;30(Suppl 1)::49S–53S.
27. Knudsen JH, Christensen NJ, Bratholm P. Lymphocyte norepinephrine and epinephrine, but not plasma catecholamines predict lymphocyte cAMP production. Life Sci 1996;59(8):639–47.
28. Chambers DA, Cohen RL, Perlman RL. Neuroimmune modulation: signal transduction and catecholamines. Neurochem Int 1993;22(2):95–110.
29. Elliott L, Brooks W, Roszman T. Inhibition of anti-CD3 monoclonal antibody-induced T-cell proliferation by dexamethasone, isoproterenol, or prostaglandin E2 either alone or in combination. Cell Mol Neurobiol 1992;12(5):411–27.
30. Bartik MM, Bauman GP, Brooks WH, Roszman TL. Costimulatory signals modulate the antiproliferative effects of agents that elevate cAMP in T cells. Cell Immunol 1994;158(1): 116–30.
31. Carlson SL, Brooks WH, Roszman TL. Neurotransmitter-lymphocyte interactions: dual receptor modulation of lymphocyte proliferation and cAMP production. J Neuroimmunol 1989;24(1–2): 155–62.
32. Soliven B, Nelson DJ. Beta-adrenergic modulation of K+ current in human T lymphocytes. J Membr Biol 1990;117(3):263–74.
33. Mosmann TR, Sad S. The expanding universe of T-cell subsets: Th1, Th2 and more. Immunol Today 1996;17(3):138–46.
34. Fearon DT, Locksley RM. The instructive role of innate immunity in the acquired immune response. Science 1996;272(5258):50–3.
35. Trinchieri G. Interleukin-12 and the regulation of innate resistance and adaptive immunity. Nat Rev Immunol 2003;3(2):133–46.
36. Elenkov IJ, Papanicolaou DA, Wilder RL, Chrousos GP. Modulatory effects of glucocorticoids and catecholamines on human interleukin-12 and interleukin-10 production: clinical implications. Proc Assoc Am Physicians 1996;108(5):374–81.
37. Panina-Bordignon P, Mazzeo D, Lucia PD, D'Ambrosio D, Lang R, Fabbri L et al. Beta2-agonists prevent Th1 development by selective inhibition of interleukin 12. J Clin Invest 1997;100(6): 1513–19.
38. Hasko G, Szabo C, Nemeth ZH, Salzman AL, Vizi ES. Stimulation of beta-adrenoceptors inhibits endotoxin-induced IL-12 production in normal and IL-10 deficient mice. J Neuroimmunol 1998;88(1–2):57–61.

39. Hetier E, Ayala J, Bousseau A, Prochiantz A. Modulation of interleukin-1 and tumor necrosis factor expression by beta-adrenergic agonists in mouse ameboid microglial cells. Exp Brain Res 1991;86:407–13.
40. Severn A, Rapson NT, Hunter CA, Liew FY. Regulation of tumor necrosis factor production by adrenaline and β-adrenergic agonists. J Immunol 1992;148:3441–5.
41. Nakamura A, Johns EJ, Imaizumi A, Abe T, Kohsaka T. Regulation of tumour necrosis factor and interleukin-6 gene transcription by beta2-adrenoceptor in the rat astrocytes. J Neuroimmunol 1998;88:144–53.
42. Koff WC, Fann AV, Dunegan MA, Lachman LB. Catecholamine-induced suppression of interleukin-1 production. Lymphokine Res 1986;5:239–47.
43. van der Poll T, Lowry SF. Epinephrine inhibits endotoxin-induced IL-1 beta production: roles of tumor necrosis factor-alpha and IL-10. Am J Physiol 1997;273(6 Pt 2):R1885–90.
44. van der Poll T, Coyle SM, Barbosa K, Braxton CC, Lowry SF. Epinephrine inhibits tumor necrosis factor-alpha and potentiates interleukin 10 production during human endotoxemia. J Clin Invest 1996;97(3):713–19.
45. Norris JG, Benveniste EN. Interleukin-6 production by astrocytes: induction by the neurotransmitter norepinephrine. J Neuroimmunol 1993;45:137–45.
46. Maimone D, Cioni C, Rosa S, Macchia G, Aloisi F, Annunziata P. Norepinephrine and vasoactive intestinal peptide induce IL-6 secretion by astrocytes: synergism with IL-1 beta and TNF alpha. J Neuroimmunol 1993;47:73–81.
47. Sanders VM, Baker RA, Ramer-Quinn DS, Kasprowicz DJ, Fuchs BA, Street NE. Differential expression of the beta2-adrenergic receptor by Th1 and Th2 clones: implications for cytokine production and B cell help. J Immunol 1997;158(9):4200–10.
48. Borger P, Hoekstra Y, Esselink MT, Postma DS, Zaagsma J, Vellenga E et al. Beta-adrenoceptor-mediated inhibition of IFN-gamma, IL-3, and GM-CSF mRNA accumulation in activated human T lymphocytes is solely mediated by the beta2-adrenoceptor subtype. Am J Respir Cell Mol Biol 1998;19(3):400–7.
49. Elenkov IJ, Hasko G, Kovacs KJ, Vizi ES. Modulation of lipopolysaccharide-induced tumor necrosis factor-α production by selective α- and β-adrenergic drugs in mice. J Neuroimmunol 1995;61(2):123–31.
50. Hasko G, Szabo C, Nemeth ZH, Salzman AL, Vizi ES. Stimulation of beta-adrenoceptors inhibits endotoxin-induced IL-12 production in normal and IL-10 deficient mice. J Neuroimmunol 1998;88:57–61.
51. Woiciechowsky C, Asadullah K, Nestler D, Eberhardt B, Platzer C, Schoning B et al. Sympathetic activation triggers systemic interleukin-10 release in immunodepression induced by brain injury. Nat Med 1998;4(7):808–13.
52. Le Tulzo Y, Shenkar R, Kaneko D, Moine P, Fantuzzi G, Dinarello CA et al. Hemorrhage increases cytokine expression in lung mononuclear cells in mice: involvement of catecholamines in nuclear factor-kappaB regulation and cytokine expression. J Clin Invest 1997;99(7):1516–24.
53. Broug-Holub E, Persoons JH, Schornagel K, Mastbergen SC, Kraal G. Effects of stress on alveolar macrophages: a role for the sympathetic nervous system. Am J Respir Cell Mol Biol 1998;19(5):842–8.
54. Mohamed-Ali V, Flower L, Sethi J, Hotamisligil G, Gray R, Humphries SE et al. Beta-Adrenergic regulation of IL-6 release from adipose tissue: in vivo and in vitro studies. J Clin Endocrinol Metab 2001; 86(12):5864–9.
55. Baumann H, Gauldie J. The acute phase response. Immunol Today 1994;15(2):74–80.
56. Murray DR, Prabhu SD, Chandrasekar B. Chronic beta-adrenergic stimulation induces myocardial proinflammatory cytokine expression. Circulation 2000;101(20):2338–41.
57. Kavelaars A, van de Pol M , Zijlstra J, Heijnen CJ. Beta 2-adrenergic activation enhances interleukin-8 production by human monocytes. J Neuroimmunol 1997;77(2):211–16.
58. Linden A. Increased interleukin-8 release by beta-adrenoceptor activation in human transformed bronchial epithelial cells. Br J Pharmacol 1996;119(2):402–6.

59. Engstad CS, Lund T, Osterud B. Epinephrine promotes IL-8 production in human leukocytes via an effect on platelets. Thromb Haemost 1999;81(1):139–45.
60. Kaplanski G, Porat R, Aiura K, Erban JK, Gelfand JA, Dinarello CA. Activated platelets induce endothelial secretion of interleukin-8 in vitro via an interleukin-1-mediated event. Blood 1993;81(10):2492–5.
61. Hasko G, Shanley TP, Egnaczyk G, Nemeth ZH, Salzman AL, Vizi ES et al. Exogenous and endogenous catecholamines inhibit the production of macrophage inflammatory protein (MIP) 1$\alpha$ via a $\beta$ adrenoceptor mediated mechanism. Br J Pharmacol 1998;125:1297–303.
62. Hellstrand K, Hermodsson S. An immunopharmacological analysis of adrenaline-induced suppression of human natural killer cell cytotoxicity. Int Arch Allergy Appl Immunol 1989;89: 334–41.
63. Whalen MM, Bankhurst AD. Effects of beta-adrenergic receptor activation, cholera toxin and forskolin on human natural killer cell function. Biochem J 1990;272:327–31.
64. Shakhar G, Ben-Eliyahu S. In vivo beta-adrenergic stimulation suppresses natural killer activity and compromises resistance to tumor metastasis in rats. J Immunol 1998;160(7):3251–8.
65. Katafuchi T, Take S, Hori T. Roles of sympathetic nervous system in the suppression of cytotoxicity of splenic natural killer cells in the rat. J Physiol (Lond) 1993;465:343–57.
66. Irwin M, Hauger R, Brown M. Central corticotropin-releasing hormone activates the sympathetic nervous system and reduces immune function: increased responsivity of the aged rat. Endocrinology 1992;131:1047–53.
67. Irwin M. Stress-induced immune suppression: role of brain corticotropin releasing hormone and autonomic nervous system mechanisms. Adv Neuroimmunol 1994;4:29–47.
68. Strausbaugh H, Irwin M. Central corticotropin-releasing hormone reduces cellular immunity. Brain Behav Immun 1992;6:11–17.
69. Vredevoe DL, Moser DK, Gan XH, Bonavida B. Natural killer cell anergy to cytokine stimulants in a subgroup of patients with heart failure: relationship to norepinephrine. Neuroimmunomodulation 1995;2(1):16–24.
70. Koff WC, Dunegan MA. Modulation of macrophage-mediated tumoricidal activity by neuropeptides and neurohormones. J Immunol 1985;135(1):350–4.
71. Koff WC, Dunegan MA. Neuroendocrine hormones suppress macrophage-mediated lysis of herpes simplex virus-infected cells. J Immunol 1986;136(2):705–9.
72. Bermudez LE, Wu M, Young LS. Effect of stress-related hormones on macrophage receptors and response to tumor necrosis factor. Lymphokine Res 1990;9:137–45.
73. Chou RC, Stinson MW, Noble BK, Spengler RN. Beta-adrenergic receptor regulation of macrophage-derived tumor necrosis factor-alpha production from rats with experimental arthritis. J Neuroimmunol 1996;67(1):7–16.
74. Suberville S, Bellocq A, Fouqueray B, Philippe C, Lantz O, Perez J et al. Regulation of interleukin-10 production by beta-adrenergic agonists. Eur J Immunol 1996;26:2601–5.
75. Miles BA, Lafuse WP, Zwilling BS. Binding of alpha-adrenergic receptors stimulates the anti-mycobacterial activity of murine peritoneal macrophages. J Neuroimmunol 1996;71:19–24.
76. Spengler RN, Allen RM, Remick DG, Strieter RM, Kunkel SL. Stimulation of $\alpha$-adrenergic receptor augments the production of macrophage-derived tumor necrosis factor. J Immunol 1990;145:1430–4.
77. Baker AJ, Fuller RW. Loss of response to beta-adrenoceptor agonists during the maturation of human monocytes to macrophages in vitro. J Leukoc Biol 1995;57(3):395–400.
78. Chou RC, Dong XL, Noble BK, Knight PR, Spengler RN. Adrenergic regulation of macrophage-derived tumor necrosis factor-alpha generation during a chronic polyarthritis pain model. J Neuroimmunol 1998;82:140–8.
79. Ignatowski TA, Gallant S, Spengler RN. Temporal regulation by adrenergic receptor stimulation of macrophage (M phi)-derived tumor necrosis factor (TNF) production post-LPS challenge. J Neuroimmunol 1996;65:107–17.
80. Foreman JC. Substance P and calcitonin gene-related peptide: effects on mast cells and in human skin. Int Arch Allergy Appl Immunol 1987;82:366–71.

81. Church MK, Lowman MA, Robinson C, Holgate ST, Benyon RC. Interaction of neuropeptides with human mast cells. Int Arch Allergy Appl Immunol 1989;88:70–8.
82. Theoharides TC, Spanos C, Pang X, Alferes L, Ligris K, Letourneau R et al. Stress-induced intracranial mast cell degranulation: a corticotropin-releasing hormone-mediated effect. Endocrinology 1995;136(12):5745–50.
83. Theoharides TC, Singh LK, Boucher W, Pang X, Letourneau R, Webster E et al. Corticotropin-releasing hormone induces skin mast cell degranulation and increased vascular permeability, a possible explanation for its proinflammatory effects. Endocrinology 1998;139(1): 403–13.
84. Lotz M, Vaughan JH, Carson DA. Effect of neuropeptides on production of inflammatory cytokines by human monocytes. Science 1988;241:1218–21.
85. Kincy-Cain T, Bost KL. Substance P-induced IL-12 production by murine macrophages. J Immunol 1997;158:2334–9.
86. Ho WZ, Stavropoulos G, Lai JP, Hu BF, Magafa V, Anagnostides S et al. Substance P C-terminal octapeptide analogues augment tumor necrosis factor-alpha release by human blood monocytes and macrophages. J Neuroimmunol 1998;82:126–32.
87. Straub RH, Cutolo M. Involvement of the hypothalamic – pituitary – adrenal/gonadal axis and the peripheral nervous system in rheumatoid arthritis: viewpoint based on a systemic pathogenetic role. Arthritis Rheum 2001;44(3):493–507.
88. Cook-Mills JM, Mokyr MB, Cohen RL, Perlman RL, Chambers DA. Neurotransmitter suppression of the in vitro generation of a cytotoxic T lymphocyte response against the syngeneic MOPC-315 plasmacytoma. Cancer Immunol Immunother 1995;40:79–87.
89. Bender BS, Bell WE, Taylor S, Scarpace PJ. Decreased sensitivity to cAMP in the in vitro generation of memory splenic cytotoxic T-lymphocytes from aged mice: role of phosphodiesterase. J Pharmacol Exp Ther 1993;264:1381–6.
90. Hatfield SM, Petersen BH, DiMicco JA. Beta adrenoceptor modulation of the generation of murine cytotoxic T lymphocytes in vitro. J Pharmacol Exp Ther 1986;239(2):460–6.
91. Sanders VM, Munson AE. Norepinephrine and the antibody response. Pharmacol Rev 1985;37: 229–48.
92. Coqueret O, Dugas B, Mencia-Huerta JM, Braquet P. Regulation of IgE production from human mononuclear cells by beta 2-adrenoceptor agonists [see comments]. Clin Exp Allergy 1995;25(4):304–11.
93. Coqueret O, Lagente V, Frere CP, Braquet P, Mencia-Huerta JM. Regulation of IgE production by beta 2-adrenoceptor agonists. Ann N Y Acad Sci 1994;725:44–9.
94. Maestroni GJ. Is hematopoiesis under the influence of neural and neuroendocrine mechanisms? Histol Histopathol 1998;13:271–4.
95. Maestroni GJ. Catecholaminergic regulation of hematopoiesis in mice [letter; comment]. Blood 1998;92:2971–3.
96. Cavallotti D, Artico M, Iannetti G, Cavallotti C. Occurrence of adrenergic nerve fibers in human thymus during immune response. Neurochem Int 2002;40(3):211–21.
97. Dunn AJ. Systemic interleukin-1 administration stimulates hypothalamic norepinephrine metabolism parallelling the increased plasma corticosterone. Life Sci 1988;43(5):429–35.
98. Lerner BH. Can stress cause disease? Revisiting the tuberculosis research of Thomas Holmes, 1949–1961. Ann Intern Med 1996;124(7):673–80.
99. Levenstein S, Ackerman S, Kiecolt-Glaser JK, Dubois A. Stress and peptic ulcer disease. JAMA 1999;281(1):10–11.
100. Cohen S, Tyrrell DA, Smith AP. Psychological stress and susceptibility to the common cold [see comments]. N Engl J Med 1991;325(9):606–12.
101. Padgett DA, Sheridan JF, Dorne J, Berntson GG, Candelora J, Glaser R. Social stress and the reactivation of latent herpes simplex virus type 1. Proc Natl Acad Sci USA 1998;95(12):7231–5.
102. Sainz B, Loutsch JM, Marquart ME, Hill JM. Stress-associated immunomodulation and herpes simplex virus infections. Med Hypotheses 2001;56(3):348–56.

103. Cole SW, Korin YD, Fahey JL, Zack JA. Norepinephrine accelerates HIV replication via protein kinase A-dependent effects on cytokine production. J Immunol 1998;161(2):610–16.
104. Cole SW, Naliboff BD, Kemeny ME, Griswold MP, Fahey JL, Zack JA. Impaired response to HAART in HIV-infected individuals with high autonomic nervous system activity. Proc Natl Acad Sci USA 2001;98(22):12695–700.
105. O'Sullivan ST, Lederer JA, Horgan AF, Chin DH, Mannick JA, Rodrick ML. Major injury leads to predominance of the T helper-2 lymphocyte phenotype and diminished interleukin-12 production associated with decreased resistance to infection [see comments]. Ann Surg 1995;222(4):482–90.
106. Elenkov IJ, Chrousos GP, Wilder RL. Neuroendocrine regulation of IL-12 and TNF-alpha/IL-10 balance. Clinical implications. Ann N Y Acad Sci 2000;917:94–105.
107. Munford RS, Tracey KJ. Is severe sepsis a neuroendocrine disease. Mol Med 2002;8(8):437–42.
108. Prass K, Meisel C, Hoflich C, Braun J, Halle E, Wolf T et al. Stroke-induced immunodeficiency promotes spontaneous bacterial infections and is mediated by sympathetic activation: Reversal by poststroke T helper cell type 1-like immunostimulation. J Exp Med 2003;198(5):725–36.
109. Wilder RL. Neuroendocrine-immune system interactions and autoimmunity. Annu Rev Immunol 1995;13:307–38.
110. Miller LE, Justen HP, Scholmerich J, Straub RH. The loss of sympathetic nerve fibers in the synovial tissue of patients with rheumatoid arthritis is accompanied by increased norepinephrine release from synovial macrophages. FASEB J 2000;14(13):2097–107.
111. Elenkov IJ, Wilder RL, Bakalov VK, Link AA, Dimitrov MA, Fisher S et al. IL-12, TNF-alpha, and hormonal changes during late pregnancy and early postpartum: implications for autoimmune disease activity during these times. J Clin Endocrinol Metab 2001;86(10):4933–8.
112. Calcagni E, Elenkov IJ. Stress system activity, innate and T helper cytokines, and susceptibility to immune-related diseases. Ann N Y Acad Sci 2006;1069:62–76.
113. Barnes P, FitzGerald G, Brown M, Dollery C. Nocturnal asthma and changes in circulating epinephrine, histamine, and cortisol. N Engl J Med 1980;303(5):263–7.
114. Colombo MP, Vagliani M, Spreafico F, Parenza M, Chiodoni C, Melani C et al. Amount of interleukin 12 available at the tumor site is critical for tumor regression. Cancer Res 1996;56(11):2531–4.
115. Chouaib S, Asselin-Paturel C, Mami-Chouaib F, Caignard A, Blay JY. The host-tumor immune conflict: from immunosuppression to resistance and destruction. Immunol Today 1997;18(10): 493–7.
116. Shakhar G, Ben-Eliyahu S. In vivo beta-adrenergic stimulation suppresses natural killer activity and compromises resistance to tumor metastasis in rats. J Immunol 1998;160:3251–8.
117. Li T, Harada M, Tamada K, Abe K, Nomoto K. Repeated restraint stress impairs the antitumor T cell response through its suppressive effect on Th1-type CD4+ T cells. Anticancer Res 1997;17(6D):4259–68.
118. Blay J, White TD, Hoskin DW. The extracellular fluid of solid carcinomas contains immunosuppressive concentrations of adenosine. Cancer Res 1997;57(13):2602–5.
119. Hoskin DW, Reynolds T, Blay J. Adenosine as a possible inhibitor of killer T-cell activation in the microenvironment of solid tumours. Int J Cancer 1994;59(6):854–5.

# *Enhancing versus Suppressive Effects of Stress on Immune Function: Implications for Immunoprotection and Immunopathology*

FIRDAUS S. DHABHAR

*Department of Psychiatry & Behavioral Sciences, Program in Immunology, Comprehensive Cancer Center, Stanford University, Stanford, CA, USA*

## ABSTRACT

It is widely believed that stress suppresses immune function and increases susceptibility to infections and cancer. Paradoxically, stress is also known to exacerbate allergic and autoimmune diseases that are proinflammatory and should theoretically be ameliorated by immunosuppression. These observations suggest that stress may have bidirectional effects on immune function, elucidation of which may have significant clinical consequences. It has recently been shown that in contrast to chronic stress that suppresses or dysregulates immune function, acute stress can be immunoenhancing. Studies have shown that acute stress experienced at the time of immune activation can affect dendritic cell, neutrophil, macrophage, and lymphocyte trafficking, maturation, and function, and augment innate as well as adaptive immune responses. Acute stress experienced prior to novel antigen exposure enhances innate immunity and memory T-cell formation and results in a significant and long-lasting immunoenhancement. Acute stress experienced during antigen re-exposure enhances secondary/adaptive immune responses. Therefore, depending on the conditions of immune activation and the nature of the immunizing antigen, acute stress may enhance the acquisition and expression of immunoprotection or immunopathology. In contrast, chronic stress dysregulates innate and adaptive immune responses by shifting the immunological balance from Type 1 to Type 2 cytokine-mediated immunity, and suppresses immunity by decreasing leukocyte numbers, trafficking, and function. Chronic stress also increases susceptibility to non-melanoma skin cancer by suppressing Type 1 cytokines and protective T cells while increasing suppressor T-cell function. We have suggested that the adaptive purpose of a physiological stress response may be to promote survival, with stress hormones and neurotransmitters serving as beacons that prepare the immune system for potential challenges (e.g., wounding or infection) perceived by the brain (e.g., detection of an attacker). However, this system may exacerbate immunopathology if the enhanced immune response is directed against innocuous or self-antigens, or dysregulated following prolonged activation during chronic stress. In view of the ubiquitous nature of stress and its significant effects on immunoprotection as well as immunopathology, it is important to further elucidate the mechanisms mediating stress–immune interactions and to meaningfully translate findings from bench to bedside.

## 1. INTRODUCTION

Psychological stress is known to suppress immune function and increase susceptibility to infections and cancer. Paradoxically, stress is also known to exacerbate allergic, autoimmune, and inflammatory diseases, which suggests that stress may enhance immune function under certain conditions. It has recently been appreciated that while chronic stress suppresses immune function, acute stress often has immunoenhancing effects [1].

One of the most under-appreciated effects of stress on the immune system is its ability to induce significant changes in leukocyte distribution in the body [2]. Importantly, these changes have significant effects on immune function in different body compartments that are either enriched or depleted of leukocytes during stress. Moreover, acute stress can affect dendritic cell, neutrophil, macrophage, and lymphocyte trafficking, maturation, or function in ways that can enhance innate as well as adaptive immunity [3–6]. Acute stress experienced prior to novel cutaneous antigen exposure increases memory T-cell formation and results in a significant and long-lasting increase in immunity [3,4,6]. Similarly, acute stress experienced during antigen re-exposure enhances secondary immune responses [7]. This suggests that depending on the conditions and compartment in which the immune response is initiated, stress can enhance the acquisition as well as expression of immunoprotection and immunopathology.

In contrast to acute stress, chronic stress suppresses or dysregulates innate and adaptive immune responses through mechanisms that involve suppression of leukocyte numbers, trafficking, and function, or changes in the Type 1–Type 2 cytokine balance [8,9]. Chronic stress has recently been shown to increase susceptibility to skin cancer by suppressing Type 1 cytokines and protective T cells while increasing suppressor T-cell function [10].

We have suggested that the primary biological purpose of an acute psycho-physiological stress response may be to promote survival, with stress hormones and neurotransmitters serving as beacons that prepare the immune system for potential challenges (e.g., wounding or infection) perceived by the brain (e.g., detection of an imminent attack) [1,2]. However, this same system may exacerbate immunopathology if the enhanced immune response is directed against innocuous or self-antigens, or if the stress-response system is over-activated as seen during chronic stress. In view of the ubiquitous nature of stress and its significant effects on immunoprotection and immunopathology, it is important to further elucidate the mechanisms mediating stress–immune interactions and to translate findings from bench to bedside.

## 2. STRESS

Although the word "stress" generally has negative connotations, stress is a familiar aspect of life, being a stimulant for some, but a burden for others. Numerous definitions have been proposed for the word stress. Each definition focuses on aspects of an internal or external challenge, disturbance, or stimulus; perception of a stimulus by an organism; or a physiological response of the organism to the stimulus [11–13]. Physical stressors have been defined as external challenges to homeostasis and psychological stressors as the "anticipation justified or not, that a challenge to homeostasis looms" [14]. An integrated definition states that stress is a constellation of events, consisting of a stimulus (stressor), which precipitates a reaction in the brain (stress perception), which activates physiologic fight or flight systems in the body (stress response) [15]. The physiologic stress response results in the release of neurotransmitters and hormones that serve as the brain's alarm signals to the body. It is often overlooked that a stress response has salubrious adaptive effects in the short run [1,3], although stress can be harmful when it is long lasting [8,13,16,17].

An important distinguishing characteristic of stress is its duration and intensity. Thus, *acute stress* has been defined as stress that lasts for a period of minutes to hours, and *chronic stress* as stress that persists for several hours per day for weeks or months [15]. The intensity of stress may be gauged by the peak levels of stress hormones, neurotransmitters, and other physiological changes such as increases in heart rate and blood pressure, and by the amount of time for which these changes persist during stress and following the cessation of stress.

It is important to bear in mind that there exist significant individual differences in the manner and extent to which stress is perceived, processed, and coped with [1]. These differences become particularly relevant while studying human subjects because stress perception, processing, and coping mechanisms can have significant effects on the kinetics and peak levels of circulating stress hormones and on the duration for which these hormone levels are elevated. The magnitude and duration of catecholamine and glucocorticoid hormone exposure in turn can have significant effects on leukocyte distribution and function [18–20].

## 3. PARADOXICAL OBSERVATIONS REGARDING THE EFFECTS OF STRESS ON IMMUNE FUNCTION

Three paradoxes present themselves when one reviews the literature examining the relationship between stress, immune function, and health: *First*, it is paradoxical that organisms would have evolved to suppress immune function at a time when an active immune response may be critical for survival, for example, under conditions of stress when an organism may be injured or infected by the actions of the stress-inducing agent (e.g., an attacking predator). *Second*, on the one hand, stress is thought to suppress immunity and increase susceptibility to infections and cancer [8,21–24], while on the other, it is thought to exacerbate inflammatory (cardiovascular disease, gingivitis, dermatitis) or autoimmune (psoriasis, arthritis, multiple sclerosis) diseases [25–34] which should be ameliorated by a suppression of immune function. *Third*, stress is known to exacerbate autoimmune and inflammatory diseases; however, stress hormones (glucocorticoids) are used clinically to treat these diseases [35]. These observations suggest that psycho-physiological stress can exert immunoenhancing as well as immunosuppressive effects.

## 4. STRESS-INDUCED CHANGES IN IMMUNE CELL DISTRIBUTION

Effective immunoprotection requires rapid recruitment of leukocytes into sites of surgery, wounding, infection, or vaccination. Immune cells circulate continuously on surveillance pathways that take them from the blood, through various organs, and back into the blood. This circulation is essential for the maintenance of an effective immune defense network [36]. The numbers and proportions of leukocytes in the blood provide an important representation of the state of distribution of leukocytes in the body and of the state of activation of the immune system. The ability of acute stress to induce changes in leukocyte distribution within different body compartments is perhaps one of the most under-appreciated effects of stress and stress hormones on the immune system [2].

Numerous studies have shown that stress and stress hormones induce significant changes in absolute numbers and relative proportions of leukocytes in the blood. In fact, changes in blood leukocyte numbers were used as a measure of stress before methods were available to directly assay the hormone [37]. Studies have also shown that glucocorticoid [38–40] and catecholamine hormones [41–46] induce rapid and significant changes in leukocyte distribution and that these hormones are the major mediators of the effects of stress. Stress-induced changes in blood leukocyte numbers have been reported in fish [47], hamsters [48], mice [49], rats [2,40,50,51], rabbits [52], horses [53], non-human primates [54], and humans [44,55–58]. This suggests that the phenomenon of stress-induced leukocyte redistribution has a long evolutionary lineage, and that perhaps it has important functional significance.

Studies in rodents have shown that stress-induced changes in blood leukocyte numbers are characterized by a significant decrease in numbers and percentages of lymphocytes and monocytes, and by an increase in numbers and percentages of neutrophils [2,50]. Flow cytometric analyses revealed that absolute numbers of peripheral blood T cells, B cells, natural killer (NK) cells, and monocytes all show a rapid and significant decrease (40%–70% lower than baseline) during stress [2]. Moreover, it has been shown that stress-induced changes in leukocyte numbers are rapidly reversed upon the cessation of stress [2]. In apparent contrast to animal studies, human studies have shown that stress can increase rather than decrease blood leukocyte numbers [56,58–61]. This apparent contradiction may be resolved by taking the following factors into consideration: First, stress-induced increases in blood leukocyte numbers are observed following stress conditions which primarily result in the activation of the sympathetic nervous system. These stressors are often of a short duration (few minutes) or relatively mild (e.g., public speaking) [56,59–61]. Second, the increase in total leukocyte numbers may be accounted for by stress- or catecholamine-induced increases in granulocytes and NK cells [43,56,59–61]. Third, stress- or pharmacologically induced increases in glucocorticoid hormones induce a significant decrease in blood lymphocyte and monocyte numbers [37,40,56,62]. Thus, stress conditions which result in a significant and sustained activation of the hypothalamic–pituitary–adrenal (HPA) axis result in a decrease in blood leukocyte numbers.

In view of the above discussion, it has been proposed that acute stress induces an initial increase followed by a decrease in blood leukocyte numbers. Stress conditions that result in activation of the sympathetic nervous system, especially conditions that induce high levels of norepinephrine, may induce an increase in circulating leukocyte numbers. These conditions may occur during the beginning of a stress response, very short-duration stress (order of minutes), mild psychological stress, or during exercise. In contrast, stress conditions that result in the activation of the HPA axis induce a decrease in circulating leukocyte numbers. These conditions often occur during the later stages of a stress response, long-duration acute stressors (order of hours), or during severe psychological, physical, or physiological stress. An elegant and interesting example in support of this hypothesis comes from Schedlowski et al. who measured changes in blood T-cell and NK-cell numbers as well as plasma catecholamine and cortisol levels in parachutists [56]. Measurements were made 2 h before, immediately after, and 1 h after the jump. Results showed a significant increase in T-cell and NK-cell numbers immediately (minutes) after the jump that was followed by a significant decrease 1 h after the jump. An early increase in plasma catecholamines preceded early increases in lymphocyte numbers, whereas the more delayed rise in plasma cortisol preceded the late decrease in lymphocyte numbers [56]. Importantly, changes in NK-cell activity and antibody-dependent cell-mediated cytotoxicity closely paralleled changes in blood NK-cell numbers, thus suggesting that changes in leukocyte numbers may be an important mediator of apparent changes in leukocyte "activity." Similarly, Rinner et al. have shown that a short stressor (1-min handling) induced an increase in mitogen-induced proliferation of T and B cells obtained from peripheral blood, while a longer stressor (2-h immobilization) induced a decrease in the same proliferative responses [63]. In another example, Manuck et al. showed that acute psychological stress induced a significant increase in blood CTL numbers only in those subjects who showed heightened catecholamine and cardiovascular reactions to stress [64].

Thus, an acute stress response may induce biphasic changes in blood leukocyte numbers. Soon after the beginning of stress (order of minutes) or during mild acute stress, or exercise, catecholamine hormones and neurotransmitters induce the body's "soldiers" (leukocytes) to exit their "barracks" (spleen, lung, marginated pool, and other organs) and enter the "boulevards" (blood vessels and lymphatics). This results in an increase in blood leukocyte numbers, the effect being most prominent for NK cells and granulocytes. As the stress response continues, activation of the HPA axis results in the release of glucocorticoid hormones which induce leukocytes to exit the blood and take position at potential "battle stations" (such as the skin, lung, gastro-intestinal and urinary-genital tracts, mucosal surfaces, and lymph nodes) in preparation for immune challenges which may be imposed by the actions of the stressor [2,7,18]. Such a redistribution of leukocytes results in a decrease in blood leukocyte numbers. Thus, acute stress may result in a redistribution of leukocytes from the barracks, through the boulevards, and to potential battle stations within the body.

Since the blood is the most accessible and commonly used compartment for human studies, it is important to carefully evaluate how changes in blood immune parameters might reflect *in vivo* immune function in the context of the specific experiments or study at hand. Moreover, since most blood collection procedures involve a certain amount of stress, since all patients or subjects will have experienced acute and chronic stress, and since many studies of psycho-physiological effects on immune function focus on stress, the effects of stress on blood leukocyte distribution become a factor of considerable importance.

Dhabhar et al. were the first to propose that stress-induced changes in blood leukocyte distribution may represent an adaptive response [50,65]. They suggested that acute stress-induced changes in blood leukocyte numbers represent a redistribution of leukocytes from the blood to organs such as the skin, draining sentinel lymph nodes, and other compartments [7,18]. They hypothesized that such a leukocyte redistribution may enhance immune function in compartments to which immune cells traffic during stress. In agreement with this hypothesis, it was demonstrated that a stress-induced redistribution of leukocytes from the blood to the skin is accompanied by a significant enhancement of skin immunity [7,65,66].

## 5. FUNCTIONAL CONSEQUENCES OF STRESS-INDUCED CHANGES IN IMMUNE CELL DISTRIBUTION

When interpreting data showing stress-induced changes in functional assays such as lymphocyte proliferation or NK activity, it may be important to bear in mind the effects of stress on the leukocyte composition of the compartment in which an immune parameter is being measured. For example, it has been shown that acute stress induces a redistribution of leukocytes from the blood to the skin and that this redistribution is accompanied by a significant enhancement of skin cell-mediated immunity [3,7]. In what might at first glance appear to be contradicting results, acute stress has been shown to suppress splenic and peripheral blood responses to T-cell mitogens [67] and splenic IgM production [68]. However, it is important to note that in contrast to the skin that is enriched in leukocytes during acute stress, peripheral blood and spleen are relatively depleted of leukocytes during acute stress [69]. This stress-induced decrease in blood and spleen leukocyte numbers may contribute to the acute stress-induced suppression of immune function in these compartments.

Moreover, in contrast to acute stress, chronic stress has been shown to suppress skin cell-mediated immunity, and a chronic stress-induced suppression of blood leukocyte redistribution is thought to be one of the factors mediating the immunosuppressive effect of chronic stress [15]. Again, in what might appear to be contradicting results, chronic stress has been shown to enhance mitogen-induced proliferation of splenocytes [70] and splenic IgM production [68]. However, the spleen is relatively enriched in T cells during chronic glucocorticoid administration, suggesting that it may also be relatively enriched in T cells during chronic stress [71], and this increase in spleen leukocyte numbers may contribute to the chronic stress-induced enhancement of immune parameters measured in the spleen.

It is also important to bear in mind that the heterogeneity of the stress-induced changes in leukocyte distribution [2] suggests that using equal numbers of leukocytes in a functional assay may not account for stress-induced changes in relative percentages of different leukocyte subpopulations in the cell suspension being assayed. For example, samples that have been equalized for absolute numbers of total blood leukocytes from control versus stressed animals may still contain different numbers of specific leukocyte subpopulations (e.g., T cells, B cells, or NK cells). Such changes in leukocyte composition may contribute to the effects of stress even in functional assays using equalized numbers of leukocytes from different treatment groups. Therefore, stress may affect immune function at a cellular level (e.g., phagocytosis, antigen presentation, killing, antibody production) and/or through leukocyte redistribution that could increase or decrease the number of cells with a specific functional capacity in the compartment being studied.

## 6. EFFECTS OF ACUTE STRESS ON LEUKOCYTE TRAFFICKING TO A SITE OF SURGERY OR IMMUNE ACTIVATION

Viswanathan and Dhabhar used a clinically relevant subcutaneously implanted surgical sponge model to elucidate the effects of stress on the kinetics, magnitude, subpopulation, and chemoattractant specificity of leukocyte trafficking to a site of immune activation or surgery [5]. Mice that were acutely stressed before subcutaneous implantation or the surgical sponge showed a two- to three-fold higher neutrophil, macrophage, NK-cell, and T-cell infiltration than non-stressed animals. Leukocyte infiltration was evident as early as 6 h and peaked between 24 and 48 h. Importantly at 72 h, sponges from non-stressed and acutely stressed mice had comparable and significantly lower leukocyte numbers indicating effective resolution of inflammation in both groups. These authors also examined the effects of stress on early (6 h) leukocyte infiltration in response to a predominantly proinflammatory cytokine, tumor necrosis factor-α (TNF-α), and lymphocyte-specific chemokine, lymphotactin (LTN). Acute stress significantly increased infiltration of macrophages, in response to saline, LTN, or TNF-α; neutrophils, only in response to TNF-α; and NK and T cells only in response to LTN. These results showed that acute stress significantly enhances the kinetics and magnitude of leukocyte infiltration into a site of immune activation or surgery in a subpopulation and chemoattractant-specific manner, with tissue damage, antigen-, or pathogen-driven chemoattractants synergizing with acute stress to further determine the specific subpopulations that are recruited [5]. Thus, depending on the primary chemoattractants driving an immune response, acute stress may selectively mobilize specific leukocyte subpopulations into sites of surgery, wounding, or inflammation. Such a stress-induced increase in leukocyte trafficking may be an important mechanism by which acute stressors alter the course of different (innate versus adaptive, early versus late, acute versus chronic) protective or pathological immune responses.

## 7. ACUTE STRESS-INDUCED ENHANCEMENT OF INNATE/PRIMARY IMMUNE RESPONSES

In view of the skin being one of the target organs to which leukocytes traffic during stress, studies were conducted to examine whether skin immunity is enhanced when immune activation/antigen exposure occurs following a stressful experience. Studies showed that acute stress experienced at the time of novel or primary antigen exposure results in a significant enhancement of the ensuing skin immune response [3]. Compared to controls, mice restrained for 2.5 h before primary immunization with keyhole limpet hemocyanin (KLH) showed a significantly enhanced immune response when re-exposed to KLH 9 months later. This immunoenhancement was mediated by an increase in numbers of memory and effector helper T cells in sentinel lymph nodes at the time of primary immunization. Further analyses showed that the early stress-induced increase in T-cell memory may have stimulated the robust increase in infiltrating lymphocyte and macrophage numbers observed months later at a novel site of antigen re-exposure. Enhanced leukocyte infiltration was driven by increased levels of the Type 1 cytokines, IL-2 and IFN-$\gamma$, and TNF-$\alpha$, observed at the site of antigen re-exposure in animals that had been stressed at the time of primary immunization. Given the importance of inducing long-lasting increases in immunological memory during vaccination, it has been suggested that the neuroendocrine stress response is nature's adjuvant that could be psychologically and/or pharmacologically manipulated to safely increase vaccine efficacy.

In a series of elegant experiments, Saint-Mezard et al. similarly showed that acute stress experienced at the time of sensitization resulted in a significant increase in the contact hypersensitivity (CHS) response [6]. These investigators showed that acute stress experienced during sensitization enhanced dendritic cell migration from skin to sentinel lymph nodes and also enhanced priming of lymph node $CD8^+$ T cells. These $CD8^+$ T cells responded in greater numbers at the site of antigen re-exposure during the recall phase of the CHS response. These studies also suggested that the effects of acute stress in this case were mediated primarily by norepinephrine [6]. Other investigators have similarly reported stress-induced enhancement of Type 1 cytokine-driven cell-mediated immunity [72–74] and Type 2 cytokine-driven humoral immunity [74–77].

Viswanathan et al. further elucidated the molecular and cellular mediators of the immunoenhancing effects of acute stress [78]. They showed that compared to non-stressed mice, acutely stressed animals showed significantly greater pinna swelling, leukocyte infiltration, and upregulated macrophage chemoattractant protein-1 (MCP-1), macrophage inflammatory protein-3$\alpha$ (MIP-3$\alpha$), IL-1$\alpha$, IL-1$\beta$, IL-6, TNF-$\alpha$, and IFN-$\gamma$ gene expression at the site of primary antigen exposure. Stressed animals also showed enhanced maturation and trafficking of dendritic cells from skin to lymph nodes, higher numbers of activated macrophages in skin and lymph nodes, increased T-cell activation in lymph nodes, and enhanced recruitment of surveillance T cells to skin. These findings showed that important interactive components of innate (dendritic cells and macrophages) and adaptive (surveillance T cells) immunity are mediators of the stress-induced enhancement of a primary immune response. Such immunoenhancement during primary immunization may induce a long-term improvement in immunologic memory resulting in subsequent augmentation of the immune response during secondary antigen exposure.

In addition to elucidating mechanisms that could be targeted to reduce stress-induced exacerbation of allergic, autoimmune, and proinflammatory reactions, the above-mentioned studies provide further support for the idea that a psycho-physiological stress response is nature's fundamental survival mechanism that could be therapeutically harnessed to augment immune function during vaccination, wound healing, or infection.

## 8. ACUTE STRESS-INDUCED ENHANCEMENT OF ADAPTIVE/SECONDARY IMMUNE RESPONSES

Studies have shown that in addition to enhancing primary cutaneous immune responses, acute stress experienced at the time of antigen re-exposure can also enhance secondary or recall responses in skin [7]. Compared to non-stressed controls, mice that were acutely stressed at the time of antigen re-exposure showed a significantly larger number of infiltrating leukocytes at the site of the immune reaction. These results demonstrated that a relatively mild behavioral manipulation can enhance an important class of immune responses that mediate harmful (allergic dermatitis) as well as beneficial (resistance to certain viruses, bacteria, and tumors) aspects of immune function.

Blecha et al. reported a similar stress-induced enhancement of CHS reactions in mice [72], and Flint et al. showed that acute stress enhanced CHS responses in both male and female mice and that immunoenhancement was partially dependent on glucocorticoid hormones [79], and a stress-induced enhancement of the elicitation phase of skin cell-mediated immunity has also been reported in hamsters [48]. Taken together, studies show that acute stress can significantly enhance the immunization/sensitization/induction as well as the re-exposure/elicitation/recall phases of skin cell-mediated immunity.

## 9. HORMONE AND CYTOKINE MEDIATORS OF STRESS-INDUCED ENHANCEMENT OF IMMUNE FUNCTION

Although much work remains to be done, to identify molecular, cellular, and physiological mechanisms mediating the adjuvant-like immunoenhancing effects of acute stress, several studies have begun to identify endocrine and immune mediators of these effects. Studies have shown that corticosterone and epinephrine are important mediators of an acute stress-induced immunoenhancement [65]. Adrenalectomy, which eliminates the glucocorticoid and epinephrine stress response, eliminated the stress-induced enhancement of skin DTH. Low-dose corticosterone or epinephrine administration significantly enhanced skin DTH [65]. In contrast, high-dose corticosterone, chronic corticosterone, or low-dose dexamethasone administration significantly suppressed skin DTH. These results suggested a novel role for adrenal stress hormones as endogenous immunoenhancing agents. They also showed that stress hormones released during a circumscribed or acute stress response may help prepare the immune system for potential challenges (e.g., wounding or infection) for which stress perception by the brain may serve as an early warning signal. Studies by Flint et al. have also suggested that corticosterone is a mediator of the stress-induced enhancement of skin CHS [79] while Saint-Mezard et al. have suggested that the adjuvant-like effects of stress on dendritic cell and $CD8^+$ T-cell migration and function are mediated by norepinephrine [6].

Studies have also examined the immunological mediators of an acute stress-induced enhancement of skin immunity. Since IFN-$\gamma$ is a critical cytokine mediator of cell-mediated immunity and delayed as well as contact hypersensitivity, studies were conducted to examine its role as a local mediator of the stress-induced enhancement of skin DTH [66]. The effect of acute stress on skin DTH was examined in wildtype and IFN-$\gamma$ receptor gene knockout mice (IFN-$\gamma R^{-/-}$) that had been sensitized with 2,4-dinitro-1-fluorobenzene (DNFB). Acutely stressed wildtype mice showed a significantly larger DTH response than non-stressed mice. In contrast, IFN-$\gamma R^{-/-}$ mice failed to show a stress-induced enhancement of skin DTH. Immunoneutralization of IFN-$\gamma$ in wildtype mice significantly reduced the stress-induced enhancement of skin DTH. In addition, an inflammatory response to direct IFN-γ-administration was significantly enhanced by acute stress. These results showed that IFN-γ is an important local mediator of a stress-induced enhancement of skin DTH [66]. In addition to IFN-γ, TNF-α, MCP-1, MIP-3α, IL-1, and IL-6 have also been associated with a stress-induced enhancement of the immunization/sensitization phase of skin cell-mediated immunity [3,78]. It is clear that further investigation is necessary in order to identify the most important molecular, cellular, and physiological mediators of a stress-induced enhancement of skin immunity.

## 10. STRESS-INDUCED SUPPRESSION OF IMMUNE FUNCTION

In contrast to acute stressors, chronic stress has been shown to suppress or dysregulate immune function (for review, see Refs [8,80–86]). Dhabhar and McEwen conducted studies designed to examine the effects of increasing the intensity and duration of acute stress as well as the transition from acute to chronic stress on skin immune function [15]. These studies showed that acute stress administered for 2 h prior to antigenic challenge significantly enhanced skin cell-mediated immunity [15]. Increasing the duration of stress from 2 to 5 h produced the same magnitude immunoenhancement. Interestingly, increasing the intensity of acute stress produced a significantly larger enhancement of the DTH response which was accompanied by increasing magnitudes of leukocyte redeployment. In contrast, these studies found suppression of the skin immune response when chronic stress exposure was begun 3 weeks before sensitization and either discontinued upon sensitization, or continued an additional week until challenge, or extended for 1 week after challenge [15]. Interestingly, acute stress-induced redistribution of peripheral blood lymphocytes was attenuated with increasing duration of stressor exposure and correlated with attenuated glucocorticoid responsivity. These results suggested that stress-induced alterations in lymphocyte redeployment may play an important role in mediating the bidirectional effects of stress on cutaneous cell-mediated immunity [15]. An association between chronic stress and reduced skin cell-mediated immunity has also been reported in human subjects [87].

Given the importance of cutaneous cell-mediated immunity in elimination of immunoresponsive tumors like squamous cell carcinoma (SCC) [88,89], Saul et al. examined the effects of chronic stress on susceptibility to ultraviolet radiation (UV)-induced SCC [10]. Mice were exposed to a minimal erythemal dose of UVB three times a week for 10 weeks. Half of the UVB-exposed mice were left non-stressed (i.e., they remained in their home cages) and the other half were chronically stressed (i.e., restrained during weeks 4–6). UV-induced tumors were measured weekly from week 11 through week 34, blood was collected at week 34, and tissues were collected at week 35. The mRNA expression of IL-12p40, IFN-γ, IL-4, IL-10, CD3ε, and CCL27/CTACK – the skin T-cell-homing chemokine – in dorsal skin was quantified using real-time polymerase chain reaction. $CD4^+$, $CD8^+$, and $CD25^+$ leukocytes were counted using immunohistochemistry and flow cytometry. Stressed mice had a shorter median time to first tumor (15 versus 16.5 weeks) and reached 50% incidence earlier than controls (15 versus 21 weeks). Stressed mice also had lower IFN-γ, CCL27/CTACK, and CD3ε gene expression and lower $CD4^+$ and $CD8^+$ T cells infiltrating within and around tumors than non-stressed mice. In addition, stressed mice had higher numbers of tumor infiltrating and circulating $CD4^+CD25^+$ suppressor cells than non-stressed mice. These studies showed that chronic stress increased susceptibility to UV-induced SCC by suppressing skin immunity, Type 1 cytokines, and protective T cells, and increasing active immunosuppression through regulatory/suppressor T cells [10].

Similarly, in human and animal studies, chronic stress has also been shown to suppress different immune parameters examples of which include CMI [90,91], antibody production [92,93], NK activity [16,94–96], leukocyte proliferation [94,95,97], skin homograft rejection [98], virus-specific T-cell and NK-cell activity [99], and antimycobacterial activity of macrophages from susceptible mouse strains [100].

## 11. SUMMARY OF FACTORS THAT DETERMINE WHETHER STRESS MAY ENHANCE OR SUPPRESS IMMUNE FUNCTION

We have proposed that several key factors may influence the direction (enhancing versus suppressive) of the effects of stress on a given immune measure [1]: (1) the effects of stress on leukocyte distribution; (2) the compartments in which the immune response occurs; (3) the duration (acute versus chronic) of exposure to stress; (4) the differential effects of physiologic versus pharmacologic concentrations of stress hormones, and the differential effects of endogenous (e.g., cortisol, corticosterone) versus synthetic (e.g., dexamethasone) hormones; and (5) the timing of stressor relative to the timecourse of the immune response. It is important to recognize that factors such as gender, genetics, age, and time of day as well as the route of administration, nature, and concentration of the antigen/pathogen may also significantly affect the relationship between stress and immune function.

## 12. THE STRESS SPECTRUM MODEL

We have proposed a definition of stress as a constellation of events (Fig.1), consisting of a stimulus (*Stressor*), which precipitates a reaction in the brain (*Stress Perception* and *Processing*), which activates physiologic fight or flight systems in the body (*Stress Response*) [15]. It is often overlooked that a stress response has salubrious adaptive effects in the short run [2,3,5,7,65,101], although stress can be harmful when it is long lasting [8,15,17,101].

In order to reconcile these seemingly contradictory effects of stress, we proposed that a stress response and its effects on immune function be viewed in the context of a *Stress Spectrum* [15,18] (Fig. 1). One region of this spectrum is characterized by *Acute Stress* or *Eustress*, that is conditions of short-duration stress that may result in immunopreparatory, or immunoenhancing, physiological conditions. An important characteristic of acute stress is a rapid physiological stress response mounted in the presence of the stressor, followed by a rapid shutdown of the response upon cessation of the stressor. The other region of the stress spectrum is characterized by *Chronic Stress* or *Distress*, that is repeated or prolonged stress which may result in dysregulation or suppression of immune function. An important characteristic of chronic stress is that the physiological response either persists long after the stressor has ceased or is activated repeatedly to result in an overall integrated increase in exposure of the organism to stress hormones. The concept of "allostatic load" has been proposed to define the "psycho-physiological wear and tear" that takes place while different biological systems work to stay within a range of equilibrium (allostasis) in response to demands placed by internal or external chronic stressors (for review, see Refs [12,17]). We suggest that conditions of high allostatic load would result in dysregulation or suppression of immune function. Importantly, a disruption of the circadian corticosterone/cortisol rhythm may be an indicator and/or mediator of distress or high allostatic load [15,102]. The Stress Spectrum also proposes that acute or chronic stress is generally superimposed on a psycho-physiological *Health Maintenance Equilibrium* (Fig. 1). The extent and

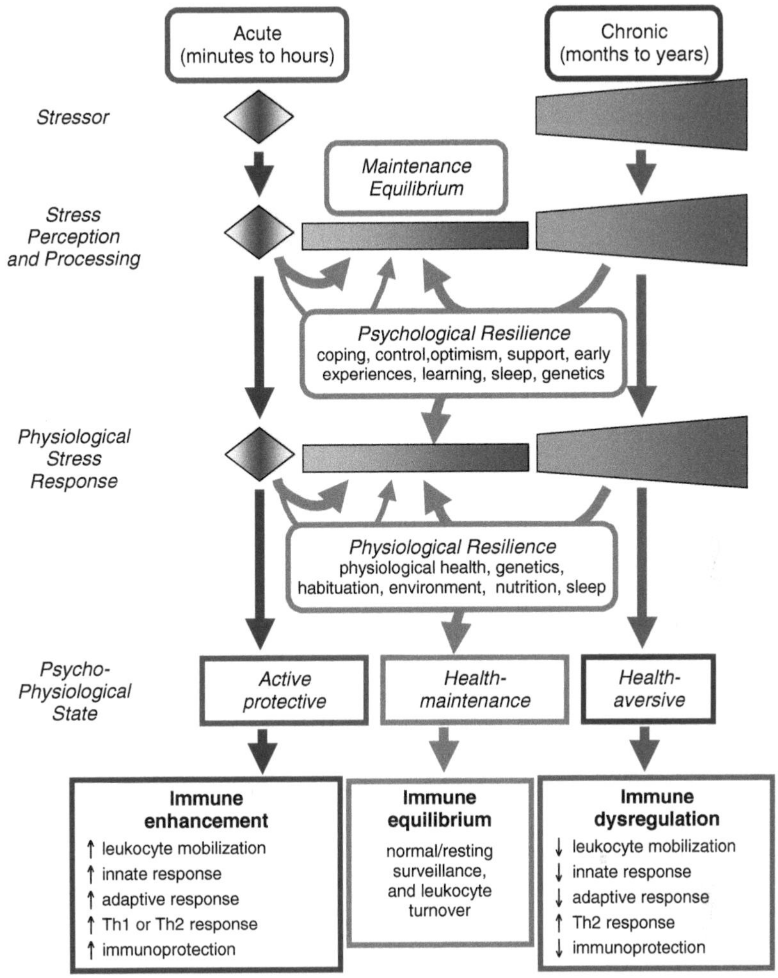


Figure 1. The stress spectrum model. We have proposed that stress be defined as a constellation of events, consisting of a stimulus (*Stressor*), that precipitates a reaction in the brain (*Stress Perception* and *Processing*), that activates physiologic fight or flight systems in the body (*Physiological Stress Response*) [15]. The duration of a physiological stress response is the critical determinant of its effects on immune function and health. The *Stressor* itself may be acute (e.g., narrowly missing being hit by a car) or chronic (e.g., caring for a chronically ill child, spouse or parent). *Stress Perception* and *Processing* by the brain are critical for determining the duration and magnitude of the *Physiological Stress Response* stimulated by any given stressor. Acute or chronic stress is generally superimposed on a psycho-physiological *Health Maintenance Steady State*. The extent and efficiency with which an organism returns to its health maintenance steady state after stress depends on *Resilience*, which we define as the capacity of psychological and interacting physiological systems to recover from challenging conditions. Factors such as coping mechanisms, sense of control, optimism, social support, early life experiences, learning, genetics, and sleep are important mediators of *Psychological Resilience*. Factors such as neuroendocrine reactivity, genetics, environment, nutrition, and sleep are important mediators of *Physiological Resilience*. Psychological resilience mechanisms are especially important in humans because they can limit the duration and magnitude of chronic stress responses. By the same token, psychogenic stressors can be particularly detrimental in human subjects because they may generate stress responses long after stressor exposure or even in the absence of physical stressors or salient threats. The *Physiological Stress Response* is the ultimate effector arm of the stress spectrum. It may consist of acute or chronic physiological activation (neurotransmitters, hormones, and their molecular, cellular, organ-level, and systemic effects) that results in *Psycho-Physiological States* that have different effects on health. Acute stress generally results in activation of mechanisms that include enhancement of immune function while chronic stress results in health-aversive conditions that result in dysregulation or suppression of immune function. The molecular mechanisms mediating conversion from positive to negative effects of stress on immune function and health are slowly beginning to emerge, and merit further investigation.

efficiency with which an organism returns to its health maintenance equilibrium after stress depends on *Resilience*, which we define as reserve capacity of psycho-physiological systems to recover from challenging conditions (Fig. 1). Factors such as coping mechanisms, sense of control, optimism, social support, early life experiences, learning, genetics, and sleep may be important mediators of *Psychological Resilience* (Fig. 1). Factors such as neuroendocrine reactivity, genetics, environment, nutrition, and sleep may be important mediators of *Physiological Resilience* (Fig. 1). The psycho-physiological basis of resilience [103] and reserve capacity are under-investigated and provide an important opportunity for future research.

The Stress Spectrum, taken together with the preceding discussion, shows that the duration, intensity/concentration, and timing of exposure to stressor-induced physiological activation (neurotransmitters, hormones, and their molecular, cellular, organ-level, and systemic effects) are critical for determining whether stress will enhance or suppress/dysregulate immune function. The model shows that the stressor itself can be acute or chronic (Fig. 1). Stress perception and processing by the brain and mechanisms mediating psychological and physiological resilience are critical for determining the duration and magnitude of the physiological stress response (Fig. 1) Psychological resilience mechanisms are especially important in humans because they can limit the duration and magnitude of chronic stress responses. Psychogenic stressors are also very important in human subjects because they can generate stress responses long after stressor exposure (e.g., post-traumatic stress disorder following a severe traumatic experience, or in a milder form, lingering anger/mood disturbance following a social altercation) or even in the absence of a physical stressor or salient threat (e.g., worrying about whether one's romantic feelings will be reciprocated). Therefore, following stressor exposure and its processing by the brain, there ensues a *Physiological Stress Response*. This response may consist of acute or chronic physiological activation (neurotransmitters, hormones, and their molecular, cellular, organ-level, and systemic effects) which results in *Psycho-Physiological States* that have different effects on overall health and immune function as shown in Fig. 1. While there is significant evidence from animal studies to support this model, it needs to be further examined in studies involving human subjects.

## 13. CONCLUSION

Stress has long been suspected to play a role in the etiology of many diseases, and numerous studies have shown that stress can be immunosuppressive and hence may be detrimental to health. Moreover, glucocorticoid stress hormones are widely regarded as being immunosuppressive, and are used clinically as anti-inflammatory agents. However, studies have shown that the acute stress response may play a critical adaptive and protective role, with stress hormones and neurotransmitters serving as messengers from the brain that prepare the immune system for potential challenges (e.g., wounding or infection) that are perceived by the brain (e.g., the detection of predator or attacker) [3,5,7,15,50]. This chapter shows that under certain conditions, stress and glucocorticoid hormones exert immunoenhancing effects. These findings need to be explored and investigated further and translated from bench to bedside.

It is important to recognize that humans as well as animals experience stress as an intrinsic part of life, and in conjunction with many standard diagnostic, clinical, and experimental manipulations. Unintended stressors may significantly affect these measures and overall health outcomes. Thus, when conducting clinical, diagnostic, or experimental procedures, it may be important to account for the effects of stress on the specific physiologic parameter or health outcome being measured. For example, it is critical to elucidate and account for the effects of stress and/or stress hormones on changes in leukocyte distribution within different body compartments. Where possible, redistribution needs to be monitored in terms of changes in absolute numbers of specific subpopulations of leukocytes. Stress-induced changes in immune cell numbers and/or function could significantly affect results (in case of experiments), diagnosis (in case of medical tests), or outcome (in case of treatment procedures and surgery).

This chapter illustrates the complex role that stress and stress hormones play as modulators and regulators of an immune response. There is evidence to show that under certain conditions stress and/or stress hormones are immunoenhancing, while under other conditions they are immunosuppressive. The following factors determine whether stress or stress hormones will enhance or suppress immune function: (1) changes in leukocyte distribution within the body; (2) the body compartments in which the immune response occurs; (3) the duration (acute versus chronic) of stress; (4) the concentration (physiologic versus pharmacologic), duration (acute versus chronic), and nature (endogenous versus synthetic) of glucocorticoid hormone exposure; and (5) the timing of stress or stress hormone exposure relative to the stage (early versus late) of an immune response. Further elucidation of the interactions between the above-mentioned factors and other nervous, endocrine, and genetic factors in mediating the effects of stress on immune function is necessary.

Due to a host of psycho-socio-political factors, stress has unfortunately become an increasing and inevitable part of people's lives. Stress is also a major factor during the diagnosis, treatment, and follow-up for most diseases. Chronic stress has been shown to dysregulate immune function and is thought to play a role in the etiology of many diseases. In contrast, it has been shown that activation of acute stress physiology may enhance protective immune responses [2–4,7]. Therefore, a determination of the physiologic mechanisms through which stress and stress hormones enhance or suppress immune responses is critically important for elucidating risk, developing preventative and therapeutic interventions, and optimizing a patient's response to treatment. The elucidation of such mechanisms would facilitate development of biomedical treatments designed to harness an individual's physiology to selectively enhance (during vaccination, wounding, infections, or cancer) or suppress (during autoimmune or inflammatory disorders) the immune response depending on what would be most beneficial for the patient.

## ACKNOWLEDGMENTS

I wish to thank current and past members of my laboratory, particularly Dr Kavitha Viswanathan, Dr Alison Saul, Kanika Ghai, Christine Daugherty, and Jean Tillie, whose work and publications are among those that are discussed in this chapter. The work described here was supported by grants from the NIH (AI48995 and CA107498) and The Dana Foundation.

## REFERENCES

1. Dhabhar FS, McEwen BS. Bidirectional effects of stress on immune function: possible explanations for salubrious as well as harmful effects. In Psychoneuroimmunology IV. Ader R, Ed.; San Diego: Elsevier, 2006; pp.723–60.
2. Dhabhar FS, Miller AH, McEwen BS, Spencer RL. Effects of stress on immune cell distribution – dynamics and hormonal mechanisms. J Immunol 1995;154:5511–27.

3. Dhabhar FS, Viswanathan K. Short-term stress experienced at the time of immunization induces a long-lasting increase in immunological memory. Am J Physiol Regul Integr Comp Physiol 2005;289(3):R738–44.
4. Viswanathan K, Daugherty C, Dhabhar FS. Stress as an endogenous adjuvant: augmentation of the immunization phase of cell-mediated immunity. Int Immunol 2005 Aug;1(8):1059–69.
5. Viswanathan K, Dhabhar FS. Stress-induced enhancement of leukocyte trafficking into sites of surgery or immune activation. Proc Natl Acad Sci USA 2005;102(16):5808–13.
6. Saint-Mezard P, Chavagnac C, Bosset S, Ionescu M, Peyron E, Kaiserlian D et al. Psychological stress exerts an adjuvant effect on skin dendritic cell functions in vivo. J Immunol 2003;171(8): 4073–80.
7. Dhabhar FS, McEwen BS. Stress-induced enhancement of antigen-specific cell-mediated immunity. J Immunol 1996;156:2608–15.
8. Glaser R, Kiecolt-Glaser JK. Stress-induced immune dysfunction: implications for health. Nat Rev Immunol 2005 Mar;5(3):243–51.
9. Ader R. Psychoneuroimmunology IV, 4th Edn. San Diego: Academic Press, 2006.
10. Saul AN, Oberyszyn TM, Daugherty C, Kusewitt D, Jones S, Jewell S et al. Chronic stress and susceptibility to skin cancer. J Nat Cancer Inst 2005;97:1760–7.
11. Goldstein DS, McEwen B. Allostasis, homeostats, and the nature of stress. Stress (Amsterdam, Netherlands) 2002;5(1):55–8.
12. McEwen BS. The end of stress as we know it. Washington, DC, Dana Press; 2002.
13. Sapolsky RM. Why zebras don't get ulcers, 3rd Edn. New York: W.H. Freeman and Company, 2004.
14. Sapolsky RM. The influence of social hierarchy on primate health. Science 2005;308:648–52.
15. Dhabhar FS, McEwen BS. Acute stress enhances while chronic stress suppresses immune function in vivo: a potential role for leukocyte trafficking. Brain Behav Immun 1997;11(4):286–306.
16. Irwin M, Patterson T, Smith TL, Caldwell C, Brown SA, Gillin CJ et al. Reduction of immune function in life stress and depression. Biol Psychiatry 1990;27:22–30.
17. McEwen BS. Protective and damaging effects of stress mediators: allostasis and allostatic load. N Engl J Med 1998;338(3):171–9.
18. Dhabhar FS, McEwen BS. Bidirectional effects of stress & glucocorticoid hormones on immune function: possible explanations for paradoxical observations. In Psychoneuroimmunology, 3rd Edn. Ader R, Felten DL, Cohen N, Eds; San Diego: Academic Press, 2001;pp.301–38.
19. Pruett SB. Quantitative aspects of stress-induced immunomodulation. Int immunopharmacol 2001;1(3):507–20.
20. Schwab CL, Fan R, Zheng Q, Myers LP, Hebert P, Pruett SB. Modeling and predicting stress-induced immunosuppression in mice using blood parameters. Toxicol Sci 2005 Jan;83(1):101–13.
21. Ben-Eliyahu S, Yirmiya R, Liebeskind JC, Taylor AN, Gale RP. Stress increases metastatic spread of a mammary tumor in rats: evidence for mediation by the immune system. Brain Behav Immun 1991;5:193–205.
22. Cohen S, Tyrrell DAJ, Smith AP. Psychological stress and susceptibility to the common cold. N Engl. J Med 1991;325:606–12.
23. Kiecolt-Glaser JK, Glaser R, Gravenstein S, Malarkey WB, Sheridan J. Chronic stress alters the immune response to influenza virus vaccine in older adults. Proc Natl Acad Sci USA 1996;93: 3043–7.
24. Sheridan JF. Stress-induced modulation of anti-viral immunity – Normal Cousins Memorial Lecture 1997. Brain Behav Immun 1998;12:1–6.
25. Solomon GF, Moos RH. Emotions, immunity and disease. Arch Gen Psychiatry 1964;11:657–69.
26. Mei-Tal V, Meyerowitz S, Engel G. Role of psychological process in a somatic disorder: multiple sclerosis. Psychosom Med 1970;32:67–86.
27. Thomason BT, Brantley PJ, Jones GN, Dyer HR, Morris JL. The relations between stress and disease activity in rheumatoid arthritis. J Behav Med 1992;15(2):215–20.
28. Amkraut AA, Solomon CF, Kraemer HC. Stress, early experience and adjuvant-induced arthritis in the rat. Psychosom Med 1971;33:203–14.

29. Pawlak C, Heiker H, Witte T, Wiese B, Heijnen CJ, Schmidt RE et al. A prospective study of daily stress and disease activity in patients with systemic lupus erythematosus. Neuroimmunomodulation 1999;6:241.
30. Al'Abadie MS, Kent GG, Gawkrodger DJ. The relationship between stress and the onset and exacerbation of psoriasis and other skin conditions. Br J Dermatol 1994;130:199–203.
31. Weigl BA. The significance of stress hormones (glucocorticoids, catecholamines) for eruptions and spontaneous remission phases in psoriasis. Int J Dermatol 2000;39(9):678–88.
32. Folks DG, Kinney FC. The role of psychological factors in dermatologic conditions. Psychosomatics 1992 Winter;33(1):45–54.
33. Ackerman KD, Heyman R, Rabin BS, Anderson BP, Houck PR, Frank E et al. Stressful life events precede exacerbations of multiple sclerosis. Psychosom Med 2002;64(6):916–20.
34. Buske-Kirschbaum A, Geiben A, Hellhammer D. Psychobiological aspects of atopic dermatitis: an overview. Psychother psychosom 2001 Jan–Feb;70(1):6–16.
35. Schleimer RP, Claman HN, Oronsky A Eds. Anti-inflammatory steroid action: basic and clinical aspects, 1st Edn. San Diego: Academic Press Inc, 1989.
36. Sprent J, Tough DF. Lymphocyte life-span and memory. Science 1994;265:1395–400.
37. Hoagland H, Elmadjian F, Pincus G. Stressful psychomotor performance and adrenal cortical function as indicated by the lymphocyte reponse. J Clin Endocrinol 1946;6:301–11.
38. Fauci AS, Dale DC. The effect of in vivo hydrocortisone on subpopulations of human lymphocytes. J Clin Invest 1974;53:240–6.
39. Fauci AS, Dale DC. The effect of hydrocortisone on the kinetics of normal human lymphocytes. Blood 1975;46:235–43.
40. Dhabhar FS, Miller AH, McEwen BS, Spencer RL. Stress-induced changes in blood leukocyte distribution – role of adrenal steroid hormones. J Immunol 1996;157:1638–44.
41. Carlson SL, Fox S, Abell KM. Catecholamine modulation of lymphocyte homing to lymphoid tissues. Brain Behav Immun 1997;11(4):307–20.
42. Benschop RJ, Oostveen FG, Heijnen CJ, Ballieux RE. Beta 2-adrenergic stimulation causes detachment of natural killer cells from cultured endothelium. Eur J Immunol 1993;23(12):3242–7.
43. Benschop RJ, Rodriguez-Feuerhahn M, Schedlowski M. Catecholamine-induced leukocytosis: early observations, current research, and future directions. Brain Behav Immun 1996;10(2):77–91.
44. Mills PJ, Ziegler MG, Rehman J, Maisel AS. Catecholamines, catecholamine receptors, cell adhesion molecules, and acute stressor-related changes in cellular immunity. Adv Pharmacol 1998;42:587–90.
45. Redwine L, Snow S, Mills P, Irwin M. Acute psychological stress: effects on chemotaxis and cellular adhesion molecule expression. Psychosom Med 2003 Jul–Aug;65(4):598–603.
46. Mills PJ, Meck JV, Waters WW, D'Aunno D, Ziegler MG. Peripheral leukocyte subpopulations and catecholamine levels in astronauts as a function of mission duration. Psychosom Med 2001 Nov–Dec;63(6):886–90.
47. Pickford GE, Srivastava AK, Slicher AM, Pang PKT. The stress response in the abundance of circulating leukocytes in the Killifish, Fundulus heteroclitus. I The cold-shock sequence and the effects of hypophysectomy. J Exp Zool 1971;177:89–96.
48. Bilbo SD, Dhabhar FS, Viswanathan K, Saul A, Yellon SM, Nelson RJ. Short day lengths augment stress-induced leukocyte trafficking and stress-induced enhancement of skin immune function. Proc Natl Acad Sci USA 2002;99(6):4067–72.
49. Jensen MM. Changes in leukocyte counts associated with various stressors. J Reticuloendothel Soc 1969;8:457–65.
50. Dhabhar FS, Miller AH, Stein M, McEwen BS, Spencer RL. Diurnal and stress-induced changes in distribution of peripheral blood leukocyte subpopulations. Brain Behav Immun 1994;8:66–79.
51. Rinder CS, Mathew JP, Rinder HM, Tracey JB, Davis E, Smith BR. Lymphocyte and monocyte subset changes during cardiopulmonary bypass: effects of aging and gender [see comments]. J Lab Clin Med 1997;129(6):592–602.
52. Toft P, Svendsen P, Tonnesen E, Rasmussen JW, Christensen NJ. Redistribution of lymphocytes after major surgical stress. Acta Anaesthesiol Scand 1993;37:245–9.

53. Snow DH, Ricketts SW, Mason DK. Hematological responses to racing and training exercise in thoroughbred horses, with particular reference to the leukocyte response. Equine Vet J 1983;15(2):149–54.
54. Morrow-Tesch JL, McGlone JJ, Norman RL. Consequences of restraint stress on natural killer cell activity, behavior, and hormone levels in Rhesus Macaques(*Macaca mulatta*). Psychoneuroendocrinology 1993;18:383–95.
55. Herbert TB, Cohen S. Stress and immunity in humans: a meta-analytic review. Psychosom Med 1993;55:364–79.
56. Schedlowski M, Jacobs R, Stratman G, Richter S, Hädike A, Tewes U et al. Changes of natural killer cells during acute psychological stress. J Clin Immunol 1993;13(2):119–26.
57. Redwine L, Mills PJ, Sada M, Dimsdale J, Patterson T, Grant I. Differential immune cell chemotaxis responses to acute psychological stress in Alzheimer caregivers compared to non-caregiver controls. Psychosom Med 2004 Sep–Oct;66(5):770–5.
58. Bosch JA, Berntson GG, Cacioppo JT, Dhabhar FS, Marucha PT. Acute stress evokes selective mobilization of T cells that differ in chemokine receptor expression: a potential pathway linking immunologic reactivity to cardiovascular disease. Brain Behav Immun 2003;17:251–9.
59. Naliboff BD, Benton D, Solomon GF, Morley JE, Fahey JL, Bloom ET et al. Immunological changes in young and old adults during brief laboratory stress. Psychosom Med 1991;53:121–32.
60. Brosschot JF, Benschop RJ, Godaert GL, Olff M, De Smet M, Heijnen CJ et al. Influence of life stress on immunological reactivity to mild psychological stress. Psychosom Med 1994;56(3): 216–24.
61. Mills PJ, Berry CC, Dimsdale JE, Ziegler MG, Nelesen RA, Kennedy BP. Lymphocyte subset redistribution in response to acute experimental stress: effects of gender, ethnicity, hypertension, and the sympathetic nervous system. Brain Behav Immun 1995;9:61–9.
62. Stein M, Ronzoni E, Gildea EF. Physiological responses to heat stress and ACTH of normal and schizophrenic subjects. Am J Psychiatry 1951;6:450–5.
63. Rinner I, Schauenstein K, Mangge H, Porta S, Kvetnansky R. Opposite effects of mild and severe stress on *in vitro* activation of rat peripheral blood lymphocytes. Brain Behav Immun 1992;6:130–40.
64. Manuck SB, Cohen S, Rabin BS, Muldoon MF, Bachen EA. Individual differences in cellular immune response to stress. Psychol Sci 1991;2(2):111–5.
65. Dhabhar FS, McEwen BS. Enhancing versus suppressive effects of stress hormones on skin immune function. Proc Natl Acad Sci USA 1999;96:1059–64.
66. Dhabhar Satoskar AR, Bluethmann H, David JR, McEwen BS. Stress-Induced Enhancement of Skin Immune Function: a role for IFN. Proc Natl Acad Sci USA 2000;97:2846–51.
67. Cunnick JE, Lysle DT, Kucinski BJ, Rabin BS. Evidence that shock-induced immune suppression is mediated by adrenal hormones and peripheral beta-adrenergic receptors. Pharmacol Biochem Behav 1990;36:645–51.
68. Zalcman FS, Anisman SH. Acute and chronic stressor effects on the antibody response to sheep red blood cells. Pharmacol Biochem Behav 1993;46(2):445–52.
69. Dhabhar FS. Stress-induced enhancement of cell-mediated immunity. Ann N Y Acad Sci 1998;840:359–72.
70. Monjan AA, Collector MI. Stress-induced modulation of the immune response. Science 1977;196:307–8.
71. Miller AH, Spencer RL, Hasset J, Kim C, Rhee R, Cira D et al. Effects of selective Type I and Type II adrenal steroid receptor agonists on immune cell distribution. Endocrinology 1994;135(5): 1934–44.
72. Blecha F, Barry RA, Kelley KW. Stress-induced alterations in delayed-type hypersensitivity to SRBC and contact sensitivity to DNFB in mice. Proc Soc Exp Biol Med 1982;169:239–46.
73. Coe CL, Lubach G, Ershler WB. Immunological consequences of maternal separation in infant primates. In Infant Stress and Coping. Lewis M, Worobey J. Eds; New York: Jossey-Bass Inc, 1989.
74. Wood PG, Karol MH, Kusnecov AW, Rabin BS. Enhancement of antigen-specific humoral and cell-mediated immunity by electric footshock stress in rats. Brain Behav Immun 1993;7:121–34.

75. Carr DJ, Woolley TW, Blalock JE. Phentolamine but not propranolol blocks the immunopotentiating effect of cold stress on antigen-specific IgM production in mice orally immunized with sheep red blood cells. Brain Behav Immun 1992;6(1):50–63.
76. Cocke R, Moynihan JA, Cohen N, Grota LJ, Ader R. Exposure to nonspecific alarm chemosignals alters immune responses in BALB/c mice. Brain Behav Immun 1993;7:36–46.
77. Persoons JHA, Berkenbosch F, Schornagel K, Thepen T, Kraal G. Increased specific IgE production in lungs after the induction of acute stress in rats. J Allergy Clin Immunol 1995;95: 765–70.
78. Viswanathan K, Daugherty C, Dhabhar FS. Stress as an endogenous adjuvant: augmentation of the immunization phase of cell-mediated immunity. Int Immunol 2005;17(8):1059–69.
79. Flint MS, Miller DB, Tinkle SS. Restraint-induced modulation of allergic and irritant contact dermatitis in male and female B6.129 mice. Brain Behav Immun 2000 Dec;14(4):256–69.
80. Sklar LS, Anisman H. Stress and cancer. Psychol Bull 1981;89(3):369–406.
81. Borysenko M, Borysenko J. Stress, behavior, and immunity: animal models and mediating mechanisms. Gen Hosp Psychiatry 1982;4:59–67.
82. Khansari DN, Murgo AJ, Faith RE. Effects of stress on the immune system. Immunol Today 1990;11(5):170–5.
83. Zwilling B. Stress affects disease outcomes. Confronted with infectious disease agents, the nervous and immune systems interact in complex ways. ASM News 1992;58(1):23–5.
84. Kort WJ. The effect of chronic stress on the immune system. Adv Neuroimmunol 1994;4:1–11.
85. Irwin M. Stress-induced immune suppression: role of brain corticotropin releasing hormone and autonomic nervous system mechanisms. Adv Neuroimmunol 1994;4:29–47.
86. Black PH. Immune system-central nervous system interactions: effect and immunomodulatory consequences of immune system mediators on the brain. Antimicrob Agents Chemother 1994;38(1):7–12.
87. Smith A. Conna U, Bennett B, Wakefield D, Hickie I, Lloyd A. The relationship between distress and the development of a primary immune response to a novel antigen. Brain Behav Immun 2004 Jan;18(1):65–75.
88. Kripke ML. Ultraviolet radiation and immunology: something new under the sun – presidential address. Cancer Res 1994;54(23):6102–5.
89. Granstein RD, Matsui MS. UV radiation-induced immunosuppression and skin cancer. Cutis 2004 Nov;74(5 Suppl):4–9.
90. Kelley KW, Greenfield RE, Evermann JF, Parish SM, Perryman LE. Delayed-type hypersensitivity, contact sensitivity, and PHA skin-test responses of heat- and cold-stressed calves. Am J Vet Res 1982;43:775–9.
91. Basso AM, Depiante-Depaoli M, Cancela L, Molina V. Seven-day variable-stress regime alters cortical β-adrenoreceptor binding and immunologic responses: reversal by imipramine. Pharmacol Biochem Behav 1993;45:665–72.
92. Edwards EA, Dean LM. Effects of crowding of mice on humoral antibody formation and protection to lethal antigenic challenge. Psychosom Med 1977;39(1):19–24.
93. Fleshner M, Laudenslager ML, Simons L, Maier SF. Reduced serum antibodies associated with social defeat in rats. Physiol Behav 1989;45(6):1183–7.
94. Bartrop R, Lazarus L, Luckhurst E, Kiloh LG, Penny R. Depressed lymphocyte function after bereavement. Lancet 1977;8016:834–6.
95. Cheng GJ, Morrow-Tesch JL, Beller DI, Levy EM, Black PH. Immunosuppression in mice induced by cold water stress. Brain Behav Immun 1990;4(4):278–91.
96. Kiecolt-Glaser JK, Garner W, Speicher C, Penn GM, Holliday J, Glaser R. Psychosocial modifiers of immunocompetence in medical students. Psychosom Med 1984;46(1):7–14.
97. Regnier JA, Kelley KW. Heat- and cold-stress suppresses in vivo and in vitro cellular immune response of chickens. Am J Vet Res 1981;42:294–9.
98. Wistar R, Hildemann WH. Effect of stress on skin transplantation immunity in mice. Science 1960;131:159–60.

99. Bonneau RH, Sheridan JF, Feng N, Glaser R. Stress-induced effects on cell-mediated innate and adaptive memory components of the murine immune response to herpes simplex virus infection. Brain Behav Immun 1991;5:274–95.
100. Brown DH, Zwilling BS. Activation of the hypothalamic-pituitary-adrenal axis differentially affects the anti-mycobacterial activity of macrophages from BCG-resistant and susceptible mice. J Neuroimmunol 1994;53:181–7.
101. Sapolsky RM, Romero LM, Munck AU. How do glucocorticoids influence stress responses? Integrating permissive, suppressive, stimulatory, and preparative actions. Endocr Rev 2000;21(1)(Feb):55–89.
102. Sephton SE, Sapolsky RM, Kraemer HC, Spiegel D. Diurnal cortisol rhythm as a predictor of breast cancer survival. J Natl Cancer Inst 2000;92(12):994–1000.
103. Charney DS. Psychobiological mechanisms of resilience and vulnerability: implications for successful adaptation to extreme stress. Am J Psychiatry 2004 Feb 1;161(2):195–216.

# III. IMMUNE–NEUROENDOCRINE FEEDBACK MECHANISMS

Edited by A. del Rey, G. P. Chrousos and H. O. Besedovsky

# *Processing of Cytokine Signals at CNS Levels: Relevance for Immune–HPA Axis Interactions*

HUGO O. BESEDOVSKY and ADRIANA DEL REY

*Department of Immunophysiology, Institute of Physiology and Pathophysiology, Marburg, Germany*

## ABSTRACT

The capacity of cytokines to elicit neural and endocrine responses, such as the stimulation of the hypothalamus–pituitary–adrenal axis (HPA), and the different pathways followed by these mediators to affect the functioning of the central nervous system (CNS) have been intensively investigated. However, less is known about how the information conveyed by cytokines is processed at central levels. Here, we discuss evidence indicating that several of the effects of proinflammatory cytokines exerted in a "healthy" brain are amplified in the CNS due to the fact that (a) peripheral cytokines, such as IL-1, have the capacity to elicit their own synthesis in the brain; and (b) a sustained increase in neuronal activity also induces the production of physiologically significant amounts of these mediators. The relevance of brain-borne interleukin (IL)-1 and IL-6, when their production is not the result of pathological events in the CNS, is discussed taking as example their effect on the stimulation of the HPA axis, and their involvement in physiologic brain mechanisms such as synaptic plasticity and memory formation, and the control of glucose homeostasis.

## 1. INTRODUCTION

The effect of hormones and neurotransmitters on immune functions has been extensively studied, and some of these effects are discussed in various chapters of this book. Conversely, there is also evidence that neuroendocrine structures change their activity in synchrony with the activation of immune cells. The present view is that, upon activation, immune cells produce factors that serve as messengers of an afferent pathway to neuroendocrine structures. It has been proposed that this afferent pathway follows either a humoral or a neural route or involves the brain vascular endothelial interphase, and there is evidence that different routes are followed depending on how the stimuli used to activate immune cells are applied and the types of inflammatory/immune responses that are elicited (see other chapters in this volume). In any case, the information conveyed to the CNS is processed at central levels since it is followed by a complex neuroendocrine response. Similarly, there is evidence that cytokines produced in the CNS are involved in brain and neuroendocrine functions, and that peripheral immune stimuli as well as neural mechanisms can control the production of cytokines in the brain. This implies that immune-derived signals and neurosensorial stimuli converge on central neuronal networks that integrate and process the information conveyed by peripheral or brain-borne cytokines. In the following, we discuss some evidence indicating that immune cell products can affect neuroendocrine and brain functions acting directly or indirectly at CNS levels. Because of the nature of this volume, special emphasis is given to the effect of cytokines on the HPA axis. Evidence that cytokines produced in the brain are involved in CNS–immune interactions is also discussed.

## 2. AN EXAMPLE OF THE RELEVANCE OF THE STIMULATION OF THE HPA AXIS DURING DISEASE

The consequences that immune-mediated neuroendocrine responses have for immunoregulation and for the course of diseases that involve the activation of the immune system are extensively addressed in other chapters of this volume (see also comments in Ref. [1]). As an example, we shall only mention recent studies showing the relevance of the activation of HPA axis during experimental infection of mice with the parasite *Trypanosoma cruzi*, the causal agent of Chagas' disease. Administration of only 100 parasites into C57BL/6J mice results in high parasitemia, thymic atrophy, characteristic heart lesions, and 100% lethality within 20 days. Corticosterone blood levels start increasing 10 days after infection and reach more than 20-fold basal levels on day 17. Interference with the effect of endogenous glucocorticoids either by blocking glucocorticoid receptors or by adrenalectomy, although preventing thymic atrophy and reducing double positive thymocytes, aggravates the disease, causes a further increase of IL-6 and tumor necrosis factor-α (TNF-α) levels in blood, and accelerates death [2]. Interference of glucocorticoid effects in infected TNF-receptor knockout mice also results in an uncontrolled production of this cytokine, even higher levels of corticosterone and an earlier death when compared to the wild type [3]. Balb/c mice are more resistant to *Trypanosoma cruzi* infection. Even inoculation of a high number of parasites results in less than 50% lethality. Balb/c mice have higher basal levels of corticosterone, and the increase in corticosterone levels following infection with *Trypanosoma cruzi* occurs earlier and reaches higher concentrations than in C57BL/6J mice. Blockade of glucocorticoid receptors or adrenalectomy in infected Balb/c mice results in full expression of the disease and 100% lethality [2]. These results provide a further example of the relevance of the activation of the HPA axis not only for the course of a disease but also for the susceptibility to its expression.

## 3. THE BRAIN AND NEUROENDOCRINE RESPONSES TO IMMUNE SIGNALS

### 3.1. Early evidence

The initial strategy leading to the demonstration that the brain and neuroendocrine structures receive and respond to signals derived from the immune system was to study whether factors released by immune cells are capable of inducing neuroendocrine effects. This approach was based on the knowledge that the immune response to innocuous antigens affects the activity of discrete brain neurons, the HPA axis, and the sympathetic nervous system. For example, the immune response to different antigens causes an increase in the rate of firing of neurons in the ventromedial hypothalamic nucleus, decreases the activity of noradrenergic brain neurons, stimulates the HPA axis, and inhibits splenic sympathetic nerve activity (for review, see Ref. [4]). To search for the immune-derived mediators that stimulate the HPA axis, the supernatants obtained from cultures of allogeneic and/or mitogen-stimulated human or murine lymphocytes were inoculated into naive rats and mice. This treatment caused a clear increase in adrenocorticotropic hormone (ACTH) and corticosterone output [5,6]. The products that mediate these effects were generically termed "glucocorticoid-increasing factors" (GIF). The fact that one million lymphocytes stimulated *in vitro* produce enough GIF to induce a several-fold increase in blood ACTH and corticosterone levels in an adult rat illustrates the potency of this material. Our attempts at purifying the product that stimulates the HPA axis, although showing that the active material was a protein of a molecular weight between 10 and 30 kDa, were not successful. Retrospectively, we interpret that this failure was due to the fact that GIF activity is, as mentioned below, shared by several cytokines. In any case, the results obtained fully confirmed our prediction that factors derived from immune cells can affect neuroendocrine mechanisms. Later, pure or recombinant immune cytokines became available, and it was possible to test their capacity to influence endocrine and neural functions.

### 3.2. Effect of immune-derived products on hypothalamic releasing hormones: the HPA axis as an example

Several cytokines, although with different potency, share the capacity to induce the same neuroendocrine effects, particularly the stimulation of the HPA axis. The large number of cytokines that can stimulate the HPA axis attests for the biologic relevance of this endocrine response.

IL-1 was the first cytokine shown to stimulate the HPA axis [7]. This effect is the most extensively studied example of the influence of cytokines on endocrine functions. Both the purified natural and the recombinant human forms of IL-1 can stimulate ACTH and corticosterone output in mice and rats. Later, it was shown that also TNF-α, different types of interferons, IL-2, IL-6, IL-11, IL-12, leukemia inhibitory factor (LIF), granulocyte macrophage colony-stimulating factor (GM-CSF), oncostatin, and stem cell factor (SCF) can affect the HPA axis. The literature regarding these studies have been extensively reviewed in Refs [4,8], in which more than 500 references on this subject can be found.

Comparative studies on the capacity of IL-1, IL-6, and TNF-α to stimulate the pituitary–adrenal axis showed that, on a molecular weight basis, IL-1 is more potent than TNF-α and IL-6 [9]. It is worth noticing that the stimulatory effect of cytokines on the HPA axis, which was originally described in animals, has been confirmed in humans for IL-1α [10,11], IL-1β [10,12], IL-2 [13–15], IL-6 [16,17], TNF-α and TNF-β [18], interferon (IFN)-α [19–22], β [23], and γ [18,24,25], and GM-CSF [26]. In humans as in laboratory animals, increased secretion of ACTH and/or cortisol/corticosterone occurs within 1 h after intravenous (i.v.) injection or within 1–4 h after subcutaneous administration of cytokines.

The effect of cytokines on the HPA axis can be exerted at different levels of this axis, namely the hypothalamus, the pituitary, and the adrenal glands. However, the main site of action of the acute stimulation of the HPA axis by IL-1 is the hypothalamus. Other structures of this axis, including the pituitary and the adrenal glands, may be affected after more prolonged exposure to the cytokine (see following chapters in this volume).

Corticotropin-releasing hormone (CRH) was the first hypothalamic releasing hormone whose production was shown to be affected by a cytokine [27–29]. IL-1, administered systemically or intracerebroventricularly (i.c.v.), stimulates CRH production in the hypothalamus. Blockade of this releasing hormone abrogates the acute effect of the cytokine observed *in vivo*. Effects of IL-1α and β, IL-2, IL-6, IL-8, TNF-α, and IFN-α on CRH release have also been reported *in vitro* (for review, see Refs [4,8]). Following the initial evidence that products from activated immune cells can affect CRH release, it was shown that cytokines can also affect the production of other hypothalamic releasing factors, such as TSH, luteinizing-hormone-releasing hormone (LHRH), and growth-hormone-releasing hormone (GhRH) [8].

There are certain contradictions between results obtained from *in vivo* and *in vitro* studies regarding the primary target of the stimulatory effect of cytokines on glucocorticoid release. Some *in vitro* studies show a direct effect on the hypothalamus, while others indicate that a stimulatory effect can be exerted directly either on the pituitary or on the adrenals. For a proper interpretation of these results, it is however necessary to consider that cytokines, acting in a paracrine way, can act as growth factors and non-specifically influence cell metabolism. A recent study indicates that conditioned medium from lymphoid cells obtained from patients with tuberculosis and stimulated with *Mycobacterium tuberculosis* antigens *in vitro* decrease dehydroepiandrosterone (DHEA) release by human adrenal cells [30]. This *in vitro* effect mimicks the marked decrease in DHEA concentrations observed in plasma of patients with pulmonary tuberculosis, and indicates that cytokines might directly affect DHEA production *in vivo* [30]. Thus, the effects as mentioned above cannot be interpreted as "classical endocrine actions" of cytokines since they might reflect effects that occur when they are locally produced in the tissues, for example, during inflammatory and autoimmune processes. Conversely, the data obtained *in vivo* do not always clarify the site of action of cytokines. In any case, both types of experiments clearly show that cytokines can serve as mediators of the bidirectional communication between the immune system and the HPA axis.

Besides cytokines, other immunologic messengers potentially capable of influencing neuroendocrine processes are histamine and serotonin, which are released during certain types of immune response [31], the pituitary hormone-like peptides that are produced by stimulated lymphoid cells [32–34], and vasopressin and oxytocin, which can be produced by thymic nurse cells [35]. Even specific antibodies might affect the endocrine system. For example, antihormone antibodies can share common characteristics with hormone receptors, and the anti-idiotypic antibody may act as the ligand, thereby behaving as an internal "image" of the hormone. This mechanism could explain the appearance of autoantihormone antibodies and antihormone-receptor antibodies during different autoimmune diseases [36,37]. Antibodies have also been shown to modify hormonal action, for example, by changing the affinity of the hormone for the carrier, by protecting hormones from enzymatic degradation, and also by stabilizing particular conformations in the molecule [38].

### 3.3. Effect of cytokines on CNS neurotransmitters and neuronal activity

We shall focus here on the effect of cytokines, particularly IL-1, on neurotransmitters and neuronal pathways involved in the activation of the HPA axis. Using a similar approach as for the identification of GIF (see Section 3.1), we found that administration of immune cell-derived conditioned media results in decreased noradrenaline (NA) content in the brain [39]. Later, using purified cytokines, it was shown that IL-1 reduces NA content and increases the ratio 3-methoxy-4-hydroxyphenylethylene glycol (MHPG)/NA, which reflects increased NA metabolism [40,41]. This effect is observed in the hypothalamus, hippocampus, brain stem, and the spinal cord. The fact that catecholaminergic fibers in the spinal cord are stimulated by IL-1 may indicate one neural pathway for the effect of this cytokine in the CNS and, conversely, that a central effect of the cytokine is conveyed to the peripheral autonomic nervous system. The stimulation of noradrenergic neurons in the CNS by IL-1 is consistent with other evidence indicating that stimulation of CRH production by catecholaminergic neurons is involved in the response of the HPA axis to IL-1 [42]. This is further supported by studies showing that the response of the HPA axis to IL-1 is blocked by surgical interruption of noradrenergic innervation of CRH-producing neurons in the paraventricular nucleus (PVN) of the hypothalamus or by NA antagonists [43–45]. This is an important point since it has been shown that noradrenergic fibers stimulate CRH and arginine-vasopressin producing neurons in the hypothalamus [46]. Studies performed *in vivo* show that IL-1 administration stimulates neurons of the PVN and supraoptic nucleus of the hypothalamus and also neurons of the stria terminalis. There is evidence that both IL-1 and lipopolysaccharide (LPS) administration at peripheral levels stimulate defined neuronal pathways as reflected by c-fos expression as a marker of neuronal activity. Among the areas affected is the PVN [47].

Another aspect of IL-1 action on the HPA axis relates to the observation by Tilders and co-workers that a single administration of IL-1 sensitizes the HPA axis to different stimuli [34]. This effect, which can last for several weeks, is paralleled by an increase in electrically evoked release of NA in the PVN [48]. These studies indicate that IL-1 can induce a long-lasting resetting of the HPA axis and in this way affects the response of this axis to stress and to products derived from activated immune cells [49].

## 4. CYTOKINE PRODUCTION AND ACTION IN THE BRAIN: RELEVANCE FOR THE ACTIVITY OF THE HPA AXIS

In the previous sections, we have discussed some evidence showing that cytokines can directly or indirectly influence brain mechanisms involved in the stimulation of the HPA axis. To this already complicated scenario, it is necessary to add another level of complexity: cytokines that can stimulate this axis are produced by brain cells and, since cytokine receptors are expressed in the CNS, their local production is likely to affect brain functions. Among the cytokines that are constitutively found or whose expression can be induced in the brain are IL-1, its natural receptor antagonist (IL-1ra), IL-2, IL-3, IL-6, IL-8, IL-10, IL-12, IL-13, TNF-α, TNF-β, IFN-α, IFN-β and γ [4,8], IL-15 [50], and IL-18 (unpublished results). In the following, we refer exclusively to the relevance of cytokine production in the brain for the stimulation of the HPA axis in the absence of overt CNS pathologies.

### 4.1. Peripheral stimulation of immune cells induces brain expression of cytokines known to activate the HPA axis

The most common approach to study the production of cytokines in the brain following stimulation of immune cells has been based on the peripheral administration of bacterial endotoxins, in particular LPS. Most reports agree that peripheral administration of LPS results in increased expression of cytokines in the brain [51–56]. However, the doses of LPS administered were rather high, may disturb the BBB [57], and affect blood and endothelial cells. We have ruled out these possibilities [58] and showed that constitutive expression of IL-1β, IL-6, and TNF-α but not of IFN-γ is detected in the brain, and that LPS administration clearly increases the expression of all four cytokines. In particular, IL-1β and IL-6 are preferentially expressed in the hypothalamus and hippocampus while TNF-α expression is more marked in the thalamus–striatum. Taken together, the results show that stimulation of peripheral immune cells induces cytokine expression in the brain. The particular pattern of regional expression of cytokines in the hypothalamus and hippocampus suggests that, during activation of the immune system, brain-borne cytokines may affect neuroendocrine mechanisms controlled in these areas.

In this context, two aspects deserve further consideration. First, certain cytokines can trigger their own production or upregulate the expression of their own receptors in the brain. For example, peripheral administration of IL-1 results in increased IL-1 gene expression in the brain [59], and peripheral administration of IL-6 induces IL-6 receptor expression in the PVN of the hypothalamus [60].

Secondly, the central expression of certain cytokines may be under neuronal control. As it will be discussed below, long-term potentiation (LTP) of synaptic activity results in IL-1 and IL-6 expression in the hippocampus [61–63]. In addition, we have recently found that the expression of IL-1β in the hypothalamus induced by peripheral administration of LPS is under noradrenergic neuronal control (unpublished results).

### 4.2. Relevance of endogenous cytokines acting at brain levels for the stimulation of the HPA axis

Several behavioral effects of peripheral induction of cytokines can be interfered by blocking cytokine receptors in the brain [64,65]. It has been reported that IL-1ra blocks the expression of CRH in the PVN of the hypothalamus that occurs at a late stage of the stimulation of the HPA axis induced by intraperitoneal (i.p.) administration of LPS [66]. This study is in line with our observation that while the initial stimulation of the HPA axis by LPS is not interfered by i.c.v. administration of IL-1ra, this treatment affects the maintenance phase of this endocrine response [67]. There is also evidence that TNF-α produced in the brain contributes to the stimulation of the HPA axis during peripheral inflammation [68]. The i.c.v. administration of either an anti-TNF antibody or soluble TNF receptors results in a significant reduction of the stimulation of the HPA axis during turpentine-induced inflammation. However, no evidence for increased TNF-α gene expression was detected. Thus, increased levels of TNF occur either as a consequence of increased translation of the gene or by stimulation of the release of pre-formed TNF-α. Alternatively, brain-borne TNF could synergize with the effect of increased brain levels of prostaglandins (or other mediators) in the stimulation of the HPA axis.

The effect of acute stress should be added to those stimuli that can affect cytokine induction in the brain. It has been reported that stress induces IL-1 gene expression in the hypothalamus [69–72] and that microinjection of IL-1ra in this area inhibits stress-induced increase in ACTH levels by as much as 50% [71]. These studies do not clarify whether IL-1 is produced in the brain as a consequence of a neural stimulus or is secondary to peripheral mechanisms affected during the response to stress. However, the fact that blockade of IL-1 receptors in the brain blunts significantly the activation of the HPA axis indicates a contribution of brain-borne IL-1 to the neuroendocrine response to stress.

The stimulation of the HPA axis induced by IL-1 is expected to result in glucose mobilization and to induce hyperglycemia since these are classic effects of glucocorticoids. Furthermore, increased levels of glucocorticoids and catecholamines are considered to be bases for the displacement of glucose to skeletal muscle that underlie the flight or fight reaction during acute stress. However, contrary to the expectance, in mice, IL-1 induces a profound and long-lasting hypoglycemia, an effect that is insulin-independent and also observed in animals with Type 2 diabetes [73]. Such effect is based on an IL-1-induced increased glucose transport and oxidation in all tissues tested but that during inflammatory/infectious diseases predominates in the sites where IL-1 is produced (for review, see Ref. [74]). However, as it will be discussed later, a central component is relevant for the maintenance of IL-1-induced hypoglycemia.

### 4.3. Increased neuronal activity induces brain expression of cytokines that can stimulate the HPA axis

The initial view that cytokine production in the brain occurs only during pathologies in the CNS has changed in the last years. There is evidence supporting the view that brain-borne cytokines are physiologic neuromodulators. This evidence derives from studies in which it was shown that increased neuronal activity results in a biologically significant increase in the production of some cytokines. The procedure used in these studies was to elicit an LTP of synaptic activity in the hippocampus *in vivo* and *in vitro*. This process appears to underlie certain forms of learning and memory, and it is triggered by a short-time (less than 1 min), high-frequency stimulation of afferent fibers. The resulting potentiation of postsynaptic neuronal activity can remain increased during hours in hippocampal slices and days in freely moving animals. During the course of this process, local IL-1 and IL-6 gene expression is increased both in slices and *in vivo*. This increase is biologically significant since interference with the action of these cytokines affects the maintenance of LTP in opposite ways: endogenous IL-1 supports this process while IL-6 tends to temporally restrict synaptic potentiation [61–63]. We can add now that also IL-1ra antagonist and IL-18 are expressed during hippocampal LTP (manuscript in preparation). These data suggest that a cytokine network is activated under conditions in which a prolonged increase in neuronal activity is elicited following a minimal manipulation of pre-synaptic fibers. The evidence, while indicating a neuromodulatory role of cytokines, also shows that these mediators are not only produced during brain pathologies or maximal neuronal stimulation. Furthermore, LTP could be operationally considered as a process in which synaptic strength is reset at a higher level and, therefore, the results indicate that cytokines are involved in this process.

## 5. CYTOKINE–NEURONAL INTERACTIONS IN THE BRAIN: RELEVANCE FOR RESETTING HOMEOSTASIS AND THE ACTIVATION OF THE HPA AXIS

It is worth noting that the same cytokines that stimulate the HPA axis when injected or induced in the periphery are expressed in the brain. If the stimulation of the HPA axis is a consequence of a peripheral event, which is then the function of the increased cytokine expression detected in parallel in the brain? Does this phenomenon just reflect "redundancy" in cytokine production? In our opinion, a likely interpretation is that the production of both peripheral and central cytokines underlies well-programmed steps of responses integrated at brain levels. Under basal conditions, the release of low amounts of cytokines by brain cells could be one of the various inputs that control the activity of neurons involved in the regulation of adaptive functions integrated at hypothalamic and limbic system levels. In conditions during which the activity of the immune system changes, peripheral cytokines and other mediators would trigger the initial step of neuroendocrine responses to immune cell stimulation. The quick neuroendocrine response observed when certain cytokines are administered peripherally may indicate that this initial step does not involve the *de novo* synthesis of cytokines in the brain. However, peripheral immune mediators, such as IL-1, increase the electrical activity of CRH-producing neurons [75] and LPS administration results in increased c-fos expression in the hypothalamus [47]. Based on our finding that an increase in neuronal activity during LTP induces cytokine expression in the brain, it is likely that the neuronal stimulation triggered by peripheral administration of cytokines would result in a comparable effect. Again, the results obtained during LTP showing that the endogenously produced IL-1 reinforces synaptic efficacy indicate that an increased local production of cytokines would maintain neuronal activation. Thus, an initial stimulation of neuronal activity triggered by peripheral cytokines could be prolonged by the local production of cytokines in the brain. A demonstration that a process such as that described for hippocampal cells does also occur in the PVN still requires direct demonstration. However, it is becoming clear that synaptic plasticity such as that reflected during LTP occurs in neurons of different areas of the brain, including the hypothalamus [76].

As already mentioned, neurosensorial stimuli, such as acute stress, which result in the activation of CRH producing neurons, also induce cytokine production in the CNS (see Section 3.2). Thus, immune- and neurosensorial stimuli converge on cytokine-producing cells in the brain. On this basis, we postulate a "relay system" based on interactions between neurons and cytokine-producing brain cells that would integrate peripheral immune- and neurosensorial signals and modulate neuroendocrine responses to these stimuli.

Cytokines produced in the brain as resultant of the different inputs to such a "relay system" may change the setpoint for the control of the neuroendocrine variables that need to be adjusted. This would explain why neuroendocrine and metabolic effects of cytokines administered systemically are more prolonged than expected from the half-life of cytokines in the circulation. The hypoglycemia induced by IL-1β [73] may serve to illustrate this point. A single low dose of IL-1 administered peripherally results in a long-lasting reduction of blood glucose levels (despite the concomitant increase in glucocorticoid and glucagon levels), and this effect can be abrogated to large extent by i.c.v. administration of IL-1ra [74,77]. Furthermore, IL-1-injected animals remain hypoglycemic for several hours even after receiving a glucose load, an effect that is also noticed in the leptin-receptor deficient db/db mice, a model of Type 2 diabetes PNAS. Central effects of IL-1 are most likely dependent on the capacity of IL-1 to induce its own production in the hypothalamus in normal and in db/db mice. As mentioned, a selective blockade of IL-1 receptors in the brain interferes to a large extent with the hypoglycemic effect of the cytokine. These results suggest that the *de novo* produced cytokine in the brain contributes to reset glucose homeostasis. Such resetting is likely orchestrated by brain-borne IL-1 through down regulation of counter-regulatory endocrine mechanisms to hypoglycemia, which are integrated at hypothalamic levels. We hypothesize that brain-borne IL-1 may alter the balance between stimulation of CRHR1 and CRHR2 in favor of last one, thus inhibiting the endocrine response to hypoglycemia [78]. The effects of IL-1 on glucose homeostasis are in our view the expression of a necessary redistribution of glucose during immune processes when this fuel has to be displaced from muscle and adipose tissue to the activated cells responsible for both innate and adaptive immunity, which are highly demanding in terms of energy.

In the following, we speculate on how this relay system hypothesis could explain the effects of cytokines on the HPA axis. This speculation is based on the following facts:

- Peripheral administration of several cytokines results in stimulation of the HPA axis. This effect lasts beyond the half-life of the injected material.
- Peripheral administration of LPS stimulates the HPA axis and induces a delayed but prolonged expression of IL-1, IL-6, TNF-α, and IFN-γ in the hypothalamus.
- Peripheral administration of IL-1 results in the expression of its own gene in the hypothalamus.
- Blockade of IL-1 receptors in the brain does not interfere with the initial stimulation of the HPA axis induced by LPS but with the maintenance of this endocrine response.
- Increased neuronal activity, as observed during LTP and following injection of pharmacological agents, results in increased IL-1 and IL-6 expression.
- Noradrenergic brain neurons control IL-1 gene expression in the hypothalamus and affect the stimulation of the HPA axis by this cytokine.

- Acute stress results in IL-1 gene expression in the hypothalamus, which contributes to the activation of the HPA axis.
- Activation of CRH neurons affects not only the HPA axis but also several other neuronal mechanisms, including the stimulation of brain noradrenergic neurons.

The quick response of the HPA axis to LPS and IL-1 administration and the kinetics of IL-1 gene expression in the brain suggest two steps in the stimulation of the HPA axis during activation of peripheral immune cells. At an early stage, the stimulation of the HPA axis may not require the presence of IL-1 in the brain since it is not affected by blockade of central IL-1 receptors. Thus, the stimulation of the HPA axis induced by LPS is most likely initiated by the quick release of mediators secondarily induced by peripheral IL-1 and other cytokines. These mediators could act either systemically at the level of brain endothelial cells and/or by stimulating afferent neural pathways to the brain that activate CRH neurons, and thus the HPA axis. In a later phase, when IL-1 gene expression is increased in the hypothalamus, the occupancy of IL-1 receptors in the CNS is needed to maintain this axis activated. According to the relay system hypothesis, the stimulation of CRH neurons by peripheral signals would trigger the gene expression of cytokines, such as IL-1 and IL-6, known to stimulate the HPA axis. The stimulation by CRH of NA neurons that control IL-1 gene expression and that are relevant to the stimulation of the HPA axis would contribute to this effect. These events would constitute an amplificatory mechanism in the stimulation of the HPA axis triggered initially by acute stimulation of peripheral immune cells. Such amplificatory mechanism based on the integration of immune-, cytokine-, and neuronal-mediated signals would initially override the negative feedback on CRH neurons mediated by increased levels of glucocorticoids. However, at a given time, the glucocorticoid feedback would start predominating and HPA axis activity would tend to return to basal levels. The expression of IL-1ra gene, which is also triggered by IL-1, could also contribute to a negative feedback. However, if immune cell activation persists, peripheral signals are expected to prolong neuronal activation and consequently also the activity of the cells that produce the cytokines. The fact that psychological and physical stresses also activate simultaneously the HPA axis and cytokine expression in the brain should now be added to this immunologically triggered circuit. Thus, the concept of a relay system based on interactions between hypothalamic PVN neurons and cytokine-producing neural cells provides a further step in the understanding of how immune and psycho-neurosensorial signals could be integrated and processed at brain levels.

Cytokines can affect a variety of processes at central levels, such as food and water intake, metabolic pathways, sleeping patterns, and behavior. The relay system hypothesis predicts that the induction of cytokines in the brain is a common mechanism in the control of these processes. Increased neuronal activity in brain areas involved in the control of different functions would result in the local induction of cytokines that would in turn affect the activity of these neurons. This possibility is experimentally testable by studying cytokine production and effects in areas of the brain in which neuronal activity is expected to increase during adaptive responses, for example neurons in the paraventricular and arcuate hypothalamic nuclei during food deprivation. The integrative role of peripheral and central cytokines in the activation of the HPA axis and resetting of homeostasis is schematically represented in Fig. 1.

In conclusion, the existence of a "relay system" that integrates immune and neuronal signals and that, by controlling cytokine production in the brain, mediates a resetting of essential physiological variables would provide an answer to critical questions such as (1) how the immune system can mediate long-lasting neuroendocrine adjustments of mechanisms controlled in different brain regions; and (2) how these adjustments are coordinated with other needs of the organism and with the response to other sensorial signals.

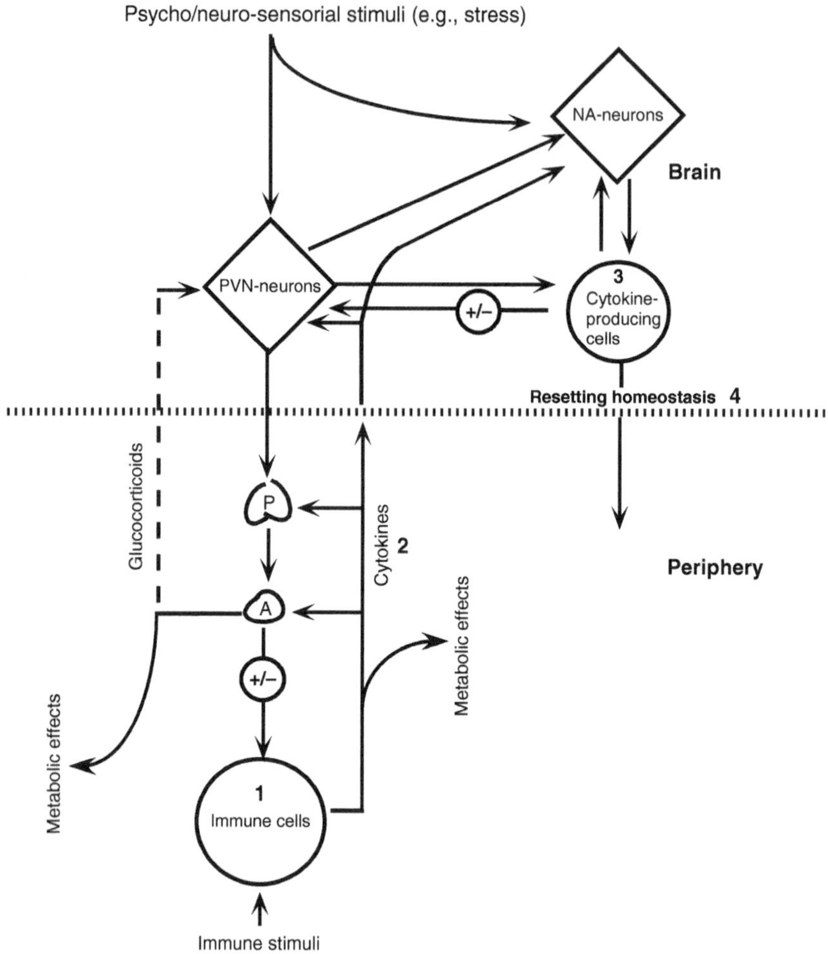


Figure 1. Role of peripheral and central cytokines in the activation of the HPA axis and resetting of homeostasis. (1) Immune cells: inflammatory cells, antigen-presenting cells (APC), and lymphoid cells; (2) cytokines and/or secondarily induced products in the periphery or in the brain; (3) cytokine-producing neural cells: glia and neurons; (4) for example, thermo- and glucoregulation; P: pituitary; A: adrenal, —— stimulation; inhibition; ± stimulation or inhibition.

## ACKNOWLEDGMENTS

This work was supported by the German Research Council (DFG, RE 1451/2-2).

## REFERENCES

1. Besedovsky HO, del Rey A. Regulating inflammation by glucocorticoids. Nat Immunol 2006; 7:537.
2. Roggero E, Perez AR, Tamae-Kakazu M, Piazzon I, Nepomnaschy I, Besedovsky HO et al. Endogenous glucocorticoids cause thymus atrophy but are protective during acute Trypanosoma cruzi infection. J Endocrinol 2006;190:495–503.

3. Perez AR, Roggero E, Nicora A, Palazzi J, Besedovsky HO, Del Rey A et al. Thymus atrophy during Trypanosoma cruzi infection is caused by an immuno-endocrine imbalance. Brain Behav Immun 2007;21:890–900.
4. Besedovsky HO, del Rey A. Immune-neuro-endocrine interactions: facts and hypotheses. Endocr Rev 1996;17:64–102.
5. Besedovsky HO, del Rey A, Sorkin E. Lymphokine-containing supernatants from con A-stimulated cells increase corticosterone blood levels. J Immunol 1981;126:385–7.
6. Besedovsky HO, Del Rey A, Sorkin E, Lotz W, Schwulera U. Lymphoid cells produce an immunoregulatory glucocorticoid increasing factor (GIF) acting through the pituitary gland. Clin Exp Immunol 1985;59:622–8.
7. Besedovsky H, del Rey A, Sorkin E, Dinarello CA. Immunoregulatory feedback between interleukin-1 and glucocorticoid hormones. Science 1986;233:652–4.
8. Turnbull AV, Rivier CL. Regulation of the hypothalamic-pituitary-adrenal axis by cytokines: actions and mechanisms of action. Physiol Rev 1999;79:1–71.
9. Besedovsky HO, del Rey A, Klusman I, Furukawa H, Monge Arditi G, Kabiersch A. Cytokines as modulators of the hypothalamus-pituitary-adrenal axis. J Steroid Biochem Mol Biol 1991;40:613–18.
10. Crown J, Jakubowski A, Kemeny N, Gordon M, Gasparetto C, Wong G et al. A phase I trial of recombinant human interleukin-1 beta alone and in combination with myelosuppressive doses of 5-fluorouracil in patients with gastrointestinal cancer. Blood 1991;78:1420–7.
11. Smith JW, 2nd, Urba WJ, Curti BD, Elwood LJ, Steis RG, Janik JE et al. The toxic and hematologic effects of interleukin-1 alpha administered in a phase I trial to patients with advanced malignancies. J Clin Oncol 1992;10:1141–52.
12. Curti BD, Urba WJ, Longo DL, Janik JE, Sharfman WH, Miller LL et al. Endocrine effects of IL-1 alpha and beta administered in a phase I trial to patients with advanced cancer. J Immunother Emphasis Tumor Immunol 1996;19:142–8.
13. Denicoff KD, Durkin TM, Lotze MT, Quinlan PE, Davis CL, Listwak SJ et al. The neuroendocrine effects of interleukin-2 treatment. J Clin Endocrinol Metab 1989;69:402–10.
14. Lissoni P, Rovelli F, Tancini G, Tisi E, Rivolta MR, Ardizzoia A et al. Inhibitory effect of interleukin-3 on interleukin-2-induced cortisol release in the immunotherapy of cancer. J Biol Regul Homeost Agents 1992;6:113–15.
15. Lissoni P, Barni S, Tisi E, Rovelli F, Pittalis S, Rescaldani R et al. In vivo biological results of the association between interleukin-2 and interleukin-3 in the immunotherapy of cancer. Eur J Cancer 1993;29A:1127–32.
16. Mastorakos G, Chrousos GP, Weber JS. Recombinant interleukin-6 activates the hypothalamic-pituitary-adrenal axis in humans. J Clin Endocrinol Metab 1993;77:1690–4.
17. Spath-Schwalbe E, Born J, Schrezenmeier H, Bornstein SR, Stromeyer P, Drechsler S et al. Interleukin-6 stimulates the hypothalamus-pituitary-adrenocortical axis in man. J Clin Endocrinol Metab 1994;79:1212–14.
18. Nolten WE, Goldstein D, Lindstrom M, McKenna MV, Carlson IH, Trump DL et al. Effects of cytokines on the pituitary-adrenal axis in cancer patients. J Interferon Res 1993;13:349–57.
19. Roosth J, Pollard RB, Brown SL, Meyer WJ 3rd. Cortisol stimulation by recombinant interferon-alpha 2. J Neuroimmunol 1986;12:311–16.
20. Muller H, Hammes E, Hiemke C, Hess G. Interferon-alpha-2-induced stimulation of ACTH and cortisol secretion in man. Neuroendocrinology 1991;54:499–503.
21. Gisslinger H, Svoboda T, Clodi M, Gilly B, Ludwig H, Havelec L et al. Interferon-alpha stimulates the hypothalamic-pituitary-adrenal axis in vivo and in vitro. Neuroendocrinology 1993;57:489–95.
22. Barbarino A, Colasanti S, Corsello SM, Satta MA, Della Casa S, Rota CA et al. Dexamethasone inhibition of interferon-alpha 2-induced stimulation of cortisol and growth hormone secretion in chronic myeloproliferative syndrome. J Clin Endocrinol Metab 1995;80:1329–32.
23. Goebel MU, Baase J, Pithan V, Exton M, Saller B, Schedlowski M et al. Acute interferon beta-1b administration alters hypothalamic-pituitary-adrenal axis activity, plasma cytokines and leukocyte distribution in healthy subjects. Psychoneuroendocrinology 2002;27:881–92.

24. Holsboer F, Stalla GK, von Bardeleben U, Hammann K, Muller H, Muller OA. Acute adrenocortical stimulation by recombinant gamma interferon in human controls. Life Sci 1988;42:1–5.
25. Spath-Schwalbe E, Porzsolt F, Digel W, Born J, Kloss B, Fehm HL. Elevated plasma cortisol levels during interferon-gamma treatment. Immunopharmacology 1989;17:141–5.
26. Zylinska K, Mucha S, Komorowski J, Korycka A, Pisarek H, Robak T et al. Influence of granulocyte-macrophage colony stimulating factor on pituitary-adrenal axis (PAA) in rats in vivo. Pituitary 1999;2:211–16.
27. Berkenbosch F, van Oers J, del Rey A, Tilders F, Besedovsky H. Corticotropin-releasing factor-producing neurons in the rat activated by interleukin-1. Science 1987;238:524–6.
28. Sapolsky R, Rivier C, Yamamoto G, Plotsky P, Vale W. Interleukin-1 stimulates the secretion of hypothalamic corticotropin-releasing factor. Science 1987;238:522–4.
29. Uehara A, Gottschall PE, Dahl RR, Arimura A. Interleukin-1 stimulates ACTH release by an indirect action which requires endogenous corticotropin releasing factor. Endocrinology 1987;121:1580–2.
30. del Rey AD, Mahuad CV, Bozza VV, Bogue C, Farroni MA, Bay ML et al. Endocrine and cytokine responses in humans with pulmonary tuberculosis. Brain Behav Immun 2007;21:171–9.
31. Bourne HR, Lichtenstein LM, Melmon KL, Henney CS, Weinstein Y, Shearer GM. Modulation of inflammation and immunity by cyclic AMP. Science 1974;184:19–28.
32. Westly HJ, Kleiss AJ, Kelley KW, Wong PK, Yuen PH. Newcastle disease virus-infected splenocytes express the proopiomelanocortin gene. J Exp Med 1986;163:1589–94.
33. Kavelaars A, Ballieux RE, Heijnen CJ. The role of IL-1 in the corticotropin-releasing factor and arginine-vasopressin-induced secretion of immunoreactive beta-endorphin by human peripheral blood mononuclear cells. J Immunol 1989;142:2338–42.
34. Schmidt ED, Janszen AW, Wouterlood FG, Tilders FJ. Interleukin-1-induced long-lasting changes in hypothalamic corticotropin-releasing hormone (CRH)-neurons and hyperresponsiveness of the hypothalamus-pituitary-adrenal axis. J Neurosci 1995;15:7417–26.
35. Dardenne M, Savino W. Control of thymus physiology by peptidic hormones and neuropeptides. Immunol Today 1994;15:518–23.
36. Cohen IRD, Elias R, Maron R, Shechter Y. Immunization to insulin generates anti-idiotypes that behave as antibodies to the insulin hormone receptors and cause diabetes mellitus. In Idiotypy in Biology and Medicine. Köhler H, Urbain J, Cazenave PA, Eds; New York: Academic Press, 1984; pp. 385–400.
37. Strosberg AD. Antiidiotypic antibodies as immunological internal images of hormones. In Idiotypy in Biology and Medicine. Köhler H, Urbain J, Cazenave PA, Eds; New York: Academic Press, 1984; pp. 365–83.
38. Aston R, Cowden WB, Ada GL. Antibody-mediated enhancement of hormone activity. Mol Immunol 1989;26:435–46.
39. Besedovsky H, del Rey A, Sorkin E, Da Prada M, Burri R, Honegger C. The immune response evokes changes in brain noradrenergic neurons. Science 1983;221:564–6.
40. Dunn AJ. Systemic interleukin-1 administration stimulates hypothalamic norepinephrine metabolism parallelling the increased plasma corticosterone. Life Sci 1988;43:429–35.
41. Kabiersch A, del Rey A, Honegger CG, Besedovsky HO. Interleukin-1 induces changes in norepinephrine metabolism in the rat brain. Brain Behav Immun 1988;2:267–74.
42. Chuluyan HE, Saphier D, Rohn WM, Dunn AJ. Noradrenergic innervation of the hypothalamus participates in adrenocortical responses to interleukin-1. Neuroendocrinology 1992;56:106–11.
43. Ovadia H, Abramsky O, Barak V, Conforti N, Saphier D, Weidenfeld J. Effect of interleukin-1 on adrenocortical activity in intact and hypothalamic deafferentated male rats. Exp Brain Res 1989;76: 246–9.
44. Weidenfeld J, Abramsky O, Ovadia H. Evidence for the involvement of the central adrenergic system in interleukin 1-induced adrenocortical response. Neuropharmacology 1989;28:1411–14.
45. Matta S, Singh J, Newton R, Sharp BM. The adrenocorticotropin response to interleukin-1 beta instilled into the rat median eminence depends on the local release of catecholamines. Endocrinology 1990;127:2175–82.

46. Helmreich DL, Itoi K, Lopez-Figueroa MO, Akil H, Watson SJ. Norepinephrine-induced CRH and AVP gene transcription within the hypothalamus: differential regulation by corticosterone. Brain Res Mol Brain Res 2001;88:62–73.
47. Ericsson A, Kovacs KJ, Sawchenko PE. A functional anatomical analysis of central pathways subserving the effects of interleukin-1 on stress-related neuroendocrine neurons. J Neurosci 1994; 14:897–913.
48. Schmidt ED, Schoffelmeer AN, De Vries TJ, Wardeh G, Dogterom G, Bol JG, et al. A single administration of interleukin-1 or amphetamine induces long-lasting increases in evoked noradrenaline release in the hypothalamus and sensitization of ACTH and corticosterone responses in rats. Eur J Neurosci 2001;13:1923–30.
49. Jansen AS, Schmidt ED, Voorn P, Tilders FJ. Substance induced plasticity in noradrenergic innervation of the paraventricular hypothalamic nucleus. Eur J Neurosci 2003;17:298–306.
50. Kurowska M, Rudnicka W, Maslinska D, Maslinski W. Expression of IL-15 and IL-15 receptor isoforms in select structures of human fetal brain. Ann N Y Acad Sci 2002;966:441–5.
51. Gatti S, Bartfai T. Induction of tumor necrosis factor-alpha mRNA in the brain after peripheral endotoxin treatment: comparison with interleukin-1 family and interleukin-6. Brain Res 1993; 624:291–4.
52. Laye S, Parnet P, Goujon E, Dantzer R. Peripheral administration of lipopolysaccharide induces the expression of cytokine transcripts in the brain and pituitary of mice. Brain Res Mol Brain Res 1994; 27:157–62.
53. Quan N, Sundar SK, Weiss JM. Induction of interleukin-1 in various brain regions after peripheral and central injections of lipopolysaccharide. J Neuroimmunol 1994;49:125–34.
54. Buttini M, Boddeke H. Peripheral lipopolysaccharide stimulation induces interleukin-1 beta messenger RNA in rat brain microglial cells. Neuroscience 1995;65:523–30.
55. Gabellec MM, Griffais R, Fillion G, Haour F. Expression of interleukin 1 alpha, interleukin 1 beta and interleukin 1 receptor antagonist mRNA in mouse brain: regulation by bacterial lipopolysaccharide (LPS) treatment. Brain Res Mol Brain Res 1995;31:122–30.
56. Van Dam AM, Bauer J, Tilders FJ, Berkenbosch F. Endotoxin-induced appearance of immunoreactive interleukin-1 beta in ramified microglia in rat brain: a light and electron microscopic study. Neuroscience 1995;65:815–26.
57. Lustig S, Danenberg HD, Kafri Y, Kobiler D, Ben-Nathan D. Viral neuroinvasion and encephalitis induced by lipopolysaccharide and its mediators. J Exp Med 1992;176:707–12.
58. Pitossi F, del Rey A, Kabiersch A, Besedovsky H. Induction of cytokine transcripts in the central nervous system and pituitary following peripheral administration of endotoxin to mice. J Neurosci Res 1997;48:287–98.
59. Hansen MK, Taishi P, Chen Z, Krueger JM. Vagotomy blocks the induction of interleukin-1beta (IL-1beta) mRNA in the brain of rats in response to systemic IL-1beta. J Neurosci 1998;18: 2247–53.
60. Vallieres L, Rivest S. Interleukin-6 is a needed proinflammatory cytokine in the prolonged neural activity and transcriptional activation of corticotropin-releasing factor during endotoxemia. Endocrinology 1999;140:3890–903.
61. Schneider H, Pitossi F, Balschun D, Wagner A, del Rey A, Besedovsky HO. A neuromodulatory role of interleukin-1beta in the hippocampus. Proc Natl Acad Sci USA 1998;95:7778–83.
62. Besedovsky H, Pitossi F, Schneider H, Balschun D, del Rey A. Cytokines in the brain as mediators of immune-neuro-endocrine interactions. J Neuroimmunol 1999;31:2.
63. Balschun D, Wetzel W, del Rey A, Pitossi F, Schneider H, Zuschratter W et al. Interleukin-6: a cytokine to forget. FASEB J 2004;18:1788–90.
64. Kent S, Bluthe RM, Kelley KW, Dantzer R. Sickness behavior as a new target for drug development. Trends Pharmacol Sci 1992;13:24–8.
65. Linthorst AC, Flachskamm C, Muller-Preuss P, Holsboer F, Reul JM. Effect of bacterial endotoxin and interleukin-1 beta on hippocampal serotonergic neurotransmission, behavioral activity, and free corticosterone levels: an in vivo microdialysis study. J Neurosci 1995;15:2920–34.

66. Kakucska I, Qi Y, Clark BD, Lechan RM. Endotoxin-induced corticotropin-releasing hormone gene expression in the hypothalamic paraventricular nucleus is mediated centrally by interleukin-1. Endocrinology 1993;133:815–21.
67. Besedovsky H, del Rey A, Rogausch H, Kabiersch A, Klusman I. Effect of cytokines integrated by the nervous system. J Neurochem 1993;61(Suppl):125.
68. Turnbull AV, Pitossi FJ, Lebrun JJ, Lee S, Meltzer JC, Nance DM et al. Inhibition of tumor necrosis factor-alpha action within the CNS markedly reduces the plasma adrenocorticotropin response to peripheral local inflammation in rats. J Neurosci 1997;17:3262–73.
69. Minami M, Kuraishi Y, Yamaguchi T, Nakai S, Hirai Y, Satoh M. Immobilization stress induces interleukin-1 beta mRNA in the rat hypothalamus. Neurosci Lett 1991;123:254–6.
70. Shintani F, Nakaki T, Kanba S, Kato R, Asai M. Role of interleukin-1 in stress responses. A putative neurotransmitter. Mol Neurobiol 1995;10:47–71.
71. Shintani F, Nakaki T, Kanba S, Sato K, Yagi G, Shiozawa M et al. Involvement of interleukin-1 in immobilization stress-induced increase in plasma adrenocorticotropic hormone and in release of hypothalamic monoamines in the rat. J Neurosci 1995;15:1961–70.
72. DeKeyser FG, Leker RR, Weidenfeld J. Activation of the adrenocortical axis by surgical stress: involvement of central norepinephrine and interleukin-1. Neuroimmunomodulation 2000;7:182–8.
73. del Rey A, Besedovsky H. Interleukin 1 affects glucose homeostasis. Am J Physiol 1987;253:R794–8.
74. del Rey A, Roggero E, Randolf A, Mahuad C, McCann S, Rettori V et al. IL-1 resets glucose homeostasis at central levels. Proc Natl Acad Sci USA 2006;103:16039–44.
75. Saphier D, Ovadia H. Selective facilitation of putative corticotropin-releasing factor-secreting neurones by interleukin-1. Neurosci Lett 1990;114:283–8.
76. Nishikawa Y, Shibata S, Watanabe S. Circadian changes in long-term potentiation of rat suprachiasmatic field potentials elicited by optic nerve stimulation in vitro. Brain Res 1995;695:158–62.
77. del Rey A, Monge-Arditi G, Besedovsky HO. Central and peripheral mechanisms contribute to the hypoglycemia induced by interleukin-1. Ann N Y Acad Sci 1998;840:153–61.
78. McCrimmon RJ, Song Z, Cheng H, McNay EC, Weikart-Yeckel C, Fan X et al. Corticotrophin-releasing factor receptors within the ventromedial hypothalamus regulate hypoglycemia-induced hormonal counterregulation. J Clin Invest 2006;116:1723–30.

*The Hypothalamus–Pituitary–Adrenal Axis*
Edited by A. del Rey, G. P. Chrousos and H. O. Besedovsky



# *Inflammatory Mediator Action on the Anterior Pituitary Gland*

DAMIANA GIACOMINI[1,3], MATIAS ACUÑA[1], ALBERTO C. NAGASHIMA[1,3], CAROLINA P. CASTRO[1], MARCELO PÁEZ-PEREDA[2], ULRICH RENNER[2], GÜNTER K. STALLA[2], and EDUARDO ARZT[1,3]

[1]*Laboratorio de Fisiología y Biología Molecular, Departamento de Fisiología, Biología Molecular y Celular, FCEN, Universidad de Buenos Aires, Argentina*
[2]*Max-Planck Institute of Psychiatry, Kraepelinstr. Munich, Germany*
[3]*Members of the Argentine National Research Council (CONICET), Argentina*

## ABSTRACT

Interleukin-1 (IL-1), IL-6, and other cytokines mainly produced by endotoxin [lipopolysaccharide (LPS)]-activated monocytes and macrophages, are known to be responsible for the acute phase and inflammatory response. These cytokines also influence the secretion of anterior pituitary hormones and play an important role in the interaction between the immune and the endocrine systems. Moreover, IL-1, IL-6, and their receptors are expressed and produced within the anterior pituitary influencing the growth and function of pituitary cells. Thus, we propose that they act as autocrine or paracrine regulators of pituitary homeostasis. In addition, intrapituitary production of IL-1 and IL-6 is also induced by LPS showing another link between the immune and the endocrine systems. The present review summarizes the actual knowledge about the actions of IL-1 and IL-6 (and related cytokines) known to regulate the anterior pituitary physiology and the involvement of LPS in modulating these intrapituitary cytokine pathways.

## 1. INTRODUCTION

The interaction between the immune and the neuroendocrine systems has been studied in various mammalian species [1]. The first studies performed established an interaction between the immune system and the adrenal activity, showing that the physiological levels of glucocorticoids are critical regulators for the organism to overcome pathological events like inflammation or infection in an optimal manner [2,3]. This interaction between the immune and endocrine systems involves cytokines and hormones. Several studies have shown that the hypothalamus is one of the neuroendocrine sites of action of cytokines. The pituitary gland is also a target for cytokine actions and constitutes an integration point among neuroendocrine and immune signals. Reciprocally, hormones modulate the function of immune cells and influence cytokine production. During the past decade, it has been demonstrated that cytokines, in addition to their role as lymphocyte messengers, act within the anterior pituitary gland regulating hormone secretion and growth of endocrine cells. Thus, some cytokines are indeed produced intrapituitary and act in a paracrine manner. In this chapter, we will focus on pituitary cytokine pathways concentrating on interleukin-1 (IL-1), IL-6, and related cytokines.

## 2. INTERLEUKIN-1

### 2.1. IL-1 regulatory action on pituitary function, in vivo studies

The stimulatory action of soluble factors produced by the immune system, which were originally nominated as glucocorticoid increasing factors (GIFs) and particularly IL-1, was first described by Besedovsky and collaborators [4–6]. The effect of this and other cytokines have been extensively studied. IL-1 is the most potent cytokine showing GIF activity. It appears to act at all levels of the hypothalamic–pituitary–adrenal (HPA) axis, stimulating the secretion of corticotrophin-releasing hormone (CRH), adrenocorticotropic hormone (ACTH), and glucocorticoids (reviewed in Refs [7–9]). In this context, many studies have been performed showing the action of cytokines in vivo in animal models in the regulation of plasma levels of pituitary hormones. We will not address this issue in depth because it is described in detail in other chapters of this book, and we will focus on the studies that show how cytokines act directly as effectors on the anterior pituitary.

Proopiomelanocortin (POMC) primary transcript RNA has been demonstrated to be increased by continuous infusion of IL-1 in Wistar rats [10]. A single intraperitoneal injection of IL-1β in rats induces an increase in POMC mRNA in the pituitary, whereas in these conditions IL-6 did not change POMC mRNA levels [11]. It has been shown that CRH can sensitize the pituitary gland to the direct ACTH-releasing activity of IL-1 [12]. Chronic infusion of IL-1 induced an increase in spontaneous in vitro secretion of β-endorphin by pituitary cells while the in vitro responses of the pituitary to CRH of animals treated in vivo with IL-1 were diminished [13].

### 2.2. IL-1 receptor expression in the anterior pituitary gland

IL-1 receptor (IL-1R) (quantitative autoradiography, binding assays, and immunocytochemistry) and IL-1R mRNA [reverse transcriptase-polymerase chain reaction (RT–PCR)] have been characterized in normal mouse and rat anterior pituitary cells and in the mouse corticotrophic tumor cell line AtT-20 (reviewed in Ref. [14]). A specific high-affinity IL-1R, whose density was not altered by treatment with lipopolysaccharide (LPS), was described in the mouse using quantitative autoradiography [15]. The expression of IL-1R in this species was also studied using competitive binding assays [16] and in situ localization, and there was a dense IL-1R binding distributed homogeneously over the anterior lobe [17]. This binding increased after treatment with glucocorticoids for 7 days [18]. In AtT-20 cells, CRH induces an increase in the density of IL-1R without altering their affinity [19]. In contrast to the stimulatory effect described above, other studies have shown that glucocorticoids also inhibit the stimulatory action of CRH over the density of IL-1R expression [20], this inhibitory effect is also observed in AtT-20 cells stimulated with CRH [21]. Both IL-1α and IL-1β act through a common receptor in the anterior pituitary [15,16,22]. Both types of receptors for IL-1 have been identified in the normal pituitary, Type 1 receptor (IL-1RI) and Type 2 receptor (IL-1RII) [23]. High density of binding sites was observed using radiolabeled recombinant IL-1β, also in the rat anterior pituitary [24]. The expression of IL-1R provides the basis for the direct action of IL-1 on pituitary cells. Interestingly, pituitary IL-1R is regulated by CRH and glucocorticoids.

### 2.3. IL-1 expression in the anterior pituitary gland

Endogenous production of IL-1 by anterior pituitary has been demonstrated. Immunoreactive IL-1β (ELISA) and its mRNA have been detected in rat anterior pituitary, and both are increased following LPS treatment [25,26]. IL-1β expression has also been detected in a series of human pituitary adenomas in vitro [27]. IL-1ra, an IL-1 endogenous competitive peptide antagonist at the IL-1R, has been demonstrated in the rat anterior pituitary by RT–PCR [28]. IL-1ra has also been detected by RT–PCR and immunofluorescence in several types of human pituitary adenomas [29]. Taken together, all these observations show that pituitary cells express the IL-1 complete system (IL-1/IL-1ra and IL-1R).

### 2.4. IL-1 action on pituitary cells

The available evidence in vitro indicates that IL-1 acts at the level of the pituitary although the exact nature of this mechanism appears to depend on the precise experimental model. In nearly identical studies utilizing pituitary monolayer cultures, the secretion of growth hormone (GH), prolactin (PRL), luteinizing hormone (LH), and follicle-stimulating hormone (FSH) has been reported not to be influenced by IL-1β [30], whereas in the other study IL-1β stimulated the secretion of GH, LH, and TSH and inhibited the secretion of PRL [31]. The secretion of PRL by normal rat pituitary cultures was also reported to be inhibited by IL-1β [32]. Bernton et al. [31] and Keher et al. [33] found IL-1β to be a potent stimulator of ACTH secretion. Uehara et al. [30], however, found only a weak stimulation of ACTH secretion by IL-1β. IL-1β has also been reported to increase ACTH secretion from perfused pituitary cells in a dose-dependent manner [34,35] as well as to have no effect on ACTH secretion [36]. A nitric oxide synthase inhibitor did not affect IL-1-induced ACTH release in pituitary rat cell cultures [37]. The conclusion we draw from these studies in primary pituitary cultures is that while the literature is not completely consistent, the prevalent opinion is that IL-1 stimulates the secretion of most hormones of the anterior pituitary directly. The exception is PRL, whose secretion appears to be inhibited by IL-1. Interestingly, PRL opposite to glucocorticoids is an immune-stimulatory hormone [38].

Despite the tumoral origin of AtT-20 cells, the studies performed on this mouse corticotrophic cell line are in agreement with those performed on the normal rat pituitary cells. IL-1β stimulates the release of β-endorphin [39–41] and ACTH [42,43] from AtT-20 cells, apparently by different signal transduction pathways. In GH3 cells, it has been shown that IL-1β increases GH secretion and synthesis through activation of MEK, p38 MAPK, and phosphoinositide 3-kinase-signaling systems [44].

In contrast to the large number of studies performed in animal models or in vitro, only a few studies have been performed in humans. However, the available data validate in humans the actions found in the different experimental models. For example, IL-1β stimulates the secretion of ACTH from pituitary cell cultures derived from humans with Cushing's disease [45].

Although the signal transduction pathways triggered by IL-1R have been widely studied [46,47] and despite the evidence that protein kinase A (PKA) pathway activation leads to ACTH secretion [48,49], there have only recently been advances on the transcription factors involved in IL-1 induction of POMC mRNA. The molecular pathways involved in CRH induction of POMC mRNA were recently clarified and involve the orphan receptor subfamily proteins Nur77, Nurr1, and NOR1 [50–53]. Induction of these proteins correlates with CRH-dependent POMC mRNA increase, and the POMC promoter sites that interact with these factors, the NurRE and NBRE sites, were characterized. The NurRE site is necessary for POMC promoter induction by CRH [52,53]. Recently, we have demonstrated that IL-1 effects on POMC mRNA induction involve Nur77 (but not Nurr1) induction and activity [54]. Stimulation at the NurRE site by IL-1 is dependent on p38 protein kinase activity. We demonstrated that Nur77 is essential for the function of IL-1 in corticotroph cells. The POMC promoter construction with a mutated NurRE site is unresponsible to IL-1 stimulation. Also the induction of POMC mRNA, stimulation of ACTH secretion, and increased transcriptional activity at the NurRE site are blocked in AtT-20N77 cells, which express a Nur77-dominant negative form. The functional relevance of Nur77 induction by IL-1 is underlined by the fact that the IL-1-CRH synergism on ACTH secretion correlates with the synergism in Nur77 transcriptional activity [54].

The involvement of NF-κB was studied in pituitaries from mice injected intraperitoneally with recombinant rat IL-1β or LPS. By immunohistochemistry, assays with an antibody directed against the p65 NF-κB subunit, the presence of an active NF-κB complex in cell nuclei was demonstrated [55]. The response to LPS was present predominantly in the nuclei of glial fibrillary acidic protein (GFAP)-positive cells of the neurohypophysis and in F4/80-labeled cells of the anterior pituitary. In contrast, NF-κB response to IL-1β was also localized in somatotroph cells [55]. Activation of NF-κB in response to IL-1β was no longer apparent in IL-1RI knock-out (KO) mice, confirming that this receptor is essential for the transduction of IL-1 signal in the pituitary [55].

Cyclooxygenase-2 (COX-2) is an immediate early response gene in macrophages and is responsible for elevated prostaglandin (PG) secretion during inflammation and immune response. This enzyme carries out catalytic activities converting free arachidonic acid into eicosanoids, high-reactive molecules. In human monocyte macrophage, elevation of COX-2 mRNA can be induced by a variety of proinflammatory stimuli such LPS and IL-1. [56]. IL-1 induces COX-2 gene expression also in human-cultured intestinal myofibroblasts [57]. Activation of NF-κB, ERK, p38, and PKC-signaling pathways are each necessary for optimal COX-2 induction [57]. IL-1β depolarizes parvocellular neurones in the paraventricular nucleus (PVN) of the hypothalamus, and these responses are abolished in the presence of a COX-2 inhibitor, indicating a dependence on PG synthesis and activation in central nervous system [58]. COX-2 plays a role in progression of colon, breast, pancreas, and lung carcinomas. Recent studies have examined the role of COX-2 expression in normal pituitaries and pituitary adenomas and have suggested a role for COX-2 in the regulation of angiogenesis in the pituitary, playing an important role in pituitary tumor development and progression [59,60]. These results leave an open question whether COX-2-signaling pathway might be involved in IL-1 inflammatory effects in the pituitary.

IL-1 also stimulates the increased glucocorticoid-induced transcriptional activity of the glucocorticoid receptor (GR) via the glucocorticoid response elements (GREs) in corticotroph cells suggesting that cytokines may modulate the sensitivity to glucocorticoid feedback [61].

Experiments in IL-1α/β KO and IL-6 KO mice have shown that IL-1α induced a normal HPA initial activation in these animals. However, after 6 h, the activation was reduced relative to wild-type mice, indicating a role for endogenous IL-α/β and IL-6 in prolonged HPA activation [62]. Moreover, induction of POMC transcript in the anterior pituitary in response to IL-1α was reduced not only in IL-1α/β KO and IL-6 KO mice but also in CRH KO mice, showing that IL-1α/β, IL-6, and CRH are all required for POMC induction during inflammation [62]. In summary, IL-1α/β induction of POMC in the anterior pituitary is through the induction of two independent pathways, CRH and IL-6 [62].

In addition to the effects on the secretion of pituitary hormones described above, cytokines can influence the growth of pituitary cells. For example, IL-1β and IL-1α inhibit anterior pituitary cell growth, an effect antagonized by IL-1ra, but do not act on lactosomatotrophic GH3 cells [63].

The direct action of IL-1 on pituitary cells contributes to the circuits by which IL-1 integrates the neuroendocrine–immune systems (Fig. 1).

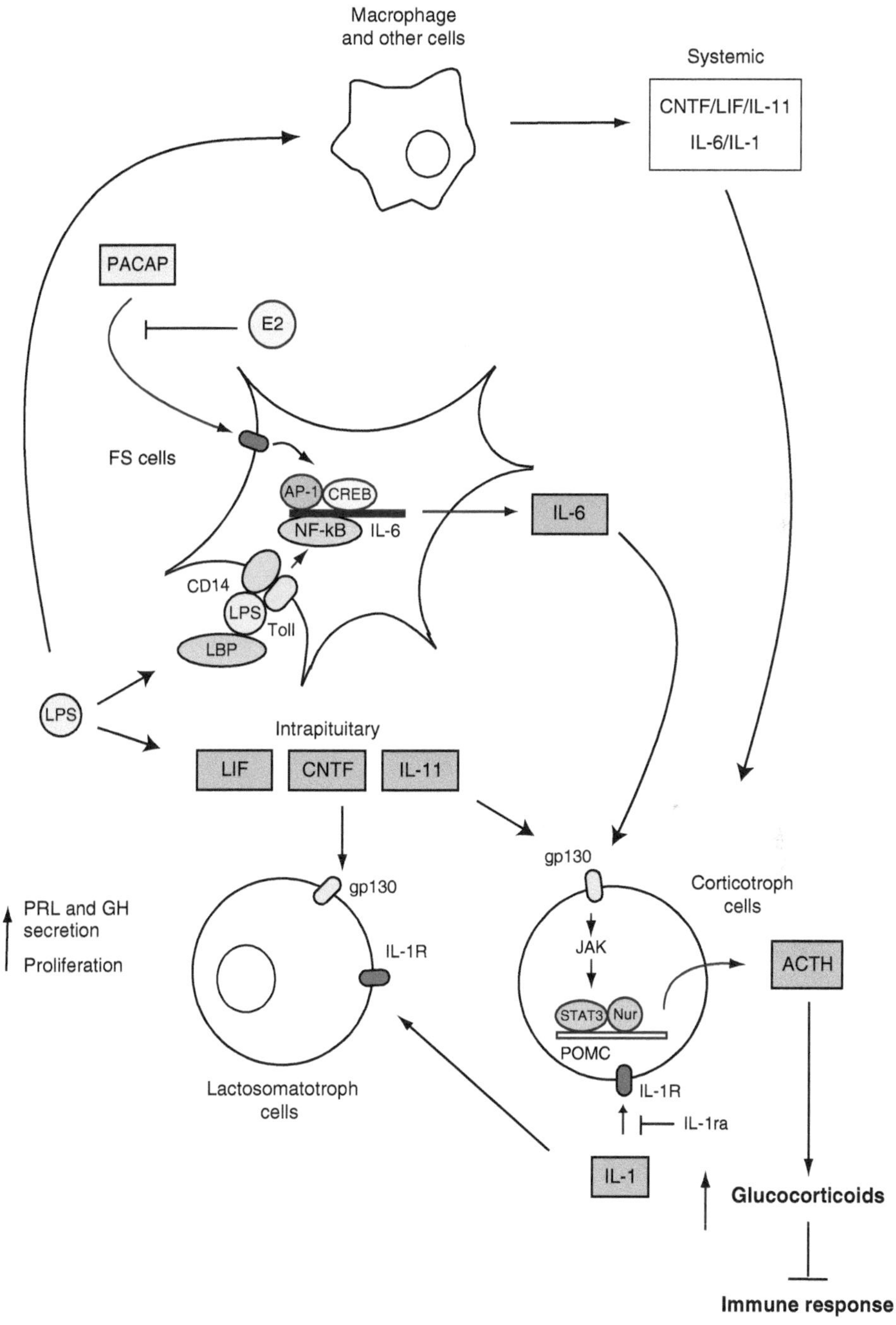


Figure 1. Inflammatory cytokine family pathways on anterior pituitary cells during acute or chronic inflammation or infection, LPS stimulates systemic or anterior pituitary (folliculostellate (FS) or other pituitary cell sources) inflammatory cytokine secretion (IL-1 and IL-6) and related gp130 cytokines (i.e., LIF), which in turn stimulate the expression of ACTH or other hormonal/functional parameters in the pituitary. ACTH secretion induces the release of glucocorticoids from the adrenals, which suppress the activated immune response. Gp130 cytokines stimulate POMC/ACTH, in contrast to CRH or IL-1, through STAT3 and provide a new and powerful mechanism for the regulation of corticotroph function.

## 3. INTERLEUKIN-6 FAMILY

This family is composed of IL-6, leukemia inhibitory factor (LIF), IL-11, oncostatin M, ciliary neurotrophic factor (CNTF), cardiotropin-1, cardiotrophin-like related cytokine, and stimulating neurotrophin-1/B cell-stimulating factor-3 cell (reviewed in Ref. [64]). Their receptors cytoplasmic domains are not required for cytokine-mediated cellular signal transduction, which is mediated by another membrane glycoprotein-denominated gp130 [65,66]. Thus, two functionally different chains compose the IL-6 family receptor complex: one cytokine-binding (alpha) chain and one signal-transducing (gp130) chain, a common signal transducer, which is also present in multiple cytokine receptors [64]. Although LIF, IL-11, and CNTF are not classic inflammatory cytokines, they are closely related to IL-6, share the signal transducer of the IL-6 receptor, and play important roles in the pituitary physiology. Thus, we will refer to them while discussing IL-6 actions on pituitary cells.

### 3.1. IL-6 and its family (gp130 cytokines) regulatory action on pituitary function, in vivo studies

IL-6 is known to influence the secretion of anterior pituitary hormones and is, therefore, considered to play an important role in the function and growth of normal and adenomatous endocrine pituitary cells [67,68].

IL-6 stimulates the release of ACTH when injected in normal rats [69], and suboptimal amounts of IL-1β and IL-6 synergize to induce an early (30–60 min) ACTH response in mice [70]. In agreement with this synergism, it has been shown that an anti-IL-6 antibody blocked the IL-1-induced increase in plasma ACTH in mice [71]. As described for IL-1β, IL-6 also induced an increase in ACTH levels at 4 h post-injection [11].

Recombinant IL-6 has been shown to activate the HPA axis in humans, at different doses, IL-6 treatment of healthy males increased dramatically and dose-dependently both ACTH and cortisol plasma concentrations [72]. In cancer patients, IL-6 induces a marked and prolonged increase in both ACTH and cortisol plasma levels [73,74].

In vivo, intraperitoneal LIF administration resulted in a fourfold increase of ACTH in mice [75], and in nonhuman primates, systemic intracarotid administration of recombinant human LIF was followed by an increase in plasma ACTH levels [76].

### 3.2. IL-6 and its family receptor expression in the anterior pituitary gland

Gp130 mRNA is expressed in human pituitaries [77]. Expression of gp130 alone, however, does not confer the binding of IL-6, but gp130 and IL-6Rα together form a high-affinity IL-6-binding site. Considering that gp130 is ubiquitously expressed, the time and place at which gp130 functions in vivo appears to be determined by spatially and chronologically regulated expression of specific cytokine-binding receptor chains or cytokines themselves [67]. This has been shown for IL-6, for which it has been shown that IL-6-specific alpha-receptors are expressed in rat anterior pituitary cells [78]. IL-6 receptor mRNA was identified by in situ hybridization and RT–PCR in the human pituitary tumor cell line, HP75, derived from a clinically nonfunctioning pituitary tumor [79].

LIF-binding sites have been demonstrated in developing human fetal pituitary and in normal and adenomatous adult human tissue [80]. LIF receptor mRNA was also demonstrated in pituitary cells by RT–PCR [81], and specific LIF-binding sites are present in murine AtT-20 cells [80]. Other gp130 cytokine receptors also appear to function in the anterior pituitary.

The mRNA for the α-chain specific for the IL-11R and CNTF-R are also expressed on folliculostellate (FS) and lactosomatotrophic GH3 cells [82,83]. Consistent with the cell line studies, the expression of these receptors was detected by northern blot in normal cells [82,83]. Further supporting a role in humans, the expression of the IL-11R has been reported by RT–PCR in corticotrophic and nonfunctioning human tumors [82], and the mRNA for the α-chain specific for the CNTF-R was detected by northern blot in tumors secreting PRL and GH and in nonfunctioning tumors [84].

The expression of gp130 itself and α-subunits for different gp130 cytokines provides the cellular and molecular basis for a direct action of these cytokines in the regulation of pituitary growth and function.

### 3.3. IL-6 and its family expression in the anterior pituitary gland

IL-6 production by anterior pituitary cells has been demonstrated by several groups [85,86], as has been the presence of IL-6 mRNA in these cells [87]. Recently, it has been shown that the human pituitary tumor cell line, HP75, synthesizes IL-6 mRNA and expresses and secretes IL-6 [79]. FS cells have been identified as the source of intrapituitary IL-6 production in the normal pituitary, whereas in pituitary adenomas IL-6 is produced by the tumor cells themselves (reviewed in Refs [14,68]). Many compounds can stimulate the production of IL-6 by these cells. The hypothalamic factor pituitary adenylate cyclase-activating polypeptide (PACAP) and the closely related vasoactive intestinal peptide (VIP) enhance the IL-6 release from normal FS cells and from a FS cell line obtained from a pituitary thyrotropic tumor, TtT/GF cells [88–90]. It has been reported that TtT/GF cells and normal FS cells also contain tumor necrosis factor (TNF)-binding sites and that TNF-α and TNF-β are able to induce the release of IL-6 [91]. Recently, we demonstrated that estradiol inhibits dose dependently PACAP-stimulated IL-6 secretion at the transcriptional level, inhibiting the AP-1/TRE transcriptional activity in TtT/GF cells [92]. The inhibition of IL-6 by estrogens in TtT/GF cells, which are important factors in the pathogenesis of lactotroph tumors, may have influences in the early stages of tumor growth, in which IL-6 has an inhibitory action.

On the contrary, estradiol not only enhanced vascular endothelial growth factor (VEGF) release in all lactotroph and lactosomatotroph human pituitary adenoma cell cultures but IL-6 secretion was also stimulated in three of five lactotroph and in all lactosomatotroph cell cultures [93]. The stimulation of IL-6 and/or VEGF secretion by estradiol in the majority of human lactotroph and lactosomatotroph adenoma cell cultures suggests that estrogens may contribute to adenoma expansion through the stimulation of these auto-/paracrine-acting adenoma progression factors [93]. The inhibition in FS cells, probably in the normal pituitary, where IL-6 has an inhibitory role, and the stimulation in the tumoral stage, where IL-6 has a stimulatory role [94], are in agreement with the permissive role of estrogens on lactotroph proliferation and further tumor formation.

IL-1 also stimulates the release of IL-6 in the pituitary in vitro [95]. Glucocorticoids inhibit the production of IL-6 from normal cells [89] and aggregate cultures [96]. Consistent with this, levels of IL-6 mRNA are increased after adrenalectomy in the rat pituitary, suggesting the presence in vivo, of negative feedback from the HPA axis end product on IL-6 production by the pituitary [97]. Further evidence for IL-6 expression and regulation has been obtained in humans. IL-6 mRNA has been detected in corticotrophic adenoma cell cultures as well as in normal human pituitaries and other adenoma types (i.e., prolactinomas, nonfunctioning adenomas, and somatotrophinomas) by in situ hybridization, immunocytochemistry, RT–PCR, and biological assay in the secreted medium (reviewed in Refs [14,98]).

Studies in the developing human fetal pituitary, in normal and adenomatous adult human tissue, and in AtT-20 cells demonstrated LIF protein and mRNA expression [80]. Furthermore, pituitary LIF and LIF-R are significantly induced in vivo in response to LPS endotoxin showing that LIF is a novel LPS-inducible proinflammatory neuroendocrine cytokine [81].

More recent studies raised the question of whether other gp130 cytokines are expressed in anterior pituitary cells. They demonstrated IL-11 mRNA in AtT-20 corticotrophic cells [82] and the expression of mRNAs for IL-11 and CNTF in FS and lactosomatotrophic cells [83].

The expression of gp130 cytokines in anterior pituitary cells points out the importance of their auto/paracrine action on pituitary cells.

### 3.4. IL-6 and its family action on the anterior pituitary

Cell culture experiments with normal primary rat pituitary cells show that picomolar concentrations of IL-6 stimulate the release of PRL, GH, and LH (reviewed in Ref. [99]). IL-6 also enhances ACTH and GH release from rat hemipituitary glands and ACTH from AtT-20 cells [100]. Yamaguchi et al. [101] confirmed that IL-6 stimulates the release of PRL and LH from rat pituitary cells and demonstrated a stimulation of FSH release. In contrast to these stimulatory effects, pretreatment of anterior pituitary cells with IL-6 inhibits both forskolin- and VIP-stimulated activation of adenylate cyclase as well as inhibiting TRH-stimulated increases in inositol phosphate turnover and intracellular calcium levels. However, IL-6 had no effect on basal levels of these intracellular second messengers [102]. The basal release of PRL is inhibited by a polyclonal antiserum to rat IL-6, a functional test showing the involvement of intrinsic IL-6 in PRL production (reviewed in Ref. [99]). IL-6 also stimulates PRL and GH release from lactosomatotrophic GH3 cells [94] and both ACTH secretion and POMC gene expression in corticotroph adenoma cell cultures [103]. This demonstration of the stimulatory action of IL-6 on human corticotroph adenoma cell function provides further evidence for a direct action of IL-6 on corticotroph pituitary cells.

Furthermore, studies in primary cell cultures of human somatotroph adenomas have shown that IL-6 stimulated GH secretion in 10 out of 11 somatotroph adenoma cultures and that the GH-stimulatory potency of IL-6 was identical, or even stronger, than that of GHRH [104]. Also, in 8 out of 11 adenoma cell cultures, IL-6 production was observed [104], suggesting that GH production might be stimulated by IL-6 in an autocrine/paracrine manner in the majority of somatotroph adenomas.

In addition to their effects on the secretion of the pituitary hormones, gp130 cytokines can influence the growth of pituitary cells. Interestingly, this regulation of pituitary cell growth by gp130 cytokines, considering the little proliferation of the pituitary gland, underlies their role among the factors controlling pituitary cell division. This is particularly important for IL-6, which stimulates the growth of TtT/GF cells [105] and pituitary tumor cells [94,106], whereas the same concentration of IL-6 inhibits the growth of normal endocrine cells [94]. In the pituitary cell line HP75, exogenous IL-6 in a low dose stimulated, whereas higher doses inhibited cell growth. This growth effect was inhibited by IL-6-blocking antibody, showing that IL-6 is an important growth regulator in HP75 cells, having an autocrine growth stimulatory effect under basal conditions [79].

Many studies have shown that IL-1 and IL-6 regulate VEGF levels in a variety of tissues. In order to study a possible relationship between VEGF and IL-1/IL-6 during pituitary pathogenesis, their levels were studied in a series of pituitary adenomas and pituitary cells. VEGF was detected in conditioned medium of HP75 cells and in 82% of the tumors tested [107]. Tumor volume and secretion of VEGF were significantly associated with levels of IL-6 and

IL-1α produced [107]. Moreover, the addition of exogenous IL-1α, but not IL-6, significantly increased VEGF production [107]. The significant associations between VEGF and the levels of IL-6 and IL-1 alpha suggest an important role for these cytokines in the development of these tumors.

In mouse pituitary primary cell culture, LIF stimulates ACTH secretion [108]. LIF stimulates the secretion of ACTH and the expression of POMC message in AtT-20 cells [80] and in primary human fetal pituitary cultures [77]. In addition, in both cell type cultures, LIF potentates the stimulatory action of CRH on ACTH secretion [77,109]. The functional importance of LIF action on ACTH secretion is further underlined by studies in LIF gene KO mice (LIFKO), in which a defect in the activation of the HPA axis was observed [110]. ACTH levels are diminished after fasting in the LIFKO animals, and replacement of LIF restores the HPA response [110], those LIFKO mice also show attenuated ACTH responses after immobilization stress [75]. The molecular pathways involved in LIF secretion of ACTH have been studied in AtT-20 cells. In these cells, LIF rapidly induces tyrosyl phosphorylation of STAT1 and STAT3, which are mediators of the gp130 signaling [111]. Overexpression of two dominant negative forms of STAT3 blocks the action of LIF in the corticotroph, confirming that LIF stimulation depends on STAT3 [112]. Using progressive 5′-deletions of POMC promoter, it has been demonstrated that the LIF-responsive POMC promoter region contains two juxtaposed sequences related to STAT3 DNA-binding motif [113]. This evidence for STAT3's direct involvement offers a new mechanistic insight for HPA axis stimulation by gp130 cytokines and suggested an explanation for the synergism between CRH and LIF in regulating ACTH secretion following exposure to events like inflammation or stress. Thus, these results indicated the importance of LIF in the neuroendocrine response to inflammatory processes. Recently, it has been demonstrated that in LIFKO, mice levels of IL-1β, IL-6, and SOCS-3 were elevated. These higher pituitary levels of proinflammatory cytokines observed in LIFKO mice and the attenuation of the HPA axis stress response resulted in a stronger inflammatory process [114]. These results indicate for the first time that, although LIF induces ACTH, SOCS-3 acts to counter-regulate the HPA axis response to inflammation.

Recently a new role of LIF during inflammation has been demonstrated. LIF maintains the HPA axis activation by decreasing GR expression and raises the possibility that LIF might contribute to the development of central glucocorticoid resistance during inflammation [115]. Also $LIFR^{-/-}$ mice present elevated pituitary GR and mineralocorticoid receptor (MR) mRNA and protein levels, indicating the importance of LIF signaling for HPA axis development [116].

Taking into account that transgenic mice overexpressing pituitary LIF develop Cushing-like syndrome with reduced PRL expression, the effects of LIF were tested on human prolactinomas and normal and tumoral rat lactotrophs. LIF treatment reduced PRL secretion in primary human prolactinoma cultures by up to 42% and in primary rat pituitary cultures by 70% and also in the rat MMQ pituitary tumor cells (derived from a PRL-secreting rat anterior pituitary adenoma) [117]. These observations indicate a role for LIF contributing to unrestrained prolactinomas PRL secretion [117].

Like IL-6 and LIF, CNTF and IL-11 also regulate pituitary function. For example, CNTF stimulates GH and PRL production from lactosomatotrophic GH3 cells [83]. IL-11 and CNTF exert a similar stimulation on GH mRNA expression in somatotrophic monolayer cell cultures from acromegalic tumors and CNTF stimulates PRL secretion in lactotrophic monolayers cell cultures from prolactinomas patients [84]. Moreover, in monolayer cell cultures from normal rat anterior pituitary, IL-11 and CNTF had no significant effect on the release of either GH or PRL and on GH mRNA. Interestingly, cell shape or cell contacts appear to be of particular importance for these signaling events. These cytokines significantly stimulated both PRL and

GH secretion when the cells were cultured in aggregates, in which the three-dimensional interaction of the cells is reconstituted. These studies show that the three-dimensional structure of the gland, in which FS cells are involved, is of critical importance for the regulatory action of gp130 cytokines in anterior pituitary cells [84]. IL-11 also stimulates the secretion of the angiogenic factor, VEGF by FS cells [83], and induces POMC expression and ACTH secretion in AtT-20 cells [82]. It also simulates the expression of SOCS-3 in AtT-20 cells, and overexpression of SOCS-3 caused a significant inhibition of IL-11-induced ACTH secretion [82] and also LIF activation of corticotrophs [118,119], indicating that the intracellular gp130 negative feedback system that shuts down gp130 signaling is active in corticotrophic cells. In addition, IL-11 and CNTF also stimulated the proliferation of FS and lactosomatotrophic GH3 cells [83]. Moreover, while all gp130 cytokines stimulate the proliferation of the MtT/SM cells, they inhibit PRL secretion by 70–80% and that of GH by 50% [120].

In concordance to gp130 cytokines action on pituitary cells growth, reduced levels of gp130 protein in GH3 cells (stably transfected with gp130 antisense cDNA) blocked cell growth and hormone secretion stimulated by CNTF and led to severely impaired in vivo tumor development in athymic *nude* mice [121], providing initial evidence supporting a link between gp130 and pituitary abnormal growth. Moreover, it is well known that the pituitary somatotrophic cell line (MtT/S) depends on the gp130 cytokine-producing TtT/GF cells, for tumorigenesis in vivo. Recently, we demonstrated the participation of gp130 cytokines in the auto-paracrine stimulation of MtT/S growth. MtT/S cells overexpressing gp130 protein (gp130-S) or with reduced gp130 levels (gp130-AS) were injected into *nude* mice. MtT/S clones respond differently depending on cell number; at high concentrations, MtT/S clones alone generated tumors equivalent in size to tumors derived from MtT/S plus TtT/GF cells. At low concentrations, MtT/S sense and control clone-generated tumors of smaller size than tumors derived from these same clones plus TtT/GF cells, showing a dependence on FS cells. In both cases, MtT/S gp130-AS clones had impaired tumor development [122]. These studies underline the importance of gp130 cytokines in proliferation and establish its role in auto-paracrine pituitary growth regulation.

IL-6 and other gp130 cytokines appear to be important mediators of immune–neuroendocrine pathway activity at the level of the pituitary gland (Fig. 1).

## 4. LPS EFFECT ON INFLAMMATORY CYTOKINE SECRETION AND ANTERIOR PITUITARY CELLS

During infection and inflammation, LPS, which is a component of gram-negative bacteria, activates cells of the immune system leading to increased levels of circulating inflammatory cytokines, which subsequently activate the HPA axis affecting hormone production. Apart from immune cells, cells from the anterior pituitary have been identified as direct targets for bacterial LPS. This immune–endocrine process is particularly important because it enhances circulating anti-inflammatory glucocorticoids.

It has been shown that LPS treatment increases pituitary IL-1β expression while decreasing IL-1R expression in the mouse brain–endocrine–immune axis, demonstrating that they are under reciprocal regulation [25]. Both IL-1 and IL-1ra mRNA increase in the pituitary, as well as in other tissues, after LPS treatment of mouse [26,123]. AtT-20 cells treatment with 1 μg/ml of LPS, in contrast to lower doses, not only increased IL-1Rs in these cells but also decreased IL-1β concentrations in the cell homogenates [124]. This LPS-induced modulation of IL-1Rs may provide a novel mechanism for the actions of LPS to alter pituitary function during endotoxemia [124].

It has been demonstrated that LPS enhances the intrapituitary secretion of IL-6 in vivo suggesting that IL-6-producing FS cells may respond to LPS; but from these in vivo studies, it is not clear whether LPS acts directly or via the stimulation and subsequent action of peripheral TNF-α or IL-1 [125]. In concordance, LPS simulates the production of IL-6 in vitro by FS cells via the specific membrane-bound CD14-Toll proteins, which represent the classical LPS receptor and the p38α mitogen-activated protein kinase (MAPK)–NF-κB pathway [126]. It has been studied whether LPS enhances ACTH secretion via paracrine-acting intrapituitary IL-6. These experiments have shown that LPS stimulated IL-6 both in monolayer and in aggregate mouse pituitary cell cultures, but only in aggregates, ACTH secretion was significantly enhanced by LPS [127]. Moreover, a neutralizing antibody against mouse IL-6 also inhibited LPS-induced ACTH secretion in aggregates [127]. This intrapituitary IL-6-mediated system represents a pituitary-specific mechanism that stimulates the HPA axis during infection and inflammation.

Besides LPS effects on normal pituitary cultures, recently it has been investigated whether Toll receptor 4 (Tlr4) is also present in normal and transformed endocrine epithelial pituitary cell types. By reverse transcriptase-polymerase chain reaction, Tlr4 mRNA expression was found in AtT-20 and HP75 pituitary epithelial tumor cell lines, whereas GH3 and alphaT3-1 cells were negative for its expression [128]. When human anterior pituitaries were analyzed, Tlr4 protein was detected in only a few epithelial cells from normal pituitaries and in 26 out of 67 human pituitary tumors [128]. Moreover, LPS had no effect on ACTH hormone secretion in Tlr4-positive AtT-20 cells but suppressed the growth of these cells in a dose-dependent manner [128]. In cell cultures of Tlr4-positive pituitary adenomas, LPS dose-dependently stimulated the production of IL-6, which in turn induce growth and hormone production in pituitary tumors [128]. The above data suggest that, during gram-negative bacteria-induced infections or inflammatory processes, LPS could affect pituitary tumor pathophysiology and progression in the subset of Tlr4-expressing pituitary adenomas.

Studies have also shown that mouse LIF mRNA is induced in vivo by intraperitoneal LPS injection, interestingly, to a greater extent than did LIF receptor mRNA [81]. These results were in concordance with an eightfold induction of murine pituitary LIF expression and elevated plasma levels of ACTH and corticosterone in mouse after intraperitoneal injection of LPS [114].

These data suggest that LPS is not only stimulating the HPA axis via the activation of macrophages/monocytes which in turn release and increase peripheral cytokines but that there is also an additional route to suppress the activated immune system. LPS induces the release of IL-6 from FS cells, which in turn act in a paracrine manner to enhance the secretion of ACTH. Moreover, LPS stimulates the secretion of intrapituitary IL-1 and LIF and both induce the release of ACTH. The increased production of ACTH would enhance glucocorticoid levels more rapidly to suppress the activated immune response and it is also related more precisely to the pituitary. During acute or chronic inflammation or infection, systemic, hypothalamic, or hypophyseal IL-1, IL-6, and gp130 cytokines, such as LIF and IL-11, act on anterior pituitary cells. This prevents overshooting by the activated immune system, which may have harmful effects, such as septic shock, and contributes to the integration of the neuroendocrine and immune systems (Fig. 1).

## ACKNOWLEDGMENTS

This work was supported by grants from the Volkswagen Foundation (I/76 803), the University of Buenos Aires, the Argentine Health Ministry, the Argentine National Research Council (CONICET), and the Agencia Nacional de Promoción Científica y Tecnológica-Argentina.

## REFERENCES


1. Besedovsky HO, del Rey A. Immune-neuro-endocrine interactions: facts and hypotheses. Endocr Rev 1996;17:64–102.
2. Besedovsky HO, Del Rey A, Sorkin E. Antigenic competition between horse and sheep red blood cells as a hormone-dependent phenomenon. Clin Exp Immunol 1979;37:106–13.
3. del Rey A, Besedovsky H, Sorkin E. Endogenous blood levels of corticosterone control the immunologic cell mass and B cell activity in mice. J Immunol 1984;133:572–5.
4. Besedovsky HO, del Rey A, Sorkin E. Lymphokine containing supernatants from con A-stimulated cells increase corticosterone blood levels. J Immunol 1981;126:385–7.
5. Besedovsky HO, del Rey A, Sorkin E, Lotz W, Schwulera U. Lymphoid cells produce an immunoregulatory glucocorticoid increasing factor (GIF) acting through the pituitary gland. Clin Exp Immunol 1985;59:622–8.
6. Besedovsky H, del Rey A, Sorkin E, Dinarello CA. Immunoregulatory feedback between interleukin-1 and glucocorticoid hormones. Science 1986;233:652–4.
7. Bateman A, Singh A, Kral T, Solomon S. The immune-hypothalamic-pituitary-adrenal axis. Endocr Rev 1989;10:92–112.
8. Besedovsky HO, del Rey A. Immune-neuroendocrine circuits: integrative role of cytokines. Front Neuroendocrinol 1992;13:61–94.
9. Hermus ARMM, Sweep CGJ. Cytokines and the hypothalamic-pituitary-adrenal axis. J Steroid Biochem Mol Biol 1990;37:867–71.
10. Parsadaniantz SM, Batsché E, Gegout-Pottie P, Terlain B, Gillet P, Netter P et al. Effects of continuous infusion of Interleukin-1β on corticotropin-releasing hormone (CRH), CRH receptors, proopiomelanocortin gene expression and secretion of corticotropin, b-endorphin and corticosterone. Neuroendocrinology 1997;65:53–63.
11. Harbuz MS, Stephanou A, Sarlis N, Lightman SL. The effects of recombinant human interleukin (IL)-1α, IL-1β or IL-6 on hypothalamus-pituitary-adrenal axis activation. J Endocrinol 1992;133:349–55.
12. Payne LC, Weigent DA, Blalock JE. Induction of pituitary sensitivity to interleukin-1: a new function for corticotropin-releasing hormone. Biochem Biophys Res Commun 1994;198:480–4.
13. Sweep CGJ, Van der Meer MJM, Hermus ARMM, Smals AGH, Van der Meer JWM, Pesman GJ et al. Chronic stimulation of the pituitary-adrenal axis in rats by interleukin-1β infusion: in vivo and in vitro studies. Endocrinology 1992;130:1153–64.
14. Arzt E, Pereda MP, Castro CP, Pagotto U, Renner U, Stalla GK. Pathophysiological role of the cytokine network in the anterior pituitary gland. Front Neuroendocrinol 1999;20:71–95.
15. Haour GH, Ban EM, Milon GM, Baran D, Fillion GM. Brain Interleukin-1 receptors: characterization and modulation after lipopolysaccharide injection. Prog Neuroendocrinimmunol 1990;3:196–204.
16. Takao T, Culp SG, Newton RC, De Souza EB. Type I interleukin-1 receptors in the mouse brain-endocrine-immune axis labeled with (125I) recombinant human interleukin-1 receptor antagonist. J Neuroimmunol 1992;41:51–60.
17. Cunningham ET, Wada E, Carter DB, Tracey DE, Battery JF, De Souza EB. In situ histochemical localization of type I interleukin-1 receptor messenger RNA in the central nervous system, pituitary, and adrenal gland of the mouse. J Neurosci 1992;3:1101–14.
18. Ban E, Marquette C, Sarrieu A, Fitzpatrick F, Fillion G, Milon G et al. Regulation of interleukin-1 receptor expression on mouse brain and pituitary by lipopolysaccharide and glucocorticoids. Neuroendocrinology 1993;58:581–7.
19. Webster EL, Tracey DE, De Souza EB. Upregulation of IL-1 receptors in mouse AtT20 pituitary tumors cells following treatment with corticotropin-releasing factor. Endocrinology 1991;129: 2796–8.
20. Takao T, Tojo C, Nishioka T, Hashimoto K, De Souza EB. Corticotropin-releasing factor treatment upregulates interleukin-1 receptors in the mouse pituitary: reversal by dexamethasone. Brain Res 1995;688:219–22.

21. Takao T, Dieterich KD, Tracey DE, De Souza EB. Cyclic AMP-dependent modulation of interleukin-1 receptors in the mouse AtT-20 pituitary tumor cell line. Brain Res 1994;656(1):177–81.
22. Matta SG, Linner KM, Sharp BM. Interleukin-1α and interleukin-1β stimulate adrenocorticotropin secretion in the rat through a similar hypothalamic receptor(s): effects of interleukin1 receptor antagonist protein. Neuroendocrinology 1993;57:14–22.
23. Parnet P, Brunke DL, Goujon E, Mainard JD, Biragyn A, Arkins S et al. Molecular identification of two types of interleukin-1 receptors in the murine pituitary gland. J Neuroendocrinol 1993;5:213–9.
24. Marquette C, Van Dam A-M, Ban E, Lanièce P, Crumeyrolle-Arias M, Fillion G et al. Rat interleukin-1β binding sites in rat hypothalamus and pituitary gland. Neuroendocrinology 1995;62:362–9.
25. Takao T, Culp SG, De Souza EB. Reciprocal modulation of interleukin-1β (IL-1β) and IL-1 receptors by lypopolysacharide (endotoxin) treatment in the mouse brain-endocrine-immune axis. Endocrinology 1993;132:1497–504.
26. Koenig JI, Snow K, Clark BD, Toni R, Cannon JG, Shaw AR et al. Intrinsic pituitary interleukin 1-β is induced by bacterial lipopolysaccharide. Endocrinology 1990;126:3053–8.
27. Green VL, Atkin L, Speirs V, Jeffreys RV, Landolt B, Mathew B et al. Cytokine expression in human anterior pituitary adenomas. Clin Endocrinol 1996;45:179–85.
28. Gatti S, Bartfai T. Induction of tumor necrosis factor-a mRNA in the brain after peripheral endotoxin treatment: comparison with interleukin-1 family and interleukin-6. Brain Res 1993;6:291–4.
29. Sauer J, Arzt E, Gumprecht H, Hopfner U, Stalla GK. Expression of interleukin-1 receptor antagonist in human pituitary adenomas in vitro. J Clin Endocrinol Metab 1994;79:1857–63.
30. Uehara A, Gillis S, Arimura A. Effests of interleukin-1 on hormone release from normal rat pituitary cells in primary culture. Nueroendocrinology 1987;45:343–7.
31. Bernton EW, Beach JE, Holaday JW, Smallridge RC, Fein HG. Release of multiple hormones by a direct action of interleukin-1 on pituitary cells. Science 1987;238:519–21.
32. Florio T, Meucci E, Landolfi E, Grimaldi M, Ventra C, Scorziello A et al. Interleukin-1 modulation of anterior pituitary function: effect on hormone release and second messenger systems. Pharmacol Res 1989;21:21–36.
33. Keher P, Turnill D, Dayer JM, Muller AF, Gaillard RC. Human recombinant interleukin-1-beta and alpha, but not recombinant tumor necrosis alpha stimulate ACTH release from rat anterior pituitary cells in vitro in a prostaglandin E2 cAMP independent manner. Neuroendocrinology 1988;48:160–6.
34. Cambronero JC, Rivas FJ, Borrel J, Gauza C. Interleukin-1-beta induces pituitary adrenocorticotropin secretion: evidence for glucocorticoid modulation. Neuroendocrinology 1992;55:648–54.
35. Beach JE, Smallridge RC, Kinzer CA, Bernton EW, Holaday JW, Fein HG. Rapid release of multiple hormones from rat pituitaries perifused with recombinant interleukin-1. Life Sci 1989;44:1–7.
36. Parsadaniantz SM, Lenoir V, Terlain B, Kerdelhue B. Lack of effect of interleukin-1α and -β, during in vitro perfusion, on anterior pituitary release of adrenocorticotropin hormone and β-endorphin in the male rat. J Neurosci Res 1993;34:315–23.
37. Hashimoto K, Nishioka T, Tojo C, Takao T. Nitric oxide plays no role in ACTH release induced by interleukin-β, corticotropin-releasing hormone, arginine vasopressin and phorbol myristate acetate in rat pituitary cell cultures. Endocr J 1995;42(3):435–9.
38. Perez Castro C, Paez Pereda M, Reul JMHM, Stalla G, Holsboer F, Arzt E. In Vivo Changes of PRL during the T-cell Dependent Immune Response, 1st Edn. Amsterdam: Elsevier Science B.V., 2002.
39. Fagarasan MO, Aiello F, Muegge K, Durum S, Axelrod J. Interleukin 1 induces b-endorphin secretion via Fos and Jun in AtT20 pituitary cells. Proc Natl Acad Sci USA 1990;87:7871–4.
40. Fagarasan MO, Bishop JF, Rinaudo MS, Axelrod J. Interleukin 1 induces early protein phosphorylation and requires only a short exposure for late induced secretion of β-endorphin in a mouse pituitary cell line. Proc Natl Acad Sci USA 1990;87:2555–9.
41. Fagarasan MO, Eskay R, Axelrod J. Interleukin 1 potentiates the secretion of β-endorphin secretion by secretagogues in a mouse pituitary cell line (AtT20). Proc Natl Acad Sci USA 1989;86:2070–3.

42. Fukata J, Usui T, Naitoh Y, Nakai Y, Imura H. Effects of recombinant human interleukin-1α, -1β, 2 and 6 on ACTH synthesis and release in the mouse pituitary tumor cell line AtT-20. J Endocrinol 1989;122:33–9.
43. Woloski BMRNJ, Smith EM, Meyer WJ, Fuller GM, Blalock JE. Corticotropin-releasing activity of monokines. Science 1985;230:1035–7.
44. Gong FY, Deng JY, Shi YF. Stimulatory effect of interleukin-1β on growth hormone gene expression and growth hormone release from rat GH3 cells. Neuroendocrinology 2005;81:217–28.
45. Malarkey WB, Zvara BJ. Interleukin-1β and other cytokines stimulate adrenocorticotropin release from cultured pituitary cells of patients with Cushing's disease. J Clin Endocrinol 1989;69:196–9.
46. Baud V, Liu ZG, Bennett B, Suzuki N, Xia Y, Karin M. Signaling by proinflammatory cytokines: oligomerization of TRAF2 and TRAF6 is sufficient for JNK and IKK activation and target gene induction via an amino-terminal effector domain. Genes Dev 1999;13:1297–308.
47. Dinarello CA. The interleukin-1 family: 10 years of discovery. FASEB J 1994;8:1314–25.
48. Gwosdow AR, O'Connell NA, Abou-Samra AB. Interleukin-1 increases protein kinase A activity by a cAMP-independent mechanism in AtT-20 cells. Am J Physiol 1994;266:E79–84.
49. Gwosdow AR, Spencer JA, O'Connell NA, Abou-Samra AB. Interleukin-1 activates protein kinase A and stimulates adrenocorticotropin hormone release from AtT20 cells. Endocrinology 1993;132:710–4.
50. Murphy EP, Coneely OM. Neuroendocrine regulation of the hypothalamic pituitary adrenal axis by the nurr1/nur77 subfamily of nuclear receptors. Mol Endocrinol 1997;11:39–47.
51. Maira M, Martens C, Philips A, Drouin J. Heterodimerization between members of the Nur subfamily of orphan nuclear receptors as a novel mechanism for gene activation. Mol Cell Biol 1999;19:7549–57.
52. Philips A, Lesage S, Gingras R, Maira MH, Gauthier Y, Hugo P et al. Novel dimeric Nur77 signaling mechanism in endocrine and lymphoid cells. Mol Cell Biol 1997;17:5946–51.
53. Kovalovsky D, Refojo D, Liberman AC, Hochbaum D, Pereda MP, Coso OA et al. Activation and induction of NUR77/NURR1 in corticotrophs by CRH/cAMP: involvement of calcium, protein kinase A, and MAPK pathways. Mol Endocrinol 2002;16:1638–51.
54. Kovalovsky D, Paez Pereda M, Labeur M, Renner U, Holsboer F, Stalla GK et al. Nur77 induction and activation are necessary for interleukin-1 stimulation of proopiomelanocortin in AtT-20 corticotrophs. FEBS Lett 2004;563:229–33.
55. Parnet P, Pousset F, Laye S. NF kappa B activation in mouse pituitary: comparison of response to interleukin-1 beta and lipopolysaccharide. Neuroendocrinology 2003;15:304–14.
56. Huang ZF, Massey JB, Via DP. Differential regulation of cyclooxygenase-2 (COX-2) mRNA stability by interleukin-1 beta (IL-1β) and tumor necrosis factor-alpha (TNF-α) in human in vitro differentiated macrophages. Biochem Pharmacol 2000;59:187–94.
57. Mifflin RC, Saada JI, Di Mari JF, Adegboyega PA, Valentich JD, Powell DW. Regulation of COX-2 expression in human intestinal myofibroblasts: mechanisms of IL-1-mediated induction. Am J Physiol Cell Physiol 2002;282:C824–34.
58. Ferri CC, Yuill EA, Ferguson AV. Interleukin-1beta depolarizes magnocellular neurons in the paraventricular nucleus of the hypothalamus through prostaglandin-mediated activation of a non selective cationic conductance. Regul Pept 2005;129:63–71.
59. Vidal S, Kovacs K, Bell D, Horvath E, Scheithauer BW, Lloyd RV. Cyclooxygenase-2 expression in human pituitary tumors. Cancer 2003;97:2814–21.
60. Onguru O, Scheithauer BW, Kovacs K, Vidal S, Jin L, Zhang S et al. Analysis of Cox-2 and thromboxane synthase expression in pituitary adenomas and carcinomas. Endocr Pathol 2004;15:17–27.
61. Costas M, Trapp T, Paez Pereda M, Sauer J, Rupprecht R, Nahmod VE et al. Molecular and functional evidence for in vitro cytokine enhancement of human and murine target cell sensitivity to glucocorticoids. J Clin Invest 1996;98:1409–16.
62. Chida D, Imaki T, Suda T, Iwakura Y. Involvement of corticotropin-releasing hormone- and interleukin (IL)-6-dependent proopiomelanocortin induction in the anterior pituitary during hypothalamic-pituitary-adrenal axis activation by IL-1alpha. Endocrinology 2005;146:5496–502.

63. Renner U, Newton CJ, Pagotto U, sauer J, Artz E, Stalla GK. Involvement of interleukin-1 and interleukin-1 receptor antagonist in rat pituitary cell growth regulation. Endocrinology 1995;136: 3186–93.
64. Kishimoto T, Akira S, Narazaki M, Taga T. Interleukin-6 family of cytokines and gp130. Blood 1995;86:1243–54.
65. Hirano T. Interleukin-6. In The Cytokine Handbook, 2nd Edn.Thomson A W, Ed.; Osaka: Academic Press, 1994; pp. 145–68.
66. Kishimoto T, Taga T, Akira S. Cytokine signal transduction. Cell 1994;76:253–62.
67. Arzt E. gp130 cytokine signaling in the pituitary gland: a paradigm for cytokine-neuro-endocrine pathways. J Clin Invest 2001;108:1729–33.
68. Renner U, Gloddek J, Paez Pereda M, Arzt E, Stalla GK. Regulation and role of intrapituitary IL-6 production by folliculostellate cells. Dom Animal Endocrinol 1998;15:353–62.
69. Naitoh Y, Fukata J, Tominaga T, Nakai Y, Tamai S, Mori K et al. Interleukin-6 stimulates the secretion of adrenocorticotropic hormone in conscious, freely-moving rats. Biochem Biophys Res Commun 1988;155:1459.
70. Perlstein RS, Mougey EH, Jackson WE, Neta R. Interleukin-1 and Interleukin-6 act synergistically to stimulate the release of adrenocorticotropic hormone in vivo. Lymphokine Cytokine Res 1991;10: 141–6.
71. Neta R, Perlstein R, Vogel SN, Ledney GD, Abrams J. Role of Interleukin 6 (IL-6) in protection from lethal irradiation and in endocrine responses to IL-1 and tumor necrosis factor. J Exp Med 1992;175:689–94.
72. Tsigos C, Papanicolaou DA, Defensor R, Mitsiadis CS, Kyrou I, Chrousos GP. Dose effects of recombinant human interleukin-6 on pituitary hormone secretion and energy expenditure. Neuroendocrinology 1997;66:54–62.
73. Mastorakos G, Chrousos GP, Weber JS. Recombinant interleukin-6 activates the hypothalamic-pituitary-adrenal axis in humans. J Clin Endocrinol Metab 1993;77:1690–4.
74. Späth-Schwalbe E, Born J, Schrezenmeier H, Bornstein SR, Stromeyer P, Drechsler S et al. Interleukin-6 stimulates the hypothalamus-pituitary-adrenocortical axis in man. J Clin Endocrinol Metabol 1994;79:1212–4.
75. Chesnokova V, Auerhamer CJ, Melmed S. Murine LIF gene disruption attenuates the hypothalamo-pituitary-adrenal axis stress response. Endocrinology 1998;139:2209–16.
76. Akita S, Conn PM, Melmed S. Leukemia inhibitory factor (LIF) induces acute adrenocorticotrophic hormone (ACTH) secretion in fetal rhesus macaque primates: a novel dynamic test of pituitary function. J Clin Endocrinol Metab 1996;81:4170–3.
77. Shimon I, Yan X, Ray DW, Melmed S. Cytokine-dependent gp130 receptor subunit regulates human fetal pituitary adrenocorticotropin hormone and growth hormone secretion. J Clin Invest 1997;100:357–63.
78. Ohmichi M, Hirota K, Koike K, Kurachi H, Ohtsuka S, Matsuzaki N et al. Binding sites for interleukin-6 in the anterior pituitary gland. Neuroendocrinology 1992;55:199–203.
79. Borg SA, Kerry KE, Baxter L, Royds JA, Jones TH. Expression of interleukin-6 and its effects on growth of HP75 human pituitary tumor cells. J Clin Endocrinol Metab 2003;88:4938–44.
80. Akita S, Webster J, Ren SG, Takino H, Said J, Zand O et al. Human and murine pituitary expression of leukemia inhibitory factor. Novel intrapituitary regulation of adrenocorticotropin hormone synthesis and secretion. J Clin Invest 1995;95:1288–98.
81. Wang Z, Ren SG, Melmed S. Hypothalamic and pituitary leukemia inhibitory factor gene expression in vivo: a novel endotoxin-inducible neuro-endocrine interface. Endocrinology 1996;137:2947–53.
82. Auernhammer CJ, Melmed S. Interleukin-11 stimulates proopiomelanocortin gene expression and adrenocorticotropin secretion in corticotroph cells: evidence for a redundant cytokine network in the hypothalamo-pituitary-adrenal axis. Endocrinology 1999;140:1559–66.
83. Perez Castro C, Nagashima AC, Pereda MP, Goldberg V, Chervin A, Largen P et al. The gp130 cytokines interleukin-11 and ciliary neurotropic factor regulate through specific receptors the function and growth of lactosomatotropic and folliculostellate pituitary cell lines. Endocrinology 2000;141:1746–53.

84. Perez Castro C, Nagashima AC, Paez Pereda M, Goldberg V, Chervín A, Carrizo G et al. Effects of the gp130 cytokines CNTF and IL-11 on pituitary cells: CNTF receptors on human pituitary adenomas and stimulation of PRL and GH secretion in normal rat anterior pituitary aggregate cultures. J Endor 2001;169:539–47.
85. Spangelo BL, MacLeod RM, Isakson PC. Production of interleukin-6 by anterior pituitary cells in vitro. Endocrinology 1990;126:582–6.
86. Vankelecom H, Carmeliet P, Van Damme J, Billiau A, Denef C. Production of Interleukin-6 by folliculo-stellate cells of the anterior pituitary gland in a histiotypic cell aggregate culture system. Neuroendocrinology 1989;49:102–6.
87. Spangelo BL, Judd AM, MacLeod RM, Goodman DW, Isakson PC. Endotoxin-induced release of interleukin-6 from rat medial basal hypothalami. Endocrinology 1990;127:1779–85.
88. Tatsumo I, Somogyvari-Vigh A, Mizumo K, Gottschall PE, Hidaka H, Arimura A. Neuropeptide regulation of interleukin-6 production from the pituitary: stimulation by pituitary adenylate cyclase activating plypeptide and calcitonin gene-related peptide. Endocrinology 1991;129:1797–804.
89. Spangelo BL, Isakson PC, MacLeod RM. Production of interleukin-6 by anterior pituitary cells is stimulated by increased intracellular adenosine 3′, 5′-monophosphate and vasoactive intestinal peptide. Endocrinology 1990;127:403–9.
90. Matsumoto H, Koyama C, Sawada T, Koike K, Hirota K, Miyake A et al. Pituitary folliculo-stellate-like cell line (TtT/GF) responds to novel hypophysiotrophic peptide (pituitary adenylate cyclase-activating peptide), showing increased adenosine 3′,5′-monophosphate and interleukin-6 secretion and cell proliferation. Endocrinology 1993;133:2150–5.
91. Kobayashi H, Fukata J, Murakami N, Usui T, Ebisui O, Muro S et al. Tumor necrosis factor receptors in the pituitary cells. Brain Res 1997;758:45–50.
92. Nagashima AC, Giacomini D, Castro CP, Pereda MP, Renner U, Stalla GK et al. Transcriptional regulation of interleukin-6 in pituitary folliculo-stellate TtT/GF cells. Mol Cell Endocrinol 2003;201:47–56.
93. Onofri C, Carbia Nagashima A, Schaaf L, Feirer M, Lohrer P, Stummer W et al. Estradiol stimulates vascular endothelial growth factor and interleukin-6 in human lactotroph and lactosomatotroph pituitary adenomas. Exp Clin Endocrinol Diabetes 2004;112:18–23.
94. Arzt E, Buric R, Stelzer G, Stalla J, Sauer J, Renner U et al. Interleukin involvement in anterior pituitary cell growth regulation: effects of interleukin-2 (IL-2) and IL-6. Endocrinology 1993;132:459–67.
95. Spangelo BL, Judd AM, Isakson PC, MacLeod RM. Interleukin-1 stimulates interleukin-6 release from rat anterior pituitary cells in vitro. Endocrinology 1991;128:2685–92.
96. Carmeliet P, Vankelecom H, Van Damme J, Billiau A, Denef C. Release of interleukin-6 from anterior pituitary cell aggregates: developmental pattern and modulation by glucocorticoids and forskolin. Neuroendocrinology 1991;53:29–34.
97. Schöbitz B, Van En Dobbelsteen M, Holsboer F, Sutanto W, De Kloet ER. Regulation of interleukin 6 gene expression in rat. Endocrinology 1993;132:1569–76.
98. Ray D, Melmed S. Pituitary cytokine and growth factor expression and action. Endocrine Review 1997;18:206–28.
99. Spangelo BL, Gorospe WC. Role of the cytokines in the neuroendocrine-immune system axis. Front Neuroendocrinol 1995;16:1–12.
100. Lyson K, McCann SM. The effect of interleukin-6 on pituitary hormone release in vivo and in vitro. Neuroendocrinology 1991;54:262–6.
101. Yamaguchi M, Matsuzaki N, Hirota K, Miyake A, Tanizawa O. Interleukin 6 possibly induced by interleukin 1 beta in the pituitary gland stimulates the release of gonadotropins and prolactin. Acta Endocrinol 1990;122:201–5.
102. Grimaldi M, Meucci O, Scorziello A, Florio T, Ventra C, De Mercato R et al. Interleukin 6 modulation of second messenger systems in anterior pituitary cells. Life Sci 1992;51:1243–8.
103. Páez Pereda M, Lohrer P, Kovalovsky D, Perez Castro C, Goldberg V, Losa M et al. Interleukin-6 is inhibited by glucocorticoids and stimulates ACTH secretion and POMC expression in human corticotroph pituitary adenomas. Exp Clin Endocr Diabetes 2000;108:202–7.

104. Thiele JO, Lohrer P, Schaaf L, Feirer M, Stummer W, Losa M et al. Functional in vitro studies on the role and regulation of interleukin-6 in human somatotroph pituitary adenomas. Eur J Endocrinol 2003;149:455–61.
105. Renner U, Gloddek J, Arzt E, Inoue K, Stalla GK. Interleukin-6 is an autocrine growth factor for folliculostellate-like TtT/GF mouse pituitary tumor cells. Exp Clin Endocrinol Diabetes 1997;105:345–52.
106. Sawada T, Koike K, Kanda Y, Ikegami H, Jikihara T, Maeda T et al. Interleukin-6 stimulates cell proliferation of rat anterior pituitary clonal cell lines in vitro. J Endocrinol Invest 1995;18:83–90.
107. Borg SA, Kerry KE, Royds JA, Battersby RD, Jones TH. Correlation of VEGF production with IL1 alpha and IL6 secretion by human pituitary adenoma cells. Eur J Endocrinol 2005;152: 293–300.
108. Stefana B, Ray DW, Melmed S. Leukemia inhibitory factor (LIF) induces differentiation of pituitary corticotroph function: a neuro-endocrine phenotypic switch. Proc Natl Acad Sci USA 1996;93:12502–6.
109. Ray DW, Ren SG, Melmed S. Leukemia inhibitory factor (LIF) stimulates proopiomelanocortin (POMC) expression in a corticotroph cell line. Role of STAT pathway. J Clin Invest 1996;97:1852–9.
110. Akita S, Malkin J, Melmed S. Disrupted murine leukemia inhibitory factor (LIF) gene attenuates adrenocorticotropic hormone (ACTH) secretion. Endocrinology 1996;137:3140–3.
111. Ray DW, Ren SG, Melmed S. Leukemia inhibitory factor (LIF) stimulates proopiomelanocortin (POMC) expression in a corticotroph cell line. Role of STAT pathway. J Clin Invest 1996;97:1852–9.
112. Bousquet C, Melmed S. Critical role for STAT3 in murine pituitary adrenocorticotropin hormone leukemia inhibitory factor signaling. J Biol Chem 1999;274:10723–30.
113. Bousquet C, Zatelli MC, Melmed S. Direct regulation of pituitary proopiomelanocortin by STAT3 provides a novel mechanism for immuno-neuroendocrine interfacing. J Clin Invest 2000;106:1417–25.
114. Chesnokova V, Kariagina A, Melmed S. Opposing effects of pituitary leukemia inhibitory factor and SOCS-3 on the ACTH axis response to inflammation. Am J Physiol Endocrinol Metab 2002;282:E1110–8.
115. Kariagina A, Zonis S, Afkhami M, Romanenko D, Chesnokova V. Leukemia inhibitory factor regulates glucocorticoid receptor expression in the hypothalamic-pituitary-adrenal axis. Am J Physiol Endocrinol Metab 2005;289:E857–63.
116. Ware CB, Kariagina A, Zonis S, Alon D, Chesnokova V. Leukemia inhibitory factor signaling is implicated in embrionic development of the HPA axis. FEBS Lett 2005;579:4465–9.
117. Ben-Shlomo A, Miklovsky I, Ren SG, Yong WH, Heaney AP, Culler MD et al. Leukemia inhibitory factor regulates prolactin secretion in prolactinoma and lactotroph cells. J Clin Endocrinol Metab 2003;88:858–63.
118. Auenhammer CJ, Chesnokova V, Gousquet C, Melmed S. Pituitary corticotroph SOCS-3: novel intracellular regulation of leukemia-inhibitory factor-mediated proopiomelanocortin gene expression and adrenocorticotropin secretion. Mol Endocrinol 1998;12:954–61.
119. Bousquet C, Susini C, Melmed S. Inhibitory roles for SHP-1 and SOCS-3 following pituitary proopiomelanocortin by leukemia inhibitory factor. J Clin Invest 1999;104:1277–85.
120. Tomida M, Yoshida U, Mogi C, Maruyama M, Goda H, Hatta Y et al. Leukemia inhibitory factor and interleukin-6 inhibit secretion of prolactin and growth hormone by rat pituitary MtT/S cells. Cytokine 2001;14:202–7.
121. Castro CP, Giacomini D, Nagashima AC, Onofri C, Graciarena M, Kobayashi K et al. Reduced expression of the cytokine transducer gp130 inhibits hormone secretion, cell growth, and tumor development of pituitary lactosomatotrophic GH3 cells. Endocrinology 2003;144:693–700.
122. Graciarena M, Carbia-Nagashima A, Onofri C, Perez-Castro C, Giacomini D, Renner U et al. Involvement of the gp130 cytokine transducer in MtT/S pituitary somatotroph tumour development in an autocrine-paracrine model. Eur J Endocrinol 2004;151:595–604.
123. Gabellec M-M, Griffais R, Fillion G, Haour F. Expression of interleukin-1α, interleukin-1β and interleukin-1 receptor antagonist mRNA in mouse brain: regulation by bacterial lipopolysaccharide (LPS) treatment. Brain Res Mol Brain Res 1995;31:122–30.

124. Kumagai C, Takao T, Matsumoto R, Asaba K, Hashimoto K. Modulation of interleukin-1 receptors followed by endotoxin lipopolysaccharide treatment in the mouse AtT-20 pituitary tumor cell line. Neuroimmunomodulation 2002;10:310–6.
125. Muramami N, Fukata J, Tsukada T, Kobayashi H, Ebisui O, Segawa H et al. Bacterial lipopolysaccharide-induced expression of interleukin-6 messenger ribonucleic acid in the rat hypothalamus, pituitary, adrenal gland, and spleen. Endocrinology 1993;133:2574–8.
126. Lohrer P, Gloddek J, Nagashima AC, Korali Z, Hopfner U, Páez Pereda M et al. Lipopolysaccharide directly stimulates the intrapituitary interleukin-6 production by folliculostellate cells via specific receptors and the p38a mitogen-activated protein kinase/nuclear factor-kB pathway. Endocrinology 2000;141:4457–65.
127. Gloddek J, Lohrer P, Stalla J, Arzt E, Stalla GK, Renner U. The intrapituitary stimulatory effect of lipopolysaccharide on ACTH secretion is mediated by paracrine-acting IL. Clin Endocrinol Diabetes 2001;6:410–15.
128. Tichomirowa M, Theodoropoulou M, Lohrer P, Schaaf L, Losa M, Uhl E et al. Bacterial endotoxin (lipopolysaccharide) stimulates interleukin-6 production and inhibits growth of pituitary tumour cells expressing the toll-like receptor 4. J Neuroendocrinol 2005;17:152–6.

Edited by A. del Rey, G. P. Chrousos and H. O. Besedovsky


# *Neuroendocrinology and the Immune Response: Inflammatory Mediator Effects on the Adrenal Gland*

ILIAS VREZAS[1], CHRISTIAN G. ZIEGLER[2], ALEXANDER W. KRUG[2], and STEFAN R. BORNSTEIN[2]

[1] *Department of Endocrinology, University Hospital of Duesseldorf, Duesseldorf, Germany*
[2] *Medical Clinic III, Carl Gustav Carus University Hospital, University of Dresden, Dresden, Germany*

## ABSTRACT

Cytokines are able to interact directly with the hypothalamic–pituitary–adrenal (HPA) axis. On the level of the adrenal, immune cells infiltrating the gland can influence adrenocortical function by secreting cytokines. On the other hand, adrenal cells themselves produce cytokines, which act on the adrenal gland. In this context, cytokines and especially interleukin (IL)-1, IL-6, and tumor necrosis factor-α (TNF-α) seem to be important regulators of the HPA axis. Especially IL-6 has a great influence on many functions, including differentiation, stimulation, and activation of immune cells, or other cells of neuroendocrine origin. In addition to cytokines interacting with adrenal function, cytokine-independent mechanisms are also responsible for a cell-to-cell-mediated immune regulation of the adrenal. Furthermore, hypothalamic hormones, including corticotropin-releasing hormone (CRH) and vasopressin have also been identified as important modulators of HPA axis in physiological as well as pathological situations. The importance of this immune-endocrine crosstalk becomes more evident in cases of autoimmune and inflammatory diseases, where an adequate adrenal stress response is decisive for the survival of the organism.

## 1. INTRODUCTION

The immune system interacts with the hypothalamic–pituitary–adrenal (HPA) axis in a systemic manner by stimulating corticotropin-releasing hormone (CRH) and adrenocorticotropic hormone (ACTH) secretion [1,2]. In addition to this systemic activation of the HPA axis, there is evidence for a local, intra-adrenal, regulation of the HPA axis with the adrenal itself being the major site for both synthesis and action of numerous cytokines [3].

The adrenal cortex of humans and rodents is extensively infiltrated by immune cells, including macrophages, lymphocytes, monocytes, mast cells, and dendritic cells [4]. Adrenocortical macrophages, when activated, are able to produce a diversity of cytokines, like interleukin (IL)-1, IL-6, and transforming growth factor (TGF)-β [5,6], which influence adrenal function in a differential, either stimulatory or inhibitory, manner. Furthermore, circulating leukocytes also exert their actions on the adrenal gland through cytokine secretion.

Adrenocortical cells have the intrinsic ability to produce cytokines themselves, as demonstrated by *zona glomerulosa* (ZG) cells producing IL-6 and tumor necrosis factor-α (TNF-α), respectively [7,8]. The inner *zona reticularis* is the main site of cytokine production in humans [9]. Furthermore, ACTH and angiotensin II induce the secretion of the cytokine, IL-6, in the adrenal cortex of rats and mice, whereas ACTH inhibits TNF-α release. This demonstrates that cytokines are expressed in a differential manner in the adrenal cortex and the release and production of cytokines are regulated selectively. Since cytokines have effects on adrenal function and are differentially regulated, they are assumed to play autocrine/paracrine roles in regulating the adrenal gland [10].

## 2. ACTION OF CYTOKINES ON THE ADRENAL GLAND

Cytokines can be defined as regulatory proteins (peptides, proteins, or glycoproteins) secreted by white blood cells and a variety of other cells. They have numerous effects on immunocompetent cells, modulate inflammatory responses, and regulate differentiation, growth, and function of different cell types. "Monokines" and "lymphokines" are commonly used terms for cytokines, describing the source of their origin: monocyte/macrophage-derived and lymphocyte-derived mediators. In contrast to hormones, which function as regulators predominantly of systemic homeostasis, cytokines are regulators of predominantly local tissue processes, in a paracrine or autocrine manner.

Cytokines not only regulate adrenocortical cell function, but also interact with chromaffin cells of the adrenal medulla, stimulating the secretion of catecholamines and neuropeptides by these cells [9]. This allows for a paracrine modulation of adrenocortical function.

Cytokines play an important role in activating the HPA axis during an immune and inflammatory reaction. The proinflammatory cytokines TNF-α, IL-1, and IL-6 are the main HPA axis-stimulating mediators in plasma [11]. TNF-α is secreted first, then IL-1, and IL-6 last. Inflammatory cytokines stimulate their own secretion from the cells that produce them. TNF-α and IL-1 stimulate IL-6 secretion [12]. However, IL-6 inhibits the production of TNF-α and IL-1, acting thus also as an anti-inflammatory cytokine [13].

### 2.1. Interleukin-1

At least three different glycoproteins constitute the IL-1 group: IL-1α and IL-1β, which are agonists and have about 25% sequence homology, and IL-1ra, which is an endogenous antagonist and lacks intrinsic biological activity [14].

The administration of IL-1β over a prolonged period of time not only elevated plasma ACTH concentrations for at least 1 week, but also increased adrenal weight [15]. IL-1 exerted both stimulatory and inhibitory actions on adrenocortical function. In human adrenal cells, perfused rat adrenal glands [16], and hypophysectomized rats [17], IL-1 stimulated glucocorticoid release. The inhibitory action of IL-1 has been shown in experiments, where IL-1 decreased angiotensin II-induced aldosterone synthesis [18]. Different factors have been suggested as mediators of the steroidogenic effects of IL-1, such as prostaglandins [16] or intra-adrenal CRH [19].

IL-1 mRNA was markedly elevated in both adrenal cortex and medulla after intravenous or intraperitoneal administration of lipopolysaccharide (LPS) [20]. The administration of IL-1β *in vivo*, however, either stimulated [19] or had no effect [21] on corticosterone secretion in hypophysectomized rats. In anesthetized rats with isolated adrenal glands, IL-1β stimulated the secretion of glucocorticoids [16]. However, high IL-1β doses have been required *in vivo* to show an effect on adrenal corticosterone secretion so that the physiological relevance of this effect remains unclear. IL-1β had no effect on basal or ACTH-stimulated production of glucocorticoids in human fetal adrenal tissue, cell, or organ culture [22]. But both IL-1α and IL-1β stimulated the secretion of glucocorticoids from rat adrenal slices [21], and from rat and human dispersed adrenal cells [17,23].

### 2.2. Interleukin-2

*In vitro* experiments on rat adrenocortical cells showed that IL-2 stimulates the secretion of corticosterone [24], along with the increase of prostaglandin $E_2$ levels and cyclic adenosine monophosphate. Furthermore, *in vivo* studies on cancer patients demonstrated that IL-2 treatment increased endothelin-1 levels within 2 h and this was followed by an increase in ACTH and cortisol within 3 h [24]. In addition to IL-2, also IL-3 and IL-6 stimulated the secretion of glucocorticoids from various adrenal cell preparations [25].

### 2.3. Interleukin-3

In addition to its important role in hematopoiesis [26], IL-3 has also been reported as a mediator in inflammatory states, particularly due to its close relationship to GM-CSF and IL-5 [27]. Additionally, IL-3 has been shown to stimulate steroidogenic enzymes in human lymphocytes and in rat granulosa cells [28]. IL-3 might also be involved in the regulation of immune-endocrine reactions, in addition to its hematopoietic and inflammatory role. Studies of Michl P et al. have further shown that the stimulatory effect of IL-3 and IL-6 on bovine adrenocortical cells might be mediated by different, cAMP-independent pathways, involving metabolites of the arachidonic system [29].

Furthermore, in cultured human adrenocortical cells, IL-3 stimulated steroidogenesis, whereas inhibition of the lipoxygenase pathway blocked the IL-3-induced increase of cortisol secretion [30].

### 2.4. Interleukin-6

IL-6 has been found to stimulate the secretory activity of the HPA axis at all three levels and the IL-6 receptor is expressed in the hypothalamus, pituitary, and adrenal gland in humans. Furthermore, IL-6 is secreted during stress; it is positively controlled by catecholamines, and negatively by glucocorticoids [31].

IL-6 activates the HPA axis by stimulating the release of cortisol and ACTH in humans. Long-term administration of IL-6 decreased plasma ACTH levels, while cortisol levels remained high [32] so that a direct action of IL-6 on the adrenal cortex can be assumed. Furthermore, IL-6 increased the secretion of cortisol and androgens by direct activation of IL-6 receptors on steroid-producing cells [33].

In addition, IL-6 stimulated corticosterone release from rat adrenocortical cells [25], alone or with ACTH. The IL-6 receptors on human adrenocortical cells have been found predominantly in the *zona reticularis* and in the inner *zona fasciculata* [33,34]. Elevated levels of IL-6 have been found in the steroid withdrawal syndrome [35], rheumatoid arthritis [36], and severe inflammatory, infectious, and traumatic states, potentially associated with inappropriate secretion of vasopressin [37] as well as in adrenal tumorigenesis [38]. IL-6 functions not only as an inflammatory cytokine but also as a hormonal factor that may be broadly connected to alterations in behavior and mood as seen in various endocrine disorders [39].

### 2.5. Interferons

IFN-α stimulated the synthesis of corticosterone in primary cultures of rat adrenal glands [40], whereas IFN-γ interfered with the differentiation and growth of the adrenal cortex.

In human fetal adrenals (HFA), IFN-γ inhibited the expression of insulin-like growth factor II [41,42]. This data suggests that TNF-α and IFN-γ may be involved in the regulation of HFA growth and differentiation via local IGF-II production.

### 2.6. Tumor necrosis factor-α

TNF-α decreased ACTH-dependent and basal cortisol production in fetal human adrenal cells. In fetal adrenal glands, TNF-α inhibited the expression of insulin-like growth factors [41,42].

TNF-α mRNA is present throughout the adrenal cortex of adult humans, particularly in steroid-producing cells. TNF-α is a quite potent inhibitory cytokine of corticotropin-induced cortisol secretion by cultured adrenocortical cells. In adult adrenocortical cells, TNF-α induced cortisol secretion, while in cultured cells and human fetal adrenal tissue, it inhibited ACTH-induced cortisol secretion [9].

### 2.7. Transforming growth factor β

TGF-β influences adrenocortical growth, differentiation, and hormone secretion. In bovine cells, TGF-β inhibits ACTH- and angiotensin II-induced steroidogenesis [43,44]. It was suggested that TGF-β also influences fetal and adult steroidogenesis by reducing the mRNA synthesis of dehydroepiandrosterone sulfotransferase [45].

## 3. SOURCE OF CYTOKINES INFLUENCING ADRENAL CELLS

In the blood stream the concentration of cytokines is too low for having a direct effect on adrenal function. Local immune cells and adrenal cells themselves are the two main sources of cytokines within the adrenal gland.

### 3.1. Cytokines produced by immune cells

Macrophages located primarily in the *zona reticularis* of the adrenal cortex possess the ability, when stimulated, to secrete cytokines, such as IL-1, IL-6, and TNF-α [4], which in turn may influence adrenocortical function. The action of cytokines produced by macrophages on adrenocortical cells can be inhibitory or stimulatory. Monocytes may stimulate cortisol production by human cells through a non-ACTH factor [17], whereas murine macrophages are able to produce factors which inhibit ACTH action on rabbit adrenocortical cells *in vitro* [46,47].

Lymphocytes produce ACTH-like substances [48] which can account for some influence on adrenal function, although their blood concentration is too low to have a significant effect. T lymphocytes regularly infiltrate adrenals of elderly patients [49]. Therefore, T lymphocyte-derived ACTH-related molecules may participate in the regulation of adrenal function.

### 3.2. Cytokines produced by adrenal cells

Adrenal cells are also capable of producing cytokines. IL-1 is produced by rat [50], mouse [50], and human adrenal cells [50]. IL-6 and TNF-α are produced by rat [8] and human adrenal cells [51,52]. IL-1 mRNA expression in human adrenals has been predominant in steroid-producing cells of the *zona reticularis* [53]. Pheochromocytomas were also found to produce IL-1 [7]. IL-6 and TNF-α were located predominantly in the *zona glomerulosa* of the rat adrenal [8]. In contrast, IL-6 mRNA in the human adrenal is mainly expressed in the *zona reticularis* and in steroid-producing cells, located in the adrenal medulla [52]. Production of TNF-α has been localized in the fetal but not in the adult human adrenal gland, using radioimmunoassay (RIA) [54], whereas TNF-α mRNA was localized in cells of the *zona reticularis* of adult adrenals by *in situ* hybridization [51].

## 4. SUMMARY

Cytokines are able to interact directly with the HPA axis. On the level of the adrenal, immune cells infiltrating the gland can influence adrenocortical function by secreting cytokines. On the other hand, adrenal cells themselves produce cytokines which act on the adrenal gland. In this context, cytokines and especially IL-1, IL-6, and TNF-α seem to be important regulators of the HPA axis. Especially in cases of autoimmune and inflammatory diseases, the importance of the immune-adrenal crosstalk becomes evident and a proper adrenal stress response might then be critical for the survival of the organism. Recently, Toll-like receptors discovered by us on adrenocortical cell lines and human adrenals are critically involved in the fine-tuned immune-endocrine crosstalk [55–57].

## REFERENCES


1. Besedovsky HO, del Rey A. Immune-neuro-endocrine interactions: facts and hypotheses. Endocr Rev 1996;17(1):64–102.
2. Chrousos GP. The hypothalamic-pituitary-adrenal axis and immune-mediated inflammation. N Engl J Med 1995;332(20):1351–62.
3. Marx C, Ehrhart-Bornstein M, Scherbaum WA, Bornstein SR. Regulation of adrenocortical function by cytokines-relevance for immune-endocrine interaction. Horm Metab Res 1998;30(6–7):416–20.
4. Gonzalez-Hernandez JA, Bornstein Ehrhart-Bornstein SRM, Geschwend JE, Adler G, Scherbaum WA. Macrophages within the human adrenal gland. Cell Tissue Res 1994;278(2):201–5.
5. Nathan CF. Secretory products of macrophages. J Clin Invest 1987;79(2):319–26.
6. Assoian RK, Fleurdelys BE, Stevenson HC, Miller PJ, Madtes DK, Raines EW et al. Expression and secretion of type beta transforming growth factor by activated human macrophages. Proc Natl Acad Sci USA 1987;84(17):6020–4.
7. Bornstein Ehrhart-Bornstein SRM, Gonzalez-Hernandez J, Schroder S,Scherbaum WA. Expression of interleukin-1 in human pheochromocytoma. J Endocrinol Invest 1996;19(10):693–8.
8. Judd AM, MacLeod RM. Differential release of tumor necrosis factor and IL-6 from adrenal zona glomerulosa cells in vitro. Am J Physiol 1995;268(1 Pt 1):E114–20.
9. Ehrhart-Bornstein M, Hinson JP, Bornstein Scherbaum SRWA, Vinson GP. Intraadrenal interactions in the regulation of adrenocortical steroidogenesis. Endocr Rev 1998;19(2):101–43.
10. Judd AM. Cytokine expression in the rat adrenal cortex. Horm Metab Res 1998;30(6–7):404–10. Review.

11. Bornstein SR, Rutkowski H, Vrezas I. Cytokines and steroidogenesis. Mol Cell Endocrinol 2004 Feb 27;215(1–2):135–41.
12. Dinarello CA. The proinflammatory cytokines interleukin-1 and tumor necrosis factor and treatment of the septic shock syndrome. J Infect Dis 1991;163(6):1177–84.
13. Burger D, Dayer JM. Inhibitory cytokines and cytokine inhibitors. Neurology 1995;45(6 Suppl 6): S39–43.
14. Hannum CH, Wilcox CJ, Arend WP, Joslin FG, Dripps DJ, Heimdal PL et al. Interleukin-1 receptor antagonist activity of a human interleukin-1 inhibitor. Nature 1990;343(6256):336–40.
15. Naito Y, Fukata J, Nakaishi S, Nakai Y, Hirai Y, Tamai S et al. Chronic effects of interleukin-1 on hypothalamus, pituitary and adrenal glands in rat. Neuroendocrinology 1990;51(6):637–41.
16. Roh MS, Drazenovich KA, Barbose JJ, Dinarello CA, Cobb CF. Direct stimulation of the adrenal cortex by interleukin. Surgery 1987;1(2):140–6.
17. Whitcomb RW, Linehan WM, Wahl LM, Knazek RA. Monocytes stimulate cortisol production by cultured human adrenocortical cells. J Clin Endocrinol Metab 1988;66(1):33–8.
18. Natarajan R, Ploszaj S, Horton R, Nadler J. Tumor necrosis factor and interleukin-1 are potent inhibitors of angiotensin-II-induced aldosterone synthesis. Endocrinology 1989;125(6):3084–9.
19. Andreis PG, Neri G, Belloni AS, Mazzocchi G, Kasprzak A, Nussdorfer GG. Interleukin-1 beta enhances corticosterone secretion by acting directly on the rat adrenal gland. Endocrinology 1991;129(1):53–7.
20. Nobel CS, Schultzberg M. Induction of interleukin-1 beta mRNA and enkephalin mRNA in the rat adrenal gland by lipopolysaccharides studied by in situ hybridization histochemistry. Neuroimmunomodulation 1995;2(2):61–73.
21. Gwosdow AR, Kumar MS, Bode HH. Interleukin 1 stimulation of the hypothalamic-pituitary-adrenal axis. Am J Physiol 1990;258(1 Pt 1):E65–70.
22. Harlin CA, Parker CR. Investigation of the effect of interleukin-1 beta on steroidogenesis in the human fetal adrenal gland. Steroids 1991;56(2):72–6.
23. O'Connell NA, Kumar A, Chatzipanteli K, Mohan A, Agarwal RK, Head C et al. Interleukin-1 regulates corticosterone secretion from the rat adrenal gland through a catecholamine-dependent and prostaglandin E2-independent mechanism. Endocrinology 1994;135(1):460–7.
24. Raab C, Weidmann E, Schmidt A, Bergmann L, Badenhoop K, Usadel KH, Haak T. The effects of interleukin-2 treatment on endothelin and the activation of the hypothalamic-pituitary-adrenal axis. Clin Endocrinol (Oxf) 1999 Jan;50(1):37–44.
25. Tominaga T, Fukata J, Naito Y, Usui T, Murakami N, Fukushima M et al. Prostaglandin-dependent in vitro stimulation of adrenocortical steroidogenesis by interleukins. Endocrinology 1991;128(1):526–31.
26. Oster W, Frisch J, Nicolay U, Schulz G. Interleukin-3. Biologic effects and clinical impact. Cancer 1991 May 15;3(10 Suppl):2712–7.
27. Hart PH, Whitty GA, Burgess DR, Hamilton JA. Regulation by interleukin-3 of human monocyte pro-inflammatory mediators. Similarities with granulocyte-macrophage colony-stimulating factor. Immunology 1990 Sep;71(1):76–82.
28. Kasson BG, Gorospe WC. Effects of interleukins 1, 2 and 3 on follicle-stimulating hormone-induced differentiation of rat granulosa cells. Mol Cell Endocrinol 1989 Mar;1(1):103–11.
29. Michl P, Beikler T, Engelhardt D, Weber MM. Interleukin-3 and interleukin-6 stimulate bovine adrenal cortisol secretion through different pathways. J Neuroendocrinol 2000 Jan;12(1):23–8.
30. Weber MM, Michl P, Auernhammer CJ, Engelhardt D. Interleukin-3 and interleukin-6 stimulate cortisol secretion from adult human adrenocortical cells. Endocrinology 1997;138(5):2207–10.
31. Papanicolaou DA, Wilder RL, Manolagas SC, Chrousos GP. The pathophysiologic roles of interleukin-6 in human disease. Ann Intern Med 1998 Jan 15;128(2):127–37. Review.
32. Spath-Schwalbe E, Born J, Schrezenmeier H, Bornstein SR, Stromeyer P, Drechsler S et al. Interleukin-6 stimulates the hypothalamus-pituitary-adrenocortical axis in man. J Clin Endocrinol Metab 1994;79(4):1212–14.

33. Path G, Bornstein SR, Ehrhart-Bornstein M, Scherbaum WA. Interleukin-6 and the interleukin-6 receptor in the human adrenal gland: expression and effects on steroidogenesis. J Clin Endocrinol Metab 1997;82(7):2343–9.
34. Path G, Bornstein SR, Spath-Schwalbe E, Scherbaum WA. Direct effects of interleukin-6 on human adrenal cells. Endocr Res 1996;22(4):867–73.
35. Papanicolaou DA, Tsigos C, Oldfield EH, Chrousos GP. Acute glucocorticoid deficiency is associated with plasma elevations of interleukin-6: does the latter participate in the symptomatology of the steroid withdrawal syndrome and adrenal insufficiency? J Clin Endocrinol Metab 1996 Jun;6(6):2303–6.
36. Crofford LJ, Sano H, Karalis K, Webster EA, Friedman TC, Chrousos GP, Wilder RL. Local expression of corticotropin-releasing hormone in inflammatory arthritis. Ann N Y Acad Sci 1995 Dec 29;771:459–71.
37. Mastorakos G, Weber JS, Magiakou MA, Gunn H, Chrousos GP. Hypothalamic-pituitary-adrenal axis activation and stimulation of systemic vasopressin secretion by recombinant interleukin-6 in humans: potential implications for the syndrome of inappropriate vasopressin secretion. J Clin Endocrinol Metab 1994 Oct;79(4):934–9.
38. Willenberg HS, Path G, Vogeli TA, Scherbaum WA, Bornstein SR. Role of interleukin-6 in stress response in normal and tumorous adrenal cells and during chronic inflammation. Ann N Y Acad Sci 2002 Jun;966:304–14.
39. Lamounier-Zepter V, Bornstein SR. Interleukin-6 and mood disorders in endocrine disease. Horm Metab Res 2005 Oct;37(10):656.
40. Gisslinger H, Svoboda T, Clodi M, Gilly B, Ludwig H, Havelec L et al. Interferon-alpha stimulates the hypothalamic-pituitary-adrenal axis in vivo and in vitro. Neuroendocrinology 1993;57(3):489–95.
41. Ilvesmaki V, Jaattela M, Saksela E, Voutilainen R. Tumor necrosis factor-alpha and interferon-gamma inhibit insulin-like growth factor II gene expression in human fetal adrenal cell cultures. Mol Cell Endocrinol 1993;91(1–2):59–65.
42. Voutilainen R, Ilvesmaki V, Ariel I, Rachmilewitz J, de Groot N, Hochberg A. Parallel regulation of parentally imprinted H19 and insulin-like growth factor-II genes in cultured human fetal adrenal cells. Endocrinology 1994;134(5):2051–6.
43. Rainey WE, Viard I, Mason JI, Cochet C, Chambaz EM, Saez JM. Effects of transforming growth factor beta on ovine adrenocortical cells. Mol Cell Endocrinol 1988;60(2–3):189–98.
44. Hotta M, Baird A. Differential effects of transforming growth factor type beta on the growth and function of adrenocortical cells in vitro. Proc Natl Acad Sci USA 1986;83(20):7795–9.
45. Parker CR, Jr, Stankovic AK Faye-Petersen O, Falany CN, Li H, Jian M. Effects of ACTH and cytokines on dehydroepiandrosterone sulfotransferase messenger RNA in human adrenal cells. Endocr Res 1998;24(3–4):669–73.
46. Mathison JC, La Forest AC, Ulevitch RJ. Properties and requirements for production of a macrophage product which suppresses steroid production by adrenocortical cells. Infect Immun 1984;45(2):360–6.
47. Mathison JC, Schreiber RD, La Forest AC, Ulevitch RJ. Suppression of ACTH-induced steroidogenesis by supernatants from LPS-treated peritoneal exudate macrophages. J Immunol 1983;130(6):2757–62.
48. Woloski BM, Smith EM, Meyer WJ, 3rd, Fuller GM, Blalock JE. Corticotropin-releasing activity of monokines. Science 1985;230(4729):1035–7.
49. Hayashi Y, Hiyoshi T, Takemura T, Kurashima C, Hirokawa K. Focal lymphocytic infiltration in the adrenal cortex of the elderly: immunohistological analysis of infiltrating lymphocytes. Clin Exp Immunol 1989;77(1):101–5.
50. Alheim K, Andersson C, Tingsborg S, Ziolkowska M, Schultzberg M, Bartfai T. Interleukin 1 expression is inducible by nerve growth factor. Proc Natl Acad Sci USA 1991;88(20):9302–6.
51. Gonzalez-Hernandez JA, Ehrhart-Bornstein M, Spath-Schwalbe E, Scherbaum WA, Bornstein SR. Human adrenal cells express tumor necrosis factor-alpha messenger ribonucleic acid: evidence for paracrine control of adrenal function. J Clin Endocrinol Metab 1996;81(2):807–13.

52. Gonzalez-Hernandez JA, Bornstein SR, Ehrhart-Bornstein M, Spath-Schwalbe E, Jirikowski G, Scherbaum WA. Interleukin-6 messenger ribonucleic acid expression in human adrenal gland in vivo: new clue to a paracrine or autocrine regulation of adrenal function. J Clin Endocrinol Metab 1994;79(5):1492–7.
53. Gonzalez-Hernandez JA, Bornstein SR, Ehrhart-Bornstein M, Gschwend JE, Gwosdow A, Jirikowski G et al. IL-1 is expressed in human adrenal gland in vivo. Possible role in a local immune-adrenal axis. Clin Exp Immunol 1995;99(1):137–41.
54. Jaattela M, Carpen O, Stenman UH, Saksela E. Regulation of ACTH-induced steroidogenesis in human fetal adrenals by rTNF-alpha. Mol Cell Endocrinol 1990;68(2–3):R31–6.
55. Bornstein SR, Ziegler CG, Krug AW, Kanczkowski W, Rettori V, McCann SM, Wirth M, Zacharowski K. The role of Toll-like receptors in the immune-adrenal crosstalk. Ann N Y Acad Sci 2006 Nov;1088:307–18. Review.
56. Tran N, Koch A, Berkels R, Boehm O, Zacharowski PA, Baumgarten G, Knuefermann P, Schott M, Kanczkowski W, Bornstein SR, Lightman SL, Zacharowski K. Toll-like receptor 9 expression in murine and human adrenal glands and possible implications during inflammation. J Clin Endocrinol Metab 2007 Jul;92(7):2773–83. Epub 2007 May 1.
57. Zacharowski K, Zacharowski PA, Koch A, Baban A, Tran N, Berkels R, Papewalis C, Schulze-Osthoff K, Knuefermann P, Zahringer U, Schumann RR, Rettori V, McCann SM, Bornstein SR. Toll-like receptor 4 plays a crucial role in the immune-adrenal response to systemic inflammatory response syndrome Proc Natl Acad Sci USA 2006 Apr 18;103(16):6392–7. Epub 2006 Apr 10.

*The Hypothalamus–Pituitary–Adrenal Axis*
Edited by A. del Rey, G. P. Chrousos and H. O. Besedovsky


# *Inflammatory Mediators Affect the Autonomic Nervous System*

RAINER H. STRAUB, MARTIN KEES, DAVID JANELE, GEORG PONGRATZ, JÜRGEN SCHÖLMERICH, and PETER HÄRLE

*Department of Internal Medicine I, University Hospital Regensburg, Regensburg, Germany*

## ABSTRACT

This chapter discusses effects of inflammatory mediators such as interleukins, tumor necrosis factor, interferons, and others on the sympathetic or parasympathetic nervous system. This includes inflammatory mediator-induced alterations at central autonomic control centers, of sympathetic and parasympathetic ganglia, of nerves and autonomic receptors, of neurotransmitter secretion from nerve terminals, of neurotransmitter receptor signaling, of peripheral neurotransmitter effects without exact localization of the site of action, of chromaffin cells, and of enzymes responsible for adrenergic or cholinergic neurotransmitter production. Although this chapter cannot be complete, it clearly demonstrates a close interaction between these inflammatory mediators and altered function of the autonomic nervous system. Particularly, the sympathetic nervous system is affected in many different ways which, in general, lead to an increase of the sympathetic nervous tone but to a loss of sympathetic innervation in inflamed tissue. At present, it is not known how inflammatory mediators exactly influence the autonomic nervous system in chronic inflammatory diseases. It is likely that this influence may lead to autonomic neuropathy and also to autonomic nervous hyper-reflexia with many deleterious consequences such as sympathetic overdrive, hypertension, arteriosclerosis, metabolic disturbances, and others. Furthermore, long-loop anti-inflammatory feedback effects of the autonomic nervous system may be altered under such circumstances, the consequences of which remain to be elucidated.

## 1. INTRODUCTION

Research during the last three decades clearly demonstrated that cytokines and other inflammatory mediators influence the autonomic nervous system in many different ways. The influences of inflammatory mediators on the autonomic nervous system are long-distance or short-distance influences [1–6]. Depending on the site of action, inflammatory mediators may locally influence central autonomic control centers, sympathetic and parasympathetic ganglia, nerves and autonomic receptors, neurotransmitter secretion from nerve terminals, neurotransmitter receptor signaling, chromaffin cells, and enzymes which produce adrenergic or cholinergic neurotransmitters. However, the same inflammatory mediators may act over long distances when the respective concentration in the blood and their half-life are high, and structures of the autonomic nervous system are accessible. This chapter focuses on inflammatory mediator-induced effects on parts of the autonomic nervous system (sympathetic, parasympathetic) which are summarized in Fig. 1. However, it does not include modulation of other systemic feedback axes such as the

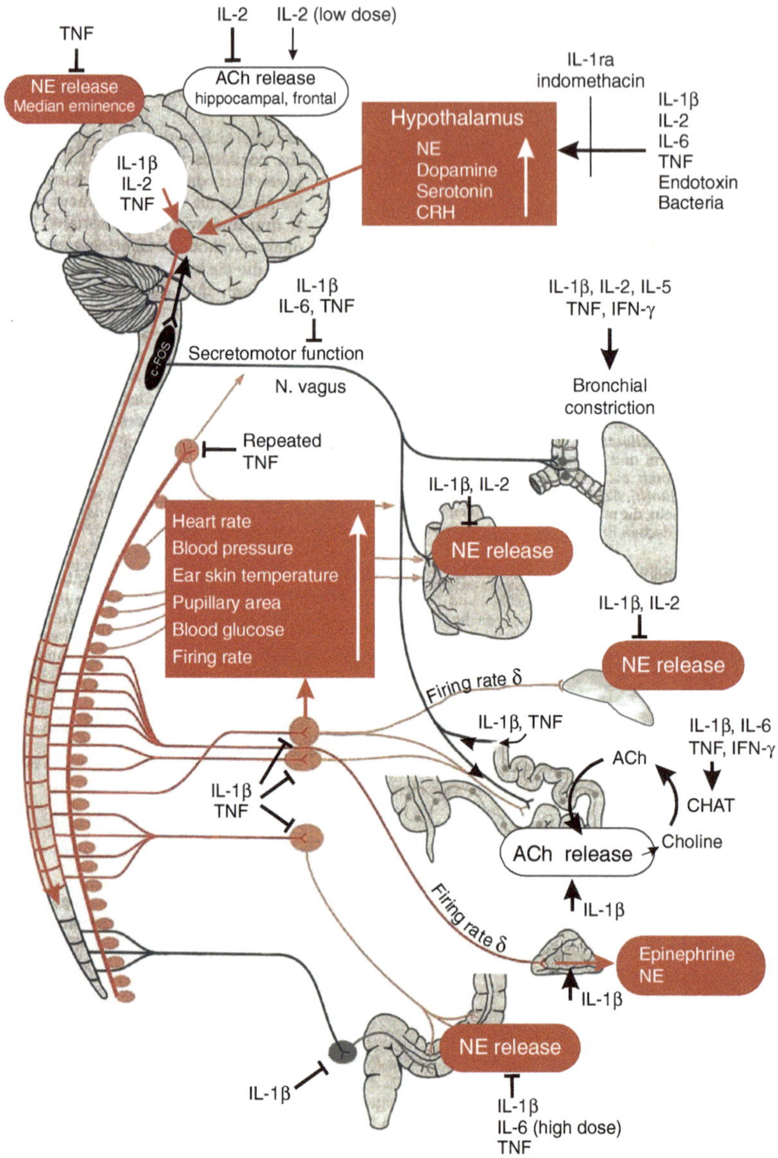


Figure 1. Effects of inflammatory mediators on sympathetic and parasympathetic nervous system. The effects appear on many different levels and the direction, whether stimulatory or inhibitory, depends on the site of action. The figure summarizes the available literature in this field. A line with an arrow (a bar) at the end indicates a stimulatory (an inhibitory) effect. Sympathetic pathways are depicted in red (light red for postganglionic) and parasympathetic pathways in black (gray for postganglionic). An increased firing rate indicated by "firing rate ↑" is induced by protein antigens or IL-1β intraperitoneal (i.p.), intravenous (i.v.), or intracerebroventricular (i.c.v.). ACh, acetylcholine; CHAT, choline acetyltransferase; CRH, corticotropin-releasing hormone; IFN, interferon; IL, interleukin; IL-1ra, IL-1 receptor antagonist; N, nervous; NE, norepinephrine; TNF, tumor necrosis factor.

hypothalamic–pituitary–adrenal (HPA) axis, the hypothalamic–pituitary–gonadal (HPG) axis, or other hormonal pathways. Furthermore, it does not demonstrate effects of neurotransmitters on immune function. These topics are covered elsewhere in this volume.

## 2. ACTIVATION OF CENTRAL AUTONOMIC NERVOUS CONTROL CENTERS

### 2.1. Central sympathetic nervous control centers

This chapter discusses direct effect of inflammatory mediators on central sympathetic nervous control centers but does not include the effects of these mediators on hormone secretion from the hypothalamus or pituitary gland, although we know that there are intimate interactions [7]. In the 1970s, different antigens were used in order to activate the central nervous control centers which finally led to adrenal corticosterone secretion [8]. In further experiments, Besedovsky et al. found that interleukin (IL)-1 is a major constituent of the stimulatory cocktail [9]. This prompted many researchers to focus on IL-1 in the following years.

Intraperitoneal (i.p.) injection of purified recombinant IL-1 into mice increased the cerebral concentration of the norepinephrine (NE) catabolite, 3-methoxy,4-hydroxyphenylethyleneglycol (MHPG), reflecting increased activity of noradrenergic neurons. This effect was dose-dependent and was largest in the hypothalamus, especially the medial division [10]. Tryptophan concentrations were also increased throughout the brain [10]. Administration of IL-1RA largely prevented the effects of IL-1α or IL-1β on increase in hypothalamic NE metabolism. IL-1RA did not prevent the increases in brain tryptophan that occurred after treatment with IL-1 [11]. It was stated that there are several independent cytokine ways to increase central NE and tryptophan increase [12]. Another study demonstrated that intracerebroventricular (i.c.v.) IL-2 is also able to induce serotonin and 5-hydroxyindoleacetic acid. The i.c.v. pretreatment with the IL-1RA showed that the effects of IL-2 on hippocampal serotonin were completely dependent on endogenous brain IL-1 [13]. Furthermore, IL-1β stimulates the release of NE and dopamine from rat hypothalamus [14,15]. One site of NE release is the paraventricular nucleus of the hypothalamus [16,17] which is an important integrating center also for the sympathetic nervous system. One may speculate whether different cytokine patterns and central nervous system (CNS) access of cytokines may stimulate distinct neurotransmitter response patterns in separate areas of the brain. This may be relevant in inflammatory disease where different autonomic and hormonal response patterns may be elicited by the CNS.

Other groups linked the IL-1-induced increase of central NE turnover to eicosanoid metabolism: The increase in NE and serotonin turnover observed 4 h following systemic administration of IL-1 was antagonized by concurrent administration of indomethacin, a potent inhibitor of cyclooxygenase [18]. Furthermore, nitric oxide (NO) was found to have neurotransmitter-like functions in hypothalamic areas which are involved in NE-modulated neurohormone secretion [19,20].

The question arises whether stimulation of central NE release also leads to peripheral increase of plasma NE. The intravenous (i.v.) or i.c.v. injection of either IL-1α or IL-1β caused a dose-related increase in plasma epinephrine and NE levels in rats [21]. This effect is due to IL-1-induced increase of hypothalamic NE release [22] and is not influenced by a cholinergic hypothalamic mechanism [23]. Interestingly, the effect is only short-lasting because continuous administration of IL-1 for 1 week showed only NE increase at day 1 but not thereafter [24]. Adaptation processes may be responsible for these changes. With respect to the HPA axis, very similar adaptation changes have been found for IL-6-induced hormone secretion [25,26]. Others found very similar NE-stimulating effects after i.p. administration of IL-1: A dose that maximally activated pituitary–adrenal activity slightly elevated venous plasma epinephrine and NE concentrations (two-fold increase) at 30 min and at 1 h after injection [27]. Even psychological immobilization stress enhances local biologically active IL-1 in the hypothalamus, and locally produced hypothalamic IL-1 plays a role in elevated monoamine release in the hypothalamus and activation of the HPA axis [28].

However, what are the mechanisms which mediate IL-1-induced increase of plasma NE? Is it the outflow from the brain or is the peripheral sympathetic nervous system activated? At the present point, it seems relatively clear that the second explanation is more likely: For example, an i.c.v. infusion of IL-1β elicited a dose-dependent increase in the electrical activity of the splenic sympathetic nerve in rats [29]. An antagonist of corticotropin-releasing factor completely abolished this effect [30]. Other studies confirmed these effects of IL-1β [31,32]. Even i.p. injection of IL-1β facilitates NE release in the spleen through activation of the sympathetic nerve, and the increased sympathetic activity is, at least in part, due to the excitation of neurons containing corticotropin-releasing hormone in the brain [33]. As a consequence, sympathetic activation leads to suppression of splenic natural killer cell activity and other splenic immune functions [34–36]. These data confirmed earlier studies of the late 1970s which demonstrated that antigen injection leads to sympathetic immunosuppression in the spleen [37].

During the 1980s, an interesting question appeared: How does i.p. injection of IL-1β activate the central autonomic nervous control centers so that an activation of the sympathetic nervous system results in immunosuppression? Many researchers believed that this would be a direct effect of blood-borne IL-1β within brain areas where the blood–brain barrier is relatively leaky such as in circumventricular organs. However, it was demonstrated that IL-1β is able to directly modulate nerve terminals in the peritoneal cavity because IL-1β-induced hyperthermia is disrupted by subdiaphragmatic vagal transection [38]. Furthermore, the central increase of NE after i.p. IL-1β is abrogated by vagotomy [39]. Similar results were found for another cytokine, tumor necrosis factor (TNF): i.p. TNF injection produces dose-dependent hyperalgesia as measured by the tailflick test, which is mediated via the induced release of IL-1β, and activation of subdiaphragmatic vagal afferents [40]. The effect of i.p. IL-1β on vagal afferents has been confirmed by others [41]. The i.p. IL-1β induces activation of many different brain areas as demonstrated by *in situ* hybridization histochemistry of immediate-early gene c-fos mRNA. This leads to coactivation of the area postrema and nucleus of the solitary tract which reflects entry into the brain and neural transduction of the peripheral signal [42]. Furthermore, IL-1β requires brainstem afferences conveyed to the hypothalamus by the ventral noradrenergic ascending bundle for the release of ACTH after intra-arterial injection of this cytokine [43].

When central sympathetic nervous control centers are activated, we observe an increase of peripheral NE. Does this lead to metabolic changes or changes of typical functions of the sympathetic nervous system in the periphery? Indeed, it was demonstrated that whole body glucose metabolism is increased after central administration of IL-1α. This effect is mediated by increased sympathoadrenal activity and an increase in pancreatic insulin and glucagon secretion, which depends on activation of central adrenoceptors [44]. The i.p. and i.c.v. IL-1β also increases heart rate and blood pressure which are markers of sympathetic activation [45–47]. This was also found in human subjects after IL-1β administration [48]. Furthermore, IL-1β, TNF, and lipopolysaccharide (LPS) induced a decrease in ear skin temperature, which is indicative of cutaneous sympathetic activation, and simultaneous inhibition of renal sympathetic nerve activity during the first phase of rising rectal temperature [49]. In addition, i.v. TNF increases pupillary area in mice – a typical marker of sympathetic activation [50].

In conclusion, the early studies of the 1970s demonstrating sympathetic activation after antigen challenge have been fully confirmed [37]. Now, many interesting activation pathways have been described which lead to long-distance stimulation of the central sympathetic nervous system (Fig. 1). From the standpoint of a clinical immunologist, sympathetic activation is an important mechanism to inhibit overshooting immune responses in the periphery. However, as a side effect, this also leads to sympathetic overdrive with an increase of systolic blood pressure and heart rate as well as metabolic changes such as increased whole body glucose metabolism. A sympathetic overdrive may be a severe problem in chronic inflammatory diseases such as systemic lupus erythematosus and inflammatory bowel disease which is discussed in Section 9.

### 2.2. Central parasympathetic nervous control centers

Effects of recombinant human IL-1β on the neuronal activities in the rat dorsal motor nucleus of the vagus were investigated by extra- and intracellular recordings in slice preparations. IL-1β mainly inhibits the vagal motoneurons in the dorsal motor nucleus, at least partly through prostaglandin synthesis [51]. This provides a mechanism that could account for the central action of IL-1β on visceral processes such as the inhibition of gastric acid secretion (Fig. 1). This was corroborated by *in vivo* studies in which i.c.v. administration of IL-1β dose dependently inhibited the gastric acid secretion induced by electrical stimulation of the vagus nerve [52]. In addition, TNF affects neurons in the nucleus of the solitary tract that are involved in vago–vagal reflex control of gastric motility [53,54]. The i.c.v. and i.p. injection of IL-1β induces high levels of c-fos mRNA in the nucleus tractus solitarius, which were 3–4 times higher in animals treated intraperitoneally compared to those treated intracerebroventricularly. The differential magnitude of the c-fos mRNA response in the nucleus tractus solitarius is consistent with vagal activation [55,56]. A similar increase of c-fos mRNA was observed after peripheral i.v. IL-6 infusion [57]. Furthermore, i.p. injection of IL-1β activated the vagal glutamatergic system in the nucleus tractus solitarius [58]. This suggests that IL-1β, IL-6, and TNF induce changes of gastrointestinal motility and secretion by modulating intrinsic vago–vagal reflex pathways during illness (Fig. 1). Furthermore, endotoxin-induced sickness behavior is linked to vagal motoneurons in the brain stem which connect behavioral and gastrointestinal pathways [59].

Since it has recently been described that vagal efferent nerve activity is immunosuppressive in endotoxin-induced shock by lowering serum TNF [60,61], the activation of the central parasympathetic nervous control centers may serve immunosuppressive effects similar to the sympathetic nervous system. Under consideration of these new data, it seems likely that general activation of many central autonomic nervous control centers is necessary to overcome overshooting of peripheral immune responses.

## 3. MODULATION OF SYMPATHETIC AND PARASYMPATHETIC GANGLIA, NERVES, AND AUTONOMIC RECEPTORS

In a more clinical situation, a soluble protein antigen, which activates antigen-specific Th2 cells and B cells *in vivo*, increases the rate of NE release and turnover in the spleen and bone marrow 18–25 h after immunization [62]. The i.v. injection of IL-1β resulted in a dose-dependent increase in the activity of the adrenal and splenic nerves, which lasted for more than 2–6 h. The activity of renal nerves showed a transient increase which was followed by a long-lasting suppression after injection of IL-1β [29]. An i.v. injection of the cyclooxygenase inhibitors ibuprofen or sodium salicylate suppressed almost completely the IL-1β-induced activity in adrenal and splenic nerves [29]. The data suggest the regional differentiation of activity in the visceral sympathetic nerves in response to IL-1β. Similar studies corroborated the IL-1β-induced increase of the nerve firing rate at the splenic nerve [63]. In an inflammatory disease state, such a differential activation of the sympathetic nervous system at different locations would be a prerequisite for a very specific influence on immune function which may depend on the above-mentioned CNS-stimulating pattern of inflammatory mediators.

IL-1β has excitatory and inhibitory actions on guinea-pig pelvic ganglion neurons. Inhibition of evoked nicotinic cholinergic fast excitatory postsynaptic potentials by IL-1β may be due to presynaptic inhibition of acetylcholine release [64]. In another study, the effect of TNF on calcium currents of cultured neurons from neonatal rat superior cervical ganglia was studied using whole-cell patch-clamp technique. The authors found that TNF-treated neurons exhibited increased calcium current density without significant alteration in the steady-state parameters of activation and availability. Thus, TNF can alter cellular functions of sympathetic neurons via modulating ionic conductance [65]. Furthermore, calcium-dependent release of catecholamines in cultured neurons from neonatal rat superior cervical ganglia is diminished after repeated stimuli with TNF which may be an interesting mechanism of adaptational processes [66]. In addition, cultured rat sympathetic neurons respond weakly to exogenous IL-6, but addition of soluble IL-6 receptor and IL-6 enhances neuronal survival in the absence of nerve growth factor (NGF). Neutralizing monoclonal antibodies against IL-6 blocks these effects [67].

In inflammatory conditions such as chronic rheumatoid arthritis, sympathetic nerve fibers were significantly decreased in synovial membrane as compared to sensory nerve fibers and as compared to sympathetic nerve fibers in patients with osteoarthritis [68]. A similar loss of sympathetic nerve fibers was observed in the spleen during adjuvant arthritis in rats [69,70]. This effect may be due to a local increase of NGF, which can induce apoptosis via the p75 neurotrophin receptor in the presence of low levels of TrkA signal activity [71–73]. Another mechanism would be an increase of local nerve repellent factors such as fibronectin and semaphorins (unpublished observation). Thus, an increase of the sympathetic tone is accompanied by a loss of sympathetic innervation in inflamed tissue.

The literature survey did not yield a high number of studies which dealt with the influence of inflammatory mediators on baroreceptor or osmoreceptor reflexes. One study investigated the effect of LPS administration on baroreceptor reflexes. LPS, which elicits the production of several cytokines, induces cardiovascular changes characterized by increased perfusion of immune organs and compensatory sympathetic vasoconstriction in other tissues. Subpyrogenic doses of LPS increased the sensitivity of the baroreceptor reflex 2 and 3 h after the i.v. administration of the endotoxin [74].

## 4. MODULATION OF NEUROTRANSMITTER SECRETION FROM AND UPTAKE BY NERVE TERMINALS

Apart from the mentioned long-distance effects of cytokines on central autonomic nervous centers, also very short-distance effects in the close proximity of nerve terminals were described. In all studies, proinflammatory cytokines, such as TNF, IL-1β, and IL-2, inhibited evoked release of NE (Fig. 1). The first experiments, which demonstrated an effect of a cytokine on NE release in the CNS, were published in the early 1990s by Elenkov et al. [75]. In these studies, TNF inhibited stimulation-evoked release of NE from noradrenergic axon terminals in the isolated rat median eminence. In contrast, TNF had no effect on the release of NE from the spleen [75]. Another study with superfusion and electrical field stimulation of rat hippocampal brain slices showed TNF-induced inhibition of NE release [76]. Furthermore, calcium-dependent release of catecholamines in cultured neurons from neonatal rat superior cervical ganglia is diminished after repeated stimuli with TNF [66]. The human neuroblastoma-derived cell-line SH-SY5Y is a model for mature postganglionic sympathetic neurons. Preincubation for 15–25 min with 60 pM IL-1β reduced potassium-evoked NE release [77].

Experiments with electrical field stimulation in longitudinal muscle–myenteric plexus or myenteric nerve varicosity preparations from jejunum of non-infected rats revealed that IL-1β suppresses NE release [78]. Preincubation of the tissue with TNF also caused a suppression of NE release stimulated by potassium chloride or electrical field stimulation. The action of TNF was time- and concentration-dependent [79]. Interestingly, 1 ng/mL of IL-6 augmented NE release, 100 ng/mL suppressed NE release, whereas 10 ng IL-6/mL had no effect. However, IL-6 plus a subthreshold concentration of IL-1β suppressed NE release from longitudinal muscle myenteric plexus [80].

Human recombinant IL-1β and IL-2 reduced stimulation-evoked NE release from isolated perfused rat spleen [81]. Human recombinant IL-6 caused no significant change in these studies. The inhibitory effect of low concentrations of IL-1β and IL-2 supports the idea that locally produced inflammatory mediators affect neuronal function in a peripheral lymphoid organ [81]. The exocytotic NE release is accompanied by a concomitant secretion of a nitric oxide-like compound which, in turn, reinforces NE release [82].

In isolated superfused mouse atria, IL-1β and TNF inhibited the stimulation-induced release of NE [83]. The effect of IL-1β was blocked by IL-1RA. IL-2 and IL-6 were ineffective in modulating release [83]. This is in contrast to another study which demonstrated that IL-2 may indirectly stimulate β-adrenergic-mediated function by triggering the presynaptic release of NE. IL-2 induces a positive inotropic response that can be blocked by the β-adrenergic antagonist propranolol but not by phentolamine [84].

Other studies focused on release of acetylcholine from nerve terminals (Fig. 1): Exogenous IL-2 significantly reduced the potassium chloride-evoked release of acetylcholine from slices of rat hippocampus. This IL-2 effect was also region-specific, such that acetylcholine release from other tissue slices (striatal, frontal cortical) was not affected [85]. Others found potentiating (at $10^{-13}$ M) and inhibitory ($10^{-9}$ M) effects of IL-2 on hippocampal acetylcholine release, which were blocked by a neutralizing IL-2-receptor antibody [86]. In addition, recombinant human IL-2 inhibited the evoked acetylcholine release from hippocampal and frontal cortical slices, but was ineffective in the parietal cortex and striatum. At very low concentrations (0.1 pM), IL-2 transiently increased hippocampal-evoked acetylcholine release, resulting in a biphasic dose–response profile [87].

In isolated guinea pig ileum, pretreatment with IL-1β for 15–60 min potentiated contractions of the ileum induced by electrical transmural stimulation [88]. Furthermore, IL-1β potentiated the electrically evoked release of acetylcholine from entire preparations of ileum, but not from longitudinal-myenteric plexus preparations and mucosa-free preparations, which is mediated by the arachidonic acid cascade [88]. In contrast, another study demonstrated that both IL-1β and IL-6 reversibly caused a presynaptic inhibition of acetylcholine release from cholinergic nerve terminals [89].

With respect to neurotransmitter uptake, NGF, which is often elevated during local immune processes, stimulates NE transport in chromaffin cells maintained in culture up to 6 days. NGF elicited approximately 60% increase in NE transporter mRNA levels in these cells, whereas glucocorticoids inhibit NE transporter function of chromaffin cells at least in part through a decrease in NE transporter mRNA [90]. Furthermore, nitric oxide (NO) and its congeners (NO(x)), including nitrosothiols, affect NE uptake by sympathetic neurons (PC-12 cells or cultured rat superior cervical ganglia) [91]. In addition, transforming growth factor (TGF)-β1 causes an increase of NE transporter mRNA levels in neural crest cells in embryonic development [92].

In conclusion, these studies clearly indicate that local cytokines can directly modulate neurotransmitter release by presynaptic heteroreceptors. The release of the sympathetic NE is blocked by typical proinflammatory cytokines. This indicates that, by a short-distance reflex loop, proinflammatory cytokines inhibit the release of an immunosuppressive neurotransmitter, an effect which may lead to local increase of these inflammatory mediators. This would lead to a more proinflammatory situation. However, studies in inflamed tissue, for example synovial membrane of patients with rheumatoid arthritis, were not carried out so that this important effect of local proinflammatory cytokines remains elusive. In the brain, modulation of NE and acetylcholine release by these inflammatory mediators may alter many important neuronal functions and behavior.

## 5. MODULATION OF NEUROTRANSMITTER RECEPTOR SIGNALING

Another location where inflammatory mediators can modulate effects of neurotransmitters is the neurotransmitter receptor itself and downstream signaling cascades (Fig. 2). It has been demonstrated that IL-1β potently inhibits the response of rat thoracic aorta to vasoconstrictor agents such as phenylephrine (α1-adrenergic agonist) [93]. Incubation of rat aortas with human monocyte-derived IL-1 or recombinant human TNF resulted in diminished aortic contraction and sensitivity to NE [94]. In another study it was demonstrated that IL-2 induced a concentration-dependent decrease of the β-adrenoceptor agonist-stimulated cAMP production in human peripheral blood mononuclear cells after a 20-h preincubation period, which was not demonstrated for IL-3 and IL-4 [95]. In a further study the same group demonstrated similar results for IL-1β, interferon (IFN)-γ, and granulocyte-macrophage colony-stimulating factor (GM-CSF) [96]. The downregulating effect of IL-2 on β-adrenoceptor signaling was confirmed by studies in T lymphocytes [97].

In cardiac muscle cells, IL-1β and TNF uncouple β-agonist-occupied receptors from adenylate cyclase which demonstrates that β-adrenoceptor or G-protein function is altered by these cytokines [98]. Others have shown that TNF can potently modulate G-protein-mediated signal transduction in rat cardiac myocytes with an increased level of membrane $G\alpha_i$ proteins, but it caused an increase in adenylate cyclase responsiveness [99]. This is in contrast to a study which demonstrated that TNF inhibits α- and β-adrenoceptor-stimulated increase in contractility and beating irregularity and impairs the impact of high extracellular calcium on contractile performance in spontaneously beating neonatal rat cardiomyocytes [100]. A similar effect on contractility of papillary muscle was demonstrated for IFN-γ in the presence of LPS. IFN-γ decreased responsiveness to β-adrenoceptor stimulation through induction of NO production [101]. These proinflammatory cytokines may exert their effects by changing calcium channels, which has been shown for IL-1β and L-type calcium channels in adult rat ventricular myocytes [102]. This indicates that proinflammatory cytokines such as IL-1β, IL-2, IFN-γ, and TNF inhibit signaling through the β-adrenoceptor. This may lead to continuously elevated levels of these cytokines because the immunosuppressive effect of NE via β-adrenoceptors is attenuated. Interestingly, in patients with juvenile chronic arthritis, the α1-adrenoceptor is upregulated which would possibly lead to a more proinflammatory situation [103].

Receptors for neurotransmitters are typically transmembrane molecules with seven membrane-spanning helices. They relay signals from extracellular stimuli to the heterotrimeric G-protein complex present on the cytoplasmic face of the membrane (Fig. 2). In this complex, the GDP-bound G-protein α subunit is prevented from interacting with downstream effector molecules because it is bound to βγ subunits and to the receptor. When the membrane receptor binds an agonistic ligand, the coupled G-protein α subunit exchanges GTP for GDP, causing it to undergo

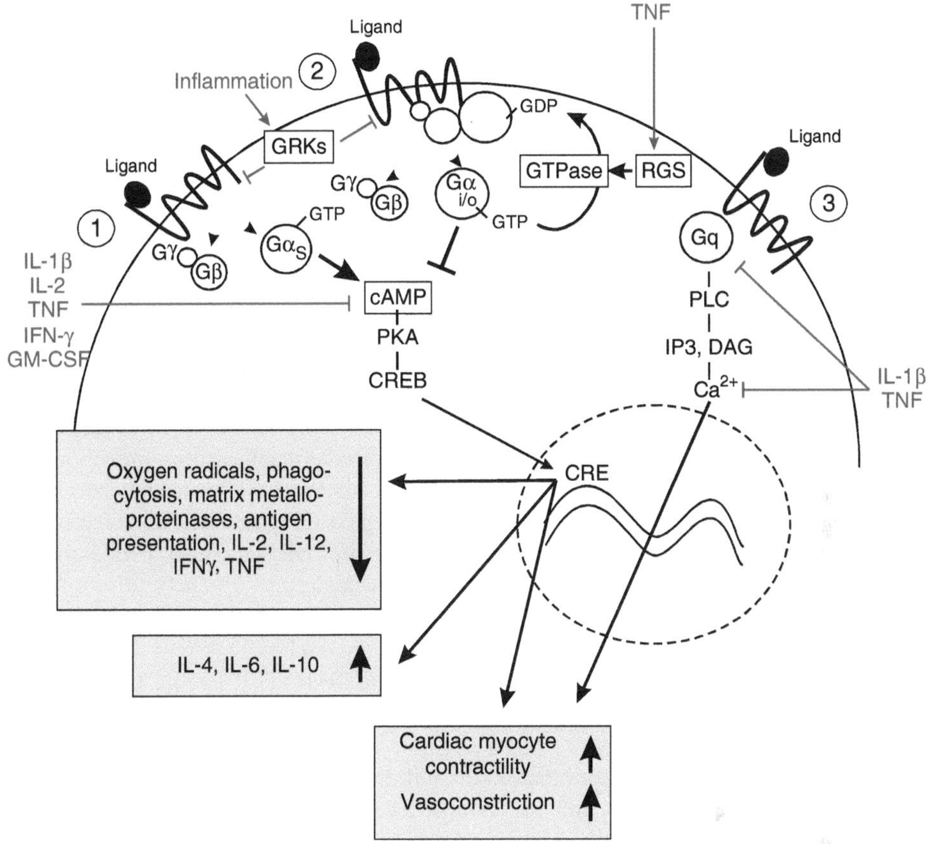


Figure 2. Effects of inflammatory mediators on neurotransmitter receptors and signaling. A line with an arrow (a bar) at the end indicates a stimulatory (an inhibitory) effect. It is known that upon ligation of a ligand to receptor ①, $G\alpha_S$ is activated which increases intracellular cyclic adenosine monophosphate (cAMP) levels (the ligand may be NE via β-adrenoceptors). This pathway signals downstream via protein kinase A (PKA) and the cAMP responsive element binding protein (CREB). Binding of a ligand to receptor ② stimulates $G\alpha_{i/o}$ which leads to decreased intracellular cAMP levels after dissociation of the G protein subunits βγ and α (the ligand may be NE via α2 adrenoceptors). Binding of NE to receptor ③ leads to activation of $G_q$ and phospholipase C (PLC) which leads to an increase of intracellular calcium via inositol-3-phosphate (IP3) and diacylglycerol (DAG) (the ligand may be NE via α1-adrenoceptors). For NE it is known that the affinities to the receptor subtypes are markedly different: low concentrations of NE preferentially stimulate receptor ② and ③, whereas high concentrations of NE stimulate both receptors. Termination of signaling is either induced by GTPase, which is activated by regulators of G protein signaling (RGS), or G-protein-coupled receptor kinases (GRKs), which leads to rapid desensitization. CRE, cAMP responsive element; G, G protein ($\alpha_S$, $\alpha_{i/0}$, β, γ or q subunits); GDP, guanosine diphosphate; GTP, guanosine triphosphate; GM-CSF, granulocyte macrophage colony-stimulating factor; IFN, interferon; IL, interleukin; TNF, tumor necrosis factor.

a conformational change and dissociate from the receptor and Gβγ subunits (Fig. 2). GTP-bound Gα engages its downstream targets and the signaling cascade continues. Hydrolysis of GTP to GDP by the GTPase activity of the Gα subunit terminates the signaling event by inducing the re-association of α with βγ (Fig. 2). The intrinsic GTPase activity of the Gα subunit is regulated by GTPase-activating proteins (GAPs). A relatively new class of GAPs consists of a large family of molecules, now known as regulators of G-protein signaling (RGS). RGS are expressed in immune cells and can be induced by treatment with TNF [104,105]. Several studies have demonstrated that different RGS proteins interact with $G\alpha_{i/o}$ or $G\alpha_q$ but not with $G\alpha_S$ [106]. Signaling through $G\alpha_{i/o}$ or $G\alpha_q$ is generally associated with a more proinflammatory response, whereas signaling through $G\alpha_S$ leads to more anti-inflammatory phenotype (summarized in Ref. [107]). Thus, TNF by inducing RGS would inhibit the pathway through $G\alpha_{i/o}$ or $G\alpha_q$ so that $G\alpha_S$ may take over and occur at higher levels leading to eventual inhibition of TNF production via elevation of cAMP when a respective neurotransmitter binds to the receptor subtypes (Fig. 2).

Another way to modulate signaling is regulation of G-protein-coupled receptor kinases (GRK) (Fig. 2): The responsiveness of G-protein-coupled neurotransmitter receptors is turned off by the GRK family (GRK-1 to GRK-6). GRKs phosphorylate receptors in an agonist-dependent manner resulting in receptor/G-protein uncoupling via subsequent binding of arrestin proteins. This is thought to be an important step in receptor desensitization [108]. In an inflammatory disease such as rheumatoid arthritis or experimental arthritis a significant decrease in GRK activity and GRK-2 protein expression was found. As a consequence, lymphocytes of patients with rheumatoid arthritis showed a significantly increased cAMP production and inhibition of TNF production by β2-adrenergic stimulation [109,110]. Thus, both intracellular mechanisms – induction of RGS and decrease of GRKs – would allow proinflammatory mediators to influence adrenergic signaling and signal transduction through other G-protein-coupled receptors. The importance for the local inflammatory process may be a downregulation of TNF and other cAMP-modulated cytokines (IFN-γ, IL-2) which needs further confirmatory studies.

## 6. MODULATION OF PERIPHERAL NEUROTRANSMITTER EFFECTS WITHOUT EXACT LOCALIZATION OF THE SITE OF ACTION

Many studies, mainly *in vivo* experiments, do not exactly mention the site of interaction between inflammatory mediators and the sympathetic or parasympathetic nervous system. These studies are summarized in the following Sections 6.1, 6.2, and 6.3.

### 6.1. Gastrointestinal function

In intestinal segments, the cholinergic contraction to electrical field stimulation was significantly increased by IL-4 treatment. The enhanced cholinergic contraction was not due to increased acetylcholine responsiveness but was dependent on the influence of leukotriene D4 [111]. Similarly, in guinea pig small intestinal submucous neurons, exposure to nanomolar concentrations of either IL-1β or IL-6 stimulated neuronal excitability [112]. The excitatory action consisted of depolarization of the membrane potential, decreased membrane conductance, and increased discharge of action potentials. Excitatory action of IL-1β was suppressed by IL-1RA. Electrical stimulation of sympathetic postganglionic axons evoked inhibitory postsynaptic potentials, and stimulation of cholinergic axons evoked nicotinic fast excitatory postsynaptic potentials. Both kinds of synaptic potentials occurred in neurons with uniaxonal morphology believed to be secretomotor neurons. Either IL-1β or IL-6 suppressed the noradrenergic inhibitory postsynaptic potentials and the fast excitatory postsynaptic potentials, and the two cytokines acted synergistically when applied in combination [112]. Suppression of the inhibitory postsynaptic potentials resulted from presynaptic inhibition of the release of NE from sympathetic nerves. The results suggest that the presence of either or both inflammatory cytokines will release the sympathetic brake from secretomotor neurons to the intestinal crypts and from nicotinic synapses in the integrative microcircuits, where NE is known to have a presynaptic inhibitory action. This, in concert with excitation of secretomotor neurons, may lead to neurogenic secretory diarrhea [112]. These stimulating effects may be completely opposite depending on the site of stimulation: For example, when effects of IL-1β on the membrane potential and synaptic transmission were examined in neurons of mammalian pelvic ganglia, IL-1β caused an initial facilitation followed by a long-lasting depression of the excitatory postsynaptic potential in rabbit pelvic ganglia. These data suggest that IL-1β presynaptically depressed the excitatory postsynaptic potentials by reducing the release of acetylcholine from the pelvic nerve terminals [113]. Others demonstrated that IL-1β is an extremely potent inhibitor of acid secretion stimulated by pentagastrin through a mechanism which is at least in part dependent on the vagus nerve and on prostaglandin synthesis [114]. In conclusion, IL-1β is able to stimulate local intestinal secretomotor neurons, whereas the same cytokine inhibits pelvic ganglia neurons and vagal motoneurons in the dorsal motor nucleus in the brain stem (Section 2.2). This indicates that IL-1β has very different effects depending on the site of action. It becomes more and more clear that the same endogenous substance may have completely different effects at remote locations. It is obvious that this fact may severely influence effects of a drug which inhibits or stimulates such a factor in a general way independent of the site of action.

### 6.2. Sepsis

Bacteria and bacteria-stimulated inflammatory mediators activate the sympathetic nervous system so that the β-adrenergic inhibitory influence on TNF secretion is increased [115]. Chemical sympathectomy attenuates this activation and increases local TNF, which subsequently leads to significantly reduced bacterial tissue burden [115]. This was corroborated in another study [116]. Injection of LPS into mice induced an increase of plasma TNF which was significantly blunted by pretreatment with a highly selective α2-adrenoreceptor antagonist (CH-38083). In contrast, LPS-induced increases in both corticosterone and IL-6 plasma levels were further increased by this α2-adrenoreceptor antagonist. Propranolol prevented the effect of α2-adrenoreceptor blockade on TNF plasma levels induced by LPS [117]. This indicates that the excessive stimulation by NE of β-adrenoceptors located on cytokine-secreting immune cells is responsible for TNF inhibition. This was corroborated by a study in spleen slices which demonstrates that TNF is mainly inhibited by β-adrenoceptor pathways [115].

### 6.3. Allergic bronchoconstriction

Incubation of isolated bronchial segments with human recombinant IL-1β for 150 min concentration-dependently decreased the contractile responses to acetylcholine. The IL-1β-induced inhibition of the contractile responses was not affected by pretreatment of tissues with indomethacin or propranolol, but it was greatly attenuated by mechanical removal of epithelium. These results suggest that IL-1β may play a protective role against bronchoconstrictor responses via epithelium-dependent mechanism [118]. In contrast, another proinflammatory mediator, TNF, increased human airway tissue responsiveness which occurs prejunctionally mediated by the parasympathetic nerve pathways [119]. In isolated atopic sensitized rabbit tracheal smooth muscle preparations, isoproterenol (β-adrenoceptor agonist)-induced relaxation was decreased by IL-1β as compared to control tissue. The impaired relaxation responses to isoproterenol were abolished in sensitized tissue which was pretreated with either IL-1RA or an IL-1β-neutralizing antibody. These observations suggest that the altered responsiveness of atopic/asthmatic-sensitized airway smooth muscle is attributed to local IL-1β [120,121]. Furthermore, exogenous administration of IL-2 and IFN-γ to atopic asthmatic serum-sensitized airway smooth muscle cells abolished enhanced constrictor responsiveness to acetylcholine and attenuated relaxation responsiveness to β-adrenoceptor stimulation with isoproterenol [122].

Inhalational challenge with antigen decreases the function of inhibitory M2 muscarinic autoreceptors on parasympathetic nerves in the lung, increasing the release of acetylcholine from the vagus nerves and potentiating vagally induced bronchoconstriction. This effect can be abrogated by a monoclonal antibody to IL-5 (TRFK-5), which inhibited the migration of eosinophils into the lungs as measured by lung lavage [123]. A similar study in guinea pigs has confirmed the importance of the allergy-promoting Th2 cytokine IL-5 [124].

## 7. MODULATION OF CHROMAFFIN CELLS

Chromaffin cells produce epinephrine in the adrenal medulla. Thus, these cells contribute to the overall sympathetic tone in the periphery. Interestingly, occurrence of IL-1-like immunoreactivity was demonstrated in the noradrenergic chromaffin cells of the rat and mouse adrenal gland [125], and there is evidence for paracrine signaling between macrophages and bovine adrenal chromaffin cells [126]. In addition, treatment of cultured bovine adrenal medullary cells with IL-1β caused an increase in accumulation of catecholamines in the cultured medium. The accumulation of catecholamines was observed in time (4–48 h)- and concentration (3–30 ng/ml)-dependent manner which was inhibited by IL-1RA [127]. In contrast, others found no increase of epinephrine secretion from porcine chromaffin cells after incubation with IL-1 or IL-2 [128]. In chromaffin cells, vasoactive intestinal polypeptide (VIP) was elevated two- to three-fold by IL-1α, while neurotensin and substance P synthesis were unaffected, and met-enkephalin levels were decreased by 25%–35%. TNF also demonstrated a neuropeptide-specific pattern of modulation of second-messenger effects on chromaffin cell neuropeptide levels similar to those seen with IL-1α [129]. These studies provide functional demonstration that inflammatory mediators modulate chromaffin cell function. The question remains whether the adrenal medulla is influenced by stimulated circulating immune cells or circulating proinflammatory mediators in the course of an inflammatory disease. Since no studies are available which directly investigated adrenal medullary tissue, this question remains to be answered.

## 8. MODULATION OF ENZYMES RESPONSIBLE FOR NEUROTRANSMITTER PRODUCTION

Classical sympathetic and parasympathetic neurotransmitters are produced in the nerve terminal of the respective nerves, whereas neuropeptides are produced in the neuronal soma. Neuropeptides are transported via the axon with a transport velocity of approx. 10 mm/h. Both classical neurotransmitters and neuropeptides are stored in nerve terminal vesicles to be released upon an action potential which enters the terminal. In neurons and in non-neuronal cells neurotransmitters are produced by the same enzymes. These enzymes may be target of inflammatory mediator influence, which is discussed in this chapter.

### 8.1. Adrenergic pathways

IL-1β stimulates tyrosine hydroxylase activity of the median eminence in a dose-dependent manner [130]. Using transgenic mice expressing TNF specifically in the CNS, it was shown that the overexpression of this cytokine reduced tyroxine hydroxylase immunoreactivity in the dorsomedial hypothalamic areas [131]. Basic fibroblast growth factor is an important autocrine/paracrine maintenance factor for adult chromaffin cells which stimulates synthesizing enzymes such as tyrosine hydroxylase and phenylethanolamine-*N*-methyltransferase of cultured chromaffin cells from young postnatal rats [132].

### 8.2. Cholinergic pathways

IFN-γ causes a dose-dependent increase in choline acetyltransferase activity in cultures of human embryonic spinal cord neurons via non-neuronal cells, probably astrocytes [133]. Following deafferentation and explantation of superior cervical sympathetic ganglia into culture, IL-1β causes an upregulation of choline acetyltransferase. TNF has a similar, though less potent, effect [134]. Furthermore, IL-6 treatment led to six-fold increase in choline acetyltransferase mRNA without a concomitant increase in choline acetyltransferase immunoreactivity [135]. In contrast, treatment with ciliary neurotrophic factor or leukemia inhibitory factor increased choline acetyltransferase mRNA levels 25-fold and resulted in intense choline acetyltransferase immunoreactivity [135]. Furthermore, activation of human mononuclear leukocytes by phytohemagglutinin induced the expression of choline acetyltransferase mRNA, and potentiated acetylcholine synthesis [136]. Choline acetyltransferase mRNA induction required more time than the induction of IL-2 mRNA. Expression of the gene encoding the vesicular acetylcholine transporter, which mediates acetylcholine transport in cholinergic neurons, was not observed in phytohemagglutinin-stimulated human mononuclear leukocytes, suggesting that the mechanisms controlling acetylcholine release from human mononuclear leukocytes differ from those in cholinergic neurons [136]. Furthermore, it has been demonstrated that purified organ resident and circulating lymphocytes, as well as various lymphoid cell lines derived from different species, exhibit choline acetyltransferase [137]. Mitogenic stimulation with phytohemagglutinin increased the acetylcholine levels in lymphoid cells as well as the release into the supernatants [138]. In conclusion, these data indicate that several proinflammatory mediators activate the choline acetyltransferase, thereby inducing acetylcholine production and secretion.

## 9. ALTERATIONS OF REFLEXES OF THE AUTONOMIC NERVOUS SYSTEM IN INFLAMMATORY DISEASES

The prototypic disease with inflammatory dysautonomia is acute inflammatory demyelinating polyneuropathy or Guillain–Barré syndrome (reviewed in Ref. [139]). Other reports of dysautonomia and polyneuropathy demonstrated inflammatory infiltrates within the autonomic nerves [140,141]. Described in the second report, patients died due to sudden cardiac death with fluctuating blood pressure, bradycardia, and ECG changes. Sudden cardiac death as a consequence of non-inflammatory autonomic neuropathy was found in diabetic subjects [142,143]. However, until the late 1970s these complications have not been envisaged in inflammatory diseases other than Guillain–Barré syndrome [144]. One of the early reports demonstrated autonomic neuropathy in patients with Sjogren's syndrome (dryness of eyes, mouth, and other mucous membranes), which superimposed on generalized neuropathy in approximately one-fourth of these patients. The nerve biopsies revealed perivascular inflammatory infiltrates [145]. Subsequent reports on autonomic nervous dysfunction in chronic inflammatory diseases were added which investigated patients with inflammatory bowel disease [146–148], Chagas' disease [149], rheumatoid arthritis [144,150–152], systemic lupus erythematosus [153–159], systemic sclerosis [157], Sjogren's syndrome [145,160,161], and others. These studies focused on hyporeflexia of the autonomic nervous system which is often equated with autonomic neuropathy.

Some reports, however, also demonstrated autonomic nervous hyper-reflexia which is accompanied by exaggerated autonomic nervous reflexes [148,162,163]. Interestingly, the two major anti-inflammatory feedback systems, the HPA axis and the sympathetic nervous system, are uncoupled in these diseases with relatively low levels of serum cortisol and an elevation of the sympathetic activity as measured by plasma neuropeptide Y [164]. Some of these response patterns may lead to a continuous sympathetic overdrive that may be harmful, particularly, for the cardiovascular system. Several studies have shown that patients with SLE are at a significantly increased risk for the development of coronary atherosclerosis [165,166], which is rarely due to arteritis. Hypertension may be the most important risk factor in these patients [167,168], as was demonstrated in hypersympathetic diabetic patients (reviewed in Ref. [169]). We note that a hypersympathetic response may be a problem of several chronic inflammatory diseases [148,170,171].

## 10. CONCLUSIONS

Due to the enormous complexity of physiology and pathophysiology, it is often difficult to draw general conclusions. Furthermore, since experiments were carried out in different species, various tissues, and under diverse conditions, a unifying formula, which may summarize the findings, will not exist. Thus, the following statements should be considered using these prerequisites:

- By circulating inflammatory mediators and vagus-stimulating cytokines, the central sympathetic nervous system is activated which leads to an increased sympathetic tone.
- In the local microenvironment of a sympathetic nerve terminal, the same inflammatory mediators inhibit NE release, which is also relevant for sympathetic ganglia.
- By circulating inflammatory mediators, central vagus neurons are activated which leads to a decrease of gastrointestinal motility and secretion, and to sickness behavior.
- In the proximity of a vagal cholinergic nerve terminal, inflammatory mediators increase acetylcholine release and choline acetyltransferase activity.
- Many direct effects on neurotransmitter receptors and signaling exist, but described effects would lead to either a more proinflammatory or anti-inflammatory local situation.

For the sympathetic and parasympathetic nervous system, inflammatory mediators have very different effects depending on the sites of action, which are often far away from each other. Such a situation may severely influence effects of a drug which inhibits or stimulates such a factor in a general way independent of the site of action.

Although in several chronic inflammatory diseases a hypersympathetic tone exists, the local situation in inflamed tissue may be completely opposite with decreased sympathetic innervation. Thus, the systemic sympathetic overdrive may harm the body (hypertension and sequelae), whereas the local immunosuppressive effects of sympathetic neurotransmitters are missing.

## REFERENCES

1. Besedovsky HO, Del Rey A. Immune-neuro-endocrine interactions. Endocr Rev 1996;17:64–102.
2. Chrousos GP. The hypothalamic-pituitary-adrenal axis and immune-mediated inflammation. N Engl J Med 1995;332:1351–62.
3. Straub RH, Westermann J, Schölmerich J, Falk W. Dialogue between CNS and immune system in lymphoid organs. Immunol Today 1998;19:409–13.
4. Elenkov IJ, Wilder RL, Chrousos GP, Vizi ES. The sympathetic nerve – an integrative interface between two supersystems: the brain and the immune system. Pharmacol Rev 2000;52:595–638.
5. Madden KS. Catecholamines, sympathetic nerves, and immunity. In Psychoneuroimmunology. Ader R, Felten DL, Cohen N, Eds; San Diego: Academic Press, 2001; pp. 197–216.
6. Webster JI, Tonelli L, Sternberg EM. Neuroendocrine regulation of immunity. Annu Rev Immunol 2002;20:125–63.
7. Calogero AE, Gallucci WT, Gold PW, Chrousos GP. Multiple feedback regulatory loops upon rat hypothalamic corticotropin-releasing hormone secretion. Potential clinical implications. J Clin Invest 1988;82:767–74.
8. Besedovsky H, Sorkin E, Keller M, Muller J. Changes in blood hormone levels during the immune response. Proc Soc Exp Biol Med 1975;150:466–70.
9. Besedovsky HO, Del Rey A, Sorkin E, Dinarello C. Immunoregulatory feedback between interleukin-1 and glucocorticoid hormones. Science 1986;233:652–4.
10. Dunn AJ. Systemic interleukin-1 administration stimulates hypothalamic norepinephrine metabolism parallelling the increased plasma corticosterone. Life Sci 1988;43:429–35.
11. Dunn AJ. The role of interleukin-1 and tumor necrosis factor alpha in the neurochemical and neuroendocrine responses to endotoxin. Brain Res Bull 1992;29:807–12.
12. Dunn AJ, Wang J, Ando T. Effects of cytokines on cerebral neurotransmission. Comparison with the effects of stress. Adv Exp Med Biol 1999;461:117–27.
13. Pauli S, Linthorst AC, Reul JM. Tumour necrosis factor-alpha and interleukin-2 differentially affect hippocampal serotonergic neurotransmission, behavioural activity, body temperature and hypothalamic-pituitary-adrenocortical axis activity in the rat. Eur. J Neurosci 1998;10:868–78.
14. Palazzolo DL, Quadri SK. Interleukin-1 stimulates catecholamine release from the hypothalamus. Life Sci 1990;47:2105–9.
15. Shintani F, Kanba S, Nakaki T, Nibuya M, Kinoshita N, Suzuki E, Yagi G, Kato R, Asai M. Interleukin-1 beta augments release of norepinephrine, dopamine, and serotonin in the rat anterior hypothalamus. J Neurosci 1993;13:3574–81.
16. MohanKumar PS, Quadri SK. Systemic administration of interleukin-1 stimulates norepinephrine release in the paraventricular nucleus. Life Sci 1993;52:1961–7.
17. Terrazzino S, Perego C, De Simoni MG. Noradrenaline release in hypothalamus and ACTH secretion induced by central interleukin-1beta. Neuroreport 1995;6:2465–8.
18. Mefford IN, Heyes MP. Increased biogenic amine release in mouse hypothalamus following immunological challenge: antagonism by indomethacin. J Neuroimmunol 1990;27:55–61.
19. Karanth S, Lyson K, McCann SM. Role of nitric oxide in interleukin 2-induced corticotropin-releasing factor release from incubated hypothalami. Proc Natl Acad Sci USA 1993;90:3383–7.
20. Costa A, Trainer P, Besser M, Grossman A. Nitric oxide modulates the release of corticotropin-releasing hormone from the rat hypothalamus in vitro. Brain Res 1993;605:187–92.
21. Rivier C, Vale W, Brown M. In the rat, interleukin-1 alpha and -beta stimulate adrenocorticotropin and catecholamine release. Endocrinology 1989;125:3096–102.
22. Matta S, Singh J, Newton R, Sharp BM. The adrenocorticotropin response to interleukin-1 beta instilled into the rat median eminence depends on the local release of catecholamines. Endocrinology 1990;127:2175–82.
23. Ohgo S, Nakatsuru K, Oki Y, Ishikawa E, Matsukura S. Stimulation by interleukin-1 (IL-1) of the release of rat corticotropin-releasing factor (CRF), which is independent of the cholinergic mechanism, from superfused rat hypothalamo-neurohypophysial complexes. Brain Res 1991;550:213–19.

24. Sweep CG, van der Meer MJ, Hermus AR, Smals AG, van der Meer JW, Pesman GJ, Willemsen SJ, Benraad TJ, Kloppenborg PW. Chronic stimulation of the pituitary-adrenal axis in rats by interleukin-1 beta infusion: in vivo and in vitro studies. Endocrinology 1992;130:1153–64.
25. Mastorakos G, Chrousos GP, Weber JS. Recombinant interleukin-6 activates the hypothalamic-pituitary-adrenal axis in humans. J Clin Endocrinol Metab 1993;77:1690–4.
26. Späth-Schwalbe E, Born J, Schrezenmeier H, Bornstein SR, Stromeyer P, Drechsler S, Fehm HL, Porzsolt F. Interleukin-6 stimulates the hypothalamus-pituitary-adrenocortical axis in man. J Clin Endocrinol Metab 1994;79:1212–14.
27. Berkenbosch F, de Goeij DE, Rey AD, Besedovsky HO. Neuroendocrine, sympathetic and metabolic responses induced by interleukin-1. Neuroendocrinology 1989;1:570–6.
28. Shintani F, Nakaki T, Kanba S, Sato K, Yagi G, Shiozawa M, Aiso S, Kato R, Asai M. Involvement of interleukin-1 in immobilization stress-induced increase in plasma adrenocorticotropic hormone and in release of hypothalamic monoamines in the rat. J Neurosci 1995;15:1961–70.
29. Niijima A, Hori T, Aou S, Oomura Y. The effects of interleukin-1 beta on the activity of adrenal, splenic and renal sympathetic nerves in the rat. J Auton Nerv Syst 1991;36:183–92.
30. Ichijo T, Katafuchi T, Hori T. Central interleukin-1 beta enhances splenic sympathetic nerve activity in rats. Brain Res Bull 1994;34:547–53.
31. Terao A, Oikawa M, Saito M. Tissue-specific increase in norepinephrine turnover by central interleukin-1, but not by interleukin-6, in rats. Am J Physiol 1994;1:R400–4.
32. Akiyoshi M, Shimizu Y, Saito M. Interleukin-1 increases norepinephrine turnover in the spleen and lung in rats. Biochem Biophys Res Commun 1990;173:1266–70.
33. Shimizu N, Hori T, Nakane H. An interleukin-1 beta-induced noradrenaline release in the spleen is mediated by brain corticotropin-releasing factor: an in vivo microdialysis study in conscious rats. Brain Behav Immun 1994;8:14–23.
34. Sundar SK, Becker KJ, Cierpial MA, Carpenter MD, Rankin LA, Fleener SL, Ritchie JC, Simson PE, Weiss JM. Intracerebroventricular infusion of interleukin 1 rapidly decreases peripheral cellular immune responses. Proc Natl Acad Sci USA 1989;86:6398–402.
35. Sundar SK, Cierpial MA, Kilts C, Ritchie JC, Weiss JM. Brain IL-1-induced immuno-suppression occurs through activation of both pituitary-adrenal axis and sympathetic nervous system by corticotropin-releasing factor. J Neurosci 1990;10:3701–6.
36. Brown R, Li Z, Vriend CY, Nirula R, Janz L, Falk J, Nance DM, Dyck DG, Greenberg AH. Suppression of splenic macrophage interleukin-1 secretion following intracerebroventricular injection of interleukin-1 beta: evidence for pituitary-adrenal and sympathetic control. Cell Immunol 1991;132:84–93.
37. Besedovsky HO, Del Rey A, Sorkin E, Da Prada M, Keller HH. Immunoregulation mediated by the sympathetic nervous system. Cell Immunol 1979;48:346–55.
38. Watkins LR, Goehler LE, Relton JK, Tartaglia N, Silbert L, Martin D, Maier SF. Blockade of interleukin-1 induced hyperthermia by subdiaphragmatic vagotomy: evidence for vagal mediation of immune-brain communication. Neurosci Lett 1995;183:27–31.
39. Ishizuka Y, Ishida Y, Kunitake T, Kato K, Hanamori T, Mitsuyama Y, Kannan H. Effects of area postrema lesion and abdominal vagotomy on interleukin-1 beta-induced norepinephrine release in the hypothalamic paraventricular nucleus region in the rat. Neurosci Lett 1997;223:57–60.
40. Watkins LR, Goehler LE, Relton J, Brewer MT, Maier SF. Mechanisms of tumor necrosis factor-alpha (TNF-alpha) hyperalgesia. Brain Res 1995;692:244–50.
41. Kapcala LP, He JR, Gao Y, Pieper JO, DeTolla LJ. Subdiaphragmatic vagotomy inhibits intra-abdominal interleukin-1 beta stimulation of adrenocorticotropin secretion. Brain Res 1996;728:247–54.
42. Brady LS, Lynn AB, Herkenham M, Gottesfeld Z. Systemic interleukin-1 induces early and late patterns of c-fos mRNA expression in brain. J Neurosci 1994;14:4951–64.
43. Barbanel G, Gaillet S, Mekaouche M, Givalois L, Ixart G, Siaud P, Szafarczyk A, Malaval F, Assenmacher I. Complex catecholaminergic modulation of the stimulatory effect of interleukin-1 beta on the corticotropic axis. Brain Res 1993;626:31–6.

44. Petit F, Jarrous A, Dickinson RD, Molina PE, Abumrad NN, Lang CH. Contribution of central and peripheral adrenergic stimulation to IL-1 alpha-mediated glucoregulation. Am J Physiol 1994;267: E49–56.
45. Morimoto K, Morimoto A, Nakamori T, Tan N, Minagawa T, Murakami N. Cardiovascular responses induced in free-moving rats by immune cytokines. J Physiol 1992;448:307–20.
46. Takahashi H, Nishimura M, Sakamoto M, Ikegaki I, Nakanishi T, Yoshimura M. Effects of interleukin-1 beta on blood pressure, sympathetic nerve activity, and pituitary endocrine functions in anesthetized rats. Am J Hypertens 1992;5:224–9.
47. Bataillard A, Sassard J. Cardiovascular effects of human recombinant interleukin-1 beta in conscious rats. Am J Physiol 1994;266:R1148–53.
48. Haefeli WE, Bargetzi MJ, Starnes HF, Blaschke TF, Hoffman BB. Evidence for activation of the sympathetic nervous system by recombinant human interleukin-1 beta in humans. J Immunother 1993;13:136–40.
49. Saigusa T. Participation of interleukin-1 and tumor necrosis factor in the responses of the sympathetic nervous system during lipopolysaccharide-induced fever. Pflugers Arch 1990;416:225–9.
50. Kurumiya S, Miura Y, Nakatsuji K, Dodo M, Hori M, Ito T. General pharmacology of recombinant human tumor necro sis factor. 2nd communication: effects on central nervous system functions. Arzneimittelforschung 1990;40:225–31.
51. Mo ZL, Katafuchi T, Hori T. Effects of IL-1 beta on neuronal activities in the dorsal motor nucleus of the vagus in rat brain slices. Brain Res Bull 1996;41:249–55.
52. Yokotani K, Okuma Y, Osumi Y. Recombinant interleukin-1 beta inhibits gastric acid secretion by activation of central sympatho-adrenomedullary outflow in rats. Eur J Pharmacol 1995;279:233–9.
53. Emch GS, Hermann GE, Rogers RC. TNF-alpha activates solitary nucleus neurons responsive to gastric distension. Am J Physiol Gastrointest Liver Physiol 2000;279:G582–6.
54. Hermann G, Rogers RC. Tumor necrosis factor-alpha in the dorsal vagal complex suppresses gastric motility. Neuroimmunomodulation 1995;2:74–81.
55. Day HE, Akil H. Differential pattern of c-fos mRNA in rat brain following central and systemic administration of interleukin-1-beta: implications for mechanism of action. Neuroendocrinology 1996;63:207–18.
56. Yang ZJ, Rao ZR, Ju G. Evidence for the medullary visceral zone as a neural station of neuroimmunomodulation. Neurosci Res 2000;38:237–47.
57. Niimi M, Wada Y, Sato M, Takahara J, Kawanishi K. Effect of continuous intravenous injection of interleukin-6 and pretreatment with cyclooxygenase inhibitor on brain c-fos expression in the rat. Neuroendocrinology 1997;66:47–53.
58. Mascarucci P, Perego C, Terrazzino S, De Simoni MG. Glutamate release in the nucleus tractus solitarius induced by peripheral lipopolysaccharide and interleukin-1 beta. Neuroscience 1998;86: 1285–90.
59. Bluthe RM, Walter V, Parnet P, Laye S, Lestage J, Verrier D, Poole S, Stenning BE, Kelley KW, Dantzer R. Lipopolysaccharide induces sickness behaviour in rats by a vagal mediated mechanism. C R Acad Sci III 1994;317:499–503.
60. Borovikova LV, Ivanova S, Zhang M, Yang H, Botchkina GI, Watkins LR, Wang H, Abumrad N, Eaton JW, Tracey KJ. Vagus nerve stimulation attenuates the systemic inflammatory response to endotoxin. Nature 2000;405:458–62.
61. Bernik TR, Friedman SG, Ochani M, DiRaimo R, Ulloa L, Yang H, Sudan S, Czura CJ, Ivanova SM, Tracey KJ. Pharmacological stimulation of the cholinergic antiinflammatory pathway. J Exp Med 2002;195:781–8.
62. Kohm AP, Tang2 Y, Sanders VM, Jones3 SB. Activation of antigen-specific CD4+ Th2 cells and B cells in vivo increases norepinephrine release in the spleen and bone marrow. J Immunol 2000;165:725–33.
63. MacNeil BJ, Jansen AH, Greenberg AH, Nance DM. Activation and selectivity of splenic sympathetic nerve electrical activity response to bacterial endotoxin. Am J Physiol 1996;270:R264–70.

64. Lin J, Krier J. Human recombinant interleukin-1 beta inhibits nicotinic transmission in neurons of guinea pig pelvic plexus ganglia. Am J Physiol 1995;269:G981–7.
65. Soliven B, Albert J. Tumor necrosis factor modulates Ca2+ currents in cultured sympathetic neurons. J Neurosci 1992;12:2665–71.
66. Soliven B, Albert J. Tumor necrosis factor modulates the inactivation of catecholamine secretion in cultured sympathetic neurons. J Neurochem 1992;58:1073–8.
67. März P, Cheng JG, Gadient RA, Patterson PH, Stoyan T, Otten U, Rose-John S. Sympathetic neurons can produce and respond to interleukin. Proc Natl Acad Sci USA 1998;6:3251–6.
68. Miller LE, Jüsten HP, Schölmerich J, Straub RH. The loss of sympathetic nerve fibers in the synovial tissue of patients with rheumatoid arthritis is accompanied by increased norepinephrine release from synovial macrophages. FASEB J 2000;14:2097–107.
69. Lorton D, Lubahn C, Felten SY, Bellinger D. Norepinephrine content in primary and secondary lymphoid organs is altered in rats with adjuvant-induced arthritis. Mech Ageing Dev 1997;94: 145–63.
70. da Silva JA, Fonseca JE, Graca L, Moita L, Carmo-Fonseca M. Reinnervation of post-arthritic joints in the rat. Clin Exp Rheumatol 1996;14:43–51.
71. Bamji SX, Majdan M, Pozniak CD, Belliveau DJ, Aloyz R, Kohn J, Causing CG, Miller FD. The p75 neurotrophin receptor mediates neuronal apoptosis and is essential for naturally occurring sympathetic neuron death. J Cell Biol 1998;140:911–23.
72. Freidin MM. Antibody to the extracellular domain of the low affinity NGF receptor stimulates p75(NGFR)–mediated apoptosis in cultural sympathetic neurons. J Neurosci Res 2001;64:331–40.
73. Majdan M, Walsh GS, Aloyz R, Miller FD. TrkA mediates developmental sympathetic neuron survival in vivo by silencing an ongoing p75NTR–mediated death signal. J Cell Biol 2001;155: 1275–85.
74. Rogausch H, Vo NT, Del Rey A, Besedovsky HO. Increased sensitivity of the baroreceptor reflex after bacterial endotoxin. Ann N Y Acad Sci 2000;917:165–8.
75. Elenkov IJ, Kovacs K, Duda E, Stark E, Vizi ES. Presynaptic inhibitory effect of TNF-alpha on the release of noradrenaline in isolated median eminence. J Neuroimmunol 1992;41:117–20.
76. Nickola TJ, Ignatowski TA, Reynolds JL, Spengler RN. Antidepressant drug-induced alterations in neuron-localized tumor necrosis factor-alpha mRNA and alpha(2)-adrenergic receptor sensitivity. J Pharmacol Exp Ther 2001;297:680–7.
77. Smith EL, Hainsworth AH. Acute effects of interleukin-1 beta on noradrenaline release from the human neuroblastoma cell line SH-SY5Y. Neurosci Lett 1998;251:89–92.
78. Hurst S, Collins SM. Interleukin-1 beta modulation of norepinephrine release from rat myenteric nerves. Am J Physiol 1993;264:G30–5.
79. Hurst SM, Collins SM. Mechanism underlying tumor necrosis factor-alpha suppression of norepinephrine release from rat myenteric plexus. Am J Physiol 1994;266:G1123–9.
80. Ruhl A, Hurst S, Collins SM. Synergism between interleukins 1 beta and 6 on noradrenergic nerves in rat myenteric plexus. Gastroenterology 1994;1077:993–1001.
81. Bognar IT, Albrecht SA, Farasaty M, Schmitt E, Seidel G, Fuder H. Effects of human recombinant interleukins on stimulation-evoked noradrenaline overflow from the rat perfused spleen. Naunyn Schmiedebergs Arch Pharmacol 1994;349:497–502.
82. Bognar IT, Albrecht SA, Farasaty M, Fuder H. Inhibition by interleukin-1 beta of noradrenaline release in rat spleen: involvement of lymphocytes, NO and opioid receptors. Naunyn Schmiedebergs Arch Pharmacol 1995;351:433–8.
83. Foucart S, Abadie C. Interleukin-1 beta and tumor necrosis factor-alpha inhibit the release of [3H]-noradrenaline from mice isolated atria. Naunyn Schmiedebergs Arch Pharmacol 1996;354:1–6.
84. Sterin-Borda L, Perez LC, Borda ES, de Bracco MM. Effects of IL-2 on the myocardium. Participation of the sympathetic system. J Mol Cell Cardiol 1996;28:2457–65.
85. Araujo DM, Lapchak PA, Chabot JG, Nair NP, Quirion R. Characterization and possible role of growth factor and lymphokine receptors in the regulation of cholinergic function in the mammalian brain. Prog Clin Biol Res 1989;317:423–36.

86. Seto D, Kar S, Quirion R. Evidence for direct and indirect mechanisms in the potent modulatory action of interleukin-2 on the release of acetylcholine in rat hippocampal slices. Br J Pharmacol 1997;120:1151–7.
87. Hanisch UK, Seto D, Quirion R. Modulation of hippocampal acetylcholine release: a potent central action of interleukin. J Neurosci 1993;2:3368–3374.
88. Mameya S, Sawa T, Taniyama K. Arachidonic acid cascade and stimulation of acetylcholine release by human recombinant interleukin-1 beta in guinea pig ileum. J Pharmacol Exp Ther 1995;275: 319–24.
89. Kelles A, Janssens J, Tack J. IL-1beta and IL-6 excite neurones and suppress cholinergic neurotransmission in the myenteric plexus of the guinea pig. Neurogastroenterol Motil 2000;12:531–8.
90. Wakade AR, Wakade TD, Poosch M, Bannon MJ. Noradrenaline transport and transporter mRNA of rat chromaffin cells are controlled by dexamethasone and nerve growth factor. J Physiol 1996;494: 67–75.
91. Kaye DM, Wiviott SD, Kobzik L, Kelly RA, Smith TW. S-nitrosothiols inhibit neuronal norepinephrine transport. Am J Physiol 1997;272:H875–83.
92. Ren ZG, Porzgen P, Zhang JM, Chen XR, Amara SG, Blakely RD, Sieber-Blum M. Autocrine regulation of norepinephrine transporter expression. Mol Cell Neurosci 2001;17:539–50.
93. Beasley D, Cohen RA, Levinsky NG. Interleukin 1 inhibits contraction of vascular smooth muscle. J Clin Invest 1989;83:331–5.
94. McKenna TM, Reusch DW, Simpkins CO. Macrophage-conditioned medium and interleukin 1 suppress vascular contractility. Circ Shock 1988;25:187–96.
95. Van Oosterhout AJ, Stam WB, Nijkamp FP. Effect of lymphokines on beta-adrenoceptor stimulation of human PBMC. Agents Actions Suppl 1990;31:163–7.
96. Van Oosterhout AJ, Stam WB, Vanderschueren RG, Nijkamp FP. Effects of cytokines on beta-adrenoceptor function of human peripheral blood mononuclear cells and guinea pig trachea. J Allergy Clin Immunol 1992;90:340–8.
97. Cazaux CA, Sterin-Borda L, Gorelik G, Cremaschi GA. Down-regulation of beta-adrenergic receptors induced by mitogen activation of intracellular signaling events in lymphocytes. FEBS Lett 1995;364:120–4.
98. Gulick T, Chung MK, Pieper SJ, Lange LG, Schreiner GF. Interleukin 1 and tumor necrosis factor inhibit cardiac myocyte beta-adrenergic responsiveness. Proc Natl Acad Sci USA 1989;86:6753–7.
99. Reithmann C, Gierschik P, Werdan K, Jakobs KH. Tumor necrosis factor alpha up-regulates Gi alpha and G beta proteins and adenylyl cyclase responsiveness in rat cardiomyocytes. Eur J Pharmacol 1991;206:53–60.
100. Muller-Werdan U, Schumann H, Fuchs R, Reithmann C, Loppnow H, Koch S, Zimny-Arndt U, He C, Darmer D, Jungblut P, Stadler J, Holtz J, Werdan K. Tumor necrosis factor alpha (TNF alpha) is cardiodepressant in pathophysiologically relevant concentrations without inducing inducible nitric oxide-(NO)-synthase (iNOS) or triggering serious cytotoxicity. J Mol Cell Cardiol 1997;29:2915–23.
101. Sun X, Delbridge LM, Dusting GJ. Cardiodepressant effects of interferon-gamma and endotoxin reversed by inhibition of NO synthase 2 in rat myocardium. J Mol Cell Cardiol 1998;30:989–97.
102. Liu SJ, Zhou W, Kennedy RH. Suppression of beta-adrenergic responsiveness of L-type Ca2+ current by IL-1beta in rat ventricular myocytes. Am J Physiol 1999;276:H141–8.
103. Roupe van der Voort C, Heijnen CJ, Wulffraat N, Kuis W, Kavelaars A. Stress induces increases in IL-6 production by leucocytes of patients with the chronic inflammatory disease juvenile rheumatoid arthritis: a putative role for alpha(1)-adrenergic receptors. J Neuroimmunol 2000;110:223–9.
104. Benzing T, Brandes R, Sellin L, Schermer B, Lecker S, Walz G, Kim E. Upregulation of RGS7 may contribute to TNF-induced changes of central nervous function. Nat Med 1999;5:913–18.
105. Moratz C, Kang VH, Druey KM, Shi CS, Scheschonka A, Murphy PM, Kozasa T, Kehrl JH. Regulator of G protein signalling 1 (RGS1) markedly impairs Gi alpha signaling responses of B lymphocytes. J Immunol 2000;164:1829–38.
106. Kehrl JH. Heterotrimeric G protein signaling: roles in immune function and fine-tuning by RGS proteins. Immunity 1998;8:1–10.

107. Straub RH, Männel DN. How the immune system puts the brain to sleep. Nat Med 1999;5:877–9.
108. Luttrell LM, Ferguson SS, Daaka Y, Miller WE, Maudsley S, Della Rocca GJ, Lin F, Kawakatsu H, Owada K, Luttrell DK, Caron MG, Lefkowitz RJ. Beta-arrestindependent formation of beta2 adrenergic receptor-Src protein kinase complexes. Science 1999;283:655–61.
109. Lombardi MS, Kavelaars A, Schedlowski M, Bijlsma JW, Okihara KL, Van de PM, Ochsmann S, Pawlak C, Schmidt RE, Heijnen CJ. Decreased expression and activity of G-protein-coupled receptor kinases in peripheral blood mononuclear cells of patients with rheumatoid arthritis. FASEB J 1999;13:715–25.
110. Lombardi MS, Kavelaars A, Cobelens PM, Schmidt RE, Schedlowski M, Heijnen CJ. Adjuvant arthritis induces down-regulation of G protein-coupled receptor kinases in the immune system. J Immunol 2001;166:1635–40.
111. Goldhill J, Morris SC, Maliszewski C, Urban JF, Jr., Funk CD, Finkelman FD, Shea-Donohue T. Interleukin-4 modulates cholinergic neural control of mouse small intestinal longitudinal muscle. Am J Physiol 1997;272:G1135–40.
112. Xia Y, Hu HZ, Liu S, Ren J, Zafirov DH, Wood JD. IL-1beta and IL-6 excite neurons and suppress nicotinic and noradrenergic neurotransmission in guinea pig enteric nervous system. J Clin Invest 1999;103:1309–16.
113. Akasu T, Tsurusaki M. Effects of interleukin-1 beta on neurons in mammalian pelvic ganglia. Kurume Med J 1999;46:143–50.
114. Wallace JL, Cucala M, Mugridge K, Parente L. Secretagogue-specific effects of interleukin-1 on gastric acid secretion. Am J Physiol 1991;261:G559–64.
115. Straub RH, Linde HJ, Männel DN, Schölmerich J, Falk W. A bacteria-induced switch of sympathetic effector mechanisms augments local inhibition of TNF-alpha and IL-6 secretion in the spleen. FASEB J 2000;14:1380–8.
116. Rice PA, Boehm GW, Moynihan JA, Bellinger DL, Stevens SY. Chemical sympathectomy increases the innate immune response and decreases the specific immune response in the spleen to infection with Listeria monocytogenes. J Neuroimmunol 2001;114:19–27.
117. Hasko G, Elenkov IJ, Kvetan V, Vizi ES. Differential effect of selective block of alpha 2-adrenoreceptors on plasma levels of tumour necrosis factor-alpha, interleukin-6 and corticosterone induced by bacterial lipopolysaccharide in mice. J Endocrinol 1995;144:457–62.
118. Tamaoki J, Yamawaki I, Takeyama K, Chiyotani A, Yamauchi F, Konno K. Interleukin-1 beta inhibits airway smooth muscle contraction via epithelium-dependent mechanism. Am J Respir Crit Care Med 1994;149:134–7.
119. Anticevich SZ, Hughes JM, Black JL, Armour CL. Induction of human airway hyperresponsiveness by tumour necrosis factor-alpha. Eur J Pharmacol 1995;284:221–5.
120. Hakonarson H, Herrick DJ, Serrano PG, Grunstein MM. Autocrine role of interleukin 1beta in altered responsiveness of atopic asthmatic sensitized airway smooth muscle. J Clin Invest 1997;99:117–24.
121. Koto H, Mak JC, Haddad EB, Xu WB, Salmon M, Barnes PJ, Chung KF. Mechanisms of impaired beta-adrenoceptor-induced airway relaxation by interleukin-1beta in vivo in the rat. J Clin Invest 1996;98:1780–7.
122. Hakonarson H, Maskeri N, Carter C, Grunstein MM. Regulation of TH1- and TH2-type cytokine expression and action in atopic asthmatic sensitized airway smooth muscle. J Clin Invest 1999;103: 1077–87.
123. Elbon CL, Jacoby DB, Fryer AD. Pretreatment with an antibody to interleukin-5 prevents loss of pulmonary M2 muscarinic receptor function in antigen-challenged guinea pigs. Am J Respir Cell Mol Biol 1995;12:320–8.
124. Adamko DJ, Yost BL, Gleich GJ, Fryer AD, Jacoby DB. Ovalbumin sensitization changes the inflammatory response to subsequent parainfluenza infection. Eosinophils mediate airway hyperresponsiveness, m(2) muscarinic receptor dysfunction, and antiviral effects. J Exp Med 1999;190: 1465–78.
125. Schultzberg M, Andersson C, Unden A, Troye-Blomberg M, Svenson SB, Bartfai T. Interleukin-1 in adrenal chromaffin cells. Neuroscience 1989;30:805–10.

126. Currie KP, Zhou Z, Fox AP. Evidence for paracrine signaling between macrophages and bovine adrenal chromaffin cell Ca(2+) channels. J Neurophysiol 2000;83:280–7.
127. Yanagihara N, Minami K, Shirakawa F, Uezono Y, Kobayashi H, Eto S, Izumi F. Stimulatory effect of IL-1 beta on catecholamine secretion from cultured bovine adrenal medullary cells. Biochem Biophys Res Commun 1994;198:81–7.
128. Lujan HJ, Mathews HL, Gamelli RL, Jones SB. Human immune cells mediate catecholamine secretion from adrenal chromaffin cells. Crit Care Med 1998;26:1218–24.
129. Eskay RL, Eiden LE. Interleukin-1 alpha and tumor necrosis factor-alpha differentially regulate enkephalin, vasoactive intestinal polypeptide, neurotensin, and substance P biosynthesis in chromaffin cells. Endocrinology 1992;130:2252–8.
130. Abreu P, Llorente E, Hernandez MM, Gonzalez MC. Interleukin-1 beta stimulates tyrosine hydroxylase activity in the median eminence. Neuroreport 1994;5:1356–8.
131. Aloe L, Fiore M. TNF-alpha expressed in the brain of transgenic mice lowers central tyroxine hydroxylase immunoreactivity and alters grooming behavior. Neurosci Lett 1997;238:65–8.
132. Unsicker K, Westermann R. Basic fibroblast growth factor promotes transmitter storage and synthesis in cultured chromaffin cells. Brain Res Dev Brain Res 1992;65:211–16.
133. Erkman L, Wuarin L, Cadelli D, Kato AC. Interferon induces astrocyte maturation causing an increase in cholinergic properties of cultured human spinal cord cells. Dev Biol 1989;132:375–88.
134. Jonakait GM. Neural-immune interactions in sympathetic ganglia. Trends Neurosci 1993;16:419–23.
135. Oh YJ, O'Malley KL. IL-6 increases choline acetyltransferase but not neuropeptide transcripts in sympathetic neurons. Neuroreport 1994;5:937–40.
136. Fujii T, Yamada S, Watanabe Y, Misawa H, Tajima S, Fujimoto K, Kasahara T, Kawashima K. Induction of choline acetyltransferase mRNA in human mononuclear leukocytes stimulated by phytohemagglutinin, a T-cell activator. J Neuroimmunol 1998;82:101–7.
137. Rinner I, Schauenstein K. Detection of choline-acetyltransferase activity in lymphocytes. J Neurosci Res 1993;35:188–91.
138. Rinner I, Kawashima K, Schauenstein K. Rat lymphocytes produce and secrete acetylcholine in dependence of differentiation and activation. J Neuroimmunol 1998;81:31–7.
139. Rostami AM. Pathogenesis of immune-mediated neuropathies. Pediatr Res 1993;33:S90–4.
140. Taubner RW, Salanova V. Acute dysautonomia and polyneuropathy. Arch Neurol 1984;41:1100–1.
141. Feiden W, Gerhard L, Borchard F. Neuritis cordis due to the acute polyneuritis of the Guillain-Barre syndrome. Virchows Arch A Pathol Anat Histopathol 1988;413:573–80.
142. Ewing DJ, Campbell IW, Clarke BF. The natural history of diabetic autonomic neuropathy. Q J Med 1980;49:95–108.
143. Ziegler D, Laux G, Dannehl K, Spuler M, Muhlen H, Mayer P, Gries FA. Assessment of cardiovascular autonomic function: age-related normal ranges and reproducibility of spectral analysis, vector analysis, and standard tests of heart rate variation and blood pressure responses. Diabet Med 1992;9:166–75.
144. Edmonds ME, Jones TC, Saunders WA, Sturrock RD. Autonomic neuropathy in rheumatoid arthritis. Br Med J 1979;2:173–5.
145. Mellgren SI, Conn DL, Stevens JC, Dyck PJ. Peripheral neuropathy in primary Sjogren's syndrome. Neurology 1989;39:390–4.
146. Lindgren S, Lilja B, Rosen I, Sundkvist G. Disturbed autonomic nerve function in patients with Crohn's disease. Scand J Gastroenterol 1991;26:361–6.
147. Lindgren S, Stewenius J, Sjolund K, Lilja B, Sundkvist G. Autonomic vagal nerve dysfunction in patients with ulcerative colitis. Scand J Gastroenterol 1993;28:638–42.
148. Straub RH, Antoniou E, Zeuner M, Gross V, Schölmerich J, Andus T. Association of autonomic nervous hyperreflexia and systemic inflammation in patients with Crohn's disease and ulcerative colitis. J Neuroimmunol 1997;80:149–57.
149. Fernandez A, Hontebeyrie M, Said G. Autonomic neuropathy and immunological abnormalities in Chagas' disease. Clin Auton Res 1992;2:409–12.
150. Leden I, Eriksson A, Lilja B, Sturfelt G, Sundkvist G. Autonomic nerve function in rheumatoid arthritis of varying severity. Scand J Rheumatol 1983;12:166–70.

151. Toussirot E, Serratrice G, Valentin P. Autonomic nervous system involvement in rheumatoid arthritis. 50 cases. J Rheumatol 1993;20:1508–14.
152. Tan J, Akin S, Beyazova M, Sepici V, Tan E. Sympathetic skin response and R-R interval variation in rheumatoid arthritis. Two simple tests for the assessment of autonomic function. Am J Phys Med Rehabil 1993;72:196–203.
153. Hoyle C, Ewing DJ, Parker AC. Acute autonomic neuropathy in association with systemic lupus erythematosus. Ann Rheum Dis 1985;44:420–4.
154. Klimiuk PS, Taylor L, Baker RD, Jayson MI. Autonomic neuropathy in systemic sclerosis. Ann Rheum Dis 1988;47:542–5.
155. Gledhill RF, Dessein PH. Autonomic neuropathy in systemic lupus erythematosus. J Neurol Neurosurg Psychiatry 1988;51:1238–40.
156. Liote F, Osterland CK. Autonomic neuropathy in systemic lupus erythematosus: cardiovascular autonomic function assessment. Ann Rheum Dis 1994;53:671–4.
157. Straub RH, Zeuner M, Lock G, Rath H, Hein R, Schölmerich J, Lang B. Autonomic and sensorimotor neuropathy in patients with systemic lupus erythematosus and systemic sclerosis. J Rheumatol 1996;23:87–92.
158. Altomonte L, Mirone L, Zoli A, Magaro M. Autonomic nerve dysfunction in systemic lupus erythematosus: evidence for a mild involvement. Lupus 1997;6:441–4.
159. Gamez-Nava JI, Gonzalez-Lopez L, Ramos-Remus C, Fonseca-Gomez MM, Cardona-Munoz EG, Suarez-Almazor ME. Autonomic dysfunction in patients with systemic lupus erythematosus. J Rheumatol 1998;25:1092–6.
160. Gemignani F, Manganelli P, Pavesi G, Marbini A. Polyneuropathy in Sjogren's syndrome. A case of prevalently autonomic neuropathy with tonic pupil and hypohidrosis. Funct Neurol 1988;3:337–48.
161. Mandl T, Jacobsson L, Lilja B, Sundkvist G, Manthorpe R. Disturbances of autonomic nervous function in primary Sjogren's syndrome. Scand J Rheumatol 1997;26:253–8.
162. Straub RH, Glück T, Zeuner M, Schölmerich J, Lang B. Association of pupillary parasympathetic hyperreflexia and systemic inflammation in patients with systemic lupus erythematosus. Br J Rheumatol 1998;37:665–70.
163. Glück T, Oertel M, Reber T, Zietz B, Schölmerich J, Straub RH. Altered function of the hypothalamic stress axes in patients with moderately active systemic lupus erythematosus. I. The hypothalamus-autonomic nervous system axis. J Rheumatol 2000;27:903–10.
164. Straub RH, Herfarth H, Falk W, Andus T, Scholmerich J. Uncoupling of the sympathetic nervous system and the hypothalamic-pituitary-adrenal axis in inflammatory bowel disease? J Neuroimmunol 2002;126:116–25.
165. Mandell BF. Cardiovascular involvement in systemic lupus erythematosus. Semin Arthritis Rheum 1987;17:126–41.
166. Ward MM. Premature morbidity from cardiovascular and cerebrovascular diseases in women with systemic lupus erythematosus. Arthritis Rheum 1999;42:338–46.
167. Seleznick MJ, Fries JF. Variables associated with decreased survival in systemic lupus erythematosus. Semin Arthritis Rheum 1991;21:73–80.
168. Ginzler EM, Felson DT, Anthony JM, Anderson JJ. Hypertension increases the risk of renal deterioration in systemic lupus erythematosus. J Rheumatol 1993;20:1694–700.
169. Julius S. Autonomic nervous dysfunction in essential hypertension. Diabetes Care 1991;14:249–59.
170. Perry F, Heller PH, Kamiya J, Levine JD. Altered autonomic function in patients with arthritis or with chronic myofascial pain. Pain 1989;39:77–84.
171. Nakajima A, Sendo W, Tsutsumino M, Koseki Y, Ichikawa N, Akama H, Taniguchi A, Terai C, Hara M, Kashiwazaki S. Acute sympathetic hyperfunction in overlapping syndromes of systemic lupus erythematosus and polymyositis. J Rheumatol 1998;25:1638–41.

# IV. THE NEUROIMMUNE BIOLOGY OF DISEASE

*The Hypothalamus–Pituitary–Adrenal Axis*
Edited by A. del Rey, G. P. Chrousos and H. O. Besedovsky



# *The Neuroendocrinology and Immunology of Critical Illness*

ISTVAN VERMES[1] and ALBERTUS BEISHUIZEN[2]

[1]*Department of Clinical Chemistry, Medical Spectrum Twente, Hospital Group, Enschede, The Netherlands*
[2]*Department of Intensive Care, VU University Medical Center, Amsterdam, The Netherlands*

## ABSTRACT

Life-threatening disease is a severe stress stimulus, which induces an acute activation of the three major regulatory systems of the body: the nervous system, the endocrine system, and the immune system. The neuroendocrine response to critical illness is a dynamic process with an acute phase (the stage of alarm reaction) and a chronic phase (the stage of resistance), both with different and multiple hormonal alterations. During the acute phase of the critical illness, the adrenocorticotrophic hormone (ACTH)-induced cortisol secretion occurs within minutes and there is a shift toward an increase in glucocorticoid and away from mineralocorticoid and androgen production in the adrenocortical cells. The growth hormone (GH) concentrations in the peripheral blood and the pulse frequency are elevated; however, the peripheral concentrations of insulin-like growth factor (IGF) and insulin-like growth factor binding protein 3 (IGFBP3) are decreased which may reflect a receptor-mediated peripheral GH resistance. Thyroid hormone metabolism is commonly affected by critical illness: Tri-iodothyronin concentration is low but thyroid-stimulating hormone (TSH) and thyroxin concentrations rise briefly and return to normal levels. During the acute alarm phase, a low serum testosterone concentration is present.

Teleologically, the acute changes in the neuroendocrine system indicate an appropriate hormonal reaction: the mobilization of fuel stores of the organism together with apparent restrains on their utilization. However, if the critical illness is prolonged or the defense response is insufficient, the hormonal response can contribute to a worsened clinical status with the patient entering the wasting syndrome. During this chronic phase, the plasma ACTH level is paradoxically low but the total cortisol concentration is still elevated. The somatotropic axis is not activated anymore, the tri-iodothyronin concentration remains low, and a temporary hypogonadotropic hypogonadism develops. Furthermore, hyperglycemia and insulin resistance accompany the wasting syndrome showing the catabolic nature of prolonged critical illness.

## 1. NEUROENDOCRINE–IMMUNE INTERACTIONS DURING CRITICAL ILLNESS

The pathogenesis of critical illness, in particular sepsis and the multiple organ dysfunction syndrome (MODS), involves a very complex interplay between the neuroendocrine and the immune systems [1]. The neuroendocrine and immune systems are interrelated via a bidirectional network through which hormones and neuropeptides affect immune functions and, in turn, immune responses are reflected in neuroendocrine changes [2,3]. Critical illness is a severe stress stimulus that disturbs the *milieu interieur* and induces homeostatic responses specific to the stimulus and generalized responses when the disturbances are prolonged and severe [4,5]. Stress can be defined as any physical circumstance that threatens to disrupt homeostasis, the response of the organism being a necessary defense mechanism directed against the stressor to maintain the homeostasis [6,7]. Conditions of extreme stress can be caused by insults such as septic or cardiogenic shock, multiple trauma, or extensive burns. The body is equipped with natural defense systems to face a great diversity of insults such as the above. The three major regulatory systems of the body – the nervous system, the endocrine system, and the immune system – protect the organism by responding to internal and external stress [6,7].

The immune–neuroendocrine response during critical illness is a dynamic process, differing between the acute and chronic phases (Fig. 1) [4,5]. The acute-phase response to critical illness is well recognized with abrupt and massive release of stress hormones including ACTH, cortisol, catecholamines, vasopressin, glucagons, and growth hormone (GH). There is a concurrent shutdown of less vital systems such as gonadal function, and anabolism is inhibited through various pathways. One manifestation of the acute response to critical illness is the induction of insulin resistance, a metabolic state in which a normal concentration of insulin produces a subnormal biological response [8–10]. These acute changes provide optimal conditions for the fight, including metabolic changes, optimal intravascular volume, and perfusion pressure. During this phase, the body provides additional metabolic substrates in the form of increased availability of glucose, amino acids, free-fatty acids (FFA), and heightened synthesis of both mitochondrial and non-mitochondrial adenosine triphosphate (ATP), which are mainly used by the central nervous system (CNS) and in establishing the host's defenses [11]. These acute alterations are adaptive and beneficial as they reduce and redirect energy consumption as well as postpone anabolism. This acute phase lasts typically from about a few hours to several days, depending on the severity of illness and is characterized by an appropriate hormonal reaction: the mobilization of fuel stores of the organism, together with apparent restrains on

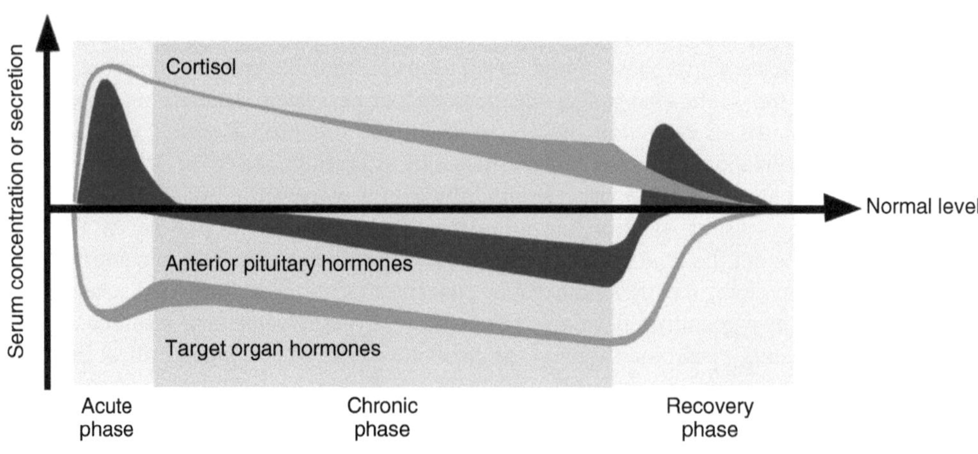


Figure 1. Hormonal changes during the course of critical illness. During the acute phase, the secretory activity of the pituitary is amplified and during the chronic phase, it is suppressed. The secretory activity of the target organs is suppressed during both phases with the exception of the adrenal cortex. The onset of recovery is characterized by the restoration of pituitary functions. Reproduced from Van den Berghe G, de Zegher F and Bouillon R (Acute and prolonged critical illness as different neuroendocrine paradigms. J Clin Endocrinol Metab 1998; 83: 1827–1834. © The Endocrine Society) with permission.

their utilization. If the process of recovery does not start within a few days, illness can become critical and often prolonged and organ system support is needed for survival [4,5].

The development of intensive care medicine changed the natural course of the severe illness-induced stress response of the organism. Patients who would previously have died during the acute phase nowadays survive, as a result of highly technological interventions. Nature has been unable to select coping mechanisms for the prolonged phase of the critical illness. On the contrary, if the critical illness is prolonged or the defense response is insufficient, the hormonal response can contribute to a worsened clinical status. In other words, the highly technological interventions in the natural course of the dying process have unmasked previously unknown clinical conditions and have in fact created a new phase in the adaptation process during critical illness: the chronic or prolonged critical illness [4,5]. This catabolic phase lasts for days and is characterized by an inappropriate hormonal response, resulting in a chronic increase in metabolic rate and a breakdown of body tissue also called the wasting syndrome [5,8]. In this phase, the hormonal profile alters substantially with concentrations of vasopressin that seem inappropriately low, onset of the sick euthyroid (T3) syndrome, low ACTH concentrations, and reduced adrenal responsiveness to ACTH. Some of these endocrine changes can themselves be the consequence of the acute phase response. The secretion of GH can be inhibited by cortisol, whereas gonadotropins are suppressed by both cortisol and prolactin. Cortisol also influences thyroid hormone metabolism [11]. The physiological hormonal homeostasis involves the balance between catabolic and anabolic hormonal responses. In the stage of exhaustion this balance is shifted in the direction of catabolic events. The body no longer uses FFA efficiently as metabolic substrate, and a large amount of proteins is lost from skeletal muscle and organs, causing impairment of vital functions. Consequently, hyperglycemia due to insulin resistance, hypoproteinemia, hypercalcemia, intracellular water and potassium depletion, and hypertriglyceridemia are present during the wasting syndrome [5,8,11].

Fortunately, the chronic phase, due to the development of intensive care medicine over the past 20 years, more often leads to the recovery phase. In this case, the homeostatic balance is redirected toward anabolic changes, normalization of hormone secretion together with restoration of peripheral feedback regulation. The importance of an intact hypothalamic–pituitary–adrenal (HPA) axis for metabolic and immunological homeostasis during critical illness is well recognized. This chapter gives an overview of the neuroendocrine responses during critical illness.

## 2. NEUROENDOCRINOLOGY OF CRITICAL ILLNESS

The neuroendocrine response to critical illness is a dynamic process with an acute phase (the stage of alarm reaction) and a chronic phase (the stage of resistance) both with different multiple hormonal alterations (Table 1). The acute phase, the fight or flight response, appears both necessary and non-injurious to the human organism. However, the chronic phase of illness sets in during more prolonged illness, with a characteristic pattern of hormonal changes in association with a catabolic state and immune paralysis, resulting in predisposition to infection and catecholamine dependency of the critically ill patient.

### 2.1. Hypothalamic–pituitary–adrenal axis

Life-threatening disease is a severe stress stimulus, which induces an acute activation of the HPA axis [7]. Activation of the HPA axis with the release of cortisol is an essential component of the general adaptation to illness and stress and contributes to the maintenance of cellular and

Table 1 Hormonal changes during critical illness

| Hormone | Acute phase | Chronic phase | Key references |
|---|---|---|---|
| *HPA axis* | | | |
| CRH | ⇑ | ⇑ | [6] |
| ACTH | ⇑ | ⇓ | [7] |
| Cortisol | ⇑ | ⇑ | [7,8] |
| DHEA(S) | ⇓⇓ | ⇓ | [9] |
| Aldosterone | ⇑ | ⇓ | [10,11] |
| CBG | ⇓⇓ | ⇓ | [12,13] |
| *Somatotropic axis* | | | |
| GH | ⇑ | ⇓ | [14] |
| IGF-I | ⇓ | ⇓ | [15,16] |
| IGFBP3 | ⇓ | ⇓ | [15] |
| *Pituitary–thyroid axis* | | | |
| TSH | N | ⇓ | [17,18] |
| TT4 | N | ⇓ | [19] |
| TT3 | ⇓ | ⇓ | [17] |
| rT3 | ⇑ | ⇑ | [20,21] |
| FT4 | N | ⇓ | [22] |
| FT3 | ⇓ | ⇓ | [22,23] |
| TBG | ? | ? | |
| *Reproductive axis* | | | |
| LH | ⇑ | ⇓ | [24] |
| FSH | N | ⇓ | [25] |
| Testosterone | ⇓ | ⇓⇓ | [24] |
| Estradiol | ⇑ | ⇑ | [26,27] |
| Prolactin | ⇑ | N | [28,29] |

organ homeostasis [12]. Stress-induced hypercorticolism is associated with augmented ACTH release, driven by corticotropin-releasing hormone (CRH), cytokines, and the noradrenergic system. Without medical intervention, the organism will pass through the three stages of the general adaptation syndrome: the stage of alarm reaction, the stage of resistance, and the stage of exhaustion [13]. The serum cortisol concentration is elevated in patients with critical illness and in the acute phase associated with enhanced ACTH (Fig. 2) [14]. The production of cortisol in the adrenocortical cells in response to ACTH occurs within minutes and there is a shift toward an increase in glucocorticoid, and away from mineralocorticoid and androgen production [15,16]. An inverse correlation between cortisol levels, illness severity, and a fatal outcome has been suggested [17–19]. There is much controversy regarding the levels of cortisol that are considered adequate in critical illness, but general agreement indicates that isolated plasma cortisol levels have limited predictive value [20].

The serum dehydroepiandrosterone (DHEA) and DHEA sulfate (DHEAS) levels are significantly decreased in these patients [17,21–23] (Fig. 3). The cortisol-binding globulin (CBG) concentration and the CBG-binding affinity are both decreased by as much as 50% [24–27], resulting in a dramatic increase in the biologically active free cortisol concentration (Fig. 4). In addition to decreased hepatic synthesis of CBG, elastase secreted by activated neutrophils at the site of inflammation has been reported to cleave CBG, resulting in preferential delivery of free cortisol to target cells at the site of inflammation.

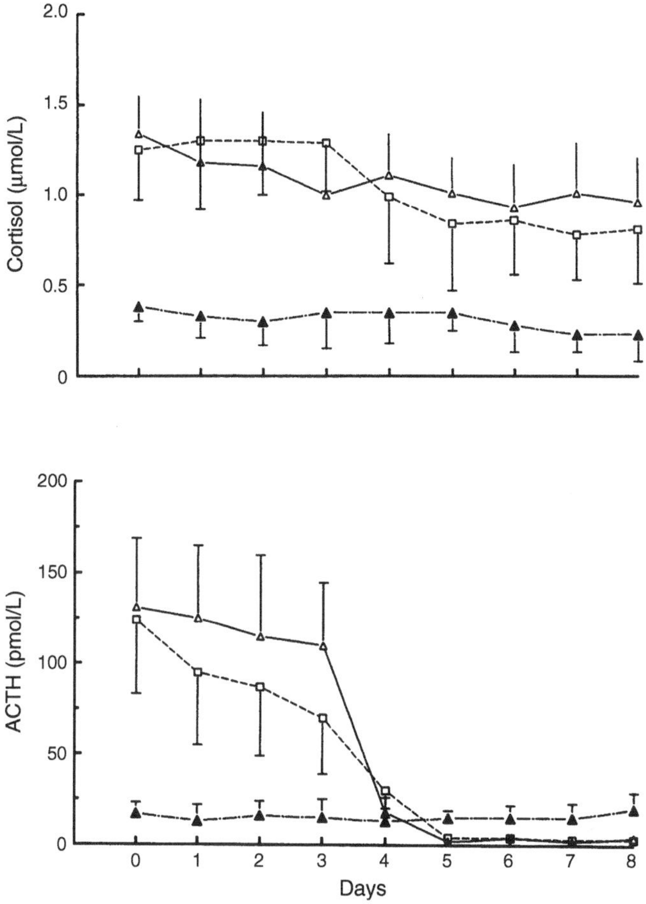


Figure 2. Plasma concentrations of immunoreactive ACTH and cortisol in patients with sepsis (△) and multiple trauma (□) and in control subjects (▲) during the first 8 days after admission to intensive care unit. Reproduced from Vermes I, Beishuizen A, Hampsink RM and Haanen C (Dissociation of plasma adrenocorticotropin and cortisol levels in critically ill patients: possible role of endothelin and atrial natriuretic hormone. J Clin Endocrinol Metab 1995; 80: 1238–42. © The Endocrine Society) with permission.

In addition, the glucocorticoid receptor (GR) shows an increased activity to free hormone during the acute phase of critical illness [28]. Proinflammatory cytokines have been shown to influence the number, expression, and function of the GR. Accordingly, the critical illness-induced changes of the biological effects of cortisol are even more pronounced than the observed increase in the plasma total cortisol concentration. Therefore, the total circulating cortisol level is a poor indicator of glucocorticoid activity at the cellular/nuclear level. Indeed, recently it was shown that during critical illness glucocorticoid secretion was markedly increased, but this increase was not discernible when only the serum total cortisol concentration was measured [29]. In this study, nearly 40% of critically ill patients with hypoproteinemia had subnormal serum total cortisol concentration, even though their adrenal function was normal. Accordingly, the measurement of plasma free cortisol or the free cortisol index is likely to be more relevant than the total plasma cortisol in the assessment of adrenal function in critical illness [30]. However, clear clinical data to prove this are currently lacking.

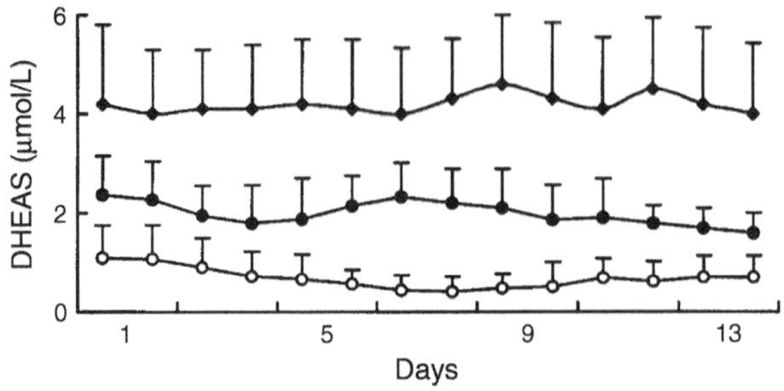


Figure 3. The serum concentration of dehydroepiandrosterone sulfate (DHEAS) in patients with septic shock (o), multiple trauma (•), and age- and sex-matched control subjects (■). Reproduced from Beishuizen A, Thijs LG, Vermes I (Decreased levels of dehydroepiandrosterone sulphate in severe critical illness: a sign of exhausted adrenal reserve? Crit care 2002; 6: 434–8) with permission.

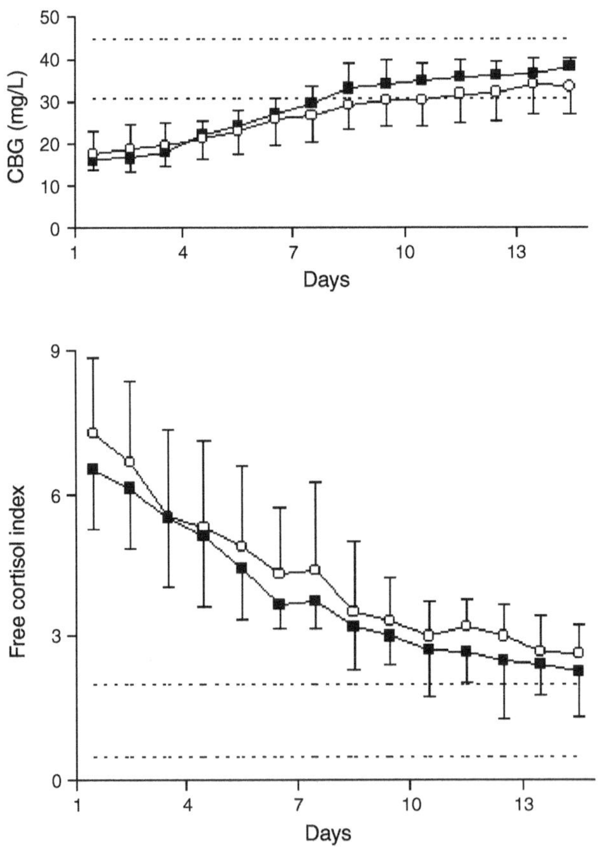


Figure 4. The time-course of cortisol binding globulin (CBG) concentrations and the free cortisol index (FCI) in patients with septic shock (□) and multitrauma (■) during the observation period of 14 days. The range of normal values, derived from healthy volunteers, is displayed between broken lines. Reproduced from Beishuizen A, Thijs LG, Vermes I (Patterns of corticosteroid-binding globulin and the free cortisol index during septic shock and multitrauma. Intensive Care Med 2001; 27: 1584–91) with permission.

Cortisol is a pluripotent hormone, acting on multiple tissues to regulate numerous aspects of metabolism, growth, and physiological functioning, being in this way essential for survival in critical illness [31,32]. Cortisol has a vital supportive role in the maintenance of vascular tone, endothelial integrity, vascular permeability, and the distribution of total body water within the vascular compartment, and potentiates the vasoconstrictory actions of catecholamines [7,12,33]. Although cortisol is required for survival, excessive amounts can be life-threatening via the well-known immunosuppressive effects of glucocorticoids and glucocorticoid-induced acute and metabolic abnormalities [4,5]. Accordingly, when critical illness is protracted, the patient enters a stress-conditioned stage of HPA adaptation. During this chronic phase of critical illness, the plasma ACTH level is low, but the total cortisol concentration is still elevated (Fig. 2). This dissociation cannot be explained by the different metabolic kinetics of ACTH and cortisol [34,35], which suggests that cortisol secretion may be driven via an alternative, non-ACTH mediated pathway [36].

There is a close bidirectional link between the HPA axis and the adrenomedullary response during stress, receiving input from the nervous and immune systems [37]. There is an intensive intermingling between the cortical steroid-producing and the medullary catecholamine-producing cells suggesting paracrine and juxtacrine interactions within the adrenal cortex [38,39]. Chromaffine cells and intramedullary immune cells are able to produce CRH and ACTH with intact biological activity, and in this way ACTH synthesized upon locally produced "tissue CRH" could stimulate adrenal glucocorticoid production in the absence of pituitary ACTH [40]. One can only speculate about the role of this non-ACTH-mediated regulation of the adrenal cortex during critical illness. The dissociation between the plasma ACTH and serum cortisol levels that we [14,41] and others [34,35,42–45] have observed during sepsis and multiple trauma suggests an important role for immunological factors to maintain the adrenocortical activation. Cytokines seem to play an important role in conditions of prolonged stress, during which CRH and ACTH are inhibited, despite continued elevation in circulating cortisol levels. Furthermore, during prolonged stress, the circadian rhythm of cortisol production is abolished, and the negative feedback from cortisol is attenuated. Failure to terminate a response to acute stress can result in the chronic stress syndrome, which causes profound endocrine, metabolic, psychiatric, and immunological alterations.

The question remains, however, whether the high total serum cortisol concentration represents sufficient glucocorticoid biological activity for the prolonged phase of conditioned stress adaptation. There are many indications in the literature that, during the prolonged phase of critical illness, a relative adrenocortical insufficiency, despite an elevated total serum cortisol level, can be present in intensive care patients [18,20,32,46–51]. The persistently decreased DHEA(S) level and the dissociation between the reduced aldosterone secretion and the elevated renin level (hyper-reninemic hypoaldosteronism) [52–54] are suggestive of an exhaustion of the adrenocortical secretory reserve capacity. In addition, the low DHEA(S) level might be an indication that the severe disease is nearing the stage of exhaustion [23,55]. The high glucocorticoid activity, together with the low DHEA(S) level, suggests an imbalance between immunosuppressive and immunostimulatory hormones of adrenocortical origin, which might be the cause of the increased susceptibility to infectious complications during the chronic phase of severe illness [56]. Increased glucocorticoid action is an essential component of the neuroendocrine adaptation during stress and therefore even minor degrees of adrenal insufficiency can be fatal in the stressed host [50]. Consequently, recognition of adrenal insufficiency in critical illness may be considered lifesaving.

Absolute adrenal insufficiency is uncommon in critically ill patients. However, relative adrenal insufficiency, when the cortisol level despite being normal or high is still considered to be functionally insufficient to control the inflammatory response, is much more common in ICU patients [57]. The definition and the diagnosis of this relative adrenocortical insufficiency continue to generate much debate [18–20,50,51,58,59]. Alternative terms have been proposed to describe dysfunction of the HPA axis such as "critical illness-related corticosteroid insufficiency" which may be defined as inadequate corticosteroid activity for the severity of the patient's illness. The diagnosis of relative adrenal insufficiency is not clear but should be suspected in any patient with septic shock who does not respond to conventional treatment. Since there are no distinguishing clinical characteristics of those with adrenal insufficiency, all patients suffering from treatment failure should be investigated [57]. The (high-dose) ACTH-stimulation test is usually used to assess adrenocortical function, but there is an area of controversy concerning the dose of the ACTH and the interpretation of cut-off values. In addition, as total cortisol is measured using ACTH testing, a proper interpretation should be supported by measuring the CBG level and measuring or calculating the free cortisol index.

Adrenal or HPA dysfunction is usually a reversible condition caused by proinflammatory mediators; however, it may also arise due to structural damage of the adrenal gland. It may have an impact on the balance between pro- and anti-inflammatory pathways and thereby influencing immune, metabolic, vascular, and organ dysfunction.

### 2.2. Somatotropic axis: GH–IGF–IGFBP

GH is a polypeptide with anabolic, immunomodulatory, and lypolytic properties. GH is released from the pituitary somatotropic cells and its secretion is under the interactive hypothalamic control of the stimulatory GH-releasing hormone (GHRH) and the inhibitory somatostatin and occurs in pulsatile fashion with diurnal variation [60]. GH action is reflected by the serum concentration of insulin-like growth factors (IGFs), which are generated by the liver and bound to specific binding proteins (IGFBPs). During the acute phase of the critical illness, both GH concentration in the peripheral blood and the pulse frequency are elevated [61–63]. This is likely to be due to the stress-induced hypothalamic activation, which could be mediated via the withdrawal of the inhibitory somatostatin tone, or via the increase of the stimulatory GHRH tone. However, the peripheral concentrations of IGF and IGFBP3 are decreased which reflect a receptor-mediated peripheral GH resistance [64–66]. It has been suggested that the reduced GH receptor expression-induced low blood IGF-I level is the primary event in this circuit, which via a decreased negative feedback inhibition evokes the pituitary release of GH during the acute phase of a critical illness [67]. This discrepancy between the circulating GH and IGF-I levels is a part of the neuroendocrine adaptation. GH induces direct lipolytic, insulin-antagonizing, and immune stimulatory effects, and in this way the body prioritizes essential substrates such as glucose, FFAs, and amino acids for survival. On the other hand, indirect IGF-I-mediated somatotropic effects are attenuated, resulting in a decrease of anabolic effects which is also beneficial for survival [67]. It has been suggested that these acute-phase responses, representing acquired peripheral resistance to GH during acute stress, are mediated by proinflammatory cytokines. Through reduced negative feedback inhibition abundant release of GH occurs. GH is thought to exert a direct action on adipocytes, resulting in a breakdown of triglycerides and suppression of circulating lipid uptake and accumulation. These changes may increase the serum levels of essential substrates such as glucose and amino acids toward survival rather than anabolism.

In patients with prolonged critical illness, the changes observed within the somatotropic axis are completely different. The total amount of GH released during the chronic phase is significantly lower compared to GH released during the acute phase, and the mean nocturnal

GH concentration is not significantly different from concentrations observed in healthy, non-stressed individuals [68,69]. In addition, the amount of GH released in pulses is reduced compared to the acute phase, but the number of pulses is still high. There is no discrepancy between GH concentration on the one hand and IGF-I and IGFBP3 concentrations on the other, during the prolonged phase, which suggests the recovery of GH-receptor sensitivity with time during critical illness. The low IGF-I level, together with other biochemical markers of impaired anabolism such as low leptin and osteocalcin concentrations, shows the catabolic nature of prolonged critical illness with a chronic increase in metabolic rate and a breakdown of body tissue [70]. One can speculate that the relative functional insufficiency of the somatotropic axis with lack of the physiological characteristics of GH secretion contributes to the pathogenesis of the wasting syndrome typical for the chronic phase of critical illness [5,71].

The underlying disease process of the alterations within the GH axis during prolonged critical illness remains incompletely understood. The pituitary gland may be failing to synthesize GH being involved in the multiple organ dysfunction syndrome (MODS). Alternatively, reduced stimulation of the somatotrophs by endogenous releasing factors or by increased bioavailability of somatostatin may account for the reduced pulsatile GH secretion. The elegant studies of Van den Berghe and colleagues clearly showed the absence of GH resistance and the capability of the pituitary gland to react on GH-secretagogues during prolonged critical illness [69,72].

IGFBP-1, which is subject to acute metabolic regulation, is high on admission to ICU and could be monitored as a potential index for hepatic insulin responsiveness. A high IGFBP-1 level on admission is related to change in nitrogen balance in the short term, and to poor outcome over the longer term.

Hyperglycemia and insulin resistance often accompany the wasting syndrome as a result from the stress response that is modulated by the HPA axis, the autonomic nervous system, and cytokines. The major metabolic alterations leading to hyperglycemia include increased glucose production secondary to enhanced glycogenolysis and gluconeogenesis as well as peripheral resistance to insulin-mediated glucose uptake [8]. Hyperglycemia during critical illness has long been considered essential to provide fuel for vital organ systems and hence was interpreted as a beneficial adaptation. Evidence is now growing against this notion as hyperglycemia is identified as an independent risk factor for adverse outcome [8]. Hyperglycemia and insulin resistance during critical illness may directly or indirectly confer a predisposition to complications such as severe infection, MODS, and death [73]. The landmark study by Van den Berghe et al. showed that by controlling the blood glucose level with intensive insulin therapy, both the morbidity and the mortality can be reduced [9,10]. This benefit concerning the mortality was much more pronounced in patients with prolonged critical illness who stayed at least 5 days in the ICU. Several potential mechanisms may explain these benefits: prevention of immune dysfunction, reduction of systemic inflammation, prevention of endothelial dysfunction, and mitochondrial dysfunctions [74–76]. Glucose control itself and not insulin infusion appeared to be responsible for the observed favorable results [77–79]. The observed benefit of intensive insulin therapy was confirmed by others [80,81] and in a recent meta-analysis [82]. However, a recent study in medical ICU patients failed to demonstrate clinical benefit in the intention-to-treat analysis, but only in the subgroup of patients who stayed in the ICU for more than 3 days [83].

In summary, stress-induced hyperglycemia may result from a combination of high levels of contra-regulatory hormones, proinflammatory cytokines, oxidative stress, β-cell dysfunction, use of corticosteroids, and from a genetic predisposition for diabetes mellitus. Intensive insulin therapy appears to be a very appealing strategy to reduce morbidity and mortality in critically ill patients. However, clinical studies are not (yet) conclusive and the mechanisms involved are not fully clear.

## 2.3. Pituitary–thyroid axis

Thyroid hormone metabolism is commonly affected by critical illness, which results in characteristic abnormalities of thyroid function known as non-thyroidal illness syndrome (NTI) or euthyroid sick syndrome [84]. The term "thyroid adaptation syndrome," however, is more suggestive for an active interplay between the hormonal and the immune systems [85]. Severe illness induces acute changes in blood thyroid hormone concentrations: the serum level of tri-iodothyronin (T3) decreases due to the peripheral inhibition of thyroxin (T4) to T3 conversion via inhibition of 1,5′-deiodinase which normally accelerates inner-ring deiodination of T4 [86,87]. The low T3 syndrome is accompanied by elevated levels of reverse T3 (rT3) due to the deficient activity of 5′-monodeiodinase [87,88]. At the same time, the concentrations of T4 and thyroid-stimulating hormone (TSH) rise briefly and return to normal, but the nocturnal TSH surge is absent [89]. The T3 concentration remains low during the acute phase and has been found to reflect the severity of the illness [67]. The discrepancy between the TSH and the T3 concentrations suggests a functional alteration of the feedback mechanism on the pituitary or the hypothalamic level. On the other hand, low concentrations of binding proteins and an inhibition of hormone binding, transport, and metabolism by elevated levels of FFAs and bilirubin have been proposed as a factor contributing to the low-T3 syndrome at the tissue level [84,90]. TNF-α, interleukin (IL)-1, and IL-6 can all produce acute stress-induced changes in thyroid hormones and may act as mediators of acute T3 syndrome.

In prolonged critical illness, the changes observed within the pituitary–thyroid axis are different. A low or low–normal TSH levels together with low T4 and T3 serum concentrations are observed in patients treated in the ICU for several weeks. The low TSH concentration is due to lack of TSH pulsatility and the loss of TSH pulse amplitude, which suggests an impaired hypothalamic regulation of the pituitary–thyroid axis [72]. Indeed, a reduced expression of the TRH gene in the hypothalamic paraventricular nuclei has been observed in patients during the chronic phase of severe illness [91]. The observed decrease in serum concentration of both thyroid hormones and TSH is not compatible with a negative feedback loop and suggests a major change in setpoint regulation of the HPA axis. In critical illness, serum T3 may even become undetectable without giving rise to an elevated concentration of serum TSH. It is currently not clearly understood whether this reflects an adaptation of the organism to illness or instead a potentially harmful condition leading to hypothyroidism at the tissue level.

One can speculate that the reduced activity of the pituitary–thyroid axis is to some extent responsible for the catabolic state of the patients during phase of critical illness. Accordingly, a rise of TSH blood concentration marks the onset of patient's recovery from the stage of resistance during critical illness [92]. Thyroid hormone levels in prolonged critical illness are inversely correlated with biochemical markers of accelerated catabolism. The low thyroid hormone levels could be either an adaptive, protective mechanism against hypercatabolism or alternatively its cause [5].

## 2.4. Reproductive axis

Profound alterations in the function of the reproductive axis occur in response to physiologic stress of many types, resulting in an inhibition of anabolic steroid secretion [93]. Teleologically, in this way the organism reduces consumption of energy and conserves substrates for more vital functions. During the acute alarm phase, a low serum testosterone concentration is present in patients with a severe illness. However, the serum luteinizing hormone (LH) level is high and the follicular stimulating hormone (FSH) blood level is normal, suggesting an acute stress-induced inhibition of Leydig cell function [94,95]. The changes in blood concentration of estrogens in patients with sepsis are more complicated. In contrast to the low androgen levels, estrogens have been found to be increased in patients with sepsis [94,96,97]. The major source of estrogens in men and in postmenopausal women is conversion by aromatization of testosterone to estradiol in muscle and in adipose tissues. An increase in aromatase activity is the main mechanism that might explain the high concentration of estrogens during sepsis [97].

The clinical consequences of these marked changes are not well understood. The decrease in testosterone is presumably accompanied by a decreased anabolic effect, possibly detrimental in major illness. These observations also raise the question as to whether a marked decrease in serum testosterone levels once offered some adaptive advantage during evolution.

Patients who are critically ill develop temporary hypogonadotropic hypogonadism when the illness is prolonged, regardless of the nature of the illness [98–101]. These patients have a normal response to gonadotropin-releasing hormone (GnRH) stimulation, and the hypogonadotropism also occurs in the presence of non-functioning gonads. These observations suggest a central, hypothalamic, and/or suprahypothalamic origin for these alterations. The compromised reproductive function occurring during chronic stress is secondary to inhibition of gonadotropin secretion induced by endogenous opioids secreted in response to endogenous CRH [102,103].

There is no doubt about the causal relationship between stress and increased pituitary prolactin release, but its biological meaning is much less clear [104]. In addition, experimental and clinical evidence supports the view that prolactin is an immunoregulating hormone which makes its changes especially interesting in intensive care patients [105,106]. In critically ill patients, prolactin concentrations have been shown to be increased in both men and women with disruption of the circadian rhythm [107]. In the chronic phase of critical illness, however, prolactin level is no longer as high as it is during the acute phase but its pulsatile release is impaired [4]. The question is whether the blunted prolactin secretion in the prolonged phase plays a role in the immune dysfunction or in the increased susceptibility to infection characterizing chronically ill patients. In patients treated in the ICU, the stress-induced changes in prolactin secretion are disturbed by dopamine administration or even endogenous dopamine. Dopamine is used as an important inotropic drug in the treatment of ICU patients and it appeared to improve short-term survival in shock states [108]. On the other hand, dopamine as a physiological prolactin-inhibiting factor can after systemic administration influence the pituitary secretion of prolactin [105]. In critically ill patients, dopamine infusion-induced hypoprolactinemia has indirect effects on cellular immunity [106]. Accordingly, one can speculate that dopamine infusion may provoke or aggravate the susceptibility to infectious complications during the chronic phase of critical illness by aggravating both T-lymphocyte dysfunction and impaired neutrophil chemotaxis [109–111].

## 3. IMMUNOLOGICAL RESPONSES DURING CRITICAL ILLNESS

A common and serious consequence of overwhelming bacterial infections is the development of septic shock and associated organ failure. Sepsis or septic shock represents an adverse systemic response to a severe infection and is usually associated with bacteremia. It is not a defined disease but a syndrome that includes fever, tachycardia, tachypnea, and hypotension. In severe cases, the hypoperfusion that is associated with hypotension or shock can lead to end-organ dysfunction or even MODS. The pathogenesis of the shock is presumed to be secondary to excessive stimulation of host cells by microbial constituents (LPS, peptidoglycan, lipoarabinomannan, lipoproteins, secreted toxins, superantigens, and bacterial DNA), which are potent activators of inflammatory cytokine synthesis [112–114]. Although activation of these cytokine cascades is an important component of the innate immune response to infection, endogenous mediators, such as tumor necrosis factor-α (TNF-α) and IL-1, contribute to the irreversible hypotension and organ damage associated with lethal septic shock [115,116]. The idea that sepsis was caused by an overwhelming reaction of the patient to invading microorganisms was supported, at least partially, on the observation that on many occasions, no clinical evidence for infection was found in patients with septic symptoms. Although the clinical appearance of gram-negative, gram-positive, and fungal sepsis is often indistinguishable, there is increasing experimental evidence that fundamental differences exist in the host response to these pathogens. The ability of a host to sense invasion by pathogenic organisms and to respond appropriately to control infection is paramount to survival. In fact, the immune system in a septic individual undergoes substantial modifications during sepsis [114]. Experimental data support the idea that an early intense inflammatory response of the immune system after infection or trauma can harm or set the stage for subsequent organ damage, but it is also well documented that, during sepsis, the innate immune system frequently loses the ability to effectively kill invading microorganisms [117]. Depending on the ability of the immune system to respond to infection, an anti-inflammatory strategy may not be helpful and could even be harmful, as in the well-known clinical trial in which a TNF-α antagonist was reported to increase mortality [118].

### 3.1. Innate and adaptive immunity

The immune response to microbial pathogens relies both on innate and adaptive components [119–122]. The innate response is largely mediated by white blood cells, such as neutrophils and macrophages. These cells phagocytose and kill the pathogens and contemporaneously coordinate additional host responses by eliciting the inflammatory response [123]. In macrophages, the pathogen is killed within the phagosome, and components of the microbe are presented to T cells, thereby activating the adaptive immune response. It is important for the macrophage to discriminate the large number of potential pathogens from itself using a restricted number of germ-line-encoded receptors that recognize conserved motifs, called pathogen-associated molecular patterns (PAMPs) [124]. Two major classes of pattern recognition receptors exist: those that mediate phagocytosis and those that lead to the activation of proinflammatory pathways [122–124]. The innate immunity is not specific to the invading pathogen and does not generate immunologic memory, and it provides rapid response acting directly on the pathogen [125]. Furthermore, the innate immune response involves an integrated multiorgan system effort by the host to not only combat microbial invasion, but also decrease tissue injury and cell death, promote recovery of the host, and reduce the likelihood of secondary or opportunistic infections [126]. The innate immune response is also aimed at containing microbial pathogenesis until the development of a more definitive acquired immune response. However, the innate immune response not only serves as an effective first line of defense against pathogenic organisms, but also, through the release of different humoral mediators casu quo, defines the nature of the acquired immune response [126].

In contrast to innate immunity, there are more highly evolved defense mechanisms, which are stimulated by exposure to infectious agents. Since this form of immunity develops in response to a particular pathogen, it has been termed "adaptive," "acquired," or "specific" immunity. The cellular components of specific immunity are primarily lymphocytes. The specific immune response can be further classified based on the components of the immune system that mediate the response: humoral immunity is mediated by B lymphocytes and cell-mediated immunity is mediated primarily by T lymphocytes. The adaptive immune response is characterized by the clonal selection of antigen-specific lymphocytes, which over time gives rise to long-lasting protection against disease [127].

### 3.2. Toll receptors in critical illness

The innate immune system has evolved as the first line of defense against invading microorganisms [128,129]. The Toll-like receptors (TLRs) are a key component of this innate immune defense, recognizing conserved patterns on microorganisms [130,131]. Ten members of the TLR family have been identified in humans, and several of them appear to recognize specific microbial products. The TLRs are essential transmembrane signaling receptors of the innate immune system that alert the host to the presence of a microbial invader by detecting the pathogen-associated molecular patterns.

The genetic variations or polymorphisms connected with TLR4 are associated with the susceptibility to sepsis or infections [128,132]. The development of antagonists for TLR proteins may provide a useful tool in counteracting the harmful proinflammatory response that complicates systemic microbial infections [133]. The classical role of gram-negative sepsis places TLR4 in a pivotal position: when stimulated by LPS or other PAMPs from microorganisms, TLR4 causes fever, shock, and death in sepsis. Mice lacking TLR4 do not develop shock and do not die when given LPS [134]. Thus, TLR4 appears to protect against rather than cause shock in sepsis. TLR4 also paradoxically protects humans from gram-negative infection. The question arises why infections are more lethal in the absence of TLR4 [131]. Sepsis and systemic inflammatory response syndrome (SIRS) seem to result from failure to contain TLR4 activation (sepsis) or from release of endogenous TLR4 activators (SIRS).

### 3.3. Complement and coagulation in critical illness

The primary soluble effector protein of the innate immune system is complement, which is activated by physical interactions with microbial pathogens or their by-products [133]. Acute inflammation, as a response to severe infection or trauma, results in a systemic activation of the coagulation system termed "disseminated intravascular coagulation" (DIC) [135,136]. The principal initiator of inflammation-induced thrombin generation is tissue factor (TF). However, more complex mechanisms may be involved in the relationship between inflammation and activation of coagulation [137]. This relationship is not unidirectional; instead, a crosstalk between the systems occurs by which activation of coagulation will also affect inflammatory activity. In particular, vascular endothelial cells seem to play a pivotal mediatory role in the coagulative response to systemic inflammation and in the crosstalk between coagulation and inflammation [137]. Endothelial cells respond to the cytokines expressed and released by activated leukocytes but can also release cytokines themselves. Furthermore, endothelial cells are able to express adhesion molecules and growth factors that may not only promote the inflammatory response further but also affect the coagulation response. However, it has recently become clear that, in addition to these mostly indirect effects of the endothelium, endothelial cells interfere directly with the initiation and regulation of fibrin formation and removal during severe infection. In fact, endothelial cells play a prominent role in all three major pathways associated with coagulopathy in sepsis: TF-mediated thrombin generation, dysfunctional anticoagulant pathways, and blocked fibrinolysis. The derangement of coagulation and fibrinolysis in sepsis is mediated by several proinflammatory cytokines, such as TNF-α, IL-1, and IL-6 [138]. The principal mediator of coagulation activation in sepsis seems to be IL-6 [139]. TNF-α indirectly influences the activation of coagulation because of its effects on IL-6, and it is the pivotal mediator of the dysregulation of the physiologic anticoagulant pathways and the fibrinolytic defect [138,140]. A number of coagulation proteases can induce proinflammatory mediators that have procoagulant effects, which may amplify the cascade that leads to DIC. Effects at the cellular level will be determined by the capacity of the coagulation inhibitors to inactivate these enzymes.

The endothelium plays a central role in all major pathways involved in the pathogenesis of hemostatic derangement during severe inflammation. Endothelial cells seem to be directly involved in the initiation and regulation of thrombin generation and the inhibition of fibrin removal. Proinflammatory cytokines are crucial in mediating these effects on endothelial cells, which themselves may also express cytokines, thereby amplifying the coagulative response.

DIC frequently complicates sepsis. Since the definition of DIC is difficult, scoring systems have been developed [136]. DIC is an acquired syndrome characterized by activation of intravascular coagulation culminating in intravascular fibrin formation and deposition in the microvasculature. Secondary fibrinolysis, or in later stages inhibition of fibrinolysis, accompanies coagulation activation. Fibrin deposition leads to a diffuse obstruction of the microcirculation resulting in progressive organ dysfunction, such as acute respiratory distress syndrome (ARDS) or renal failure. Several trials have investigated coagulation-inhibitors in septic patients. In only one trial, a clear benefit was found: activated protein C (APC) led to significant decreased mortality compared to placebo and was subsequently approved by the FDA [141]. This beneficial effect on outcome was prominent in patients with a high risk of death and multiple organ failure. APC is a powerful anti-inflammatory molecule capable of inhibiting cytokine production in monocytes and reducing adhesive interactions between neutrophils and endothelial cells. However, the exact mechanisms by which APC improves survival cannot be explained at this time.

### 3.4. Pro- and anti-inflammatory cytokines

Cytokines are a family of protein mediators of both innate and specific immunity. In general, cytokines are synthesized in response to inflammatory or antigenic stimuli and act locally. Certain cytokines may be produced in sufficient quantity to circulate and exert endocrine actions. Thus, cytokines serve many functions that are critical to host defense against pathogens and provide links between specific and innate immunity. Cytokines serve to initiate the inflammatory response and to define the magnitude and the nature of the acquired immune response. The response of critically ill patients to their injury and/or invading pathogens is dependent, in large part, on the pattern of cytokines production. The immunologic response of critically ill patients can vary from a strongly proinflammatory response, characterized by increased production of TNF-$\alpha$, IL-1, interferon (IFN)-$\gamma$, and IL-12, to one predominantly of anergy, characterized by increased production of $T_H2$ cytokines, like IL-10 and IL-4. Cytokines are the primary communicators of the innate immune system. In innate immunity, the effector cytokines are mostly produced by mononuclear phagocytes and natural killer (NK) cells; on the other hand, most cytokines in specific immunity are produced by activated T lymphocytes. Three cytokines in particular, TNF-$\alpha$, IL-1, and IL-6, appear to play central roles in initiating the innate immune response. These three cytokines not only play critical roles in regulating the innate immune response, but also are a key to the development and propagation of acquired immune responses, particularly $T_H1$ responses. This overlap in functions of TNF-$\alpha$, IL-1, and IL-6 emphasizes how the innate and acquired immune responses are tightly interrelated and regulated during critical illness. Cytokines initiate their action by binding to specific receptors on the surface of the target cell, which may be either the same cell that secretes the cytokine (autocrine) or an adjacent (paracrine) or a distant cell that is stimulated through the circulation (endocrine).

Critically ill patients are often anergic and show an increased susceptibility to invading microorganisms through inhibition of the cellular, humoral, and phagocytic immune system [142]. Typically, after severe trauma or in critically ill patients, an early proinflammatory response is followed by a more sustained anti-inflammatory response. However, the temporal relationship between a proinflammatory and anti-inflammatory cytokine response has not been fully delineated, and patients may move temporally and repeatedly between proinflammatory and anti-inflammatory states. The former response has been called SIRS, whereas the latter has been more recently defined as CARS (compensatory anti-inflammatory response syndrome) involving the increased expression of anti-inflammatory cytokines [143]. Although the timing for the development of a SIRS or CARS response will vary, there is general consensus that the interaction between the proinflammatory and anti-inflammatory mediators ultimately determines the severity of immune dysfunction and the outcome of the patient. Ultimately, either a balance is achieved and homeostasis is restored or the proinflammatory (TNF-$\alpha$, IL-1, IL-6) and anti-inflammatory (IL-4, IL-10) mediators, respectively, take the lead, causing SIRS or CARS and ultimately resulting in MODS.

Early studies unequivocally demonstrated that in response to the systemic administration of either endotoxin or gram-negative bacteria the initial cytokine response was decidedly proinflammatory, characterized by an early release of TNF-$\alpha$, IL-1, and later of IFN-$\gamma$ [144,145]. Similar findings could also be in seen in patients with acute meningococcal sepsis [146]. Early studies with septic and burn patients, respectively, also showed a correlation between the magnitude of the plasma IL-1, IL-6, and TNF-$\alpha$ response with outcome [147,148]. Furthermore, differences were seen in proinflammatory cytokine patterns in trauma versus septic shock patients. In patients with septic shock, both TNF-$\alpha$ and IL-6 concentrations were elevated compared with trauma patients, in whom only IL-6 was increased. Non-surviving septic patients maintained higher TNF-$\alpha$ levels throughout the study period compared with trauma patients [149]. Walley et al. [150] demonstrated that the level of pro- (TNF-$\alpha$) and anti-inflammatory cytokine (IL-10) production was dependent on the severity of sepsis and that severe sepsis or septic shock, a largely unopposed proinflammatory mediator response, results in death.

However, the frequency of septic patients manifesting an exaggerated systemic proinflammatory cytokine response may be less than originally anticipated from the primate and rodent models. In the IL-1 receptor antagonist (IL-1ra) trials for septic patients, $<10\%$ of these individuals had measurable TNF-$\alpha$ in their circulation, and considerably fewer had detectable IL-1$\beta$ [151]. Rather, a much larger proportion of these patients had elevated concentrations of cytokine inhibitors (IL-1ra, p75) and anti-inflammatory cytokines (IL-10). In additional studies, increased levels of IL-10 and other anti-inflammatory cytokines in trauma patients correlated with the severity of the trauma and with an increased risk of developing complications (ARDS, sepsis) [152,153]. The main sources for IL-10 production after injury in both humans and in the murine model are $CD4^+$ $T_H2$ cells, $CD8^+$ T cells, and monocytes/macrophages [154]. This reduced appearance of proinflammatory cytokines is indicative of an anergic state, when macrophages become unresponsive in terms of antigen presentation and proinflammatory cytokine production [154,155], which is in part induced by IL-10 and increased macrophage apoptosis [123]. After hemorrhagic shock, marked depression of both specific and non-specific immunity has been reported. Not only has it been reported that the ability of lymphocytes to proliferate and produce important growth/differentiation factors, such as IL-2 and IFN-$\gamma$, is decreased, but also that macrophage/monocyte responses, such as cytokine release, phagocytosis, and cell surface receptor expression, are markedly depressed [156].

### 3.5. Immunodepression in critically ill patients

Sepsis and multiple organ failure are leading causes of morbidity and mortality in critically ill patients. Alterations in the patient's immune system, with an excessive systemic response (SIRS) on one hand and paralysis of cell-mediated immunity on the other, appear to be key elements in the pathogenesis of MODS and susceptibility to infection [157,158]. SIRS has also been described as a malignant form of intravascular inflammation, rather than merely the overexpression of pro- and anti-inflammatory substances [159]. The primary effector organ for cell injury in SIRS is the activated immunocyte, including polymorphonuclear leukocytes (PMNs), monocyte–macrophages, and lymphocytes. These cells can induce organ system dysfunction across different vascular beds at sites remote from the initial stimuli. Thus, their immune function is both protective and damaging. On the proximal limb of immunocompetence, identification of foreign antigen and the development of antibodies reflect essential initiating aspects of cell-mediated immunity. The cell surface molecule engaged in antigen presentation is the human leukocyte antigen-DR (HLA-DR) moiety. SIRS induces a profound decline in HLA-DR expression, which is consistent with the immune depression, indicating that SIRS is associated with a reduced ability of the host to initiate an immunological response to new antigen [159,160].

In the state of CARS, monocytes are often deactivated resulting in reduced antigen presentation and decreased production of proinflammatory cytokines. The exact cause of this deactivation is not fully understood. Anti-inflammatory cytokines, such as TGF-β, IL-4, and IL-10, and to some extent, pro-inflammatory cytokines, such as IL-1 or IL-8, can induce this response. In general, the cellular machinery becomes altered in terms of its ability to respond to subsequent inflammatory challenges. The induction of immune suppression centers on control of final intracellular signal promoting genetic expression of proinflammatory species [159]. The primary promotor of the proinflammatory response is nuclear factor-κB (NF-κB), and in the endotoxin tolerance model NF-κB dysfunction develops [161]. Either dysfunctional NF-κB or Inhibitor-κB-alpha (IκBa) excess may reduce the release of TNF-α and IL-1. Furthermore, the heat-shock protein (hsp) system, a well-established anti-inflammatory system that is non-specifically induced by a variety of oxidative stresses, may become exhausted and may also contribute to the immunological changes in critical illness [162,163]. The mechanism of hyporesponsiveness to TNF-α in critical illness is probably complex and may represent as much cell exhaustion as downregulation.

Multiple immune cell types are dysfunctional in critically ill patients. Impaired monocyte antigen presentation and inflammatory cytokine production have been correlated to the development of organ failure and mortality. Circumstantial evidence suggests that PMN function is important in determining outcome from sepsis. PMN activation also has the real potential to be detrimental to the host. It can be highly destructive in both acute lung injury and septic states. PMN activation and their ability to respond to inflammatory mediators appear beneficial in some clinical settings and harmful in others. In general, conditions associated with a systemic inflammatory state that are devoid of active systemic bacterial infection do better with suppression of proinflammatory pathways. However, localized infectious processes (e.g., pneumonia, meningitis, peritonitis) fare better with PMN upregulation, even if this is associated with a generalized systemic proinflammatory response.

### 3.6. The role of MIF as immunoneuroendocrine mediator in critical illness

Macrophage migration inhibitory factor (MIF) has features of a cytokine, a hormone, and an enzyme [164], and although its exact mechanism of action is still incompletely understood, it has become well established that MIF is a critical mediator of the innate and acquired immune response [165]. MIF has been proposed to be the physiologic counter-regulator of glucocorticoid action within the immune system [166]. In this role, the position of MIF within the cytokine cascade is to act in concert with glucocorticoids to control both the "setpoint" and the magnitude of the inflammatory response. In addition to overriding the immunosuppressive effects of glucocorticoids, it is clear that MIF has a direct proinflammatory role in inflammatory diseases, such as sepsis [167].

Further studies in rodents indicated that the pituitary release of MIF was an integral part of the host's systemic stress response. When mice received an intraperitoneal injection of LPS, there was a dramatic fall in the pituitary content of MIF, a concomitant increase in plasma levels of MIF, and a gradual elevation of MIF mRNA expression in pituitary tissues [168,169]. Circulating MIF levels in animals were also observed to rise 3–4 h after exposure to handling-induced stress, similar to the more classically described stress-related increases in circulating ACTH and glucocorticoid levels [169]. MIF was, thus, "rediscovered" as a pituitary-derived mediator of systemic stress responses [170]. In addition to being secreted from the pituitary, MIF was also found to be released from immune/inflammatory cells as a consequence of glucocorticoid stimulation, and is in fact the only cytokine that is stimulated instead of inhibited by glucocorticoids [169]. Monocytes and macrophages contain large quantities of preformed MIF that were readily released in response to stimulation with LPS, gram-positive exotoxins (such as toxic shock syndrome toxin-1), and proinflammatory cytokines (TNF-$\alpha$, IFN-$\gamma$) [171]. On release from macrophages, MIF can exert potent autocrine and paracrine effects, promoting cell activation and proinflammatory cytokine release and overriding glucocorticoid action at the site of inflammation [169,172].

Treatment with recombinant MIF exacerbates LPS-induced toxicity, whereas treatment with anti-MIF neutralizing antibodies rescues mice from lethal endotoxic shock [168] MIF also has been shown to play an important role in the pathogenesis of sepsis caused by gram-negative bacteria [173]. Septic mice were still protected from lethality when anti-MIF treatment was initiated as late as 8 h after the onset of infection. This is of great importance in the clinical situation because treatment of septic shock in humans is always initiated after symptoms are expressed and the infection is well established. Consistent with observations that gram-positive exotoxins induce secretion of MIF from T cells and macrophages [174], neutralization of MIF in mice treated with toxic shock syndrome toxin-1 protects animals from toxic shock lethality.

Several studies have provided clinical evidence for systemically elevated MIF expression during sepsis [173]. We serially measured serum MIF, cortisol, plasma ACTH, TNF-$\alpha$, and IL-6 in 40 critical care patients during a period of 14 days or until discharge or death [41]. On day 1, MIF levels were significantly elevated in septic shock patients compared with multitrauma patients and normal controls. Furthermore, the time-course of MIF expression in serum paralleled that of cortisol in the septic shock patients. A significant correlation also was observed between elevated MIF levels at admission and occurrence of death. Interestingly, MIF levels were not elevated in non-septic, multitrauma patients. These data were complemented by Joshi et al. [175], who reported elevated MIF levels in multitrauma patients that correlated with positive tests for bacterial cultures in blood, urine, sputum, or at the wound site.

### 3.7. Immune activation and the HPA axis in critical illness

Infections and immune challenges can activate the HPA axis, but is this a specific response or merely the reaction of the organism to a disturbance of its homeostasis? Work by Besedovsky and others implicates cytokines produced by the immune system as mediators of the HPA response, which suggests that it is specific. The initial observation was that supernatants of immune cells challenged *in vitro* with mitogens, such as Con A, had the ability to activate the HPA axis. The active factor synthesized and secreted in response to the mitogens was suggested to be IL-1 [176]. It is notable that IL-1 is a considerably more potent activator of ACTH and glucocorticoid secretion than CRF itself. It is important to distinguish the HPA activation that occurs at different stages in the immune response. There is an acute HPA response to immune stimulation that occurs within the first few hours. This acute response is observed following LPS administration [177]. It seems likely that this early response is related to cytokine secretion. The mechanism of the activation of the HPA axis by immune stimuli has been the subject of intense investigation. Speculation has largely focused on the role of IL-1. Evidence exists for IL-1-induced activation of the axis at every level. However, the bulk of the evidence strongly favors the need for hypothalamic CRF. Deafferentation of the hypothalamus and lesions of the paraventricular nucleus prevent the ACTH response to IL-1, and hypophysectomy prevents the effect of IL-1 and LPS [178]. Moreover, antibodies to CRF attenuate or block the ACTH and corticosterone responses to IL-1 [179]. IL-1 is not the only cytokine that can affect the HPA axis; other cytokines, such as IL-6 and TNF-α, can have similar effects [180,181]. Interestingly, the production of TNF-α and IL-1 is inhibited by glucocorticoids so that the HPA activation elicited by the cytokines provides feedback regulation of cytokine synthesis [182]. Cytokines, such as IL-1 and IL-6, are clearly potent activators of the HPA axis, but also exert a variety of other physiological effects [178]. Although our current understanding of the system is primitive, it may be important to distinguish local effects from systemic effects. Circulating concentrations of catecholamines and steroids are probably adequate to exert physiological effects, and this appears to be true also for cytokines. Glucocorticoids and catecholamines predominantly inhibit immune responses, whereas the peptides (such as neuropeptides) are largely facilitators. When the organism is threatened, the systemic activity of the glucocorticoids to limit immune responses may be important to depress immune activity to prevent undesirable autoimmune actions. By contrast, peptides could facilitate immune responses in small areas close to the site of their release, for example, in an area of inflammation induced by infection or tissue damage. Catecholamines occupy an intermediate position, existing in sufficient concentrations to have systemic actions but not having broad access to tissues and having relatively short durations of action, except when chronically elevated. Such an arrangement would permit focusing of the activation of immune response in local areas of inflammation, while preventing potentially damaging autoimmune actions that could be triggered by widespread activation.

Immune activation is often accompanied by profound alterations in neuroendocrine functions and is associated with increased activity of the HPA axis [183,184]. There is growing evidence that this activation is involved in the acute-phase response caused by infection and immunological reactions [184]. However, the mechanisms and crucial site for the stimulating effects of these immune signals on the HPA axis are not well understood. Studies demonstrating the effects of hypothalamic lesions or pharmacological blockade of CRH release in order to interfere with LPS-induced pituitary–adrenal hormone secretion suggest that the primary locus of action may be in the hypothalamus [6].

One of the most important built-in mechanisms that limit tissue damage is the integrated response to endotoxic shock by the nervous, immune, and endocrine systems [185]. Peripheral IL-1 and TNF-α activate the HPA axis via both neural and humoral pathways, resulting in a dramatic increase in plasma levels of glucocorticoids [186]. Either adrenalectomy or hypophysectomy markedly sensitizes animals to endotoxin lethality, indicating a key role of the HPA axis in preventing endotoxin shock [187]. Injection of selective glucocorticoid receptor antagonists into animals also results in a heightened state of sensitivity to endotoxin [188], whereas administration of glucocorticoids in animals challenged by endotoxin [189] or in humans with clinical sepsis affords significant protection [49,190,191]. Thus, the actions of endogenous glucocorticoids seem essential for the survival of animals challenged by endotoxin.

## 4. CONCLUDING REMARKS

Critical illness is a severe stress stimulus that disturbs the milieu interior and induces homeostatic responses specific to the stimulus and generalized responses when the disturbances are prolonged and severe. The immune–neuroendocrine response during critical illness is a dynamic process, differing between the acute and chronic phases. This acute phase lasts typically from about a few hours to several days, depending on the severity of the illness and is characterized by an appropriate hormonal reaction: the mobilization of fuel stores of the organism, together with apparent restrains on their utilization. During the acute phase of critical illness, the secretory activity of the pituitary gland is stimulated, whereas anabolic target organ hormones are inactivated. Due to the development of intensive care medicine, patients who would previously have died during the acute phase nowadays survive and enter the chronic or prolonged phase of the critical illness. This prolonged phase lasts for days and is characterized by an inappropriate hormonal response, resulting in a chronic increase in metabolic rate and a breakdown of body tissue. The secretory activity of the pituitary gland is uniformly inhibited in relation to reduced levels of target organ hormones. These hormonal changes suggest an imbalance between immunosuppressive and immunostimulatory hormones which might be the cause of the increased susceptibility to infectious complications during the chronic phase of severe illness. The suppressed pituitary activity allows the respective target organ hormones to decrease, resulting in a restored balance between the catabolic and anabolic hormonal responses. This recovery phase is characterized by restored sensitivity of the pituitary gland to reduced feedback control. The distinction of acute and prolonged critical illness as different entities with regard to the immunoneuroendocrine adaptation may be helpful in further understanding of the pathogenesis of critical illness and the targeting of therapeutic intervention.

## REFERENCES

1. Beishuizen A, Vermes I, Haanen C. Endogenous mediators in sepsis and septic shock. Adv Clin Chem 1998;33:55–131.
2. Reichlin S. Neuroendocrine-immune interactions. N Engl J Med 1993;329:1246–53.
3. Basedowsky HO, DelRey A. Immune-neuro-endocrine interactions: facts and hypotheses. Endocr Rev 1996;17:64–102.
4. Van den Berghe G, de Zegher F, Bouillon R. Acute and prolonged critical illness as different neuroendocrine paradigms. J Clin Endocrinol Metab 1998;83:1827–34.
5. Van den Berghe G. Dynamic neuroendocrine responses to critical illness. Front Neuroendocrinol 2002;23:370–91.
6. McEwen BS. Protective and damaging effects of stress mediators. N Engl J Med 1998;338:171–9.
7. Habib KE, Gold PW, Chrousos GP. Neuroendocrinology of stress. Endocrinol Metab Clin North Am 2001;30:695–728.
8. Mizock BA. Alterations in fuel metabolism in critical illness: hyperglycaemia. Baillière's Best Pract Res Clin Endocrinol Metab 2001;15:533–51.
9. Van den Berghe G, Wouters P, Weekers F, Verwaest C, Bruyninckx F, Schets M, Vlasselaers D, Ferdinande P, Lauwers P, Bouillon R. Intensive insulin therapy in critically ill patients. N Engl J Med 2001;345:1359–67.

10. Van den Berghe G, Wilmer A, Hermans G, Meersseman W, Wouters PJ, Milants I, VanWijngaerden E, Bobbaers H, Bouillon R. Intensive insulin therapy in the medical ICU. N Engl J Med 2006;354:449–61.
11. Singer M, DeSantis V, Vitale D, Jeffcoate W. Multiorgan failure is an adaptive, endocrine-mediated, metabolic response to overwhelming systemic inflammation. Lancet 2004;364:545–8.
12. Ligtenberg JJ, Girbes ARJ, Beentjes JAM, Tulleken JE, van der Werf TS, Zijlstra JG. Hormones in the critically ill patient: to intervene or not to intervene? Intensive Care Med 2001;27:1567–77.
13. Selye J. A syndrome produced by diverse nocuous agents. Nature 1936;138:32.
14. Vermes I, Beishuizen A, Hampsink RM, Haanen C. Dissociation of plasma adrenocorticotropin and cortisol levels in critically ill patients: possible role of endothelin and atrial natriuretic hormone. J Clin Endocrinol Metab 1995;80:1238–42.
15. Parker LN, Levin ER, Lifrak ET. Evidence for adrenocortical adaptation to severe illness. J Clin Endocrinol Metab 1985;60:947–52.
16. Manglik S, Flores E, Lubarsky L, Fernandez F, Chibber VL, Tayek JA. Glucocorticoid insufficiency in patients who present to the hospital with severe sepsis: a prospective clinical trial. Crit Care Med 2003;31:1668–75.
17. Marx C, Petros S, Bornstein R, Weise M, Wendt M, Menschikowski M, Engelmann L, Hoffken G. Adrenocortical hormones in survivors and nonsurvivors of severe sepsis: diverse time course of dehydroepiandrosterone, dehydroepiadrosterone-sulfate, and cortisol. Crit Care Med 2003;31:1382–8.
18. Rai R, Cohen J, Venkatesh B. Assessment of adrenocortical function in the critically ill. Crit Care Resusc 2004;6:123–9.
19. Dimopoulou I, Tsagarakis S, Kouyialis AT, Roussou P, Assithianakis S, Christoforaki M, Illias I, Sakas DE, Thalassinos N, Roussos C. Hypothalamic-pituitary-adrenal axis dysfunction in critically ill patients with traumatic brain injury: incidence, pathophysiology, and relationship to vasopressor dependence and peripheral interleukin-6 levels. Crit Care Med 2004;32:404–8.
20. Siraux V, DeBacker D, Yalavatti G, Mélot C, Gervy C, Mockel J, Vincent J-L. Relative adrenal insufficiency in patients with septic shock: comparison of low-dose and conventional corticotropin tests. Crit Care Med 2005;33:2479–86.
21. Van den Berghe G, de Zegher F, Wouters P, Schetz M, Verwaest C, Ferdinande P, Lauwers P. Dehydroepiandrosterone sulphate in critical illness: effect of dopamine. Clin Endocrinol 1994;41:563–9.
22. Folan MM, Stone RA, Pittenger AL, Stoffel JA, Hess MM, Kroboth PD. Dehydroepiandrosterone, dehydroepiandrosterone-sulfate, and cortisol concentrations in intensive care unit patients. Crit Care Med 2001;29:965–70.
23. Beishuizen A, Thijs LG, Vermes I. Decreased levels of dehydroepiandrosterone sulphate in severe critical illness: a sign of exhausted adrenal reserve? Crit Care 2002;6:434–38.
24. Savu L, Zouaghi H, Carli A, Nunez EA. Serum depletion of corticosteroid binding activities, an early marker of human septic shock. Biochem Biophys Res Commun 1981;102:411–19.
25. Schlechte JA, Hamilton D. The effect of glucocorticoids on glucocorticoid binding globulin. Clin Endocrinol 1991;27:197–203.
26. Garrel DR. Corticosteroid-binding globulin during inflammation and burn injury: nutritional modulation and clinical implications. Horm Res 1996;45:245–51.
27. Beishuizen A, Thijs LG, Vermes I. Pattern of corticosteroid-binding globulin and the free cortisol index during septic shock and multitrauma. Intensive Care Med 2001;27:1584–91.
28. Molijn GJ, Spek JJ, VanUffelen JCJ, DeJong FH, Brinkmann AO, Bruining HA, Lamberts SWJ, Koper JW. Differential adaptation of glucocorticoid sensitivity of peripheral blood mononuclear leukocytes in patients with sepsis or septic shock. J Clin Endocrinol Metab 1995;80:1799–803.
29. Hamrahian AH, Oseni T, Arafah BM. Measurement of serum free cortisol in critically ill patients. N Engl J Med 2004;350:1629–38.
30. Loriaux L. Glucocorticoid therapy in the intensive care unit. N Engl J Med 2004;350:1601–2.
31. Munck A, Guyre PM, Holbrook NJ. Physiological functions of glucocorticoids in stress and their relations to pharmacologic actions. Endocr Rev 1984;5:25–44.

32. Lamberts SWJ, Bruining HA, DeJong FH. Corticosteroid therapy in severe illness. N Engl J Med 1997;337:1285–92.
33. Watsky J, Leinung MC. The adrenocortical response to critical illness. In Contemporary Endocrinology: Endocrinology of Critical Disease. Ober KP, Ed.; Totowa: Humana Press Inc, 1996; pp. 123–36.
34. Reincke M, Allolio B, Würth G, Winkelman W. The hypothalamic-pituitary-adrenal axis in critical illness: response to dexamethasone and corticotropin releasing hormone. J Clin Endocrinol Metab 1993;77:151–6.
35. Siegel LM, Grinspoon SK, Garvey GJ, Bilezikian JP. Sepsis and adrenal function. Trends Endocrinol Metab 1994;5:324–8.
36. Bornstein SR, Chrousos GP. Adrenocorticotropin (ACTH) and non-ACTH-mediated regulation of the adrenal cortex: neural and immune inputs. J Clin Endocrinol Metab 1999;84:1729–36.
37. Ehrhart-Bornstein M, Hinson JP, Bornstein SR, Scherbaum WA, Vinson GP. Intraadrenal interactions in the regulation of adrenocortical steroidogenesis. Endocr Rev 1998;19:101–43.
38. Bornstein SR, Ehrhart-Bornstein M, Usadel H, Böckmann M, Scherbaum WA. Morphological evidence for a close interaction of chromaffin cells with cortical cells within the adrenal gland. Cell Tissue Res 1991;265:1–9.
39. Bornstein SR, González-Hernández JA, Ehrhart-Bornstein M, Adler G, Scherbaum WA. Intimate contact of chromaffin and cortical cells within the human adrenal gland forms the cellular basis for important intraadrenal interactions. J Clin Endocrinol Metab 1994;78:225–32.
40. Chrousos GP. The hypothalamic-pituitary-adrenal axis and immune-mediated inflammation. N Engl J Med 1995;332:1351–62.
41. Beishuizen A, Thijs LG, Haanen C, Vermes I. Macrophage migration inhibitory factor and hypothalamo-pituitary-adrenal function during critical illness. J Clin Endocrinol Metab 2001; 86:2811–16.
42. Drucker D, Shandling M. Variable adrenocortical function in acute medical illness. Crit Care Med 1985;13:477–9.
43. Naito Y, Fukata J, Tamai S, Seo N, Nakai Y, Mori K, Imura H. Biphasic changes in hypothalamo-pituitary-adrenal function during the early recovery period after major abdominal surgery. J Clin Endocrinol Metab 1991;73:111–17.
44. Legakis I, Saramantis A, Voros D, Chalevelakis G, Tolis G. Dissociation of ACTH, β-endorphin and cortisol in graded sepsis. Horm Metab Res 1998;30:570–4.
45. Loisa P, Rinne T, Kaukinen S. Adrenocortical function and multiple organ failure in severe sepsis. Acta Anaesthesiol Scand 2002;46:145–51.
46. Kidess AI, Caplan RH, Reynertson RH, Wickus GG, Goodnough DE. Transient corticotropin deficiency in critical illness. Mayo Clin Proc 1993;68:435–41.
47. Beishuizen A, Vermes I, Hylkema BS, Haanen C. Relative eosinophilia and functional adrenal insufficiency in critically ill patients. Lancet 1999;353:1675–6.
48. Beishuizen A, Vermes I. Relative eosinophilia (Thorn test) as a bioassay to judge the clinical relevance of cortisol values during severe stress. J Clin Endocrinol Metab 1999;84:3400.
49. Annane D, Sébille V, Charpentier C, Bollaert P-E, Francois B, Korach J-M, Capellier G, Cohen Y, Azoulay E, Troché G, Chaumet-Riffaut P, Bellissant E. Effect of treatment with low doses of hydrocortisone and fludrocortisone on mortality in patients with septic shock. JAMA 2002; 288:862–71.
50. Marik PE, Zaloga GP. Adrenal insufficiency in the critically ill. A new look at an old problem. Chest 2002;122:1784–96.
51. Cooper MS, Stewart PM. Corticosteroid insufficiency in acutely ill patients. N Engl J Med 2003;348:727–34.
52. Zipser RD, Davenport MW, Martin KL, Tuck ML, Warner NE, Swinney RR, Devis CL, Horton R. Hyperrenemic hypoaldosteronism in the critically ill: a new entity. J Clin Endocrinol Metab 1981;53:867–73.
53. Findling JW, Waters VO, Raff H. The dissociation of renin and aldosterone during critical illness. J Clin Endocrinol Metab 1987;64:592–5.

54. Raff H, Findling JW. Aldosterone control in critically ill patients: ACTH, metoclopropamide, and atrial natriuretic peptide. Crit Care Med 1990;18:915–20.
55. Oelkers W. Dehydroepiandrosterone for adrenal insufficiency. N Engl J Med 1999;341:1073–4.
56. Vermes I, Besihuizen A. The hypothalamic-pituitary-adrenal response to critical illness. Baillière's Best Pract Res Clin Endocrinol Metab 2001;15:495–511.
57. Beishuizen A, Thijs LG. Relative adrenal failure in intensive care: an identifiable problem requiring treatment? Baillières Best Pract Res Clin Endocrinol Metab 2001;15:513–31.
58. Marik PE. Unraveling the mystery of adrenal failure in the critical ill. Crit Care Med 2004;32:596–7.
59. Annane D. Low-dose adrenocorticotropic hormone test is not ready for routine adrenal function testing in the intensive care unit. Crit Care Med 2005;33:2688–9.
60. Devesa J, Lima L, Tresguerres JAF. Neuroendocrine control of growth hormone secretion in humans. Trends Endocrinol Metab 1992;3:175–83.
61. Ross R, Miell J, Freeman E, Jonest J, Matthews D, Preece M, Buchman C. Critically ill patients have high basal growth hormone levels with attenuated oscillatory activity associated with low levels of insulin-growth factor-I. Clin Endocrinol 1991;35:47–54.
62. Voerman HJ, Strack van Schijndel RJM, Groeneveld ABJ, De Boer H, Nauta JP, Van der Veen EA, Thijs LG. Effect of recombinant human growth hormone in patients with severe sepsis. Ann Surg 1992;216:648–55.
63. Voerman HJ, Groeneveld ABJ, De Boer H, Strack van Schijndel RJM, Nauta JP, Van der Veen EA, Thijs LG. Time course and variability of the endocrine and metabolic response to severe sepsis. Surgery 1993;114:951–9.
64. Hermansson M, Wickelgren RB, Hammerqvist F, Bjarnason R, Wennström I, Wernerman J, Carlsson B, Carlsson LMS. Measurement of human growth hormone receptor messenger ribonucleic acid by a quantitative polymerase chain reaction-based assay: demonstration of reduced expression after elective surgery. J Clin Endocrinol Metab 1997;82:421–8.
65. Baxter RC, Hawker FH, To C, Stewart PM, Holman SR. Thirty day monitoring of insulin-like growth factors and their binding proteins in intensive care unit patients. Growth Horm IGF Res 1998;8: 455–63.
66. Rodriguez-Arnao J, Yarwood Y, Ferguson C, Miell J, Hinds CJ, Ross RJ. Reduction in circulating IGF-I and hepatic IGF-I mRNA levels after ceacal ligation and puncture are associated with differential regulation of hepatic IGF-binding protein-1, -2 and -3 mRNA levels. J Endocrinol 1996;151:287–92.
67. Van den Berghe G. The neuroendocrine response to stress is a dynamic process. Baillières Best Pract Res Clin Endocrinol Metab 2001;15:405–19.
68. Van den Berghe G, de Zegher F, Veldhuis JD, Wouters P, Wouters M, Verbruggen W, Schetz M, Verwaest C, Lauwers P, Bouillon R, Bowers CY. The somatotropic axis in critical illness: effect of continuous GHRH and GHRP-2 infusion. J Clin Endocrinol Metab 1997;82:590–9.
69. Van den Berghe G, de Zegher F, Baxter RC, Veldhuis JD, Wouters P, Schetz M, Verwaest C, Van der Vorst E, Lauwers P, Bouillon R, Bowers CY. Neuroendocrinology of prolonged critical illness: effect of continuous thyrotropin-releasing hormone infusion and its combination with growth hormone-secretagogues. J Clin Endocrinol Metab 1998;83:309–19.
70. Van den Berghe G, Wouters P, Weekers F, Mohan S, Baxter R, Veldhuis J, Bowers CY, Bouillon R. Reactivation of pituitary hormone release and metabolic improvement by infusion of growth hormone releasing peptide and thyrotropin-releasing hormone in patients with protracted critical illness. J Clin Endocrinol Metab 1999;84:1311–23.
71. Baxter RC. Changes in the IGF-IGFBP axis in critical illness. Baillières Best Pract Res Clin Endocrinol Metab 2001;15:421–34.
72. Van den Berghe G, DeZegher F, Veldhuis JD, Wouters P, Gouwy S, Stockman W, Weekers M, Schetz M, Lauwers P, Bouillon R, Bowers CY. Thyrotropin and prolactin release in prolonged critical illness: dynamics of spontaneous secretion and effects of growth hormone secretagogues. Clin Endocrinol 1997;47:599–612.
73. Mesotten D, Van den Berghe G. Clinical potential of insulin therapy in critically ill patients. Drugs 2003;63:625–36.

74. Van den Berghe G. How does blood glucose control with insulin save lives in intensive care? J Clin Invest 2004;114:187–95.
75. Vanhorebeek I, DeVos R, Mesotten M, Wouters PJ, De Wolf-Peeters C, Van den Berghe G. Protection of hepatocyte mitochondrial ultrastructure and function by strict blood glucose control with insulin in critically ill patients. Lancet 2005;365:53–9.
76. Ellger B, Debaveye Y, Vanhorebreek I, Langouche L, Giulietti A, Van Etten E, Herijgers P, Mathieu C, van den Berghe G. Survival benefits of intensive insulin therapy in critical illness. Impact of maintaining normoglycaemia versus glycemia-independent. Actions of insulin. Diabetes 2006; 55:1096–105.
77. Finney SJ, Zekveld C, Elia A, Evans TW. Glucose control and mortality in critically ill patients. JAMA 2003;290:2041–7.
78. Van der Voort PH, Feenstra RA, DeHeide L, Boerma EC, Van den Horst CC. Intravenous glucose intake independently related to intensive care unit and hospital mortality: an argument for glucose toxicity in critically ill patients. Clin Endocrinol 2006;64:141–5.
79. Esposito K, Nappo F, Marfella R, Giugliano G, Giugliano F, Ciotola M, Quagliaro L, Ceriello A, Giugliano D. Inflammatory cytokine concentrations are acutely increased by hyperglycemia in humans. Circulation 2002;106:2067–72.
80. Krinsley JS. Effect of an intensive glucose management protocol on the mortality of critically ill adult patients. Mayo Clin Proc 2004;79:992–1000.
81. Grey NJ, Perdrizet GA. Reduction of nosocomial infections in surgical intensive care unit by strict glycemic control. Endocr Pract 2004;10(Suppl 2):46–52.
82. Pittas AG, Siegel RD, Lau J. Insulin therapy for critically ill hospitalized patients: a meta-analysis of randomized controlled trials. Arch Intern Med 2004;164:2005–11.
83. Van den Berghe G, Wilmer A, Hermans G, Meersseman W, Wouters PJ, Milants I, Van Wijngaerden E, Bobbaers H, Bouillon R. Intensive insulin therapy in the medical ICU. N Engl J Med 2006;354: 449–51.
84. De Groot LJ. Dangerous dogmas in medicine: the nonthyroideal illness syndrome. J Clin Endocrinol Metab 1999;84:151–64.
85. Fliers E, Alkemade A, Wiersinga WM. The hypothalamic-pituitary-thyroid axis in critical illness. Baillières Best Pract Res Clin Endocrinol Metab 2001;15:453–64.
86. Chopra IJ, Chua Tec GN, Mead JF, Huang TS, Beredo A, Solomon DH. Relationship between serum free fatty acid and thyroid binding inhibitor in nonthyroidal illness. J Clin Endocrinol Metab 1985;60:980–4.
87. Kaptein EM, Robinson WJ, Grieb DA, Nicoloff JT. Peripheral serum thyroxine, triidothyronine and reverse triiodothyronine kinetics in the low thyroxine state of acute non-thyroidal illness: a non-compartemental analysis. J Clin Invest 1982;53:764–71.
88. Richmand DA, Molitch ME, O'Donnell TF. Altered thyroid hormone levels in bacterial sepsis: the role of nutritional adequacy. Metabolism 1980;29:936–9.
89. Romijn JA, Wiersinga WM. Decreased nocturnal surge of thyrotropin in nonthyroidal illness. J Clin Endocrinol Metab 1990;70:35–42.
90. Docter R, Krenning EP, DeJong M, Hennemann G. The sick euthyroid syndrome: changes in thyroid hormone serum parameters and hormone metabolism. Clin Endocrinol 1993;39:499–518.
91. Fliers E, Guldenaar SEF, Wiersinga WM, Swaab DF. Decreased hypothalamic thyrotropin-releasing hormone gene expression in patients with nonthyroidal illness. J Clin Endocrinol Metab 1997; 82:4032–6.
92. Bacci V, Schussler GC, Kaplan TC. The relationship between serum triiodothyronine and thyrotropin during systemic illness. J Clin Endocrinol Metab 1982;54:1229–35.
93. Spratt DI. Altered gonadal steroidogenesis in critical illness: is treatment with anabolic steroids indicated? Baillières Best Pract Res Clin Endocrinol Metab 2001;15:479–94.
94. Luppa P, Muncker R, Nagel D, Weber W, Engelhardt D. Serum androgens in intensive care patients: correlation with clinical findings. Clin Endocrinol 1991;34:305–10.
95. Dong Q, Hawker F, McWilliam D et al. Circulating immunoreactive inhibin and testosterone levels in patients with critical illness. Clin Endocrinol 1992;36:399–404.

96. Fourrier F, Jallot A, Leclerc L, Jourdain M, Racadot A, Chagnon JL, Rime A, Chopin C. Sex steroid hormones in circulatory shock, sepsis syndrome, and septic shock. Circ Shock 1994;43:171–8.
97. Christeff N, Benassayag C, Carli-Vielle C, Carli A, Nunez EA. Elevated estrogen and reduced testosterone levels in the serum of male septic shock patients. J Steroid Biochem 1988;29:435–40.
98. Wolf PD, Hamill RW, McDonald JV, Lee LA, Kelly M. Transient hypogonadotropic hypogonadism caused by critical illness. J Clin Endocrinol Metab 1985;60:444–50.
99. Gebhart SSP, Watts NB, Clark RV, Umpierrez G, Sgoutas D. Reversible impairment of gonadotrophin secretion in critical illness. Arch Intern Med 1989;149:1637–41.
100. Spratt DI, Bigos ST, Beitins I, Cox P, Longcope C, Orav J. Both hyper- and hypogonadotropic hypogonadism occur transiently in acute illness: bio- and immunoreactive gonadotropins. J Clin Endocrinol Metab 1992;75:1562–70.
101. Van den Berghe G, De Zegher F, Lauwers P, Veldhuis JD. Luteinizing hormone secretion and hypoandrogenemia in critically ill men: effect of dopamine. Clin Endocrinol 1994;41:563–9.
102. Rivier C, Vale W. Influence of corticotropin-releasing factor on reproductive function in the rat. Endocrinology 1984;114:914–21.
103. Barbarion A, De Marinis L, Tofani A, Casa SD, D'Amico C, Mancini A, Corsello SM, Sciuto R, Barini A. Corticotropin-releasing hormone inhibition of gonadotropin release and the effect of opioid blockade. J Clin Endocrinol Metab 1989;68:523–8.
104. Gala RR. The physiology and mechanisms of the stress-induced changes in prolactin secretion in the rat. Life Sci 1990;46:1407–20.
105. Murphy WJ, Rui H, Longo DL. Effects of growth hormone and prolactin: immune development and function. Life Sci 1995;57:1–14.
106. Russell DH. New aspects of prolactin and immunity: a lymphocyte-derived prolactin-like product and nuclear protein kinase C activation. Trends Pharmacol Sci 1989;10:40–4.
107. Noel GL, Suh HK, Stone SJG, Frantz AE. Human prolactin and growth hormone release during surgery and other conditions of stress. J Clin Endocrinol Metab 1972;35:840–51.
108. Goldberg LI. Dopamine-clinical use of an endogenous catecholamine. N Engl J Med 1974;291: 707–10.
109. Van den Berghe G, De Zegher F, Lauwers P. Dopamine suppresses pituitary function in infants and children. Crit Care Med 1994;22:1747–53.
110. Devins SS, Miller A, Herndon BL et al. Effects of dopamine on T-lymphocyte proliferative responses and serum prolactin concentration in critically ill patients. Crit Care Med 1992;20:1644–9.
111. De Zegher F, Van den Berghe G, Devlieger H, Eggermont E, Veldhuis JD. Dopamine inhibits growth hormone and prolactin secretion in human newborn. Pediatr Res 1993;34:642–5.
112. Annane D, Belissant E, Cavaillon J. Septic shock. Lancet 2006;365:63–78.
113. Hotchkiss RS, Karl IE. The pathophysiology and treatment of sepsis. N Engl J Med 2003;348: 138–50.
114. Riedemann NC, Guo R-F, Ward PA. The enigma of sepsis. J Clin Invest 2003;112:460–7.
115. Tracey KJ, Fong Y, Hesse DG, Manogue KR, Lee AT, Kuo GC et al. Anti-cachectin/TNF monoclonal antibodies prevent septic shock during lethal bacteraemia. Nature 1987;330:662–4.
116. Beutler B, Cerami A. Cachectin (tumor necrosis factor): a macrophage hormone governing cellular metabolism and inflammatory response. Endocr Rev 1988;9:57–66.
117. Riedemann RC, Guo RF, Ward PA. Novel strategies for the treatment of sepsis. Nat Med 2003;9:517–24.
118. Fisher CJ, Agosti JM, Opal SM, Lowry SF, Balk RA, Sadoff JC, Abraham E, Schein RMH, Benjamin E. Treatment of septic shock with the tumor necrosis factor receptor: Fc fusion protein. The soluble TNF receptor sepsis study group. N Engl J Med 1996;334:1697–702.
119. Martins PS, Brunialti MK, da Luz Fernandes M, Martos LS et al. Bacterial recognition and induced cell activation in sepsis. Endocr Metab Immune Disord Drug Targets 2006;6:183–91.
120. Brunn GJ, Platt JL. The etiology of sepsis: turned inside out. Trends Mol Med 2006;12:10–16.
121. Hoffmann JA, Kafatos FC, Janeway CA, Ezekowitz RA. Phylogenetic perspectives in innate immunity. Science 1999;284:1313–18.

122. Hornef MW, Wick MJ, Rhen M, Normark S. Bacterial strategies for overcoming host innate and adaptive immune responses. Nat Immunol 2002;3:1033–40.
123. Aderem A, Underhill DM. Mechanisms of phagocytosis in macrophages. Annu Rev Immunol 1999;17:593–623.
124. Janeway CA Jr, Medzhitov R. Introduction: the role of innate immunity in the adaptive immune response. Semin Immunol 1998;10:349–50.
125. Janeway CA Jr, Medzhitov R. Innate immune recognition. Annu Rev Immunol 2002;20:197–216.
126. Fearon DT, Locksley RM. The instructive role of innate immunity in the acquired immune response. Science 1996;272:50–3.
127. Janeway CA Jr. The immune system evolved to discriminate infectious nonself from noninfectious self. Immunol Today 1992;13:11–16.
128. Lien E, Ingalls RR. Toll-like receptors. Crit Care Med 2002;30:S1–11.
129. Beutler B, Poltorak A. Sepsis and evolution of the innate immune response. Crit Care Med 2001; 29:S2–6.
130. Ozinsky A, Underhill DM, Fontenot JD, Hajjar AM, Smith KD, Wilson CB et al. The repertoire for pattern recognition of pathogens by the innate immune system is defined by cooperation between Toll-like receptors. Proc Natl Acad Sci USA 2000;97:13766–71.
131. Cristofaro P, Opal S. The Toll-like receptors and their role in septic shock. Expert Opin Ther Targets 2003;7:603–12.
132. Kiechl S, Lorenz E, Reindl M, Wiedermann CJ, Oberhollenzer F, Bonora E et al. Toll-like receptor 4 polymorphisms and atherogenesis. N Engl J Med 2002;347:185–92.
133. Opal SM, Huber CE. Bench-to-bedside review: Toll-like receptors and their role in septic shock. Crit Care 2002;6:125–36.
134. Vogel SN, Fisher CJ, Hansen CT, Rosenstreich RL. Characterization of a congenitally LPS-resistant, athymic mouse strain. J Immunol 1979;122:619–22.
135. Levi M, de Jonge E, Meijers J. The diagnosis of disseminated intravascular coagulation. Blood Rev 2002;16:217–23.
136. Zeerleder S, Hack E, Wuillemin WA. Disseminated intravascular coagulation in sepsis. Chest 2005;128:2864–75.
137. Marshall JC. Inflammation, coagulopathy, and the pathogenesis of multiple organ dysfunction syndrome. Crit Care Med 2001;29:S99–106.
138. Levi M, van der Poll T, ten Cate H, van Deventer SJ. The cytokine-mediated imbalance between coagulant and anticoagulant mechanisms in sepsis and endotoxaemia. Eur J Clin Invest 1997;27:3–9.
139. van der Poll T, Levi M, Hack CE, ten Cate H, van Deventer SJ, Eerenberg AJ et al. Elimination of interleukin 6 attenuates coagulation activation in experimental endotoxemia in chimpanzees. J Exp Med 1994;179:1253–9.
140. van der Poll T, Levi M, van Deventer SJ, ten Cate H, Haagmans BL, Biemond BJ et al. Differential effects of anti-tumor necrosis factor monoclonal antibodies on systemic inflammatory responses in experimental endotoxemia in chimpanzees. Blood 1994;83:446–51.
141. Bernard GR, Vincent JL, Laterre PF et al. Efficacy and safety of recombinant human activated protein C for severe sepsis. N Engl J Med 2001;344:699–709.
142. Ertel W, Keel M, Neidhardt R, Steckholzer U, Kremer JP, Ungethuem U et al. Inhibition of the defense system stimulating interleukin-12 interferon-gamma pathway during critical illness. Blood 1997;89:1612–20.
143. Bone RC. Sir Isaac Newton, sepsis, SIRS, and CARS. Crit Care Med 1996;24:1125–8.
144. Hesse DG, Tracey KJ, Fong Y, Manogue KR, Palladino MA Jr, Cerami A et al. Cytokine appearance in human endotoxemia and primate bacteremia. Surg Gynecol Obstet 1988;166:147–53.
145. Fong Y, Tracey KJ, Moldawer LL, Hesse DG, Manogue KB, Kenney JS et al. Antibodies to cachectin/tumor necrosis factor reduce interleukin 1 beta and interleukin 6 appearance during lethal bacteremia. J Exp Med 1989;170:1627–33.
146. Waage A, Brandtzaeg P, Halstensen A, Kierulf P, Espevik T. The complex pattern of cytokines in serum from patients with meningococcal septic shock. Association between interleukin 6, interleukin 1, and fatal outcome. J Exp Med 1989;169:333–8.

147. Casey LC, Balk RA, Bone RC. Plasma cytokine and endotoxin levels correlate with survival in patients with the sepsis syndrome. Ann Intern Med 1993;119:771–8.
148. Drost AC, Burleson DG, Cioffi WG Jr, Jordan BS, Mason AD Jr, Pruitt BA Jr. Plasma cytokines following thermal injury and their relationship with patient mortality, burn size, and time postburn. J Trauma 1993;35:335–9.
149. Martin C, Boisson C, Haccoun M, Thomachot L, Mege JL. Patterns of cytokine evolution (tumor necrosis factor-alpha and interleukin-6) after septic shock, hemorrhagic shock, and severe trauma. Crit Care Med 1997;25:1813–19.
150. Walley KR, Lukacs NW, Standiford TJ, Strieter RM, Kunkel SL. Balance of inflammatory cytokines related to severity and mortality of murine sepsis. Infect Immun 1996;64:4733–8.
151. Pruitt JH, Welborn MB, Edwards PD, Harward TR, Seeger JW, Martin TD et al. Increased soluble interleukin-1 type II receptor concentrations in postoperative patients and in patients with sepsis syndrome. Blood 1996 15;87:3282–8.
152. Rogy MA, Coyle SM, Oldenburg HS, Rock CS, Barie PS, Van Zee KJ et al. Persistently elevated soluble tumor necrosis factor receptor and interleukin-1 receptor antagonist levels in critically ill patients. J Am Coll Surg 1994;178:132–8.
153. Neidhardt R, Keel M, Steckholzer U, Safret A, Ungethuem U, Trentz O et al. Relationship of interleukin-10 plasma levels to severity of injury and clinical outcome in injured patients. J Trauma 1997;42:863–70.
154. Ayala A, Perrin MM, Kisala JM, Ertel W, Chaudry IH. Polymicrobial sepsis selectively activates peritoneal but not alveolar macrophages to release inflammatory mediators (interleukins-1 and -6 and tumor necrosis factor). Circ Shock 1992;36:191–9.
155. Ertel W, Kremer JP, Kenney J, Steckholzer U, Jarrar D, Trentz O et al. Downregulation of proinflammatory cytokine release in whole blood from septic patients. Blood 1995;85:1341–7.
156. Bogdan C, Vodovotz Y, Nathan C. Macrophage deactivation by interleukin 10. J Exp Med 1991;174:1549–55.
157. Wichmann MW, Ayala A, Chaudry IH. Severe depression of host immune functions following closed-bone fracture, soft-tissue trauma, and hemorrhagic shock. Crit Care Med 1998;26:1372–8.
158. Volk HD, Reinke P, Docke WD. Clinical aspects: from systemic inflammation to "immunoparalysis". Chem Immunol 2000;74:162–77.
159. Adrie C, Pinsky MR. The inflammatory balance in human sepsis. Intensive Care Med 2000;26: 364–75.
160. Volk HD, Reinke P, Krausch D, Zuckermann H, Asadullah K, Muller JM et al. Monocyte deactivation – rationale for a new therapeutic strategy in sepsis. Intensive Care Med 1996;22:S474–81.
161. Ziegler-Heitbrock HW. Molecular mechanism in tolerance to lipopolysaccharide. J Inflamm 1995;45:13–26.
162. Jaattela M, Wissing D. Heat-shock proteins protect cells from monocyte cytotoxicity: possible mechanism of self-protection. J Exp Med 1993;177:231–6.
163. Schroeder S, Bischoff J, Lehmann LE, Hering R, von Spiegel T, Putensen C et al. Endotoxin inhibits heat shock protein 70 (HSP70) expression in peripheral blood mononuclear cells of patients with severe sepsis. Intensive Care Med 1999;25:52–7.
164. Swope MD, Lolis E. Macrophage migration inhibitory factor: cytokine, hormone, or enzyme? Rev Physiol Biochem Pharmacol 1999;139:1–32.
165. Baugh JA, Bucala R. Macrophage migration inhibitory factor. Crit Care Med 2002;30:S27–35.
166. Calandra T, Bucala R. Macrophage migration inhibitory factor (MIF): a glucocorticoid counter-regulator within the immune system. Crit Rev Immunol 1997;17:77–88.
167. Bernhagen J, Calandra T, Bucala R. The emerging role of MIF in septic shock and infection. Biotherapy 1994;8:123–7.
168. Bernhagen J, Calandra T, Mitchell RA, Martin SB, Tracey KJ, Voelter W et al. MIF is a pituitary-derived cytokine that potentiates lethal endotoxaemia. Nature 1993;365:756–9.
169. Calandra T, Bernhagen J, Metz CN, Spiegel LA, Bacher M, Donnelly T et al. MIF as a glucocorticoid-induced modulator of cytokine production. Nature 1995;377:68–71.

170. Bucala R. MIF rediscovered: cytokine, pituitary hormone, and glucocorticoid-induced regulator of the immune response. FASEB J 1996;10:1607–13.
171. Calandra T, Bernhagen J, Mitchell RA, Bucala R. The macrophage is an important and previously unrecognized source of macrophage migration inhibitory factor. J Exp Med 1994;179:1895–902.
172. Bacher M, Metz CN, Calandra T, Mayer K, Chesney J, Lohoff M et al. An essential regulatory role for macrophage migration inhibitory factor in T-cell activation. Proc Natl Acad Sci USA 1996;93:7849–54.
173. Calandra T, Echtenacher B, Roy DL, Pugin J, Metz CN, Hultner L et al. Protection from septic shock by neutralization of macrophage migration inhibitory factor. Nat Med 2000;6:164–70.
174. Calandra T, Spiegel LA, Metz CN, Bucala R. Macrophage migration inhibitory factor is a critical mediator of the activation of immune cells by exotoxins of Gram-positive bacteria. Proc Natl Acad Sci USA 1998;95:11383–8.
175. Joshi PC, Poole GV, Sachdev V, Zhou X, Jones Q. Trauma patients with positive cultures have higher levels of circulating macrophage migration inhibitory factor (MIF). Res Commun Mol Pathol Pharmacol 2000;107:13–20.
176. Besedovsky H, del Rey A, Sorkin E, Dinarello CA. Immunoregulatory feedback between interleukin-1 and glucocorticoid hormones. Science 1986;23:652–4.
177. Dunn AJ. Endotoxin-induced activation of cerebral catecholamine and serotonin metabolism: comparison with interleukin-1. J Pharmacol Exp Ther 1992;261:964–9.
178. Dunn AJ. Cytokine activation of the HPA axis. Ann N Y Acad Sci 2000;917:608–17.
179. Uehara A, Gottschall PE, Dahl RR, Arimura A. Interleukin-1 stimulates ACTH release by an indirect action which requires endogenous corticotropin releasing factor. Endocrinology 1987;121:1580–2.
180. Dunn AJ. The role of interleukin-1 and tumor necrosis factor alpha in the neurochemical and neuroendocrine responses to endotoxin. Brain Res Bull 1992;29:807–12.
181. Bethin KE, Vogt SK, Muglia LJ. Interleukin-6 is an essential, corticotropin-releasing hormone-independent stimulator of the adrenal axis during immune system activation. Proc Natl Acad Sci USA 2000;97:9317–22.
182. Zuckerman SH, Shellhaas J, Butler LD. Differential regulation of lipopolysaccharide-induced interleukin 1 and tumor necrosis factor synthesis: effects of endogenous and exogenous glucocorticoids and the role of the pituitary-adrenal axis. Eur J Immunol 1989;19:301–5.
183. Schreiber W, Pollmacher T, Fassbender K, Gudewill S, Vedder H, Wiedemann K et al. Endotoxin- and corticotropin-releasing hormone-induced release of ACTH and cortisol. A comparative study in men. Neuroendocrinology 1993;58:123–8.
184. Rivier C. Influence of immune signals on the hypothalamic-pituitary axis of the rodent. Front Neuroendocrinol 1995;16:151–82.
185. Sternberg EM, Hill JM, Chrousos GP, Kamilaris T, Listwak SJ, Gold PW et al. Inflammatory mediator-induced hypothalamic-pituitary-adrenal axis activation is defective in streptococcal cell wall arthritis- susceptible Lewis rats. Proc Natl Acad Sci USA 1989;86:2374–8.
186. Maier SF, Goehler LE, Fleshner M, Watkins LR. The role of the vagus nerve in cytokine-to-brain communication. Ann N Y Acad Sci 1998;840:289–300.
187. Bertini R, Bianchi M, Ghezzi P. Adrenalectomy sensitizes mice to the lethal effects of interleukin 1 and tumor necrosis factor. J Exp Med 1988;167:1708–12.
188. Hawes AS, Rock CS, Keogh CV, Lowry SF, Calvano SE. In vivo effects of the antiglucocorticoid RU 486 on glucocorticoid and cytokine responses to Escherichia coli endotoxin. Infect Immun 1992;60:2641–7.
189. Pitcairn M, Schuler J, Erve PR, Holtzman S, Schumer W. Glucocorticoid and antibiotic effect on experimental gram-negative bacteremic shock. Arch Surg 1975;110:1012–15.
190. Bollaert PE, Charpentier C, Levy B, Debouverie M, Audibert G, Larcan A. Reversal of late septic shock with supraphysiologic doses of hydrocortisone. Crit Care Med 1998;26:645–50.
191. Briegel J, Forst H, Haller M, Schelling G, Kilger E, Kuprat G et al. Stress doses of hydrocortisone reverse hyperdynamic septic shock: a prospective, randomized, double-blind, single-center study. Crit Care Med 1999;27:723–32.

Edited by A. del Rey, G. P. Chrousos and H. O. Besedovsky


# *Neuroendocrinology of Inflammatory Disorders*

JEANETTE I. WEBSTER MARKETON[1] and ESTHER M. STERNBERG[2]

[1] *Institute for Behavioral Medicine Research and Division of Pulmonary, Allergy, Critical Care and Sleep Medicine, Department of Internal Medicine, The Ohio State University, Columbus, OH, USA*
[2] *Section on Neuroendocrine Immunology and Behavior, National Institute of Mental Health, National Institutes of Health, Department of Health and Human Services, Bethesda, MD, USA*

## ABSTRACT

The central nervous system (CNS) and immune systems regulate each other; the immune system is regulated by the CNS through hormonal signals and neuronal pathways and conversely the CNS is regulated by the immune system through cellular pathways and molecular signals including cytokines, chemokines, and interleukins (ILs). The CNS regulates the immune system through several routes: the hypothalamic–pituitary–adrenal (HPA) axis, the autonomic nervous system, and the peripheral nervous system. This neuroendocrine regulation of immune function through resultant glucocorticoid release is essential for survival from stress, infection, or inflammatory diseases. Glucocorticoids function through the glucocorticoid receptor (GR) to elicit multiple effects on immune cells and molecules. This chapter will mainly focus on the regulation of the immune response by the neuroendocrine system and the implications for inflammatory disease. Interruptions of this regulatory loop at multiple levels predispose and enhance inflammatory diseases and cause glucocorticoid resistance in a number of diseases.

## 1. INTRODUCTION

Glucocorticoids have been known to influence the immune system for many years and have been used in the treatment of inflammatory diseases since the 1940s. In fact, the Nobel Prize was awarded to Kendall, Reichstein, and Hench for this discovery in 1950 [1]. The pharmacological effects of glucocorticoids on many aspects of immune cell function have since been extensively researched [2,3] although it was not until recently that the fact that glucocorticoids play an essential physiological role in regulation of the immune system [4] in health and disease was fully appreciated. Fundamental to our understanding of the pathogenesis of inflammatory disease and ultimately for the development of effective therapies for such diseases is an understanding of the physiological mechanisms involved in glucocorticoid secretion and their regulation of the immune system both under normal and disease conditions.

This physiological regulation of the immune system by glucocorticoids is, in fact, only one part of an extensive regulatory network between the central nervous system (CNS), the neuroendocrine system and immune system. Nerve pathways, hormonal cascades, and cellular interactions form a network of connections that allows the CNS to regulate the immune system locally, regionally, and systemically at sites of inflammation, in immune organs, and through hormonal routes. Similarly, the immune system also regulates the CNS. During inflammation, the symptoms of sickness behavior and fever are induced by cytokines produced at the inflammatory site that signal the brain [5,6]. Other cells of the brain, such as glia, neurons, and macrophages, also express cytokines, which have been shown to play a role both in neuronal cell death [7–10] and in survival [11,12]. In several neurodegenerative diseases, such as neuro-AIDS, Alzheimer's, multiple sclerosis (MS), stroke, and nerve trauma, cytokine-mediated neuronal cell death is thought to play an important role [3–16].

Cytokines produced in the periphery can also function as hormones and can stimulate the CNS by several mechanisms. They can cross the blood–brain barrier (BBB) at leaky points, for example at the organum vasculosum lamina terminalis (OVLT) or median eminence, or may be actively transported across the BBB in small amounts [7]. They can, additionally, rapidly signal the CNS through the vagus nerve [8–20]. After binding to receptors on endothelial cells, cytokines can activate second messengers, such as nitric oxide (NO) and prostaglandins, which influence the brain [8,21–23]. Although a full understanding of this communication between the CNS and the immune systems is important, it will not be the main focus of this chapter, and for further information on this regulation see our earlier review in Ref. [24]. This chapter will focus only on the regulation of the immune system by the CNS, mainly through the neuroendocrine system.

There are two major mechanisms through which the CNS regulates the immune system at a systemic and regional level: (1) the production of glucocorticoids by the hormonal stress response and (2) the release of noradrenalin by the autonomic nervous system. The CNS can also regulate the immune system locally by release of neuropeptides such as substance P and locally produced corticotrophin-releasing hormone (CRH) from the peripheral nerves. In general, locally released neuropeptides are pro-inflammatory. This latter mechanism is not the focus of this chapter, and for further information on this refer to Ref. [24].

The hypothalamic–pituitary–adrenal (HPA) axis is the main regulator of the glucocorticoid effect on the immune system (Fig. 1) The HPA axis comprises of the following endocrine structures, the paraventricular nucleus (PVN) of the hypothalamus, the anterior pituitary gland located at the base of the brain, and the adrenal glands. CRH secreted from the PVN into the hypophyseal portal blood supply stimulates the expression of adrenocorticotropin hormone (ACTH) in the anterior pituitary gland. ACTH secreted into the systemic circulation then induces the expression and release of glucocorticoids by the adrenal glands.

The HPA axis is regulated from both within the CNS and from the periphery. The HPA axis is negatively regulated at the hypothalamic and pituitary level by its end product, glucocorticoids. Other factors such as neurotransmitters and neuropeptides of the sympathetic nervous system, and other neuropeptides, such as arginine vasopressin (AVP) as well as cytokines, also regulate the HPA axis [25–28]. Negative regulation of CRH occurs by ACTH and by itself, as well as by other neuropeptides and neurotransmitters in the brain, for example, γ-aminobutyric acid–benzodiazepines (GABA–BDZ) and opioid peptide systems. The serotonergic, cholinergic, and histaminergic systems positively regulate CRH [29–34].

This finely tuned regulatory system between the neuroendocrine and the immune systems is essential for health. Enhanced susceptibility to infection, inflammatory, or autoimmune disease results from imbalances of this system caused by disturbances at any level of the HPA axis that alter glucocorticoid action. Excessive amounts of circulating glucocorticoids and overall suppression of immune responses result from over-stimulation of the HPA axis and generally leads to enhanced susceptibility to, or severity of, infection, whereas lower circulating levels of glucocorticoids caused by under-stimulation is associated with susceptibility to inflammation.

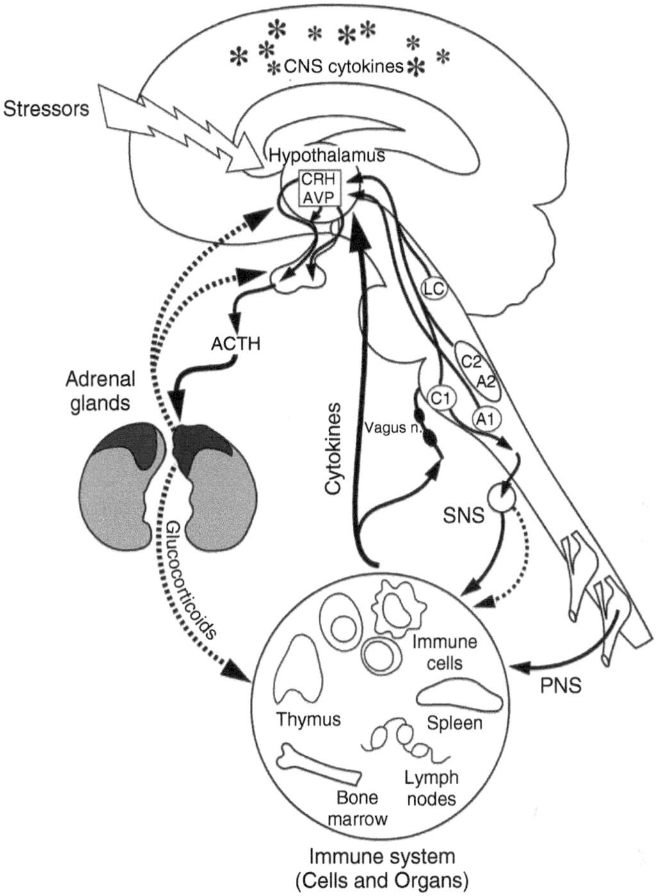


Figure 1. Two-way communication between the immune and the nervous system. The immune system and the central nervous system (CNS) communicate with and regulate each other. Cytokines made in the periphery signal the CNS by several routes. When activated, the hypothalamus releases corticotrophin-releasing hormone (CRH) and arginine vasopressin (AVP) that together stimulate the pituitary gland to release adrenocorticotrophin (ACTH). This, in turn, stimulates the adrenal cortex to release glucocorticoids. These hormones suppress immunity at a molecular, cellular, and organ level. Other routes by which the CNS regulates immunity are through the sympathetic nervous system (SNS) and peripheral nervous system (PNS). Stress activates the HPA axis to release these immunosuppressive hormones.

Dysregulation at the molecular level results in molecular-based glucocorticoid resistance and results in enhanced inflammation. Therefore, understanding this CNS and neuroendocrine system regulation of the immune system at the systemic, anatomical, and molecular levels will aid our understanding of the pathogenesis of inflammatory/autoimmune and infectious diseases and improve treatment and diagnosis of predisposition to these illnesses.

## 2. GLUCOCORTICOID MODULATION OF THE IMMUNE SYSTEM

Two receptors for glucocorticoids exist: the glucocorticoid receptor (GR) and the mineralocorticoid receptor (MR). Corticosterone has a higher affinity for MR than for GR. Thus, at low glucocorticoid levels, the preferential receptor is MR, and GR is occupied only at high stress levels of glucocorticoids [35]. MR and GR are capable of binding as heterodimers to DNA [36,37] and are co-expressed in some cell types suggesting possible involvement in gene transcription or repression. In the brain, a pro-active function has been suggested for MR in the maintenance of homeostasis, whereas a reactive role has been suggested for GR in the recovery from disturbance [38]. In immune cells, the primary receptor for glucocorticoids is GR. The expression of 11β-hydroxysteroid dehydrogenase, an enzyme responsible for the conversion of steroids from the active form, for example, cortisol and corticosterone, into an 11-keto inactive form, for example, cortisone and 11-dehydrocortisone, also regulates the availability of glucocorticoids. This enzyme exists in two forms: Type 1, which catalyzes the regeneration of active glucocorticoids from the inactive 11-keto form; and Type 2, which catalyzes the reverse reaction, that is, inactivation of glucocorticoids to the inert 11-keto form [39–41].

### 2.1. Mechanism of action of the GR

GR (NR3C1) is a member of the steroid and thyroid hormone receptor superfamily along with the progesterone, estrogen, mineralocorticoid, and thyroid receptors and essentially is a ligand-dependent transcription factor that mediates the end point tissue effect of glucocorticoids [42]. These receptors all share a similar three-domain structure (Fig. 2), consisting of the N-terminal transactivation domain, the central DNA-binding domain (DBD), and the C-terminal domain ligand-binding domain (LDB). The LBD is also involved in transactivation, dimerization, and hsp90 binding as well as ligand binding [43,44].

In plasma, glucocorticoids circulate associated with cortisol-binding globulin (CBG) or albumin. Glucocorticoids enter the cell by passive diffusion although evidence exists for active transport out of the cell [45]. In the inactivated state, GR is located in the cytoplasm in a multiprotein complex containing hsp90 and immunophilins. This is thought to hold the GR in a conformation that is accessible to the ligand. Upon activation by ligand binding, GR dissociates from this complex, translocates to the nucleus, and binds as a homodimer to its hormone response element (HRE) or glucocorticoid response element (GRE). The bound GR homodimer then modulates gene expression by interacting either directly or via cofactors with the basal transcription machinery (Fig. 3) GR can either up-regulate or down-regulated its target genes [43,46,47]. Down-regulation can occur via a negative glucocorticoid response element (nGRE), for example, the bovine prolactin gene and the pro-opiomelanocortin (POMC) gene [48–50] but

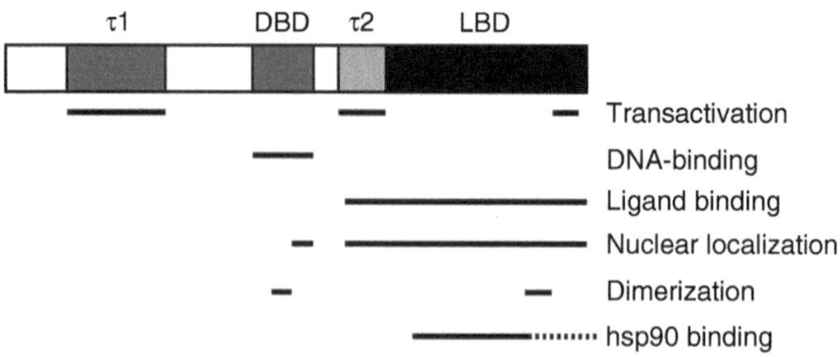


Figure 2. Molecular structure of the glucocorticoid receptor. The areas associated with the functions of transactivation, DNA binding, ligand binding, nuclear localization, dimerization, and hsp90 binding are shown.

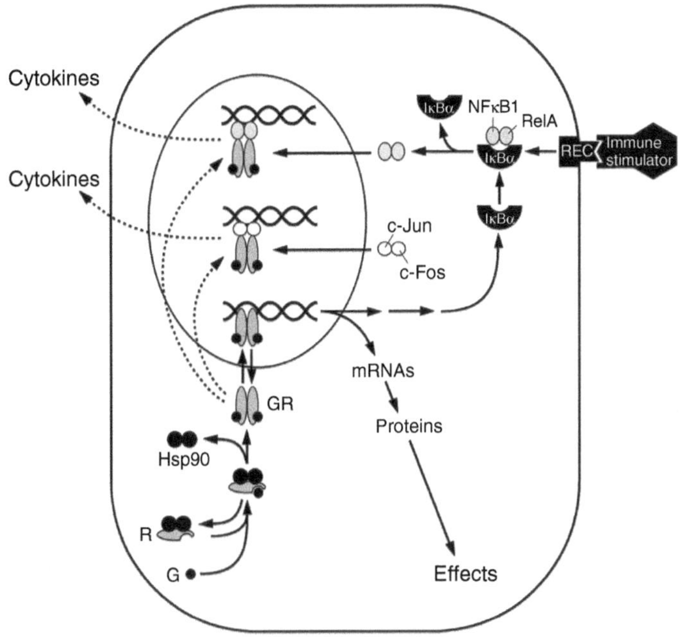


Figure 3. Molecular mechanisms of glucocorticoid effects on immune cell function. Schematic diagram of the mechanism of action of the glucocorticoid receptor including interactions with the NF-κB and AP-1 pathways. Dotted lines represent repressive pathways.

mostly occurs by interaction with other transcription factors, such as activator protein-1 (AP-1) and nuclear factor kappa B (NF-κB) [46,51–54]. It should be pointed out that the reverse can also occur, that is, AP-1 and NFκB are capable of repressing GR function [55,56].

Originally, two clones for GR were found that differed in the C-terminus [57]. The second is a splice variant of the full-length GR and was termed GRβ. It lacks the last 50 amino acids but instead it has a unique 15 amino acid C-terminus. GRβ is located in the nucleus regardless of ligand status and is also found in the cytoplasm complexed to hsp90. It is unable to bind ligand or activate gene transcription. It has been suggested to act as a dominant negative receptor in vitro by the formation of transcriptionally inactive heterodimers with GRα [58–64], but this mechanism is still under dispute as other studies have shown no effect of GRβ on GRα-mediated transactivation or transrepression [65–68].

### 2.2. Effects of glucocorticoids – pharmacological versus physiological

It is important to recognize that pharmacological and physiological doses or forms of glucocorticoids exert different effects. Modulation of transcription of genes involved in the inflammatory response is achieved by physiological doses of glucocorticoids, whereas total suppression of the inflammatory response occurs as a response to pharmacological doses [69]. Synthetic glucocorticoids, such as dexamethasone, and natural glucocorticoids, such as hydrocortisone, also elicit different immune responses. For example, consistent with the greater affinity of dexamethasone than hydrocortisone for GR, dexamethasone exerts a greater suppression on interleukin-12 (IL-12) than hydrocortisone [70].

## 2.3. Glucocorticoid effects on immune molecules

A wide variety of immune cell functions and expression of immune molecules are regulated by glucocorticoids through the molecular mechanisms described above. Glucocorticoids modulate gene expression of many immune-related molecules, including cytokines, adhesion molecules, chemoattractants, inflammatory mediators, and other inflammatory molecules (Table 1). (For comprehensive reviews see Ref. [71].)

### *2.3.1. Cytokines*

The transcription of many cytokines is modulated by glucocorticoids. In general, pro-inflammatory cytokines, such as IL-1 [72–74], IL-2 [75], IL-5 [76–79] IL-6 [80–83], IL-8 [82,84–86], IL-11 [87], IL-12 [88–91], tumor necrosis factor α (TNF-α) [92–94], interferon (IFNγ) [75,95], and granulocyte macrophage colony-stimulating factor (GM-CSF) [82,96,97], are suppressed while anti-inflammatory cytokines, IL-4 [78,98,99] and IL-10 [95,100], are up-regulated.

Glucocorticoids can also modulate cytokine expression and signaling by de-stabilization of mRNA, for example, IL-1β [72–74], IL-5 [79], IL-8 [85] and IL-11 [87]; modulation of receptors and decoy receptor expression, for example, IL-1R II [101–104] and IL-12R [88,89,91]; or regulation of translation, e.g., TNF-α [92,93]. The signaling pathways of these cytokines can also be inhibited by glucocorticoids, for example dexamethasone inhibits IL-2 signaling via the Jak-STAT cascade [105]. Although the anti-inflammatory cytokines are generally thought to be up-regulated by glucocorticoids, there are some cell and dose-dependant effects. IL-10 is up-regulated by pharmacological doses but suppressed by physiological doses of glucocorticoids [95,100], whereas IL-4 is induced in the lymph nodes and spleen of mice in response to physiological concentrations of glucocorticoids [106,107]. However, down-regulation of IL-4 expression in T cells in response to high stress levels of dexamethasone has been shown [75].

Table 1. Summary of immune-related genes regulated by glucocorticoids

| Increase | Decrease |
|---|---|
| *Cytokines*<br>IL-4, IL-10, IL-RI, IL-1 receptor antagonist | *Cytokines*<br>IL-1, IL-2, IL-3, IL-4, IL-5, IL-6, IL-8, IL-11, IL-12, IL-13, TNF-α, GM-CSF, IFNγ |
| | *Chemokines*<br>RANTES, Eotaxin, MIPα, MCP-1, MCP-2, MCP-3, CINC/gro |
| *Inflammatory mediators*<br>Lipocortin 1 (annexin 1), SLPI, MKP-1 | *Inflammatory mediators*<br>Prostaglandins, NO, NOS II, iNOS, COX-2, $cPLA_2$ |
| | *Adhesion molecules*<br>ICAM-1, ELAM-1, VCAM-1, E-selectin, L-selectin |
| *Receptors*<br>$\beta_2$-adrenoceptor<br>IκBα | *Receptors*<br>IL-2R, $NK_1R$, $NK_2R$, IL-4Rα |

This glucocorticoid modulation of cytokine expression causes a shift from a Th1 (cellular immunity) to a Th2 (humoral immunity) pattern of immunity. Other neural factors, such as the sympathetic neuropeptides, noradrenalin and adrenalin, can induce a Th1–Th2 shift [88]. A shift toward Th1 immunity is characteristic of some autoimmune diseases, such as rheumatoid arthritis, MS, and Type 1 diabetes mellitus, whereas in systemic lupus erythematosus (SLE), there is a shift toward Th2-mediated [88,108,109]. In cases of excessive glucocorticoid production, for example, in animal models with a hyperactive HPA axis [Fischer (F344/N) rats] or in women in the third trimester of pregnancy, a relative resistance to Th1-associated autoimmune diseases has been described. Conversely, animals lacking glucocorticoids or exhibiting a hypoactive HPA axis [Lewis (LEW/N) rats] are susceptible to Th1-associated autoimmune diseases [109]. For example, LEW/N rats are susceptible to the Th1-mediated autoimmune disease, experimental allergic encephalomyelitis (EAE). Spontaneous recovery is correlated with an increase in glucocorticoids, whereas adrenalectomy results in fatal exacerbation of the disease that can be reversed by administration of glucocorticoids, indicating the shift from Th1 to Th2 immunity [110].

#### *2.3.2. Cell adhesion molecules*

Glucocorticoids down-regulate the proteins involved in the attraction and adhesion of leukocytes to areas of inflammation. The expression of intracellular adhesion molecule-1 (ICAM-1) [55,111–115], endothelial-leukocyte adhesion molecule-1 (ELAM-1) [111], vascular adhesion molecule-1 (VCAM-1) [112,116,117], E-selectin [118,119], and L-selectin [120,121] are inhibited by dexamethasone.

### 2.4. Chemoattractants

Chemoattractants for neutrophils, eosinophils, and monocytes are important for the accumulation of these cells at sites of inflammation. Glucocorticoids suppress IL-1β induction of cytokine-induced neutrophil chemoattractant (CINC)/gro [122], TNF-α, and IL-1β induction of eotaxin [123] and down-regulate RANTES (regulated upon activation of normal T cell expressed and secreted) [124] and monocyte chemoattractant protein-1 (MCP-1), MCP-2, and MCP-3 [125–128].

### 2.5. Inflammatory mediators

Glucocorticoids suppress the production of inflammatory mediators including prostaglandins and NO. Glucocorticoids repress cytosolic phospholipase $A_2$ ($PLA_2$) and cyclooxygenase-2 (COX-2) mRNA levels, key enzymes in the synthesis of prostaglandins from arachadonic acid [129–131]. Repression of prostaglandin synthesis is also a result of glucocorticoid induction of lipocortin 1 (or annexin 1), which inhibits arachadonic acid release [132–134]. Glucocorticoids have been shown to repress the cytokine induction of NO synthase II (NOS II) [135] and inhibit transcription of the inducible nitric oxide synthase (iNOS) gene [136–138], thereby suppressing NO production.

### 2.6. Other inflammatory response factors

Glucocorticoids regulate receptors involved in regulation of the immune system. Transcription of the $\beta_2$-adrenoreceptor, which is involved in the adrenergic control of the immune system, is increased by glucocorticoids [139–141], transcription of the $NK_1$-receptor, the receptor for substance P [142], and the $NK_2$-receptor [143] are down-regulated by glucocorticoids.

## 3. ROLE OF ENDOGENOUS GLUCOCORTICOIDS IN PROTECTION FROM INFLAMMATION AND REGULATION OF IMMUNE FUNCTION

Circumstantial evidence exists that physiological fluctuations in glucocorticoid levels, such as during exercise and with circadian rhythm, are associated with altered immune function, for example changes in cytokine levels and production by leukocytes [144–147]. However, the strongest evidence that endogenous glucocorticoids are essential physiological regulators of the immune response and inflammatory/autoimmune disease comes from animal models.

HPA axis intervention, either surgically or pharmacologically, in rodent strains has been shown to alter the course and severity of inducible autoimmune/inflammatory disease. In F344/N rats, injection of streptococcal cell walls together with administration of the glucocorticoid antagonist RU486 is associated with development of arthritis and high mortality [148]. Adrenalectomy of rats increases the mortality rate from approximately 50 to 100% after infection with *Salmonella typhimurium* [149]. Similarly, in mice, adrenalectomy results in increased lethality following murine cytomegalovirus (MCMV) infection [150], Shiga toxin [151], and anthrax lethal toxin [152]. The importance of an intact HPA axis to protect against septic shock is indicated by these high rates of mortality, within 12–24 h, following exposure to a wide range of antigenic, pro-inflammatory, or infectious stimuli.

Reconstitution of the HPA axis, pharmacologically with glucocorticoids or surgically by intracerebral fetal hypothalamic tissue transplantation in animal models, attenuates inflammatory disease. Arthritis and carrageenan inflammation are significantly attenuated in LEW/N rats either by low-dose dexamethasone treatment [148] or by intracerebroventricular transplantation of F344/N hypothalamic tissue [153]. Adrenalectomized mice were protected from MCMV lethality by dexamethasone [150]. Immunization with myelin basic protein in LEW/N rats induces EAE. Initial immunization results in transient paralysis followed by some recovery along with an increase in endogenous corticosterone production that is essential for survival. Interruption of this endogenous production by adrenalectomy results in mortality. Replacement of basal corticosterone levels by a subcutaneous steroid implant does not reduce mortality. However, replacement of a dose equivalent to the EAE-induced corticosterone level results in increased survival rates and EAE development similar to control animals. Complete remission of the disease is achieved if replacement of even higher corticosterone levels is used [110]. Such experiments showing that interruption of the HPA axis predisposes to enhanced inflammation, while reconstitution attenuates autoimmune/inflammatory disease, suggests a role of the HPA axis in intensity of human diseases.

## 4. ANIMAL MODELS OF INFLAMMATORY DISEASE

Understanding the pathogenesis of autoimmune/inflammatory disease has been greatly aided by the use of animal models. A blunted HPA axis response is present in some animal species predisposed to autoimmune disease. These include the obese strain (OS) chicken (a model for autoimmune thyroiditis) [154], certain mouse lupus (SLE) models (MRL strain but not NZB/NZW) [155,156], and the inbred rat strain, LEW/N rats, that are prone to a large variety of autoimmune/inflammatory diseases the pattern of which is related to the particular antigen or pro-inflammatory stimulus to which they are exposed (Table 2) For a review of these and other animal models of inflammatory diseases refer to Jafarian-Tehrani and Sternberg [157] and Tonelli et al. [158].

Table 2. Inflammatory/autoimmune diseases correlated with a dysfunctional HPA axis in humans and in some animal models

| Inflammatory/Autoimmune disease | Reference |
|---|---|
| *Chicken* | |
| Thyroiditis | 154 |
| Scleroderma | 170 |
| *Mice* | |
| Systemic lupus erythematosus (SLE) | 155, 156 |
| *Rat* | |
| Arthritis | 148, 160–162 |
| Experimental allergic encephalomyelitis (EAE) | 159 |
| Septic shock | 148 |
| Inflammation | 153 |
| *Human* | |
| Rheumatoid arthritis | 180–186 |
| Systemic lupus erythematosus (SLE) | 182, 187 |
| Sjögren's syndrome | 182, 188, 189 |
| Dermatitis | 190–192 |
| Fibromyalgia | 182, 194, 198 |
| Chronic fatigue syndrome | 193–197 |
| Multiple sclerosis | 199–202 |
| Inflammatory bowel disease | 203, 204 |

A natural genetically uniform system, such as inbred rat strains, in which altered neuroendocrine responsiveness is associated with differential susceptibility and resistance to autoimmune/inflammatory disease, can be used to test the role of neuroendocrine regulation of various aspects of immunity. Inbred LEW/N rats are highly susceptible to development of a wide range of autoimmune/inflammatory diseases in response to a variety of antigenic or pro-inflammatory stimuli. For example, upon immunization with myelin basic protein, LEW/N rats develop EAE [159], in response to carrageenan, heat-killed *Mycobacterium tuberculosis*, or streptococcal cell walls, they develop arthritis [148,160–162]. Their largely histocompatible counterparts, F344/N rats, are relatively resistant to these same illnesses after exposure to the same dose of antigens. These two strains also show differences in HPA axis responsiveness, in comparison to outbred rats, the inflammatory-susceptible LEW/N exhibit a blunted response, whereas inflammatory-resistant F344/N rats have an excessive HPA response [148,162–165]. Differences in the expression of molecules involved in the HPA axis and glucocorticoid action such as hypothalamic CRH [162], POMC [164], CBG [163], and GR expression and activation [163,166,167], have been shown in these two rat strains. Glucocorticoids also play a role in protection from septic shock. Blockade of the GR by the antagonist RU486 leads to rapid death of F344/N rats co-administered with lipopolysaccharide (LPS) [148] and also to exacerbation of endotoxemia effects in rats [168].

OS chickens that develop spontaneous thyroiditis, a model for Hashimoto's thyroiditis, also exhibit a blunted HPA axis [154]. These chickens also have decreased free corticosterone levels because of an increase in CBG [169]. UCD-200 chickens, a model for human scleroderma, exhibit a hyporesponsive adrenal gland to ACTH [170]. MRP lupus-prone mice also exhibit such blunted HPA axis responses [155,156]. However, other strains such as NZB do not. This supports the notion that multiple genetic factors, including but not solely HPA axis responsiveness, contribute to overall autoimmune disease susceptibility.

## 5. DISRUPTIONS IN THE HPA AXIS OR GLUCOCORTICOID RESPONSE LEADING TO GLUCOCORTICOID RESISTANCE AND EXACERBATION OF HUMAN INFLAMMATORY DISEASE

However, associations as described above between a blunted HPA axis and an inflammatory/autoimmune disease do not prove cause and effect. In order to do this, intervention studies must be performed. Furthermore, many other factors likely contribute to such illnesses in humans. In animal and human genetic studies, genes in over 20 different regions on 15 different chromosomes determine susceptibility and resistance to inflammatory arthritis [171–174]. Many of these genes are related to immunity but many are not and still more are unknown. Amongst these factors, HPA axis responsiveness is one variable that contributes to overall susceptibility and resistance to such complex autoimmune diseases.

Amongst individuals, there is considerable degree of variance of HPA axis responsiveness. Intra-individual HPA axis responses, even in normal healthy individuals, are considerably variable and normal individuals can be sub-grouped depending on the response of their HPA axis to stimuli into high or low responders [175,176].

The anti-inflammatory immunosuppressive effects of glucocorticoids at the molecular level through GR is an important mechanism by which activation of the HPA axis regulates these immune responses and severity of expression of resultant disease. The HPA axis or glucocorticoid response can be disrupted at many levels, such as at the hypothalamus, pituitary, or adrenals, with changes in the expression of CRH, ACTH, or cortisol, respectively, or with changes in the sensitivity of the system to stimuli. At the target tissue, there are then many steps involved in the induction of gene activation by cortisol. These include entry of the hormone into the cell, binding to GR, dimerization, translocation of GR to the nucleus and interaction with cofactors and the basal transcription machinery to modulate gene expression. Interruption of any of these stages in this pathway could result in a defective HPA axis or glucocorticoid response resulting in either a reduction in glucocorticoid production or glucocorticoid resistance and ultimately leading to enhanced autoimmune/inflammatory disease susceptibility. The effectiveness of exogenous glucocorticoid therapy will be dependent on the specific defect. For further information on glucocorticoid resistance syndromes, see the recent review by Charmandari et al. [177].

It is noteworthy that many autoimmune/inflammatory diseases such as SLE, rheumatoid arthritis, and MS are more common in women than in men. Modulation of immunity by sex hormones, particularly estrogen, has been extensively researched, and there are many interactions between the HPA and the hypothalamic-pituitary-gonadal (HPG) axes. Although an important area of research, this subject will not be reviewed here (for further reading, see Jansson and Holmdahl [178] and Crofford et al. [179]).

### 5.1. HPA axis

It is often difficult to determine in patient populations precisely where a problem lies within the HPA axis. Many of the regulatory components cannot be directly measured in blood, making their analysis virtually impossible. Levels of ACTH and cortisol in peripheral blood are used as an indirect measure of CRH responsiveness since CRH cannot be measured directly in peripheral venous blood. Low levels of glucocorticoids or a low glucocorticoid response to HPA axis stimulation resulting from a blunted HPA axis have been implicated in a number of inflammatory diseases, such as rheumatoid arthritis [180–186], SLE [182,187], Sjogren's syndrome [182,188,189], allergic asthma and atopic skin disease [190–192], chronic fatigue syndrome (CSF) [193–197], fibromyalgia [182,194,198], MS [199–202], and inflammatory and irritable bowel syndrome [203,204]. Conversely, in chronic stress situations, such as those experienced by care-givers of Alzheimer's patients, students taking examinations, couples during marital conflict, and Army Rangers undergoing extreme exercise, there is excessive stimulation of the HPA axis with resultant chronically elevated glucocorticoid levels, which is associated with an enhanced susceptibility to viral infection, prolonged wound healing, or decreased antibody production after vaccination [205–210]. While in chronic stress situations, baseline measures of HPA axis function can provide evidence for elevated hormonal levels, it is important to note that baseline measure alone does not provide sufficient evidence for hypo-HPA function. In these cases, stimulation studies must be performed using exogenous hormones (ovine CRH, AVP, and ACTH) [195,201], physical stress (exercise) [145,196], or psychosocial stress (public speaking and mental arithmetic) [190,192,196]. Insulin hypoglycemia is also a potent method to stimulate the HPA axis [187].

A hyporesponsive HPA axis has been described in patients with rheumatoid arthritis. Such patients exhibit an under-responsive axis to IL-6, IL-1, or CRH stimulation together with loss of the cortisol circadian rhythm [181,211–213]. In untreated patients, it has been shown that although there is hypersecretion of ACTH, there is no corresponding increase in adrenal production of cortisol suggesting a defect in adrenal responsiveness [212,214,215].

Both CFS and fibromyalgia have also been associated with dysfunction of the HPA axis. In CFS, patients exhibit reduced free urinary cortisol compared to controls [195,216–220]. The defect appears to lie at the pituitary or hypothalamus as adult patients exhibit low ACTH and cortisol production in response to oCRH but normal basal levels [194,195,198]. In another study, no differences were observed in basal or CRH-stimulated cortisol and ACTH levels between adolescent CFS and control-matched subjects, but a relative resistance to glucocorticoids with regard to T-cell proliferation was observed. The authors attribute this difference in HPA axis responsiveness to the age of the patients [221]. In other studies using insulin-induced hypoglycemia, an increase in ACTH but not cortisol in CFS patients compared to controls was observed [222]. However, another group studying CFS patients, also without depression, did show reduced ACTH responses to stimulation [220]. Similar to CFS, patients with fibromyalgia show low 24-h free urinary cortisol and loss of diurnal cortisol secretion [179,223,224]. An abnormal HPA axis response to stimulation by oCRH was observed with a reduced secretion of cortisol but an exaggerated ACTH response [223]. A blunted cortisol secretion to exogenous CRH, insulin-induced hypoglycemia [225], or to physical exercise [226] has also been described. However, this deficiency appears to be at the level of the adrenal gland, as no difference exists in ACTH response to oCRH in fibromyalgia patients or controls [223]. In some cases, using exogenous CRH or different stimuli of the HPA axis, even higher ACTH response levels have been shown in fibromyalgia patients compared to controls [225,227].

MS has also been associated with dysregulation of the HPA axis. In one study, a central up-regulation of the HPA axis was described but with some degree of adrenal insensitivity. A normal stress response was described in these patients [202]. We have shown a relative blunting of ACTH response in AVP stimulation in the context of high basal cortisol [201]. Other studies support the association of dysregulations of the HPA axis correlating with activity of MS symptoms [199,200,202,228].

Patients with Sjogren's syndrome also exhibit lower basal HPA axis activity compared to controls as well as hyporesponsiveness of the HPA axis following oCRH stimulation or infusion of CRF, which is due to hyporesponsiveness at the level of the hypothalamus or pituitary as they exhibit a blunted ACTH response [188,189]. In SLE, studies have shown that SLE patients exhibit an attenuated cortisol response to insulin-induced hypoglycemia compared to controls [187,229]. Thus, a wide range of autoimmune/inflammatory diseases in humans are associated with a blunted HPA axis (Table 2).

### 5.2. Adrenal glands

Isolated glucocorticoid deficiency (IGD) is an autosomal recessive disorder characterized by adenocortical but not mineralocorticoid deficiency. Patients have low, undetectable cortisol levels that upon treatment with exogenous ACTH do not increase and high endogenous ACTH levels. This condition has been shown to be due to 17 individual point mutation and frameshifts in the receptor for ACTH [230–232]. Thus, defective ACTH receptors render the adrenal gland incapable of sensing the high ACTH levels that continue to be secreted by the pituitary, and despite the high levels of ACTH, little or no cortisol is produced. Novel mutations in the ACTH receptor have also been described in a patient with ACTH hypersensitivity syndrome, exhibiting normal cortisol levels but low undetectable ACTH levels [233]. Some patients with IGD have been reported to develop autoimmune diseases, such as organ-specific autoimmunity and autoimmune-mediated hypothroidism [234,235]. Adrenal insensitivity or a defect in adrenal responsiveness has also been described to occur in rheumatoid arthritis [212,214] and MS patients [202].

### 5.3. Cortisol-binding globulin

CBG is the major protein-binding cortisol in the blood and therefore limits the amount of free cortisol available. The availability of cortisol can be further affected by changes in the expression of this protein or in its binding capacity. The increased expression of CBG in some patients with long-standing Crohn's disease has been suggested to limit the bioavailability of glucocorticoids, resulting in the partial or complete resistance to steroids that has been described [236]. Likewise, an increase in CBG levels has also been described in CFS patients [198].

### 5.4. 11β-hydroxysteroid dehydrogenase

As described earlier, this enzyme catalyzes the conversion between active and inactive glucocorticoid [39,40] and thus differences in circulating or tissue glucocorticoid concentrations could result from changes in this enzyme [237]. An impairment in the conversion of the inactive cortisone to the active cortisol resulting in a drop in plasma cortisol levels was noted in obese men, indicating an impairment in the type 1 11β-hydroxysteroid dehydrogenase [238]. In ulcerative colitis, a decrease in 11β-hydroxysteroid dehydrogenase mRNA has also been demonstrated [239].

### 5.5. Glucocorticoid receptor

Familial glucocorticoid resistance, a hereditary condition, is caused by mutation of the GR resulting in decreased number of receptors, decreased affinity, or stability of the receptor or a decrease in translocation of the receptor to the nucleus. Currently, the molecular defects of eight families and one sporadic case have been determined and include six different point mutations in the LBD [240–247], one in the DNA-binding region [246,248] and a deletion in the LBD [249]. (For a review, see Charmandari et al. [177].) Another mutant has been described, which prevented GR-glucocorticoid receptor-interacting protein 1 (GRIP1) binding [250]. A phase shift mutation in GR has been described in patients with lupus nephritis [251], and a mutation in the LBD has been found in patients with lupus [252].

Mutations of GR are not the only mechanism by which changes in the gene can affect sensitivity to glucocorticoids. Increased sensitivity to glucocorticoids has been associated with the GR polymorphisms N363S [253–255] and BclI [256]. However, decreased peripheral sensitivity in a normal population has also been associated with N363S [257]. The polymorphism ER22/23RK has been associated with reduced sensitivity to glucocorticoids [255,258,259]. However, in other populations, these polymorphisms could not be correlated with differences in glucocorticoid sensitivity [257,260,261]. The polymorphism ER22/23EK and N363S have been shown to affect GR transactivation but not transrepression [255]. However, in another study, ER22/23EK was shown to affect GR-mediated transrepression but not transactivation [262].

GR contains several phosphorylation sites of unclear function. However, mutation of these phosphorylation sites results in reduced GR transactivation at a minimal promoter and reduces the stability of the GR protein [263]. Such changes in phosphorylation status of GR could have profound effects on GR function and maybe one method by which glucocorticoid resistance occurs. Other defects in the steps in the pathway leading to gene activation by glucocorticoids could result in glucocorticoid resistance, as suggested by some patients with glucocorticoid resistance who have no detectable mutation in GR [264].

Variations in the ratios of GRα and the proposed negative regulator of GR function, GRβ, could be associated with apparent glucocorticoid resistance, both initial and acquired. A higher expression of GRβ has been associated with glucocorticoid-resistant asthma [265–273]. Cytokines induce GRβ expression [274], and consequently, as inflammatory disease progresses exacerbation or development of glucocorticoid-resistant disease states could be caused by cytokine-induced GRβ expression [269]. Increased GRβ expression has also been shown in rheumatoid arthritis [275–277], Cushing's syndrome [278], Crohn's disease [279], ulcerative colitis [280,281], inflammatory bowel disease [282], chronic lymphocytic leukemia [283], septic shock [277], ankylosing spondylitis [284], and nasal polyposis disease [285,286]. Another splice variant, GRγ, has an insertion of three bases in the DBD of GR. This insertion decreases the transactivation activity of GR [287]. However, it has been suggested not to play a role in glucocorticoid sensitivity [288]. These studies suggest that a variety of pre- and post-translational changes in both forms of the GR are associated with autoimmune/inflammatory diseases.

### 5.6. Cofactors

To date, a mutation in a cofactor has not been found associated with glucocorticoid resistance. However, in a recent study into the glucocorticoid hypersensitive state associated with HIV-1 infection, a HIV-1 accessory protein, virion-associated protein (Vpr) that is able to act as a transcriptional activator to aid the replication of the HIV-1 virus, has been shown to contain a nuclear receptor-binding motif (LXXLL). Vpr binds directly to GR and co-operatively enhances transcription via SRC1a and p300/CBP cofactors [289,290]. A cofactor defect has been suggested to be the cause of resistance to multiple steroids that has been described in two sisters [291,292]. Smoking has been shown to cause oxidative stress, and this can reduce the expression of histone deacetylases (HDACs). In chronic obstructive pulmonary disease, a glucocorticoid-resistant disease, there is a decrease in HDAC2 [293,294].

### 5.7. Transport proteins

Glucocorticoids diffuse into cells although evidence exists for an active transport out of the cells, thereby adding yet another step in the process that can modulate glucocorticoid sensitivity by affecting the intracellular concentration of the ligand. Multidrug resistance (MDR) proteins, members of the ATP-binding cassette (ABC) family of transmembrane transporters [295], have been suggested to be involved in this active transport of glucocorticoids out of cells, and there is now increasing evidence from cell systems and animal studies that these proteins can indeed transport glucocorticoids [296–300]. An increased expression of MDR1 has been shown in patients with inflammatory bowel disease who have failed medical therapy [301–305], in patients with rheumatoid arthritis [306], Crohn's disease [305], and SLE [307], suggesting a role of these transport proteins in poor prognosis of these diseases and unresponsive glucocorticoid therapy. The novel orphan receptor pregnane X receptor [PXR; also known as steroid and xenobiotic receptor (SXR)] has been identified as one of the components involved in the up-regulation of these transporter proteins [308–311]. Glucocorticoids are amongst the many ligands that activate PXR [312], possibly representing another mechanism by which acquired glucocorticoid resistance might develop.

## 6. INFECTION

Bacterial and viral infections may also play a role in modulation of GR signaling. We have shown that the anthrax lethal toxin non-competitively represses GR signaling by preventing GR–DNA binding [313,314]. There are small pieces of evidence that other bacterial or viral infections or proteins may modify GR signaling (reviewed in Ref. [315]). Probably, the most well-studied viral protein is the HIV Vpr protein that was discussed earlier [289,290]. Modulation of nuclear hormone receptor signaling by bacterial or viral infections or proteins could be another mechanism by which glucocorticoid resistance occurs and may play a role in inflammatory disease etiology.

## 7. CONCLUSION

In summary, multiple neuro-anatomical, hormonal, and molecular mechanisms exist by which the CNS regulates immune function and plays a role in susceptibility to and pathogenesis of autoimmune/inflammatory disease. The HPA axis and glucocorticoids, the final effector endpoint of the HPA axis, and their regulation of immunity and involvement in severity of disease have been the major focus of this chapter. There are many potential mechanisms at the gene, protein, receptor, signaling and cell function levels where dysregulation could lead to disease. The potential mechanisms for pathogenesis of autoimmune disease(s) resulting from disruptions in these interactions is large as there are many different hormonal and nerve pathways involved in the regulation of immunity. Nonetheless, new insights into the bi-directional regulation of these systems through a thorough understanding of all levels by which the CNS and immune systems communicate and the disruptions in these communications that lead to disease will ultimately provide new avenues of therapy.

## REFERENCES


1. Hench P. Effects of cortisone in the rheumatic diseases. Lancet 1950;2:483–4.
2. Adcock IM. Glucocorticoids: new mechanisms and future agents. Curr Allergy Asthma Rep 2003;3:249–57.
3. Barnes PJ. Anti-inflammatory actions of glucocorticoids: molecular mechanisms. Clin Sci 1998;94:557–72.
4. Cupps TR, Fauci AS. Corticosteroid-mediated immunoregulation in man. Immunol Rev 1982;65:133–55.
5. Dantzer R. Cytokine-induced sickness behaviour: a neuroimmune response to activation of innate immunity. Eur J Pharmacol 2004;500:399–411.
6. Watkins LR, Maier SF. Implications of immune-to-brain communication for sickness and pain. Proc Natl Acad Sci USA 1999;96:7710–13.
7. Benveniste EN. Cytokine actions in the central nervous system. Cytokine Growth Factor Rev 1998;9:259–75.
8. Gibson RM, Rothwell NJ, Le Feuvre RA. CNS injury: the role of the cytokine IL-1. Vet J 2004;168:230–7.
9. Hopkins SJ, Rothwell NJ. Cytokines and the nervous system. I: expression and recognition. Trends Neurosci 1995;18:83–8.
10. Patel HC, Boutin H, Allan SM. Interleukin-1 in the brain: mechanisms of action in acute neurodegeneration. Ann N Y Acad Sci 2003;992:39–47.
11. Brenneman DE, Schultzberg M, Bartfai T, Gozes I. Cytokine regulation of neuronal survival. J Neurochem 1992;58:454–60.
12. Strle K, Zhou JH, Shen WH, Broussard SR, Johnson RW, Freund GG, Dantzer R, Kelley KW. Interleukin-10 in the brain. Crit Rev Immunol 2001;21:427–49.
13. Benveniste EN, Nguyen VT, O'Keefe GM. Immunological aspects of microglia: relevance to Alzheimer's disease. Neurochem Int 2001;39:381–91.
14. Feuerstein GZ, Wang X, Barone FC. The role of cytokines in the neuropathology of stroke and neurotrauma. Neuroimmunomodulation 1998;5:143–59.
15. Rosenberg PB. Clinical aspects of inflammation in Alzheimer's disease. Int Rev Psychiatry 2005;17:503–14.
16. Zhao B, Schwartz JP. Involvement of cytokines in normal CNS development and neurological diseases: recent progress and perspectives. J Neurosci Res 1998;52:7–16.
17. Banks WA, Kastin AJ, Broadwell RD. Passage of cytokines across the blood-brain barrier. Neuroimmunomodulation 1995;2:241–8.
18. Dantzer R, Konsman JP, Bluthe RM, Kelley KW. Neural and humoral pathways of communication from the immune system to the brain: parallel or convergent. Auton Neurosci 2000;85:60–5.
19. Goehler LE, Gaykema RP, Hansen MK, Anderson K, Maier SF, Watkins LR. Vagal immune-to-brain communication: a visceral chemosensory pathway. Auton Neurosci 2000;85:49–59.
20. Maier SF, Goehler LE, Fleshner M, Watkins LR. The role of the vagus nerve in cytokine-to-brain communication. Ann N Y Acad Sci 1998;840:289–300.
21. Elmquist JK, Scammell TE, Saper CB. Mechanisms of CNS response to systemic immune challenge: the febrile response. Trends Neurosci 1997;20:565–70.
22. Gaillard RC. Cytokines in the neuroendocrine system. Int Rev Immunol 1998;17:181–216.
23. McCann SM, Kimura M, Karanth S, Yu WH, Rettori V. Nitric oxide controls the hypothalamic-pituitary response to cytokines. Neuroimmunomodulation 1997;4:98–106.
24. Sternberg EM. Neural regulation of innate immunity: a coordinated nonspecific host response to pathogens. Nat Rev Immunol 2006;6:318–28.
25. Haddad JJ, Saade NE, Safieh-Garabedian B. Cytokines and neuro-immune-endocrine interactions: a role for the hypothalamic-pituitary-adrenal revolving axis. J Neuroimmunol 2002;133:1–19.
26. Rivest S. How circulating cytokines trigger the neural circuits that control the hypothalamic-pituitary-adrenal axis. Psychoneuroendocrinology 2001;26:761–88.

27. Scott LV, Dinan TG. Vasopressin and the regulation of hypothalamic-pituitary-adrenal axis function: implications for the pathophysiology of depression. Life Sci 1998;62:1985–98.
28. Volpi S, Rabadan-Diehl C, Aguilera G. Vasopressinergic regulation of the hypothalamic pituitary adrenal axis and stress adaptation. Stress 2004;7:75–83.
29. Bugajski J, Gadek-Michalska A, Bugajski AJ. Nitric oxide and prostaglandin systems in the stimulation of hypothalamic-pituitary-adrenal axis by neurotransmitters and neurohormones. J Physiol Pharmacol 2004;55:679–703.
30. Giordano R, Pellegrino M, Picu A, Bonelli L, Balbo M, Berardelli R, Lanfranco F, Ghigo E, Arvat E. Neuroregulation of the hypothalamus-pituitary-adrenal (HPA) axis in humans: effects of GABA-, mineralocorticoid-, and GH-Secretagogue-receptor modulation. Scientific WorldJournal 2006;6:1–11.
31. Hanley NR, Van de Kar LD. Serotonin and the neuroendocrine regulation of the hypothalamic – pituitary-adrenal axis in health and disease. Vitam Horm 2003;66:189–255.
32. Knigge U, Warberg J. The role of histamine in the neuroendocrine regulation of pituitary hormone secretion. Acta Endocrinol 1991;124:609–19.
33. Kovacs KJ, Miklos IH, Bali B. GABAergic mechanisms constraining the activity of the hypothalamo-pituitary-adrenocortical axis. Ann N Y Acad Sci 2004;1018:466–76.
34. Watts AG. Glucocorticoid regulation of peptide genes in neuroendocrine CRH neurons: a complexity beyond negative feedback. Front Neuroendocrinol 2005;26:109–30.
35. Funder JW. Glucocorticoid receptors. J Steroid Biochem Mol Biol 1992;43:389–94.
36. Liu W, Wang J, Sauter NK, Pearce D. Steroid receptor heterodimerization demonstrated in vitro and in vivo. Proc Natl Acad Sci USA 1995;92:12480–4.
37. Trapp T, Holsboer F. Heterodimerization between mineralocorticoid and glucocorticoid receptors increases the functional diversity of corticosteroid action. Trends Pharmacol Sci 1996;17:145–9.
38. De Kloet ER. Hormones and the stressed brain. Ann N Y Acad Sci 2004;1018:1–15.
39. Draper N, Stewart PM. 11beta-hydroxysteroid dehydrogenase and the pre-receptor regulation of corticosteroid hormone action. J Endocrinol 2005;186:251–71.
40. Seckl JR. 11beta-hydroxysteroid dehydrogenases: changing glucocorticoid action. Curr Opin Pharmacol 2004;4:597–602.
41. Tomlinson JW, Walker EA, Bujalska IJ, Draper N, Lavery GG, Cooper MS, Hewison M, Stewart PM. 11beta-hydroxysteroid dehydrogenase type 1: a tissue-specific regulator of glucocorticoid response. Endocr Rev 2004;25:831–66.
42. Nuclear Receptors Nomenclature Committee. A unified nomenclature system for the nuclear receptor superfamily. Cell 1999;97:161–3.
43. Aranda A, Pascual A. Nuclear hormone receptors and gene expression. Physiol Rev 2001;81:1269–304.
44. Evans RM. The steroid and thyroid hormone receptor superfamily. Science 1988;240:889–95.
45. Ballard PL. Delivery and transport of glucocorticoids to target cells. Monogr Endocrinol 1979;12:25–48.
46. Hayashi R, Wada H, Ito K, Adcock IM. Effects of glucocorticoids on gene transcription. Eur J Pharmacol 2004;500:51–62.
47. Kumar R, Thompson EB. Gene regulation by the glucocorticoid receptor: structure: function relationship. J Steroid Biochem Mol Biol 2005;94:383–94.
48. Dostert A, Heinzel T. Negative glucocorticoid receptor response elements and their role in glucocorticoid action. Curr Pharm Des 2004;10:2807–16.
49. Drouin J, Sun YL, Chamberland M, Gauthier Y, De Lean A, Nemer M, Schmidt TJ. Novel glucocorticoid receptor complex with DNA element of the hormone-repressed POMC gene. EMBO J 1993;12:145–56.
50. Subramaniam N, Cairns W, Okret S. Studies on the mechanism of glucocorticoid-mediated repression from a negative glucocorticoid response element from the bovine prolactin gene. DNA Cell Biol 1997;16:153–63.
51. Adcock IM, Caramori G. Cross-talk between pro-inflammatory transcription factors and glucocorticoids. Immunol Cell Biol 2001;79:376–84.

52. Karin M, Chang L. AP-1-glucocorticoid receptor crosstalk taken to a higher level. J Endocrinol 2001;169:447–51.
53. Scheinman RI, Gualberto A, Jewell CM, Cidlowski JA, Baldwin AS. Characterization of mechanisms involved in transrepression of NF-kappa B by activated glucocorticoid receptors. Mol Cell Biol 1995;15:943–53.
54. Smoak KA, Cidlowski JA. Mechanisms of glucocorticoid receptor signaling during inflammation. Mech Ageing Dev 2004;125:697–706.
55. Caldenhoven E, Liden J, Wissink S, Van de Stolpe A, Raaijmakers J, Koenderman L, Okret S, Gustafsson JA, Van der Saag PT. Negative cross-talk between RelA and the glucocorticoid receptor: a possible mechanism for the antiinflammatory action of glucocorticoids. Mol Endocrinol 1995;9:401–12.
56. Schule R, Rangarajan P, Kliewer S, Ransone LJ, Bolado J, Yang N, Verma IM, Evans RM. Functional antagonism between oncoprotein c-Jun and the glucocorticoid receptor. Cell 1990;62:1217–26.
57. Encio IJ, Detera-Wadleigh SD. The genomic structure of the human glucocorticoid receptor. J Biol Chem 1991;266:7182–8.
58. Bamberger CM, Bamberger AM, de Castro M, Chrousos GP. Glucocorticoid receptor beta, a potential endogenous inhibitor of glucocorticoid action in humans. J Clin Invest 1995;95:2435–41.
59. Charmandari E, Chrousos GP, Ichijo T, Bhattacharyya N, Vottero A, Souvatzoglou E, Kino T. The human glucocorticoid receptor (hGR) beta isoform suppresses the transcriptional activity of hGRalpha by interfering with formation of active coactivator complexes. Mol Endocrinol 2005;19:52–64.
60. Fruchter O, Kino T, Zoumakis E, Alesci S, De Martino M, Chrousos G, Hochberg Z. The human glucocorticoid receptor (GR) isoform {beta} differentially suppresses GR {alpha}-induced transactivation stimulated by synthetic glucocorticoids. J Clin Endocrinol Metab 2005;90:3505–9.
61. Lewis-Tuffin LJ, Cidlowski JA. The physiology of human glucocorticoid receptor beta (hGRbeta) and glucocorticoid resistance. Ann N Y Acad Sci 2006;1069:1–9.
62. Oakley RH, Jewell CM, Yudt MR, Bofetiado DM, Cidlowski JA. The dominant negative activity of the human glucocorticoid receptor beta isoform. Specificity and mechanisms of action. J Biol Chem 1999;274:27857–66.
63. Oakley RH, Sar M, Cidlowski JA. The human glucocorticoid receptor beta isoform. Expression, biochemical properties, and putative function. J Biol Chem 1996;271:9550–9.
64. Vottero A, Chrousos GP. Glucocorticoid Receptor beta: view I. Trends Endocrinol Metab 1999;10:333–8.
65. Brogan IJ, Murray IA, Cerillo G, Needham M, White A, Davis JR. Interaction of glucocorticoid receptor isoforms with transcription factors AP-1 and NF-kappaB: lack of effect of glucocorticoid receptor beta. Mol Cell Endocrinol 1999;157:95–104.
66. Carlstedt-Duke J. Glucocorticoid Receptor beta: view II. Trends Endocrinol Metab 1999;10:339–42.
67. de Lange P, Koper JW, Brinkmann AO, de Jong FH, Lamberts SW. Natural variants of the beta isoform of the human glucocorticoid receptor do not alter sensitivity to glucocorticoids. Mol Cell Endocrinol 1999;153:163–8.
68. Hecht K, Carlstedt-Duke J, Stierna P, Gustafsson J, Bronnegard M, Wikstrom AC. Evidence that the beta-isoform of the human glucocorticoid receptor does not act as a physiologically significant repressor. J Biol Chem 1997;272:26659–64.
69. Munck A, Guyre PM, Holbrook NJ. Physiological functions of glucocorticoids in stress and their relation to pharmacological actions. Endocr Rev 1984;5:25–44.
70. Visser J, van Boxel-Dezaire A, Methorst D, Brunt T, de Kloet ER, Nagelkerken L. Differential regulation of interleukin-10 (IL-10) and IL-12 by glucocorticoids in vitro. Blood 1998;91:4255–64.
71. Webster JI, Tonelli L, Sternberg EM. Neuroendocrine regulation of immunity. Annu Rev Immunol 2002;20:125–63.
72. Goujon E, Laye S, Parnet P, Dantzer R. Regulation of cytokine gene expression in the central nervous system by glucocorticoids: mechanisms and functional consequences. Psychoneuroendocrinology 1997;22:S75–80.

73. Lee SW, Tsou AP, Chan H, Thomas J, Petrie K, Eugui EM, Allison AC. Glucocorticoids selectively inhibit the transcription of the interleukin 1 beta gene and decrease the stability of interleukin 1 beta mRNA. Proc Natl Acad Sci USA 1988;85:1204–8.
74. Schmidt M, Pauels HG, Lugering N, Lugering A, Domschke W, Kucharzik T. Glucocorticoids induce apoptosis in human monocytes: potential role of IL-1 beta. J Immunol 1999;163:3484–90.
75. Moynihan JA, Callahan TA, Kelley SP, Campbell LM. Adrenal hormone modulation of type 1 and type 2 cytokine production by spleen cells: dexamethasone and dehydroepiandrosterone suppress interleukin-2, interleukin-4, and interferon-gamma production in vitro. Cell Immunol 1998; 184:58–64.
76. Jee YK, Gilmour J, Kelly A, Bowen H, Richards D, Soh C, Smith P, Hawrylowicz C, Cousins D, Lee T, Lavender P. Repression of interleukin-5 transcription by the glucocorticoid receptor targets GATA3 signaling and involves histone deacetylase recruitment. J Biol Chem 2005;280:23243–50.
77. Mordvinov VA, Kok CC, Arthaningtyas E, Schwenger GT, Cristow A, Sanderson CJ. Dexamethasone suppresses human interleukin-5 gene promoter. Bull Exp Biol Med 2005;140:80–2.
78. Sewell WA, Scurr LL, Orphanides H, Kinder S, Ludowyke RI. Induction of interleukin-4 and interleukin-5 expression in mast cells is inhibited by glucocorticoids. Clin Diagn Lab Immunol 1998;5:18–23.
79. Staples KJ, Bergmann MW, Barnes PJ, Newton R. Evidence for post-transcriptional regulation of interleukin-5 by dexamethasone. Immunology 2003;109:527–35.
80. De Bosscher K, Schmitz ML, Vanden Berghe W, Plaisance S, Fiers W, Haegeman G. Glucocorticoid-mediated repression of nuclear factor-kappaB-dependent transcription involves direct interference with transactivation. Proc Natl Acad Sci USA 1997;94:13504–9.
81. Ray A, Prefontaine KE. Physical association and functional antagonism between the p65 subunit of transcription factor NF-kappa B and the glucocorticoid receptor. Proc Natl Acad Sci USA 1994;91:752–6.
82. Tobler A, Meier R, Seitz M, Dewald B, Baggiolini M, Fey MF. Glucocorticoids downregulate gene expression of GM-CSF, NAP-1/IL-8, and IL-6, but not of M-CSF in human fibroblasts. Blood 1992;79:45–51.
83. Vanden Berghe W, Vermeulen L, De Wilde G, De Bosscher K, Boone E, Haegeman G. Signal transduction by tumor necrosis factor and gene regulation of the inflammatory cytokine interleukin-6. Biochem Pharmacol 2000;60:1185–95.
84. Bruun JM, Pedersen SB, Richelsen B. Regulation of interleukin 8 production and gene expression in human adipose tissue in vitro. J Clin Endocrinol Metab 2001;86:1267–73.
85. Chang. MM, Juarez M, Hyde DM, Wu R. Mechanism of dexamethasone-mediated interleukin-8 gene suppression in cultured airway epithelial cell. Am J Physiol Lung Cell Mol Physiol 2001; 280:L107–115.
86. Kwon OJ, Au BT, Collins PD, Baraniuk JN, Adcock IM, Chung KF, Barnes PJ. Inhibition of interleukin-8 expression by dexamethasone in human cultured airway epithelial cells. Immunology 1994;81:389–94.
87. Wang J, Zhu Z, Nolfo R, Elias JA. Dexamethasone regulation of lung epithelial cell and fibroblast interleukin-11 production. Am J Physiol 1999;276:L175–85.
88. Elenkov IJ, Webster EL, Torpy DJ, Chrousos GP. Stress, corticotropin-releasing hormone, glucocorticoids, and the immune/inflammatory response: acute and chronic effects. Ann N Y Acad Sci 1999;876:1–11.
89. Franchimont D, Galon J, Gadina M, Visconti R, Zhou Y, Aringer M, Frucht DM, Chrousos GP, O'Shea JJ. Inhibition of Th1 immune response by glucocorticoids: dexamethasone selectively inhibits IL-12-induced Stat4 phosphorylation in T lymphocytes. J Immunol 2000;164:1768–74.
90. Ma W, Gee K, Lim W, Chambers K, Angel JB, Kozlowski M, Kumar A. Dexamethasone inhibits IL-12p40 production in lipopolysaccharide-stimulated human monocytic cells by down-regulating the activity of c-Jun N-terminal kinase, the activation protein-1, and NF-kappa B transcription factors. J Immunol 2004;172:318–30.
91. Wu CY, Wang K, McDyer JF, Seder RA. Prostaglandin E2 and dexamethasone inhibit IL-12 receptor expression and IL-12 responsiveness. J Immunol 1998;161:2723–30.

92. Joyce DA, Gimblett G, Steer JH. Targets of glucocorticoid action on TNF-alpha release by macrophages. Inflamm Res 2001;50:337–40.
93. Joyce DA, Steer JH, Abraham LJ. Glucocorticoid modulation of human monocyte/macrophage function: control of TNF-alpha secretion. Inflamm Res 1997;46:447–51.
94. Steer JH, Ma DT, Dusci L, Garas G, Pedersen KE, Joyce DA. Altered leucocyte trafficking and suppressed tumour necrosis factor alpha release from peripheral blood monocytes after intra-articular glucocorticoid treatment. Ann Rheum Dis 1998;57:732–7.
95. Verhoef CM, van Roon JA, Vianen ME, Lafeber FP, Bijlsma JW. The immune suppressive effect of dexamethasone in rheumatoid arthritis is accompanied by upregulation of interleukin 10 and by differential changes in interferon gamma and interleukin 4 production. Ann Rheum Dis 1999;58:49–54.
96. Adkins KK, Levan TD, Miesfeld RL, Bloom JW. Glucocorticoid regulation of GM-CSF: evidence for transcriptional mechanisms in airway epithelial cells. Am J Physiol 1998;275:L372–8.
97. Smith PJ, Cousins DJ, Jee YK, Staynov DZ, Lee TH, Lavender P. Suppression of granulocyte-macrophage colony-stimulating factor expression by glucocorticoids involves inhibition of enhancer function by the glucocorticoid receptor binding to composite NF-AT/activator protein-1 elements. J Immunol 2001;167:2502–10.
98. Kunicka JE, Talle MA, Denhardt GH, Brown M, Prince LA, Goldstein G. Immunosuppression by glucocorticoids: inhibition of production of multiple lymphokines by in vivo administration of dexamethasone. Cell Immunol 1993;149:39–49.
99. Yoshikawa H, Nakajima Y, Tasaka K. Glucocorticoid suppresses autocrine survival of mast cells by inhibiting IL-4 production and ICAM-1 expression. J Immunol 1999;162:6162–70.
100. Hodge S, Hodge G, Flower R, Han P. Methyl-prednisolone up-regulates monocyte interleukin-10 production in stimulated whole blood. Scand J Immunol 1999;49:548–53.
101. Colotta F, Re F, Muzio M, Bertini R, Polentarutti N, Sironi M, Giri JG, Dower SK, Sims JE, Mantovani A. Interleukin-1 type II receptor: a decoy target for IL-1 that is regulated by IL-4. Science 1993;261:472–5.
102. Muller B, Peri G, Doni A, Perruchoud AP, Landmann R, Pasqualini F, Mantovani A. High circulating levels of the IL-1 type II decoy receptor in critically ill patients with sepsis: association of high decoy receptor levels with glucocorticoid administration. J Leukoc Biol 2002;72:643–9.
103. Pousset F, Cremona S, Dantzer R, Kelley KW, Parnet P. Dexamethasone up-regulates type II IL-1 receptor in mouse primary activated astrocytes. J Neurochem 2001;76:901–9.
104. Re F, Muzio M, De Rossi M, Polentarutti N, Giri JG, Mantovani A, Colotta F. The type II "receptor" as a decoy target for interleukin 1 in polymorphonuclear leukocytes: characterization of induction by dexamethasone and ligand binding properties of the released decoy receptor. J Exp Med 1994;179:739–43.
105. Bianchi M, Meng C, Ivashkiv LB. Inhibition of IL-2-induced Jak-STAT signaling by glucocorticoids. Proc Natl Acad Sci USA 2000;97:9573–8.
106. Daynes RA, Araneo BA. Contrasting effects of glucocorticoids on the capacity of T cells to produce the growth factors interleukin 2 and interleukin 4. Eur J Immunol 1989;19:2319–25.
107. Daynes RA, Araneo BA, Dowell TA, Huang K, Dudley D. Regulation of murine lymphokine production in vivo. III. The lymphoid tissue microenvironment exerts regulatory influences over T helper cell function. J Exp Med 1990;171:979–96.
108. Crane IJ, Forrester JV. Thl and Th2 lymphocytes in autoimmune disease. Crit Rev Immunol 2005;25:75–102.
109. Elenkov IJ. Glucocorticoids and the Th1/Th2 balance. Ann N Y Acad Sci 2004;1024:138–46.
110. MacPhee IA, Antoni FA, Mason DW. Spontaneous recovery of rats from experimental allergic encephalomyelitis is dependent on regulation of the immune system by endogenous adrenal corticosteroids. J Exp Med 1989;169:431–45.
111. Cronstein BN, Kimmel SC, Levin RI, Martiniuk F, Weissmann G. A mechanism for the antiinflammatory effects of corticosteroids: the glucocorticoid receptor regulates leukocyte adhesion to endothelial cells and expression of endothelial-leukocyte adhesion molecule 1 and intercellular adhesion molecule 1. Proc Natl Acad Sci USA 1992;89:9991–5.

112. Ito A, Miyake M, Morishita M, Ito K, Torii S, Sakamoto T. Dexamethasone reduces lung eosinophilia, and VCAM-1 and ICAM-1 expression induced by Sephadex beads in rats. Eur J Pharmacol 2003;468:59–66.
113. Liden J, Rafter I, Truss M, Gustafsson JA, Okret S. Glucocorticoid effects on NF-kappaB binding in the transcription of the ICAM-1 gene. Biochem Biophys Res Commun 2000;273:1008–14.
114. Shirasaki H, Watanabe K, Kanaizumi E, Sato J, Konno N, Narita S, Himi T. Effect of glucocorticosteroids on tumour necrosis factor-alpha-induced intercellular adhesion molecule-1 expression in cultured primary human nasal epithelial cells. Clin Exp Allergy 2004;34:945–51.
115. Wissink S, van de Stolpe A, Caldenhoven E, Koenderman L, van der Saag PT. NF-kappa B/Rel family members regulating the ICAM-1 promoter in monocytic THP-1 cells. Immunobiology 1997;198:50–64.
116. Atsuta J, Plitt J, Bochner BS, Schleimer RP. Inhibition of VCAM-1 expression in human bronchial epithelial cells by glucocorticoids. Am J Respir Cell Mol Biol 1999;20:643–50.
117. Kang H, Wei EQ, Yang XH, Zhang WP, Shen JZ. VCAM-1 expression, eosinophil infiltration, and pharmacological modulation in rat allergic airway inflammation. Acta Pharmacol Sin 2002; 23:157–61.
118. Ray KP, Farrow S, Daly M, Talabot F, Searle N. Induction of the E-selectin promoter by interleukin 1 and tumour necrosis factor alpha, and inhibition by glucocorticoids. Biochem J 1997;328:707–15.
119. Ray KP, Searle N. Glucocorticoid inhibition of cytokine-induced E-selectin promoter activation. Biochem Soc Trans 1997;25:189S.
120. Nakagawa M, Bondy GP, Waisman D, Minshall D, Hogg JC, van Eeden SF. The effect of glucocorticoids on the expression of L-selectin on polymorphonuclear leukocyte. Blood 1999;93:2730–7.
121. Weber PS, Toelboell T, Chang LC, Tirrell JD, Saama PM, Smith GW, Burton JL. Mechanisms of glucocorticoid-induced down-regulation of neutrophil L-selectin in cattle: evidence for effects at the gene-expression level and primarily on blood neutrophils. J Leukoc Biol 2004;75:815–27.
122. Ohtsuka T, Kubota A, Hirano T, Watanabe K, Yoshida H, Tsurufuji M, Iizuka Y, Konishi K, Tsurufuji S. Glucocorticoid-mediated gene suppression of rat cytokine-induced neutrophil chemoattractant CINC/gro, a member of the interleukin-8 family, through impairment of NF-kappa B activation. J Biol Chem 1996;271:1651–9.
123. Lilly CM, Nakamura H, Kesselman H, Nagler-Anderson C, Asano K, Garcia-Zepeda EA, Rothenberg ME, Drazen JM, Luster AD. Expression of eotaxin by human lung epithelial cells: induction by cytokines and inhibition by glucocorticoids. J Clin Invest 1997;99:1767–73.
124. Wingett D, Forcier K, Nielson CP. Glucocorticoid-mediated inhibition of RANTES expression in human T lymphocytes. FEBS Lett 1996;398:308–11.
125. Matsuo H, Tamura M, Kabashima N, Serino R, Tokunaga M, Shibata T, Matsumoto M, Aijima M, Oikawa S, Anai H, Nakashima Y. Prednisolone inhibits hyperosmolarity-induced expression of MCP-1 via NF-kappaB in peritoneal mesothelial cells. Kidney Int 2006;69:736–46.
126. Park SK, Yang WS, Han NJ, Lee SK, Ahn H, Lee IK, Park JY, Lee KU, Lee JD. Dexamethasone regulates AP-1 to repress TNF-alpha induced MCP-1 production in human glomerular endothelial cells. Nephrol Dial Transplant 2004;19:312–19.
127. Poon M, Liu B, Taubman MB. Identification of a novel dexamethasone-sensitive RNA-destabilizing region on rat monocyte chemoattractant protein 1 mRNA. Mol Cell Biol 1999;19:6471–8.
128. Pype JL, Dupont LJ, Menten P, Van Coillie E, Opdenakker G, Van Damme J, Chung KF, Demedts MG, Verleden GM. Expression of monocyte chemotactic protein (MCP)-1, MCP-2, and MCP-3 by human airway smooth-muscle cells. Modulation by corticosteroids and T-helper 2 cytokines. Am J Respir Cell Mol Biol 1999;21:528–36.
129. Dolan-O'keefe M, Nick HS. Inhibition of cytoplasmic phospholipase A2 expression by glucocorticoids in rat intestinal epithelial cells. Gastroenterology 1999;116:855–64.
130. Newton R, Kuitert LM, Slater DM, Adcock IM, Barnes PJ. Cytokine induction of cytosolic phospholipase A2 and cyclooxygenase-2 mRNA is suppressed by glucocorticoids in human epithelial cells. Life Sci 1997;60:67–78.

131. Perkins DJ, Kniss DA. Tumor necrosis factor-alpha promotes sustained cyclooxygenase-2 expression: attenuation by dexamethasone and NSAIDs. Prostaglandins 1997;54:727–43.
132. Buckingham JC. Fifteenth Gaddum Memorial Lecture December 1994. Stress and the neuroendocrine-immune axis: the pivotal role of glucocorticoids and lipocortin 1. Br J Pharmacol 1996;118:1–19.
133. Buckingham JC, Flower RJ. Lipocortin 1: a second messenger of glucocorticoid action in the hypothalamo-pituitary-adrenocortical axis. Mol Med Today 1997;3:296–302.
134. Flower RJ, Rothwell NJ. Lipocortin-1: cellular mechanisms and clinical relevance. Trends Pharmacol Sci 1994;15:71–6.
135. Kleinert H, Euchenhofer C, Ihrig-Biedert I, Forstermann U. Glucocorticoids inhibit the induction of nitric oxide synthase II by down-regulating cytokine-induced activity of transcription factor nuclear factor-kappa B. Mol Pharmacol 1996;49:15–21.
136. De Vera ME, Taylor BS, Wang Q, Shapiro RA, Billiar TR, Geller DA. Dexamethasone suppresses iNOS gene expression by upregulating I-kappa B alpha and inhibiting NF-kappa B. Am J Physiol 1997;273:G1290–6.
137. Matsumura M, Kakishita H, Suzuki M, Banba N, Hattori Y. Dexamethasone suppresses iNOS gene expression by inhibiting NF-kappaB in vascular smooth muscle cells. Life Sci 2001;69:1067–77.
138. Shinoda J, McLaughlin KE, Bell HS, Swaroop GR, Yamaguchi S, Holmes MC, Whittle IR. Molecular mechanisms underlying dexamethasone inhibition of iNOS expression and activity in C6 glioma cells. Glia 2003;42:68–76.
139. Collins S, Bolanowski MA, Caron MG, Lefkowitz RJ. Genetic regulation of beta-adrenergic receptors. Annu Rev Physiol 1989;51:203–15.
140. Mak JC, Nishikawa M, Barnes PJ. Glucocorticosteroids increase beta 2-adrenergic receptor transcription in human lung. Am J Physiol 1995;268:L41–6.
141. Mak JC, Nishikawa M, Shirasaki H, Miyayasu K, Barnes PJ. Protective effects of a glucocorticoid on downregulation of pulmonary beta 2-adrenergic receptors in vivo. J Clin Invest 1995;96: 99–106.
142. Ihara H, Nakanishi S. Selective inhibition of expression of the substance P receptor mRNA in pancreatic acinar AR42J cells by glucocorticoids. J Biol Chem 1990;265:22441–5.
143. Katsunuma T, Mak JC, Barnes PJ. Glucocorticoids reduce tachykinin NK2 receptor expression in bovine tracheal smooth muscle. Eur J Pharmacol 1998;344:99–106.
144. DeRijk R, Michelson D, Karp B, Petrides J, Galliven E, Deuster P, Paciotti G, Gold PW, Sternberg EM. Exercise and circadian rhythm-induced variations in plasma cortisol differentially regulate interleukin-1 beta (IL-1 beta), IL-6, and tumor necrosis factor-alpha (TNF alpha) production in humans: high sensitivity of TNF alpha and resistance of IL-6. J Clin Endocrinol Metab 1997;82:2182–91.
145. DeRijk RH, Petrides J, Deuster P, Gold PW, Sternberg EM. Changes in corticosteroid sensitivity of peripheral blood lymphocytes after strenuous exercise in humans. J Clin Endocrinol Metab 1996;81:228–35.
146. Singh A, Zelazowska EB, Petrides JS, Raybourne RB, Sternberg EM, Gold PW, Deuster PA. Lymphocyte subset responses to exercise and glucocorticoid suppression in healthy men. Med Sci Sports Exerc 1996;28:822–8.
147. Zelazowska EB, Singh A, Raybourne RB, Sternberg EM, Gold PW, Deuster PA. Lymphocyte subpopulation expression in women: effect of exercise and circadian rhythm. Med Sci Sports Exerc 1997;29:467–73.
148. Sternberg EM, Hill JM, Chrousos GP, Kamilaris T, Listwak SJ, Gold PW, Wilder RL. Inflammatory mediator-induced hypothalamic-pituitary-adrenal axis activation is defective in streptococcal cell wall arthritis-susceptible Lewis rats. Proc Natl Acad Sci USA 1989;86:2374–8.
149. Edwards CK 3rd, Yunger LM, Lorence RM, Dantzer R, Kelley KW. The pituitary gland is required for protection against lethal effects of Salmonella typhimurium. Proc Natl Acad Sci USA 1991;88:2274–7.
150. Ruzek MC, Pearce BD, Miller AH, Biron CA. Endogenous glucocorticoids protect against cytokine-mediated lethality during viral infection. J Immunol 1999;162:3527–33.

151. Gomez SA, Fernandez GC, Vanzulli S, Dran G, Rubel C, Berki T, Isturiz MA, Palermo MS. Endogenous glucocorticoids attenuate Shiga toxin-2-induced toxicity in a mouse model of haemolytic uraemic syndrome. Clin Exp Immunol 2003;131:217–24.
152. Moayeri M, Webster JI, Wiggins JF, Leppla SH, Sternberg EM. Endocrine perturbation increases susceptibility of mice to anthrax lethal toxin. Infect Immun 2005;73:4238–44.
153. Misiewicz B, Poltorak M, Raybourne RB, Gomez M, Listwak S, Sternberg EM. Intracerebroventricular transplantation of embryonic neuronal tissue from inflammatory resistant into inflammatory susceptible rats suppresses specific components of inflammation. Exp Neurol 1997;146:305–14.
154. Wick G, Sgonc R, Lechner O. Neuroendocrine-immune disturbances in animal models with spontaneous autoimmune diseases. Ann N Y Acad Sci 1998;840:591–8.
155. Hu Y, Dietrich H, Herold M, Heinrich PC, Wick G. Disturbed immuno-endocrine communication via the hypothalamo-pituitary-adrenal axis in autoimmune disease. Int Arch Allergy Immunol 1993;102:232–41.
156. Lechner O, Hu Y, Jafarian-Tehrani M, Dietrich H, Schwarz S, Herold M, Haour F, Wick G. Disturbed immunoendocrine communication via the hypothalamo-pituitary-adrenal axis in murine lupus. Brain Behav Immun 1996;10:337–50.
157. Jafarian-Tehrani M, Sternberg EM. Animal models of neuroimmune interactions in inflammatory diseases. J Neuroimmunol 1999;100:13–20.
158. Tonelli L, Webster JI, Rapp KL, Sternberg E. Neuroendocrine responses regulating susceptibility and resistance to autoimmune/inflammatory disease in inbred rat strains. Immunol Rev 2001; 184:203–11.
159. Stefferl A, Linington C, Holsboer F, Reul JM. Susceptibility and resistance to experimental allergic encephalomyelitis: relationship with hypothalamic-pituitary-adrenocortical axis responsiveness in the rat. Endocrinology 1999;140:4932–8.
160. Aksentijevich S, Whitfield HJ, Young WS, Wilder RL, Chrousos GP, Gold PW, Sternberg EM. Arthritis-susceptible Lewis rats fail to emerge from the stress hyporesponsive period. Brain Res Dev Brain Res 1992;65:115–18.
161. Harbuz MS, Windle RJ, Jessop DS, Renshaw D, Ingram CD, Lightman SL. Differential effects of psychological and immunological challenge on the hypothalamo-pituitary-adrenal axis function in adjuvant-induced arthritis. Ann N Y Acad Sci 1999;876:43–52.
162. Sternberg EM, Young WS, Bernardini R, Calogero AE, Chrousos GP, Gold PW, Wilder RL. A central nervous system defect in biosynthesis of corticotropin-releasing hormone is associated with susceptibility to streptococcal cell wall-induced arthritis in Lewis rats. Proc Natl Acad Sci USA 1989;86:4771–5.
163. Dhabhar FS, McEwen BS, Spencer RL. Stress response, adrenal steroid receptor levels and corticosteroid-binding globulin levels-a comparison between Sprague-Dawley, Fischer 344 and Lewis rats. Brain Res 1993;616:89–98.
164. Moncek F, Kvetnansky R, Jezova D. Differential responses to stress stimuli of Lewis and Fischer rats at the pituitary and adrenocortical level. Endocr Regul 2001;35:35–41.
165. Wilder RL, Calandra GB, Garvin AJ, Wright KD, Hansen CT. Strain and sex variation in the susceptibility to streptococcal cell wall-induced polyarthritis in the rat. Arthritis Rheum 1982;25:1064–72.
166. Dhabhar FS, Miller AH, McEwen BS, Spencer RL. Differential activation of adrenal steroid receptors in neural and immune tissues of Sprague Dawley, Fischer 344, and Lewis rats. J Neuroimmunol 1995;56:77–90.
167. Smith CC, Omeljaniuk RJ, Whitfield HJ, Aksentijevich S, Fellows MQ, Zelazowska E, Gold PW, Sternberg EM. Differential mineralocorticoid (type 1) and glucocorticoid (type 2) receptor expression in Lewis and Fischer rats. Neuroimmunomodulation 1994;1:66–73.
168. Fan J, Gong XQ, Wu J, Zhang YF, Xu RB. Effect of glucocorticoid receptor (GR) blockade on endotoxemia in rats. Circ Shock 1994;42:76–82.
169. Wick G, Hu Y, Schwarz S, Kroemer G. Immunoendocrine communication via the hypothalamo-pituitary-adrenal axis in autoimmune diseases. Endocr Rev 1993;14:539–63.

170. Brezinschek HP, Gruschwitz M, Sgonc R, Moormann S, Herold M, Gershwin ME, Wick G. Effects of cytokine application on glucocorticoid secretion in an animal model for systemic scleroderma. J Autoimmun 1993;6:719–33.
171. Griffiths MM, Remmers EF. Genetic analysis of collagen-induced arthritis in rats: a polygenic model for rheumatoid arthritis predicts a common framework of cross-species inflammatory/autoimmune disease loci. Immunol Rev 2001;184:172–83.
172. Listwak S, Barrientos RM, Koike G, Ghosh S, Gomez M, Misiewicz B, Sternberg EM. Identification of a novel inflammation-protective locus in the Fischer rat. Mamm Genome 1999;10:362–5.
173. Remmers EF, Longman RE, Du Y, O'Hare A, Cannon GW, Griffiths MM, Wilder RL. A genome scan localizes five non-MHC loci controlling collagen-induced arthritis in rats. Nat Genet 1996;14:82–5.
174. Wilder RL, Griffiths MM, Cannon GW, Caspi R, Gulko PS, Remmers EF. Genetic factors involved in central nervous system/immune interactions. Adv Exp Med Biol 2001;493:59–67.
175. Huizenga NA, Koper JW, de Lange P, Pols HA, Stolk RP, Grobbee DE, de Jong FH, Lamberts SW. Interperson variability but intraperson stability of baseline plasma cortisol concentrations, and its relation to feedback sensitivity of the hypothalamo-pituitary-adrenal axis to a low dose of dexamethasone in elderly individuals. J Clin Endocrinol Metab 1998;83:47–54.
176. Petrides JS, Gold PW, Mueller GP, Singh A, Stratakis C, Chrousos GP, Deuster PA. Marked differences in functioning of the hypothalamic-pituitary-adrenal axis between groups of men. J Appl Physiol 1997;82:1979–88.
177. Charmandari E, Kino T, Chrousos GP. Familial/sporadic glucocorticoid resistance: clinical phenotype and molecular mechanisms. Ann N Y Acad Sci 2004;1024:168–81.
178. Jansson L, Holmdahl R. Estrogen-mediated immunosuppression in autoimmune diseases. Inflamm Res 1998;47:290–301.
179. Crofford LJ. The hypothalamic-pituitary-adrenal stress axis in fibromyalgia and chronic fatigue syndrome. Z Rheumatol 1998;57:67–71.
180. Cash JM, Crofford LJ, Gallucci WT, Sternberg EM, Gold PW, Chrousos GP, Wilder RL. Pituitary-adrenal axis responsiveness to ovine corticotropin releasing hormone in patients with rheumatoid arthritis treated with low dose prednisone. J Rheumatol 1992;19:1692–6.
181. Chikanza IC, Chrousos G, Panayi GS. Abnormal neuroendocrine immune communications in patients with rheumatoid arthritis. Eur J Clin Invest 1992;22:635–7.
182. Crofford LJ. The hypothalamic-pituitary-adrenal axis in the pathogenesis of rheumatic diseases. Endocrinol Metab Clin North Am 2002;31:1–13.
183. Cutolo M, Sulli A, Villaggio B, Seriolo B, Accardo S. Relations between steroid hormones and cytokines in rheumatoid arthritis and systemic lupus erythematosus. Ann Rheum Dis 1998;57:573–7.
184. Eijsbouts AM, Murphy EP. The role of the hypothalamic-pituitary-adrenal axis in rheumatoid arthritis. Baillieres Best Pract Res Clin Rheumatol 1999;13:599–613.
185. Gutierrez MA, Garcia ME, Rodriguez JA, Mardonez G, Jacobelli S, Rivero S. Hypothalamic-pituitary-adrenal axis function in patients with active rheumatoid arthritis: a controlled study using insulin hypoglycemia stress test and prolactin stimulation. J Rheumatol 1999;26:277–81.
186. Neeck G, Kluter A, Dotzlaw H, Eggert M. Involvement of the glucocorticoid receptor in the pathogenesis of rheumatoid arthritis. Ann N Y Acad Sci 2002;966:491–5.
187. Gutierrez MA, Garcia ME, Rodriguez JA, Rivero S, Jacobelli S. Hypothalamic-pituitary-adrenal axis function and prolactin secretion in systemic lupus erythematosus. Lupus 1998;7:404–8.
188. Johnson EO, Moutsopoulos HM. Neuroendocrine manifestations in Sjogren's syndrome. Relation to the neurobiology of stress. Ann N Y Acad Sci 2000;917:797–808.
189. Johnson EO, Vlachoyiannopoulos PG, Skopouli FN, Tzioufas AG, Moutsopoulos HM. Hypofunction of the stress axis in Sjogren's syndrome. J Rheumatol 1998;25:1508–14.
190. Buske-Kirschbaum A, Geiben A, Hollig H, Morschhauser E, Hellhammer D. Altered responsiveness of the hypothalamus-pituitary-adrenal axis and the sympathetic adrenomedullary system to stress in patients with atopic dermatitis. J Clin Endocrinol Metab 2002;87:4245–51.
191. Buske-Kirschbaum A, Hellhammer DH. Endocrine and immune responses to stress in chronic inflammatory skin disorders. Ann N Y Acad Sci 2003;992:231–40.

192. Buske-Kirschbaum A, Jobst S, Psych D, Wustmans A, Kirschbaum C, Rauh W, Hellhammer D. Attenuated free cortisol response to psychosocial stress in children with atopic dermatitis. Psychosom Med 1997;59:419–26.
193. Cleare AJ. The neuroendocrinology of chronic fatigue syndrome. Endocr Rev 2003;24:236–52.
194. Demitrack MA. Neuroendocrine correlates of chronic fatigue syndrome: a brief review. J Psychiatr Res 1997;31:69–82.
195. Demitrack MA, Dale JK, Straus SE, Laue L, Listwak SJ, Kruesi MJ, Chrousos GP, Gold PW. Evidence for impaired activation of the hypothalamic-pituitary-adrenal axis in patients with chronic fatigue syndrome. J Clin Endocrinol Metab 1991;73:1224–34.
196. Gaab J, Huster D, Peisen R, Engert V, Heitz V, Schad T, Schurmeyer TH, Ehlert U. Hypothalamic-pituitary-adrenal axis reactivity in chronic fatigue syndrome and health under psychological, physiological, and pharmacological stimulation. Psychosom Med 2002;64:951–62.
197. Neeck G, Crofford LJ. Neuroendocrine perturbations in fibromyalgia and chronic fatigue syndrome. Rheum Dis Clin North Am 2000;26:989–1002.
198. Demitrack MA, Crofford LJ. Evidence for and pathophysiologic implications of hypothalamic-pituitary-adrenal axis dysregulation in fibromyalgia and chronic fatigue syndrome. Ann N Y Acad Sci 1998;840:684–97.
199. Gold SM, Mohr DC, Huitinga I, Flachenecker P, Sternberg EM, Heesen C. The role of stress-response systems for the pathogenesis and progression of MS. Trends Immunol 2005;26:644–52.
200. Huitinga I, Erkut ZA, van Beurden D, Swaab DF. The hypothalamo-pituitary-adrenal axis in multiple sclerosis. Ann N Y Acad Sci 2003;992:118–28.
201. Michelson D, Stone L, Galliven E, Magiakou MA, Chrousos GP, Sternberg EM, Gold PW. Multiple sclerosis is associated with alterations in hypothalamic-pituitary-adrenal axis function. J Clin Endocrinol Metab 1994;79:848–53.
202. Wei T, Lightman SL. The neuroendocrine axis in patients with multiple sclerosis. Brain 1997;120:1067–76.
203. Niess JH, Monnikes H, Dignass AU, Klapp BF, Arck PC. Review on the influence of stress on immune mediators, neuropeptides and hormones with relevance for inflammatory bowel disease. Digestion 2002;65:131–40.
204. Straub RH, Herfarth H, Falk W, Andus T, Scholmerich J. Uncoupling of the sympathetic nervous system and the hypothalamic-pituitary-adrenal axis in inflammatory bowel disease? J Neuroimmunol 2002;126:116–25.
205. Glaser R, Kiecolt-Glaser JK. Stress-associated immune modulation: relevance to viral infections and chronic fatigue syndrome. Am J Med 1998;105:35S–42S.
206. Glaser R, Kiecolt-Glaser JK. Stress-induced immune dysfunction: implications for health. Nat Rev Immunol 2005;5:243–51.
207. Kiecolt-Glaser JK, Loving TJ, Stowell JR, Malarkey WB, Lemeshow S, Dickinson SL, Glaser R. Hostile marital interactions, proinflammatory cytokine production, and wound healing. Arch Gen Psychiatry 2005;62:1377–84.
208 Rozlog LA, Kiecolt-Glaser JK, Marucha PT, Sheridan JF, Glaser R. Stress and immunity: implications for viral disease and wound healing. J Periodontol 1999;70:786–92.
209. Vedhara K, Cox NK, Wilcock GK, Perks P, Hunt M, Anderson S, Lightman SL, Shanks NM. Chronic stress in elderly carers of dementia patients and antibody response to influenza vaccination. Lancet 1999;353:627–31.
210. Yang EV, Glaser R. Stress-associated immunomodulation and its implications for responses to vaccination. Expert Rev Vaccines 2002;1:453–9.
211. Harbuz M. Neuroendocrine function and chronic inflammatory stress. Exp Physiol 2002;87:519–25.
212. Gudbjornsson B, Skogseid B, Oberg K, Wide L, Hallgren R. Intact adrenocorticotropic hormone secretion but impaired cortisol response in patients with active rheumatoid arthritis. Effect of glucocorticoids. J Rheumatol 1996;23:596–602.
213. Neeck G, Federlin K, Graef V, Rusch D, Schmidt KL. Adrenal secretion of cortisol in patients with rheumatoid arthritis. J Rheumatol 1990;17:24–9.

214. Crofford LJ, Kalogeras KT, Mastorakos G, Magiakou MA, Wells J, Kanik KS, Gold PW, Chrousos GP, Wilder RL. Circadian relationships between interleukin (IL)-6 and hypothalamic-pituitary-adrenal axis hormones: failure of IL-6 to cause sustained hypercortisolism in patients with early untreated rheumatoid arthritis. J Clin Endocrinol Metab 1997;82:1279–83.
215. Cutolo M, Sulli A, Pizzorni C, Secchi ME, Soldano S, Seriolo B, Straub RH, Otsa K, Maestroni GJ. Circadian rhythms: glucocorticoids and arthritis. Ann N Y Acad Sci 2006;1069:289–99.
216. Jerjes WK, Peters TJ, Taylor NF, Wood PJ, Wessely S, Cleare AJ. Diurnal excretion of urinary cortisol, cortisone, and cortisol metabolites in chronic fatigue syndrome. J Psychosom Res 2006;60:145–53.
217. Cleare AJ, Blair D, Chambers S, Wessely S. Urinary free cortisol in chronic fatigue syndrome. Am J Psychiatry 2001;158:641–3.
218. Cleare AJ, Bearn J, Allain T, McGregor A, Wessely S, Murray RM, O'Keane V. Contrasting neuroendocrine responses in depression and chronic fatigue syndrome. J Affect Disord 1995;34:283–9.
219. Poteliakhoff A. Adrenocortical activity and some clinical findings in acute and chronic fatigue. J Psychosom Res 1981;25:91–5.
220. Scott LV, Dinan TG. Urinary free cortisol excretion in chronic fatigue syndrome, major depression and in healthy volunteers. J Affect Disord 1998;47:49–54.
221. Kavelaars A, Kuis W, Knook L, Sinnema G, Heijnen CJ. Disturbed neuroendocrine-immune interactions in chronic fatigue syndrome. J Clin Endocrinol Metab 2000;85:692–6.
222. Bearn J, Allain T, Coskeran P, Munro N, Butler J, McGregor A, Wessely S. Neuroendocrine responses to d-fenfluramine and insulin-induced hypoglycemia in chronic fatigue syndrome. Biol Psychiatry 1995;37:245–52.
223. Crofford LJ, Pillemer SR, Kalogeras KT, Cash JM, Michelson D, Kling MA, Sternberg EM, Gold PW, Chrousos GP, Wilder RL. Hypothalamic-pituitary-adrenal axis perturbations in patients with fibromyalgia. Arthritis Rheum 1994;37:1583–92.
224. McCain GA, Tilbe KS. Diurnal hormone variation in fibromyalgia syndrome: a comparison with rheumatoid arthritis. J Rheumatol Suppl 1989;19:154–7.
225. Griep EN, Boersma JW, de Kloet ER. Altered reactivity of the hypothalamic-pituitary-adrenal axis in the primary fibromyalgia syndrome. J Rheumatol 1993;20:469–74.
226. van Denderen JC, Boersma JW, Zeinstra P, Hollander AP, van Neerbos BR. Physiological effects of exhaustive physical exercise in primary fibromyalgia syndrome (PFS): is PFS a disorder of neuroendocrine reactivity? Scand J Rheumatol 1992;21:35–7.
227. Griep EN, Boersma JW, Lentjes EG, Prins AP, van der Korst JK, de Kloet ER. Function of the hypothalamic-pituitary-adrenal axis in patients with fibromyalgia and low back pain. J Rheumatol 1998;25:1374–81.
228. Schumann EM, Kumpfel T, Then Bergh F, Trenkwalder C, Holsboer F, Auer DP. Activity of the hypothalamic-pituitary-adrenal axis in multiple sclerosis: correlations with gadolinium-enhancing lesions and ventricular volume. Ann Neurol 2002;51:763–7.
229. Rovensky J, Blazickova S, Rauova L, Jezova D, Koska J, Lukac J, Vigas M. The hypothalamic-pituitary response in SLE. Regulation of prolactin, growth hormone and cortisol release. Lupus 1998;7:409–13.
230. Tsigos C. Isolated glucocorticoid deficiency and ACTH receptor mutations. Arch Med Res 1999;30:475–80.
231. Tsigos C, Tsiotra P, Garibaldi LR, Stavridis JC, Chrousos GP, Raptis SA. Mutations of the ACTH receptor gene in a new family with isolated glucocorticoid deficiency. Mol Genet Metab 2000;71:646–50.
232. Wu SM, Stratakis CA, Chan CH, Hallermeier KM, Bourdony CJ, Rennert OM, Chan WY. Genetic heterogeneity of adrenocorticotropin (ACTH) resistance syndromes: identification of a novel mutation of the ACTH receptor gene in hereditary glucocorticoid deficiency. Mol Genet Metab 1998;64:256–65.
233. Hiroi N, Yakushiji F, Shimojo M, Watanabe S, Sugano S, Yamaguchi N, Miyachi Y. Human ACTH hypersensitivity syndrome associated with abnormalities of the ACTH receptor gene. Clin Endocrinol 1998;48:129–34.

234. Burke CW. Adrenocortical insufficiency. Clin Endocrinol Metab 1985;14:947–76.
235. Kageyama Y. A case of isolated ACTH deficiency who developed autoimmune-mediated hypothyroidism and impaired water diuresis during glucocorticoid replacement therapy. Endocr J 2000;47:667–74.
236. Mingrone G, DeGaetano A, Pugeat M, Capristo E, Greco AV, Gasbarrini G. The steroid resistance of Crohn's disease. J Investig Med 1999;47:319–25.
237. Pretorius E, Wallner B, Marx J. Cortisol resistance in conditions such as asthma and the involvement of 11beta-HSD-2: a hypothesis. Horm Metab Res 2006;38:368–76.
238. Rask E, Olsson T, Soderberg S, Andrew R, Livingstone DE, Johnson O, Walker BR. Tissue-specific dysregulation of cortisol metabolism in human obesity. J Clin Endocrinol Metab 2001;86:1418–21.
239. Takahashi KI, Fukushima K, Sasano H, Sasaki I, Matsuno S, Krozowski ZS, Nagura H. Type II 11beta-hydroxysteroid dehydrogenase expression in human colonic epithelial cells of inflammatory bowel disease. Dig Dis Sci 1999;44:2516–22.
240. Karl M, Lamberts SW, Koper JW, Katz DA, Huizenga NE, Kino T, Haddad BR, Hughes MR, Chrousos GP. Cushing's disease preceded by generalized glucocorticoid resistance: clinical consequences of a novel, dominant-negative glucocorticoid receptor mutation. Proc Assoc Am Physicians 1996;108:296–307.
241. Vottero A, Kino T, Combe H, Lecomte P, Chrousos GP. A novel, C-terminal dominant negative mutation of the GR causes familial glucocorticoid resistance through abnormal interactions with p160 steroid receptor coactivators. J Clin Endocrinol Metab 2002;87:2658–67.
242. Kino T, Stauber RH, Resau JH, Pavlakis GN, Chrousos GP. Pathologic human GR mutant has a transdominant negative effect on the wild-type GR by inhibiting its translocation into the nucleus: importance of the ligand-binding domain for intracellular GR trafficking. J Clin Endocrinol Metab 2001;86:5600–8.
243. Charmandari E, Kino T, Souvatzoglou E, Vottero A, Bhattacharyya N, Chrousos GP. Natural glucocorticoid receptor mutants causing generalized glucocorticoid resistance: molecular genotype, genetic transmission, and clinical phenotype. J Clin Endocrinol Metab 2004;89:1939–49.
244. Mendonca BB, Leite MV, de Castro M, Kino T, Elias LL, Bachega TA, Arnhold IJ, Chrousos GP, Latronico AC. Female pseudohermaphroditism caused by a novel homozygous missense mutation of the GR gene. J Clin Endocrinol Metab 2002;87:1805–9.
245. Hurley DM, Accili D, Stratakis CA, Karl M, Vamvakopoulos N, Rorer E, Constantine K, Taylor SI, Chrousos GP. Point mutation causing a single amino acid substitution in the hormone binding domain of the glucocorticoid receptor in familial glucocorticoid resistance. J Clin Invest 1991;87:680–6.
246. Ruiz M, Lind U, Gafvels M, Eggertsen G, Carlstedt-Duke J, Nilsson L, Holtmann M, Stierna P, Wikstrom AC, Werner S. Characterization of two novel mutations in the glucocorticoid receptor gene in patients with primary cortisol resistance. Clin Endocrinol 2001;55:363–71.
247. Malchoff DM, Brufsky A, Reardon G, McDermott P, Javier EC, Bergh CH, Rowe D, Malchoff CD. A mutation of the glucocorticoid receptor in primary cortisol resistance. J Clin Invest 1993;91:1918–25.
248. Charmandari E, Kino T, Ichijo T, Zachman K, Alatsatianos A, Chrousos GP. Functional characterization of the natural human glucocorticoid receptor (hGR) mutants hGRalphaR477H and hGRalphaG679S associated with generalized glucocorticoid resistance. J Clin Endocrinol Metab 2006;91:1535–43.
249. Karl M, Lamberts SW, Detera-Wadleigh SD, Encio IJ, Stratakis CA, Hurley DM, Accili D, Chrousos GP. Familial glucocorticoid resistance caused by a splice site deletion in the human glucocorticoid receptor gene. J Clin Endocrinol Metab 1993;76:683–9.
250. Kunz S, Sandoval R, Carlsson P, Carlstedt-Duke J, Bloom JW, Miesfeld RL. Identification of a novel glucocorticoid receptor mutation in budesonide-resistant human bronchial epithelial cells. Mol Endocrinol 2003;17:2566–82.
251. Jiang T, Liu S, Tan M, Huang F, Sun Y, Dong X, Guan W, Huang L, Zhou F. The phase-shift mutation in the glucocorticoid receptor gene: potential etiologic significance of neuroendocrine mechanisms in lupus nephritis. Clin Chim Acta 2001;313:113–17.

252. Lee YM, Fujiwara J, Munakata Y, Ishii T, Sugawara A, Kaku M, Kokubun S, Sasaki T, Funato T. A mutation of the glucocorticoid receptor gene in patients with systemic lupus erythematosus. Tohoku J Exp Med 2004;203:69–76.
253. Huizenga NA, Koper JW, De Lange P, Pols HA, Stolk RP, Burger H, Grobbee DE, Brinkmann AO, De Jong FH, Lamberts SW. A polymorphism in the glucocorticoid receptor gene may be associated with and increased sensitivity to glucocorticoids in vivo. J Clin Endocrinol Metab 1998;83:144–51.
254. Bonifati DM, Witchel SF, Ermani M, Hoffman EP, Angelini C, Pegoraro E. The glucocorticoid receptor N363S polymorphism and steroid response in Duchenne dystrophy. J Neurol Neurosurg Psychiatry 2006;77:1177–9.
255. Russcher H, Smit P, van den Akker EL, van Rossum EF, Brinkmann AO, de Jong FH, Lamberts SW, Koper JW. Two polymorphisms in the glucocorticoid receptor gene directly affect glucocorticoid-regulated gene expression. J Clin Endocrinol Metab 2005;90:5804–10.
256. Panarelli M, Holloway CD, Fraser R, Connell JM, Ingram MC, Anderson NH, Kenyon CJ. Glucocorticoid receptor polymorphism, skin vasoconstriction, and other metabolic intermediate phenotypes in normal human subjects. J Clin Endocrinol Metab 1998;83:1846–52.
257. van Winsen LL, Hooper-van Veen T, van Rossum EF, Polman CH, van den Berg TK, Koper JW, Uitdehaag BM. The impact of glucocorticoid receptor gene polymorphisms on glucocorticoid sensitivity is outweighted in patients with multiple sclerosis. J Neuroimmunol 2005;167:150–6.
258. Russcher H, van Rossum EF, de Jong FH, Brinkmann AO, Lamberts SW, Koper JW. Increased expression of the glucocorticoid receptor-A translational isoform as a result of the ER22/23EK polymorphism. Mol Endocrinol 2005;19:1687–96.
259. van Rossum EF, Koper JW, Huizenga NA, Uitterlinden AG, Janssen JA, Brinkmann AO, Grobbee DE, de Jong FH, van Duyn CM, Pols HA, Lamberts SW. A polymorphism in the glucocorticoid receptor gene, which decreases sensitivity to glucocorticoids in vivo, is associated with low insulin and cholesterol levels. Diabetes 2002;51:3128–34.
260. Koper JW, Stolk RP, de Lange P, Huizenga NA, Molijn GJ, Pols HA, Grobbee DE, Karl M, de Jong FH, Brinkmann AO, Lamberts SW. Lack of association between five polymorphisms in the human glucocorticoid receptor gene and glucocorticoid resistance. Hum Genet 1997;99:663–8.
261. Tissing WJ, Meijerink JP, den Boer ML, Brinkhof B, van Rossum EF, van Wering ER, Koper JW, Sonneveld P, Pieters R. Genetic variations in the glucocorticoid receptor gene are not related to glucocorticoid resistance in childhood acute lymphoblastic leukemia. Clin Cancer Res 2005;11:6050–6.
262. van den Akker EL, Russcher H, van Rossum EF, Brinkmann AO, de Jong FH, Hokken A, Pols HA, Koper JW, Lamberts SW. Glucocorticoid receptor polymorphism affects transrepression but not transactivation. J Clin Endocrinol Metab 2006;91:2800–3.
263. Webster JC, Jewell CM, Bodwell JE, Munck A, Sar M, Cidlowski JA. Mouse glucocorticoid receptor phosphorylation status influences multiple functions of the receptor protein. J Biol Chem 1997;272:9287–93.
264. Huizenga NA, de Lange P, Koper JW, de Herder WW, Abs R, Kasteren JH, de Jong FH, Lamberts SW. Five patients with biochemical and/or clinical generalized glucocorticoid resistance without alterations in the glucocorticoid receptor gene. J Clin Endocrinol Metab 2000;85:2076–81.
265. Bergeron C, Fukakusa M, Olivenstein R, Lemiere C, Shannon J, Ernst P, Martin JG, Hamid Q. Increased glucocorticoid receptor-beta expression, but not decreased histone deacetylase 2, in severe asthma. J Allergy Clin Immunol 2006;117:703–5.
266. Christodoulopoulos P, Leung DY, Elliott MW, Hogg JC, Muro S, Toda M, Laberge S, Hamid QA. Increased number of glucocorticoid receptor-beta-expressing cells in the airways in fatal asthma. J Allergy Clin Immunol 2000;106:479–84.
267. Goleva E, Li LB, Eves PT, Strand MJ, Martin RJ, Leung DY. Increased glucocorticoid receptor beta alters steroid response in glucocorticoid-insensitive asthma. Am J Respir Crit Care Med 2006;173:607–16.
268. Hamid QA, Wenzel SE, Hauk PJ, Tsicopoulos A, Wallaert B, Lafitte JJ, Chrousos GP, Szefler SJ, Leung DY. Increased glucocorticoid receptor beta in airway cells of glucocorticoid-insensitive asthma. Am J Respir Crit Care Med 1999;159:1600–4.

269. Leung DY, Hamid Q, Vottero A, Szefler SJ, Surs W, Minshall E, Chrousos GP, Klemm DJ. Association of glucocorticoid insensitivity with increased expression of glucocorticoid receptor beta. J Exp Med 1997;186:1567–74.
270. Sousa AR, Lane SJ, Cidlowski JA, Staynov DZ, Lee TH. Glucocorticoid resistance in asthma is associated with elevated in vivo expression of the glucocorticoid receptor beta-isoform. J Allergy Clin Immunol 2000;105:943–50.
271. Strickland I, Kisich K, Hauk PJ, Vottero A, Chrousos GP, Klemm DJ, Leung DY. High constitutive glucocorticoid receptor beta in human neutrophils enables them to reduce their spontaneous rate of cell death in response to corticosteroids. J Exp Med 2001;193:585–93.
272. Tliba O, Cidlowski JA, Amrani Y. CD38 expression is insensitive to steroid action in cells treated with tumor necrosis factor-alpha and interferon-gamma by a mechanism involving the up-regulation of the glucocorticoid receptor beta isoform. Mol Pharmacol 2006;69:588–96.
273. Torrego A, Pujols L, Roca-Ferrer J, Mullol J, Xaubet A, Picado C. Glucocorticoid receptor isoforms alpha and beta in in vitro cytokine-induced glucocorticoid insensitivity. Am J Respir Crit Care Med 2004;170:420–5.
274. Webster JC, Oakley RH, Jewell CM, Cidlowski JA. Proinflammatory cytokines regulate human glucocorticoid receptor gene expression and lead to the accumulation of the dominant negative beta isoform: a mechanism for the generation of glucocorticoid resistance. Proc Natl Acad Sci USA 2001;98:6865–70.
275. Chikanza IC. Mechanisms of corticosteroid resistance in rheumatoid arthritis: a putative role for the corticosteroid receptor beta isoform. Ann N Y Acad Sci 2002;966:39–48.
276. Derijk RH, Schaaf MJ, Turner G, Datson NA, Vreugdenhil E, Cidlowski J, de Kloet ER, Emery P, Sternberg EM, Detera-Wadleigh SD. A human glucocorticoid receptor gene variant that increases the stability of the glucocorticoid receptor beta-isoform mRNA is associated with rheumatoid arthritis. J Rheumatol 2001;28:2383–8.
277. Goecke A, Guerrero J. Glucocorticoid receptor beta in acute and chronic inflammatory conditions: clinical implications. Immunobiology 2006;211:85–96.
278. Hagendorf A, Koper JW, de Jong FH, Brinkmann AO, Lamberts SW, Feelders RA. Expression of the human glucocorticoid receptor splice variants alpha, beta, and P in peripheral blood mononuclear leukocytes in healthy controls and in patients with hyper- and hypocortisolism. J Clin Endocrinol Metab 2005;90:6237–43.
279. Towers R, Naftali T, Gabay G, Carlebach M, Klein A, Novis B. High levels of glucocorticoid receptors in patients with active Crohn's disease may predict steroid resistance. Clin Exp Immunol 2005;141:357–62.
280. Honda M, Orii F, Ayabe T, Imai S, Ashida T, Obara T, Kohgo Y. Expression of glucocorticoid receptor beta in lymphocytes of patients with glucocorticoid-resistant ulcerative colitis. Gastroenterology 2000;118:859–66.
281. Zhang H, Ouyang Q, Wen ZH, Fiocchi C, Liu WP, Chen DY, Li FY. Significance of glucocorticoid receptor expression in colonic mucosal cells of patients with ulcerative colitis. World J Gastroenterol 2005;11:1775–8.
282. Orii F, Ashida T, Nomura M, Maemoto A, Fujiki T, Ayabe T, Imai S, Saitoh Y, Kohgo Y. Quantitative analysis for human glucocorticoid receptor alpha/beta mRNA in IBD. Biochem Biophys Res Commun 2002;296:1286–94.
283. Shahidi H, Vottero A, Stratakis CA, Taymans SE, Karl M, Longui CA, Chrousos GP, Daughaday WH, Gregory SA, Plate JM. Imbalanced expression of the glucocorticoid receptor isoforms in cultured lymphocytes from a patient with systemic glucocorticoid resistance and chronic lymphocytic leukemia. Biochem Biophys Res Commun 1999;254:559–65.
284. Lee CK, Lee EY, Cho YS, Moon KA, Yoo B, Moon HB. Increased expression of glucocorticoid receptor beta messenger RNA in patients with ankylosing spondylitis. Korean J Intern Med 2005;20:146–51.
285. Hamilos DL, Leung DY, Muro S, Kahn AM, Hamilos SS, Thawley SE, Hamid QA. GRbeta expression in nasal polyp inflammatory cells and its relationship to the anti-inflammatory effects of intranasal fluticasone. J Allergy Clin Immunol 2001;108:59–68.

286. Pujols L, Mullol J, Benitez P, Torrego A, Xaubet A, de Haro J, Picado C. Expression of the glucocorticoid receptor alpha and beta isoforms in human nasal mucosa and polyp epithelial cells. Respir Med 2003;97:90–6.
287. Rivers C, Levy A, Hancock J, Lightman S, Norman M. Insertion of an amino acid in the DNA-binding domain of the glucocorticoid receptor as a result of alternative splicing. J Clin Endocrinol Metab 1999;84:4283–6.
288. Stevens A, Donn R, Ray D. Regulation of glucocorticoid receptor gamma (GRgamma) by glucocorticoid receptor haplotype and glucocorticoid. Clin Endocrinol 2004;61:327–31.
289. Kino T, Gragerov A, Kopp JB, Stauber RH, Pavlakis GN, Chrousos GP. The HIV-1 virion-associated protein Vpr is a coactivator of the human glucocorticoid receptor. J Exp Med 1999;189:51–62.
290. Kino T, Gragerov A, Slobodskaya O, Tsopanomichalou M, Chrousos GP, Pavlakis GN. Human immunodeficiency virus type 1 (HIV-1) accessory protein Vpr induces transcription of the HIV-1 and glucocorticoid-responsive promoters by binding directly to p300/CBP coactivators. J Virol 2002;76:9724–34.
291. New MI, Nimkarn S, Brandon DD, Cunningham-Rundles S, Wilson RC, Newfield RS, Vandermeulen J, Barron N, Russo C, Loriaux DL, O'Malley B. Resistance to several steroids in two sisters. J Clin Endocrinol Metab 1999;84:4454–64.
292. New MI, Nimkarn S, Brandon DD, Cunningham-Rundles S, Wilson RC, Newfield RS, Vandermeulen J, Barron N, Russo C, Loriaux DL, O'Malley B. Resistance to multiple steroids in two sisters. J Steroid Biochem Mol Biol 2001;76:161–6.
293. Barnes PJ. How corticosteroids control inflammation: quintiles Prize Lecture 2005. Br J Pharmacol 2006;148:245–54.
294. Barnes PJ, Ito K, Adcock IM. Corticosteroid resistance in chronic obstructive pulmonary disease: inactivation of histone deacetylase. Lancet 2004;363:731–3.
295. Borst P, Elferink RO. Mammalian ABC transporters in health and disease. Annu Rev Biochem 2002;71:537–92.
296. Kralli A, Yamamoto KR. An FK506-sensitive transporter selectively decreases intracellular levels and potency of steroid hormones. J Biol Chem 1996;271:17152–6.
297. Medh RD, Lay RH, Schmidt TJ. Agonist-specific modulation of glucocorticoid receptor-mediated transcription by immunosuppressants. Mol Cell Endocrinol 1998;138:11–23.
298. Meijer OC, de Lange EC, Breimer DD, de Boer AG, Workel JO, de Kloet ER. Penetration of dexamethasone into brain glucocorticoid targets is enhanced in mdr1A P-glycoprotein knockout mice. Endocrinology 1998;139:1789–93.
299. Schinkel AH, Wagenaar E, van Deemter L, Mol CA, Borst P. Absence of the mdr1a P-Glycoprotein in mice affects tissue distribution and pharmacokinetics of dexamethasone, digoxin, and cyclosporin A. J Clin Invest 1995;96:1698–705.
300. Webster JI, Carlstedt-Duke J. Involvement of multidrug resistance proteins (MDR) in the modulation of glucocorticoid response. J Steroid Biochem Mol Biol 2002;82:277–88.
301. Farrell RJ, Kelleher D. Glucocorticoid resistance in inflammatory bowel disease. J Endocrinol 2003;178:339–46.
302. Farrell RJ, Murphy A, Long A, Donnelly S, Cherikuri A, O'Toole D, Mahmud N, Keeling PW, Weir DG, Kelleher D. High multidrug resistance (P-glycoprotein 170) expression in inflammatory bowel disease patients who fail medical therapy. Gastroenterology 2000;118:279–88.
303. Hirano T, Onda K, Toma T, Miyaoka M, Moriyasu F, Oka K. MDR1 mRNA expressions in peripheral blood mononuclear cells of patients with ulcerative colitis in relation to glucocorticoid administration. J Clin Pharmacol 2004;44:481–6.
304. Ho GT, Soranzo N, Nimmo ER, Tenesa A, Goldstein DB, Satsangi J. ABCB1/MDR1 gene determines susceptibility and phenotype in ulcerative colitis: discrimination of critical variants using a gene-wide haplotype tagging approach. Hum Mol Genet 2006;15:797–805.
305. Urcelay E, Mendoza JL, Martin MC, Mas A, Martinez A, Taxonera C, Fernandez-Arquero M, Diaz-Rubio M, de la Concha EG. MDR1 gene: susceptibility in Spanish Crohn's disease and ulcerative colitis patients. Inflamm Bowel Dis 2006;12:33–7.

306. Llorente L, Richaud-Patin Y, Diaz-Borjon A, Alvarado de la Barrera C, Jakez-Ocampo J, de la Fuente H, Gonzalez-Amaro R, Diaz-Jouanen E. Multidrug resistance-1 (MDR-1) in rheumatic autoimmune disorders. Part I: increased P-glycoprotein activity in lymphocytes from rheumatoid arthritis patients might influence disease outcome. Joint Bone Spine 2000;67:30–9.
307. Diaz-Borjon A, Richaud-Patin Y, Alvarado de la Barrera C, Jakez-Ocampo J, Ruiz-Arguelles A, Llorente L. Multidrug resistance-1 (MDR-1) in rheumatic autoimmune disorders. Part II: increased P-glycoprotein activity in lymphocytes from systemic lupus erythematosus patients might affect steroid requirements for disease control. Joint Bone Spine 2000;67:40–8.
308. Geick A, Eichelbaum M, Burk O. Nuclear receptor response elements mediate induction of intestinal MDR1 by rifampin. J. Biol Chem 2001;276:14581–7.
309. Johnson DR, Klaassen CD. Regulation of rat multidrug resistance protein 2 by classes of prototypical microsomal enzyme inducers that activate distinct transcription pathways. Toxicol Sci 2002; 67:182–9.
310. Kast HR, Goodwin B, Tarr PT, Jones SA, Anisfeld AM, Stoltz CM, Tontonoz P, Kliewer S, Willson TM, Edwards PA. Regulation of multidrug resistance-associated protein 2 (ABCC2) by the nuclear receptors pregnane X receptor, farnesoid X-activated receptor, and constitutive androstane receptor. J Biol Chem 2002;277:2908–15.
311. Kauffmann HM, Pfannschmidt S, Zoller H, Benz A, Vorderstemann B, Webster JI, Schrenk D. Influence of redox-active compounds and PXR-activators on human MRP1 and MRP2 gene expression. Toxicology 2002;171:137–46.
312. Moore LB, Parks DJ, Jones SA, Bledsoe RK, Consler TG, Stimmel JB, Goodwin B, Liddle C, Blanchard SG, Willson TM, Collins JL, Kliewer SA. Orphan nuclear receptors constitutive androstane receptor and pregnane X receptor share xenobiotic and steroid ligands. J Biol Chem 2000;275:15122–7.
313. Webster JI, Tonelli LH, Moayeri M, Simons SS, Leppla SH, Sternberg EM. Anthrax lethal factor represses glucocorticoid and progesterone receptor activity. Proc Natl Acad Sci USA 2003;100:5706–11.
314. Webster JI, Sternberg EM. Anthrax lethal toxin represses glucocorticoid receptor (GR) transactivation by inhibiting GR-DNA binding in vivo. Mol Cell Endocrinol 2005;241:21–31.
315. Webster JI, Sternberg EM. Role of the hypothalamic-pituitary-adrenal axis, glucocorticoids and glucocorticoid receptors in toxic sequelae of exposure to bacterial and viral products. J Endocrinol 2004;181:207–21.

*The Hypothalamus–Pituitary–Adrenal Axis*
Edited by A. del Rey, G. P. Chrousos and H. O. Besedovsky


# *Glucocorticoid Resistance in Inflammatory Diseases*

DENIS P. FRANCHIMONT[1] and GEORGE P. CHROUSOS[2]

[1]*Division of Gastroenterology – Montreal General Hospital, Mcgill University, Montreal, Canada*
[2]*First Department of Pediatrics, Aghia Sophia Childrens Hospital, University of Athens, Athens, Greece*

## ABSTRACT

The process of inflammation, the disease itself, and the genetic background may all influence the response of patients to glucocorticoids (GCs). GC-dependent patients respond normally to GC but relapse once GCs are withdrawn or at dose tapering. A GC-resistant patient does not respond to GCs. The clinical response to GC can be correlated with in vitro leukocyte sensitivity to glucocorticoids. GC resistance in inflammatory diseases is not genetically determined.

Cytokines, such as interleukin (IL)-1β, IL-6, tumor necrosis factor alpha (TNF-α), interferon-γ (IFNγ), IL-2, IL-4, and IL-13, modulate glucocorticoid receptor (GR) numbers and binding affinity with parallel changes of GC sensitivity. Certain cytokine-stimulated transcription factors may interact in the nucleus with ligand-activated GRα and prevent ligand-activated GRα either from binding to glucocorticoid-responsive elements (GREs) or from exerting its transcriptional actions. Cytokines may directly target the GR through post-translational mechanisms. Phosphorylation and dephosphorylation of the receptor may participate in the ligand activation/inactivation, recycling, and turnover of the receptor.

Excessive nuclear factor (NF)-κB, activator protein-1 (AP-1), signal transducers and activators of transcription (STAT) expression, and activity in an inflammatory site might inhibit GC activity, contributing to a decrease in GC sensitivity. The combinatorial interactions of co-activators and co-repressors with the nuclear receptors may explain the cell and tissue selectivity of nuclear receptor actions. Such co-regulators could undergo post-translational modifications and integrate multiple signals from different signaling pathways such as the mitogen-activated protein kinase (MAPK) pathway.

The expression of 11β-hydroxysteroid dehydrogenase, which regulates intracellular active GC metabolite concentration, may be modulated by cytokines, enhancing or decreasing GC sensitivity. The multi-drug resistance-1 (MDR-1) gene product *P*-glycoprotein 170 is a membrane-based drug-efflux pump, which transports GCs out of the cells, thereby lowering intracellular active GC concentration.

Finally, alterations of post-receptor mechanisms in the GC-signaling pathway and transcription machinery may account for resistance in some patients with familial GC resistance.

## 1. INTRODUCTION

Despite the broad use of glucocorticoids (GCs) in inflammatory diseases (Table 1), a major question remains unanswered, that is, why some patients with inflammatory diseases respond

Table 1 Glucocorticoid therapy in inflammatory/autoimmune diseases, allergic diseases, and infectious diseases

| Inflammatory diseases | Allergic diseases | Infectious diseases |
|---|---|---|
| Rheumatoid arthritis | Allergic rhinitis | Septic shock |
| Connective diseases | Atopic dermatitis | Acute respiratory |
| Lupus, Sjogren's syndrome, sclerodermia, and dermatomyositis | Eczema | Distress syndrome |
| Inflammatory bowel diseases | Asthma | Meningitis |
| Autoimmune hepatitis | Anaphylactic shock | |
| Multiple sclerosis | Churg-Strauss syndrome | |
| Glomerulonephritis and nephritis | Hypereosinophilia | |
| Autoimmune thrombopenia | | |
| Vasculitis | | |
| Wegener granulomatosis | | |
| Sarcoidosis | | |

well to GCs while others are refractory or need chronic GC therapy to maintain their disease in remission. Indeed, some patients are steroid-sensitive (50%–75%), while others are steroid-dependent (30%–35%) or steroid-resistant (SR) (10%–15%). Most steroid-dependent or SR patients develop a cushingoid appearance or metabolic dysfunction, which suggests that their poor GC sensitivity is limited to the immune system rather than involving non-immune tissues. Recent studies have revealed that both the process of inflammation or the disease itself and the genetic and constitutional background of the patient may influence his or her variable response to GCs [1–4].

## 2. MECHANISMS OF CELL- AND TISSUE-SPECIFIC GC RESISTANCE

The clinical response to GCs can sometimes be correlated with in vitro leukocyte sensitivity to GCs. Several mechanisms of tissue GC resistance have been described (Fig. 1). Cytokines can decrease GC sensitivity of tissues by interfering with the GC-signaling pathway. GC sensitivity seems to be modulated by cytokines in various in vitro studies. Some cytokines, such as interleukin-1β (IL-1β), IL-6, TNF-α, IFNγ, IL-2, IL-4, and IL-13, modulate GC receptor numbers and binding affinity with parallel changes of GC sensitivity [5–11]. While glucocorticoid receptor alpha (GRα) is the classic receptor that binds to GCs and transduces their biological activities, GRβ does not bind to GCs and exerts weak dominant negative activity on GR-mediated genomic actions [12]. Cytokines such as IL-2 and IL-4 have been shown to enhance GRβ expression contributing to a decreased GC response following exposure to these cytokines [13–15]. Very interestingly, neutrophils demonstrate higher constitutive GRβ expression than peripheral blood mononuclear cells (PBMCs), and this expression is further enhanced after IL-8 exposure [16]. This may explain the relative resistance of neutrophils to GC-induced cell death and their enhanced survival in the presence of GCs during inflammation. Also, certain cytokine-stimulated transcription factors may interact in the nucleus with ligand-activated GRα and prevent ligand-activated GRα either from binding to glucocorticoid-responsive elements (GREs) or from exerting its transcriptional actions. Excessive nuclear factor (NF)-κB, activator protein-1 (AP-1), and signal transducers and activators of transcription (STAT) expression and activity in an inflammatory site might inhibit GC activity, contributing to a decrease in GC sensitivity [17–19]. NF-κB is a key regulator of the transcriptional activity of many cytokines.

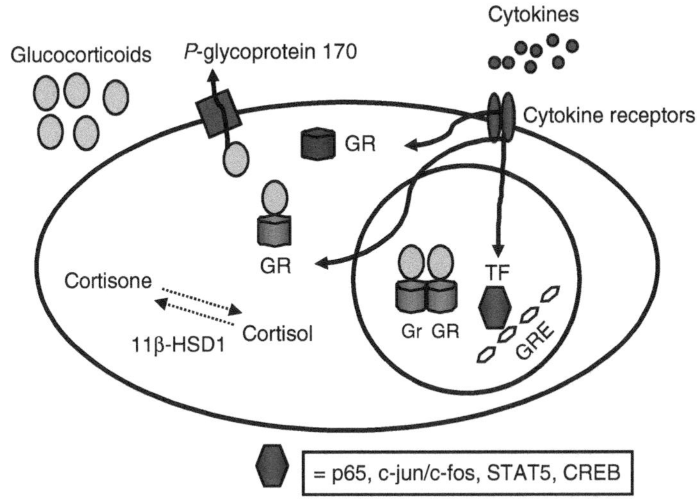


Figure 1. Mechanisms of glucocorticoid resistance. GR, glucocorticoid receptor; GRE, glucocorticoid responsive element; 11β-HSD, hydroxysteroid dehydrogenase; TF, transcription factor; STAT, signal transducer and activator of transcription; CREB, CAMP-responsive element-binding protein, p65, nuclear factor-κb subunit.

NF-κB can be viewed as an intracellular amplification mechanism that exacerbates/sustains chronic inflammatory processes. Indeed, cytokines activate NF-κB leading to a positive feed-forward cycle at an inflammatory site [3,20]. Enhanced activation and concentration of p65 inside the nucleus inhibits the ability of GCs to transactivate their GRE-complementary DNA sequence [17,21]. Type 2 cytokines exert their cellular effects through the Jak/Stat-signaling pathways. The transcription factor Stat 5 of the Type 2 cytokine receptors, such as IL-2 Rα and IL-7 Rα, directly interacts with the GR so that the Stat 5/GRα complex strongly suppresses the response of a GRE-containing promoter to GCs [19]. Cytokines may also directly target the GR through post-translational mechanisms. Phosphorylation and dephosphorylation of the receptor may participate in the ligand activation/inactivation, recycling, and turnover of the receptor. Mitogen-activated protein kinase (MAPK) family members, extracellular regulated kinases (ERKs), c-Jun N-terminal kinase (JNK), and p38 MAPK (p38) phosphorylate the GR and inhibit GR-mediated transcriptional activation [22,23]. This could represent an early repression effect of mitogenic and pro-inflammatory signals on the expression of GR-dependent genes. These examples help explain the "acquired" cytokine-induced GC resistance during inflammation.

Ligand-activated GRs regulate the transcription of responsive genes by forming complexes with the recently discovered co-regulators (i.e., co-activators or co-repressors) of transcription and several chromatin modulators. These molecules not only alter chromatin structure but also enhance or inhibit the transduction of the transcriptional signal of the ligand-activated GRα to the general transcription complex, which includes RNA polymerase II (RNPII) and its ancillary factors [24,25]. The combinatorial interactions of co-activators and co-repressors with the nuclear receptors may explain the cell and tissue selectivity of nuclear receptor actions [26]. Their relative levels may vary from tissue to tissue and can be stoichiometrically limiting between nuclear receptors and transcription factors in a particular tissue. A change in this flexible mix could lead to hypersensitivity and/or resistance to steroid hormones with hormonal and/or tissue predilection. It is fascinating that these co-regulators not only are enhancing or repressing nuclear receptor transcription but could also undergo post-translational modifications and integrate multiple signals from different signaling pathways such as the MAP kinase pathway.

Finally, the expression and activity of 11β-hydroxysteroid dehydrogenase, a key enzyme in the regulation of intracellular GC active metabolite concentration, may be modulated by cytokines, enhancing or decreasing GC sensitivity [27]. Recently, there has been a growing interest in the multi-drug resistance-1 (MDR-1) gene product *P*-glycoprotein 170, which is a membrane-based drug efflux pump that transports MDR substrates, such as GCs, out of the cells, thereby lowering their intracellular active concentration. While independent of disease activity, *P*-glycoprotein 170 may play an important role in the response of inflammatory bowel disease and rheumatoid arthritis patients to GC therapy [28,29].

## 3. GLUCOCORTICOID RESISTANCE IN INFLAMMATORY, ALLERGIC, AND AUTOIMMUNE DISEASES

GC resistance in inflammatory diseases does not appear to be related to primary, genetically determined GC resistance even though several mutations and polymorphisms have been described in familial GC resistance or in the normal population [4,30–33]. Differential GC-dependent transcriptional activation and repression have been observed among human GR variants associated with general GC resistance [34]. The dexamethasone-induced repression of transcription from elements in the promoter of the intercellular adhesion molecule-1 via NF-κB seemed in fact more efficient for the D641V-variant GR than for the wild-type GR; however, the patients with this mutation did not have clinically significant immunosuppression. Five GR cDNA polymorphisms have been reported in the normal population, which were at variance with the original sequence reported by Hollenberg et al. [35,36]. In a population of 216 healthy subjects, a reduced response in a short dexamethasone suppression test as a marker of relative GC insensitivity was observed in 20 otherwise healthy persons. However, this relative GC insensitivity was not associated with any of these polymorphisms [37]. Nevertheless, this selected sample of the general population could represent the lowest section of a normal distribution of GC sensitivity, defined by differences in the GR signal transduction system unrelated to GR polymorphisms. Finally, alterations of post-receptor mechanisms in the GC-signaling pathway and transcription machinery may account for resistance in some patients with familial GC resistance [38].

An appropriate definition of dependence and resistance is necessary to tackle this major inflammatory disease issue with relevant clinical research protocols. A GC-dependent patient responds normally to GCs but relapses once GCs are completely withdrawn and/or at dose tapering. A GC-resistant patient does not respond to GCs. These two concepts are often misused and are sometimes confusing.

### 3.1. Asthma

GC resistance in asthma is often associated with a decreased inhibition of PBMC proliferation and cytokine secretion by GCs. In these patients, GC resistance appears related to alterations of T-cell GR density and binding affinity and/or to changes in post-receptor mechanisms while GR density and binding affinity in mononuclear cell subpopulation remain unaffected [39–41]. Hyperactivation of T lymphocytes with overexpression of IL-2R and HLA-DR, reported by many groups in GC-resistant patients, suggests inflammation-induced GC resistance [39,40]. Two different populations of SR asthma patients have been described: Type 1, defined by decreased GR-binding affinity; and Type 2, defined by decreased GR number per cell [15]. Type 1 SR asthma was shown to be reversible and secondary to inflammation, namely IL-2 and IL-4 dependent, while Type 2 SR asthma appeared to be genetically determined [15,41]. Resistance to GC therapy could also be due to a primary defect in the GR signal transduction pathway. However, no consistent GR polymorphisms have been reported to date in the hGR cDNA from GC-sensitive and corticoSR asthma patients [42–44]. The ability of the GR to bind to GREs seems impaired in T cells from SR asthma patients because of a reduced number of receptors available for binding to DNA [45]. Indeed, increased level of AP-1 DNA binding with increased basal and PMA-stimulated protein levels of c-fos was recognized in GC-resistant patients compared with GC-sensitive asthma patients. Pre-treatment of PBMCs from GC-resistant asthma patients with c-fos antisense oligonucleotides enhanced GR–DNA binding activity, suggesting that overexpression of c-fos was responsible for the decreased GR–DNA binding observed in these patients [42,46]. Finally, recent studies have suggested that this decreased binding to DNA could be related to a markedly increased hGRβ/hGRα ratio such as observed in peripheral blood and bronchial lavage cells from these patients. These hGRβ/hGRα ratio changes appear reversible and cytokine inducible [13,14,42,46,47].

### 3.2. Rheumatoid arthritis

The 50% decrease of GR number per cell reported once in patients with rheumatoid arthritis did not appear to influence the in vitro GC sensitivity of PBMCs from these patients [13,14,42,46–49]. However, leukocyte MDR-1 expression might affect steroid requirements for maintenance of remission in patients with lupus erythematosus and influences the disease outcome in patients with rheumatoid arthritis [28].

### 3.3. Inflammatory bowel diseases

There is a lack of evidence of any change of GR number and binding affinity in the mucosa and/or in peripheral mononuclear cells from patients with inflammatory bowel disease [50]. Moreover, in contrast to GC resistance, dependence in Crohn's disease did not appear to be related to altered GC sensitivity but rather due to an excessive, although sensitive, ongoing inflammatory process [51]. Interestingly, GC resistance observed in ulcerative colitis is correlated with a poor GC sensitivity of proliferating peripheral blood lymphocytes [52,53]. This poor GC response could be explained by a higher detection rate of hGRβ expression in these GC-resistant patients [54]. HGRβ seemed to change over time in these patients and its expression appeared reversible. Finally, high constitutive MDR (*P*-glycoprotein 170) expression has been shown to be associated with poor medical response to GCs in both Crohn's disease and ulcerative colitis patients [29].

### 3.4. Septic shock and acute respiratory distress syndrome

Loss of ability to down-regulate proinflammatory cytokine production is an early event in the pathophysiologic course of lethal acute respiratory distress syndrome (ARDS) [55,56]. Endogenous GCs do not seem always effective in suppressing life-threatening systemic inflammation even though the degree of cortisolemia frequently correlates with severity of illness and mortality rate [57,58]. Failure to suppress inflammation could be due to inadequacy of, and/or tissue resistance to, the levels and duration of endogenous cortisol elevations [59]. Recent randomized studies have shown that prolonged exogenous GC administration – at doses that are clearly pharmacologic for normal persons – compensate adequately for the inability of target organs to respond to GCs, restore GC anti-inflammatory action, and therefore normalize GC sensitivity [60]. Improvers had declining inflammatory cytokine levels over time, and cellular findings included a progressive rise in all aspects of GRα-mediated activity and a concomitant reduction in NF-κB-mediated activity (regulated inflammatory response). By contrast, non-improvers had persistent and exaggerated elevation in plasma inflammatory cytokine levels over time, and cellular findings included only a modest increase in GRα-mediated activity and a progressive escalation in NF-κB activation over time (dysregulated inflammatory response) [60]. Indeed, the systemic inflammation-induced GC resistance observed in patients with ARDS or septic shock returned to normal with *prolonged* GC treatment at moderate doses [59,61,62]. Thus, excessive NF-κB activation could explain the resistance to GCs observed in septic shock.

## 4. PHARMACOGENETICS AND GENOMICS OF GLUCOCORTICOID THERAPY

### 4.1. Predicting glucocorticoid sensitivity, dependence, and resistance

Physicians face major clinical dilemmas on a day-to-day basis. When treating a patient with GCs, they cannot predict whether this patient will be suffering from a GC-dependent or GC-resistant inflammatory disease rather than a GC-sensitive disease. First, they must be aware of the concomitant factors that may interfere with the diagnosis of GC dependence and resistance, because GC-sensitive disease may be mistaken for a GC-dependent disease. These concomitant factors are disease-related complications, such as infections, taking certain drugs, environmental factors, inadequate doses or excessive catabolism of GCs, inefficient administration routes, and compliance of the patient. Second, physicians may consider the disease type, behavior, and severity when using GCs to treat patients. These relevant parameters are important because of the long-term adverse effects of GCs. As far as the disease type, there is a remarkable difference in the prevalence of steroid-dependence and resistance even in similar inflammatory diseases (i.e., Crohn's disease and ulcerative colitis). Also, a clinical response to GCs is to be expected from a specific disease behavior or clinical pattern [63,64].

Today, predicting GC response would be possible on the basis of analysis of host genetic factors. There are approximately 1.4 million single-nucleotide polymorphisms (SNPs) across the human genome with around 50–60,000 SNPs that may account for the variability observed in response to treatment with GCs. There are two complementary methods to look for SNPs associated with drug responsiveness. Wide genome screening of the SNPs enables to design a SNPs map for each patient and perform linkage disequilibrium studies in responders and non-responders. The second-approach analyses candidate SNP of selected genes known to be involved in drug or disease pathway (*Association studies*). For example, this second approach was used by several groups who have observed an increased prevalence of two 308 single base pair polymorphisms of the TNF-α promoters (TNF1 and TNF2) among steroid-sensitive and steroid-dependent patients with autoimmune hepatitis or Crohn's disease [65,66]. However, analyzing candidate SNPs may provide biased associations giving that several SNPs from a single gene interact with each other to produce a drug's response phenotype. Thus, combinations of SNPs will provide a much more predictable drug response. In the near future, pharmacogenetics will allow physicians to have access to the genetic and constitutional factors that may influence treatment response.

Although clinical and genetic information will be available before treating the patient with GCs, clinical evaluation of disease response to treatment is vague and not always accurate. Therefore, the biological profile and dynamics of the disease will offer a precise picture of the disease course and response. Microarray technologies and proteomics offer the opportunity to analyze global gene and protein expression in a timely manner. For example, the effect of GCs on the gene expression profile of PBMCs from healthy donors with *DNA chip microarray* has been recently realized [67]. This immediate picture of global gene regulation allows profiling of the GC-mediated pharmacological effects. Also, specific molecular markers can be targeted for quick analysis of disease response given the magnitude or pattern of their regulation. Many potential molecular markers, such as the scavenger receptor CD163, have been identified and could be used for clinical application to predict response to GCs. Again, pharmacogenomics and proteomics may represent an invaluable tool in the decision and assessment of treatment response.

## REFERENCES


1. Bamberger CM, Schulte HM, Chrousos GP. Molecular determinants of glucocorticoid receptor function and tissue sensitivity to glucocorticoids. Endocr Rev 1996;17:245–61.
2. Barnes PJ. Molecular mechanisms of corticosteroids in allergic diseases. Allergy 2001;56:928–36.
3. Barnes PJ, Greening AP, Crompton GK. Glucocorticoid resistance in asthma. Am J Respir Crit Care Med 1995;152:S125–40.
4. Chrousos GP, Castro M, Leung DY, Webster E, Kino T, Bamberger C et al. Molecular mechanisms of glucocorticoid resistance/hypersensitivity. Potential clinical implications. Am J Respir Crit Care Med 1996;154:S39–43.
5. Costas M, Trapp T, Pereda MP, Sauer J, Rupprecht R, Nahmod VE et al. Molecular and functional evidence for in vitro cytokine enhancement of human and murine target cell sensitivity to glucocorticoids. TNF-alpha priming increases glucocorticoid inhibition of TNF-alpha-induced cytotoxicity/ apoptosis. J Clin Invest 1996;98:1409–16.
6. Franchimont D, Martens H, Hagelstein MT, Louis E, Dewe W, Chrousos GP et al. Tumor necrosis factor alpha decreases, and interleukin-10 increases, the sensitivity of human monocytes to dexamethasone: potential regulation of the glucocorticoid receptor. J Clin Endocrinol Metab 1999; 84:2834–9.
7. Kam JC, Szefler SJ, Surs W, Sher ER, Leung DY. Combination IL-2 and IL-4 reduces glucocorticoid receptor-binding affinity and T cell response to glucocorticoids. J Immunol 1993;151:3460–6.
8. Rakasz E, Gal A, Biro J, Balas G, Falus A. Modulation of glucocorticosteroid binding in human lymphoid, monocytoid and hepatoma cell lines by inflammatory cytokines interleukin (IL)-1 beta, IL-6 and tumour necrosis factor (TNF)-alpha. Scand J Immunol 1993;37:684–9.
9. Salkowski CA, Vogel SN. IFN-gamma mediates increased glucocorticoid receptor expression in murine macrophages. J Immunol 1992;148:2770–7.
10. Spahn JD, Landwehr LP, Nimmagadda S, Surs W, Leung DY, Szefler SJ. Effects of glucocorticoids on lymphocyte activation in patients with steroid-sensitive and steroid-resistant asthma. J Allergy Clin Immunol 1996;98:1073–9.
11. Spahn JD, Szefler SJ, Surs W, Doherty DE, Nimmagadda SR, Leung DY. A novel action of IL-13: induction of diminished monocyte glucocorticoid receptor-binding affinity. J Immunol 1996; 157:2654–9.
12. Bamberger CM, Bamberger AM, de Castro M, Chrousos GP. Glucocorticoid receptor beta, a potential endogenous inhibitor of glucocorticoid action in humans. J Clin Invest 1995;95:2435–41.
13. Leung DY, Chrousos GP. Is there a role for glucocorticoid receptor beta in glucocorticoid-dependent asthmatics? Am J Respir Crit Care Med 2000;162:1–3.
14. Leung DY, Hamid Q, Vottero A, Szefler SJ, Surs W, Minshall E et al. Association of glucocorticoid insensitivity with increased expression of glucocorticoid receptor beta. J Exp Med 1997;186:1567–74.
15. Sher ER, Leung DY, Surs W, Kam JC, Zieg G, Kamada AK et al. Steroid-resistant asthma. Cellular mechanisms contributing to inadequate response to glucocorticoid therapy. J Clin Invest 1994;93:33–9.

16. Strickland I, Kisich K, Hauk PJ, Vottero A, Chrousos GP, Klemm DJ et al. High constitutive glucocorticoid receptor beta in human neutrophils enables them to reduce their spontaneous rate of cell death in response to corticosteroids. J Exp Med 2001;193:585–93.
17. Ray A, Prefontaine KE. Physical association and functional antagonism between the p65 subunit of transcription factor NF-kappa B and the glucocorticoid receptor. Proc Natl Acad Sci USA 1994;91:752–6.
18. Schule R, Rangarajan P, Kliewer S, Ransone LJ, Bolado J, Yang N et al. Functional antagonism between oncoprotein c-Jun and the glucocorticoid receptor. Cell 1990;62:1217–26.
19. Stocklin E, Wissler M, Gouilleux F, Groner B. Functional interactions between Stat5 and the glucocorticoid receptor. Nature 1996;383:726–8.
20. McKay LI, Cidlowski JA. Molecular control of immune/inflammatory responses: interactions between nuclear factor-kappa B and steroid receptor-signaling pathways. Endocr Rev 1999;20:435–59.
21. Wissink S, van Heerde EC, vand der Burg B, van der Saag PT. A dual mechanism mediates repression of NF-kappaB activity by glucocorticoids. Mol Endocrinol 1998;12:355–63.
22. Irusen E, Matthews JG, Takahashi A, Barnes PJ, Chung KF, Adcock IM. p38 Mitogen-activated protein kinase-induced glucocorticoid receptor phosphorylation reduces its activity: role in steroid-insensitive asthma. J Allergy Clin Immunol 2002;109:649–57.
23. Rogatsky I, Logan SK, Garabedian MJ. Antagonism of glucocorticoid receptor transcriptional activation by the c-Jun N-terminal kinase. Proc Natl Acad Sci USA 1998;95:2050–5.
24. Glass CK, Rosenfeld MG. The coregulator exchange in transcriptional functions of nuclear receptors. Genes Dev 2000;14:121–41.
25. McKenna NJ, Lanz RB, O'Malley BW. Nuclear receptor coregulators: cellular and molecular biology. Endocr Rev 1999;20:321–44.
26. Szapary D, Huang Y, Simons SS, Jr. Opposing effects of corepressor and coactivators in determining the dose-response curve of agonists, and residual agonist activity of antagonists, for glucocorticoid receptor-regulated gene expression. Mol Endocrinol 1999;13:2108–21.
27. Escher G, Galli I, Vishwanath BS, Frey BM, Frey FJ. Tumor necrosis factor alpha and interleukin 1beta enhance the cortisone/cortisol shuttle. J Exp Med 1997;186:189–98.
28. Diaz-Borjon A, Richaud-Patin Y, Alvarado de la Barrera C, Jakez-Ocampo J, Ruiz-Arguelles A, Llorente L. Multidrug resistance-1 (MDR-1) in rheumatic autoimmune disorders. Part II: Increased P-glycoprotein activity in lymphocytes from systemic lupus erythematosus patients might affect steroid requirements for disease control. Joint Bone Spine 2000;67:40–8.
29. Farrell RJ, Murphy A, Long A, Donnelly S, Cherikuri A, O'Toole D et al. High multidrug resistance (P-glycoprotein 170) expression in inflammatory bowel disease patients who fail medical therapy. Gastroenterology 2000;118:279–88.
30. Chrousos GP, Detera-Wadleigh SD, Karl M. Syndromes of glucocorticoid resistance. Ann Intern Med 1993;119:1113–24.
31. Lamberts SW, Koper JW, Biemond P, den Holder FH, de Jong FH. Cortisol receptor resistance: the variability of its clinical presentation and response to treatment. J Clin Endocrinol Metab 1992;74:313–21.
32. Lamberts SW, Koper JW, de Jong FH. Familial and iatrogenic cortisol receptor resistance. J Steroid Biochem Mol Biol 1992;43:385–8.
33. Lamberts SW, Poldermans D, Zweens M, de Jong FH. Familial cortisol resistance: differential diagnostic and therapeutic aspects. J Clin Endocrinol Metab 1986;63:1328–33.
34. de Lange P, Koper JW, Huizenga NA, Brinkmann AO, de Jong FH, Karl M et al. Differential hormone-dependent transcriptional activation and -repression by naturally occurring human glucocorticoid receptor variants. Mol Endocrinol 1997;11:1156–64.
35. Hollenberg SM, Weinberger C, Ong ES, Cerelli G, Oro A, Lebo R et al. Primary structure and expression of a functional human glucocorticoid receptor cDNA. Nature 1985;318:635–41.
36. Huizenga NA, Koper JW, De Lange P, Pols HA, Stolk RP, Burger H et al. A polymorphism in the glucocorticoid receptor gene may be associated with and increased sensitivity to glucocorticoids in vivo. J Clin Endocrinol Metab 1998;83:144–51.

37. Koper JW, Stolk RP, de Lange P, Huizenga NA, Molijn GJ, Pols HA et al. Lack of association between five polymorphisms in the human glucocorticoid receptor gene and glucocorticoid resistance. Hum Genet 1997;99:663–8.
38. Huizenga NA, de Lange P, Koper JW, de Herder WW, Abs R, Kasteren JH et al. Five patients with biochemical and/or clinical generalized glucocorticoid resistance without alterations in the glucocorticoid receptor gene. J Clin Endocrinol Metab 2000;85:2076–81.
39. Corrigan CJ. Glucocorticoid-resistant asthma. T-lymphocyte defects. Am J Respir Crit Care Med 1996;154:53–5.
40. Corrigan CJ, Brown PH, Barnes NC, Tsai JJ, Frew AJ, Kay AB. Glucocorticoid resistance in chronic asthma. Peripheral blood T lymphocyte activation and comparison of the T lymphocyte inhibitory effects of glucocorticoids and cyclosporin A. Am Rev Respir Dis 1991;144:1026–32.
41. Lane SJ, Lee TH. Glucocorticoid receptor characteristics in monocytes of patients with corticosteroid-resistant bronchial asthma. Am Rev Respir Dis 1991;143:1020–4.
42. Adcock IM, Lane SJ, Brown CR, Lee TH, Barnes PJ. Abnormal glucocorticoid receptor-activator protein 1 interaction in steroid-resistant asthma. J Exp Med 1995;182:1951–8.
43. Lane SJ, Arm JP, Staynov DZ, Lee TH. Chemical mutational analysis of the human glucocorticoid receptor cDNA in glucocorticoid-resistant bronchial asthma. Am J Respir Cell Mol Biol 1994; 11:42–8.
44. Lane SJ, Lee TH. Mechanisms of corticosteroid resistance in asthmatic patients. Int Arch Allergy Immunol 1997;113:193–5.
45. Adcock IM, Lane SJ, Brown CR, Peters MJ, Lee TH, Barnes PJ. Differences in binding of glucocorticoid receptor to DNA in steroid-resistant asthma. J Immunol 1995;154:3500–5.
46. Lane SJ, Adcock IM, Richards D, Hawrylowicz C, Barnes PJ, Lee TH. Corticosteroid-resistant bronchial asthma is associated with increased c-fos expression in monocytes and T lymphocytes. J Clin Invest 1998;102:2156–64.
47. Leung DY, de Castro M, Szefler SJ, Chrousos GP. Mechanisms of glucocorticoid-resistant asthma. Ann N Y Acad Sci 1998;840:735–46.
48. Schlaghecke R, Beuscher D, Kornely E, Specker C. Effects of glucocorticoids in rheumatoid arthritis. Diminished glucocorticoid receptors do not result in glucocorticoid resistance. Arthritis Rheum 1994;37:1127–31.
49. Schlaghecke R, Kornely E, Wollenhaupt J, Specker C. Glucocorticoid receptors in rheumatoid arthritis. Arthritis Rheum 1992;35:740–4.
50. Schottelius A, Wedel S, Weltrich R, Rohde W, Buttgereit F, Schreiber S et al. Higher expression of glucocorticoid receptor in peripheral mononuclear cells in inflammatory bowel disease. Am J Gastroenterol 2000;95:1994–9.
51. Franchimont D, Louis E, Dupont P, Vrindts-Gevaert Y, Dewe W, Chrousos G et al. Decreased corticosensitivity in quiescent Crohn's disease: an ex vivo study using whole blood cell cultures. Dig Dis Sci 1999;44:1208–15.
52. Corrigan CJ. Glucocorticoid unresponsiveness in severe ulcerative colitis. Gut 1999;45:324–5.
53. Hearing SD, Norman M, Probert CS, Haslam N, Dayan CM. Predicting therapeutic outcome in severe ulcerative colitis by measuring in vitro steroid sensitivity of proliferating peripheral blood lymphocytes. Gut 1999;45:382–8.
54. Honda M, Orii F, Ayabe T, Imai S, Ashida T, Obara T et al. Expression of glucocorticoid receptor beta in lymphocytes of patients with glucocorticoid-resistant ulcerative colitis. Gastroenterology 2000; 118:859–66.
55. Meduri GU, Headley S, Kohler G, Stentz F, Tolley E, Umberger R et al. Persistent elevation of inflammatory cytokines predicts a poor outcome in ARDS. Plasma IL-1 beta and IL-6 levels are consistent and efficient predictors of outcome over time. Chest 1995;107:1062–73.
56. Meduri GU, Headley S, Tolley E, Shelby M, Stentz F, Postlethwaite A. Plasma and BAL cytokine response to corticosteroid rescue treatment in late ARDS. Chest 1995;108:1315–25.
57. Briegel J, Forst H, Hellinger H, Haller M. Contribution of cortisol deficiency to septic shock. Lancet 1991;338:507–8.

58. Reincke M, Allolio B, Wurth G, Winkelmann W. The hypothalamic-pituitary-adrenal axis in critical illness: response to dexamethasone and corticotropin-releasing hormone. J Clin Endocrinol Metab 1993;77:151–6.
59. Meduri GU, Kanangat S. Glucocorticoid treatment of sepsis and acute respiratory distress syndrome: time for a critical reappraisal. Crit Care Med 1998;26:630–3.
60. Meduri GU, Tolley EA, Chrousos GP, Stentz F. Prolonged methylprednisolone treatment suppresses systemic inflammation in patients with unresolving acute respiratory distress syndrome: evidence for inadequate endogenous glucocorticoid secretion and inflammation-induced immune cell resistance to glucocorticoids. Am J Respir Crit Care Med 2002;165:983–91.
61. Bollaert PE, Charpentier C, Levy B, Debouverie M, Audibert G, Larcan A. Reversal of late septic shock with supraphysiologic doses of hydrocortisone. Crit Care Med 1998;26:645–50.
62. Molijn GJ, Spek JJ, van Uffelen JC, de Jong FH, Brinkmann AO, Bruining HA et al. Differential adaptation of glucocorticoid sensitivity of peripheral blood mononuclear leukocytes in patients with sepsis or septic shock. J Clin Endocrinol Metab 1995;80:1799–803.
63. Franchimont DP, Louis E, Croes F, Belaiche J. Clinical pattern of corticosteroid dependent Crohn's disease. Eur J Gastroenterol Hepatol 1998;10:821–5.
64. Lindgren SC, Flood LM, Kilander AF, Lofberg R, Persson TB, Sjodahl RI. Early predictors of glucocorticosteroid treatment failure in severe and moderately severe attacks of ulcerative colitis. Eur J Gastroenterol Hepatol 1998;10:831–5.
65. Czaja AJ, Cookson S, Constantini PK, Clare M, Underhill JA, Donaldson PT. Cytokine polymorphisms associated with clinical features and treatment outcome in type 1 autoimmune hepatitis. Gastroenterology 1999;117:645–52.
66. Louis E, Peeters M, Franchimont D, Seidel L, Fontaine F, Demolin G et al. Tumour necrosis factor (TNF) gene polymorphism in Crohn's disease (CD): influence on disease behaviour. Clin Exp Immunol 2000;119:64–8.
67. Galon J, Franchimont D, Hiroi N, Frey G, Boettner A, Ehrhart-Bornstein M et al. Gene profiling reveals unknown enhancing and suppressive actions of glucocorticoids on immune cells. FASEB J 2002;16:61–71.

# *Glucocorticosteroids for Asthma: From Genes to the Clinic*

PETER J. BARNES

*Department of Thoracic Medicine, National Heart and Lung Institute, Imperial College, London, UK*

## ABSTRACT

Glucocorticoids are the most effective controllers of asthma and other allergic diseases. They suppress inflammation mainly by switching off multiple activated inflammatory genes through reversing histone acetylation via the recruitment of histone deacetylase-2 (HDAC2). Through suppression of airway inflammation, inhaled glucocorticoids reduce airway hyperresponsiveness and control asthma symptoms. Inhaled glucocorticoids are now first-line therapy for all patients with persistent asthma, controlling asthma symptoms and preventing exacerbations. Inhaled long-acting $\beta_2$-agonists added to inhaled glucocorticoids further improve asthma control and are commonly given as combination inhalers, which improve compliance and control asthma at lower doses of glucocorticoids. Inhaled glucocorticoids, which are absorbed form the lungs into the systemic circulation, have negligible systemic side effects at the doses most patients require. Systemic glucocorticoids are used in the treatment of acute exacerbations of asthma and as a maintenance treatment in patients with severe asthma that is not controlled by maximum inhaled therapy. Oral steroids have numerous metabolic and endocrine side effects and the lowest dose needed to control the disease should be used.

## 1. INTRODUCTION

Glucocorticosteroids (also known as glucocorticoids, corticosteroids, and steroids) are by far the most effective controllers used in the treatment of asthma and the only drugs that can effectively suppress the characteristic inflammation in asthmatic airways. They are also highly effective in the treatment of rhinitis and other allergic diseases. After discussing the mechanism of action and pharmacology of glucocorticoids, I will then describe their use in the treatment of asthma.

## 2. MECHANISMS OF ACTION

There have been major advances in understanding the molecular mechanisms whereby glucocorticoids suppress inflammation, based on recent developments in understanding the fundamental mechanisms of gene transcription [1,2]. Glucocorticoids activate and suppress many genes relevant to understanding their action in asthma and other allergic diseases (Table 1).

Table 1 Effect of glucocorticoids on gene transcription

*Increased transcription*

- Lipocortin-1
- $\beta_2$-Adrenoceptor
- Secretory leukocyte inhibitory protein
- IκB-α (inhibitor of NF-κB)
- Anti-inflammatory or inhibitory cytokines
  IL-10, IL-12, IL-1 receptor antagonist
- Mitogen-activated protein kinase phosphatase-1 (MKP-1, inhibits MAP kinase pathways)

*Decreased transcription*

- Inflammatory cytokines
  IL-2, IL-3, IL-4, IL-5, IL-6, IL-11, IL-13, IL-15, TNFα, GM-CSF, SCF
- Chemokines
  IL-8, RANTES, MIP-1α, eotaxin
- Inducible nitric oxide synthase (iNOS)
- Inducible cyclo-oxygenase (COX-2)
- Inducible phospholipase $A_2$ ($cPLA_2$)
- Endothelin-1
- Neurokinin ($NK_1$)-, bradykinin ($B_2$)-receptors
- Adhesion molecules (ICAM-1,VCAM-1)

## 2.1. Cellular effects

At a cellular level, glucocorticoids reduce the numbers of inflammatory cells in the airways, including eosinophils, T lymphocytes, mast cells, and dendritic cells (Fig. 1). These effects of glucocorticoids are produced through inhibiting the recruitment of inflammatory cells into the airway by suppressing the production of chemotactic mediators and adhesion molecules and by inhibiting the survival in the airways of inflammatory cells, such as

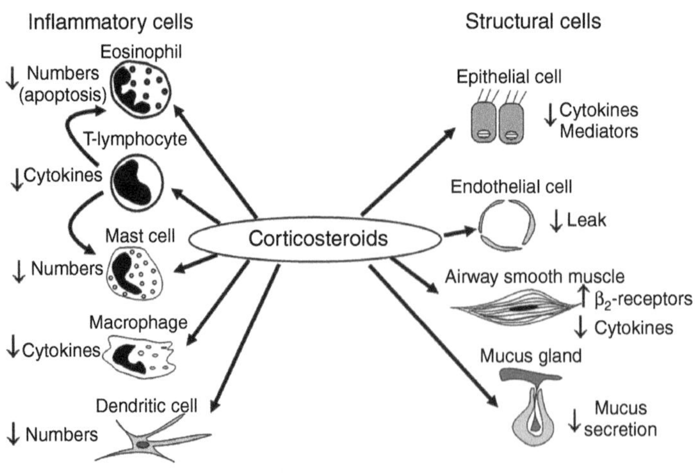


Figure 1. Cellular effect of glucocorticoids.

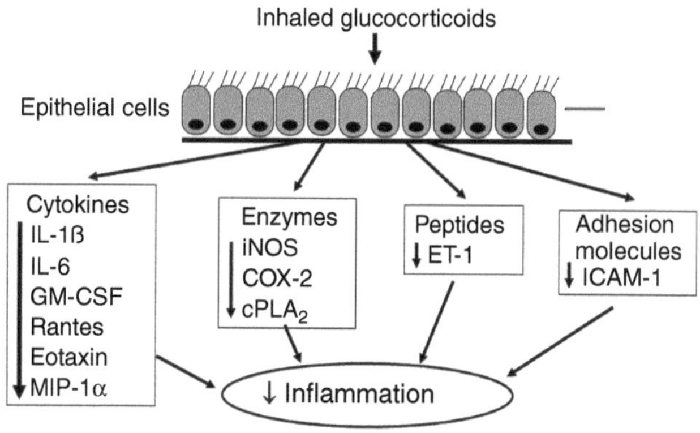


Figure 2. Inhaled glucocorticoids may inhibit the transcription of several "inflammatory" genes in airway epithelial cells and thus reduce inflammation in the airway wall. NF-κB, nuclear factor-κB; AP-1, activator protein-1; GM-CSF, granulocyte-macrophage colony-stimulating factor; IL-1, interleukin-1; iNOS, inducible nitric oxide synthase; NO, nitric oxide; COX-2, inducible cyclooxygenase; $cPLA_2$, cytoplasmic phospholipase $A_2$; PG, prostaglandin; ET, endothelin; ICAM, intercellular adhesion molecule.

eosinophils, T-lymphocytes, and mast cells. Epithelial cells may be the major cellular target for inhaled glucocorticoids, which are the mainstay of modern asthma management. Inhaled glucocorticoids suppress many activated inflammatory genes in airway epithelial cells (Fig. 2). Epithelial integrity is restored by regular inhaled glucocorticoids. The suppression of mucosal inflammation is relatively rapid with a significant reduction in eosinophils detectable within 6 h and associated with reduced airway hyperresponsiveness [3,4]. Reversal of airway hyperresponsiveness may take several months to reach a plateau, probably reflecting recovery of structural changes in the airway [5].

## 2.2. Glucocorticoid receptors

Glucocorticoids diffuse across the cell membrane and bind to glucocorticoid receptors (GRs) in the cytoplasm [2]. There is only one form of GR that binds glucocorticoids termed GRα. GRβ is an alternatively spliced form of GR that interacts with DNA but not with glucocorticoids, so may act as a dominant negative inhibitor of glucocorticoid action by interfering with the binding of GR to DNA [6]. Whether GRβ is involved in steroid resistance in asthma is controversial. Activated GRs rapidly translocate to the nucleus where they produce their molecular effects. A pair of GRs (GR dimer) bind to glucocorticoid response elements in the promoter region of steroid-responsive genes, and this interaction switches on (and sometimes switches off) gene transcription (Fig. 3). Examples of genes that are activated by glucocorticoids include genes encoding $\beta_2$-adrenergic receptors and the anti-inflammatory proteins secretory leukoprotease inhibitor and mitogen-activated protein kinase phosphatase-1 (MKP-1) that inhibits MAP kinase pathways (Fig. 4). These effects may contribute to the anti-inflammatory actions of glucocorticoids [7,8]. GR interaction with negative GREs may suppress gene transcription, and it is thought that this may be important in mediating many of the side effects of glucocorticoids. For example, glucocorticoids inhibit the expression of osteocalcin that is involved in bone synthesis [9].

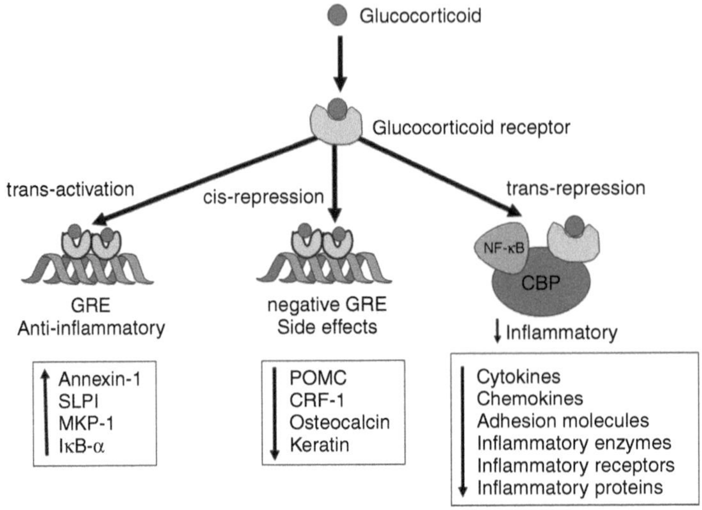


Figure 3. Glucocorticoids may regulate gene expression in several ways. Glucocorticoids enter the cell to bind to glucocorticoid receptors in the cytoplasm that translocate to the nucleus. GR homodimers bind to glucocorticoid-response elements (GREs) in the promoter region of steroid-sensitive genes, which may encode anti-inflammatory proteins. Less commonly, GR homodimers interact with negative GREs to suppress genes, particularly those linked to side effects of glucocorticoids. Nuclear GR also interact with coactivator molecules, such as CREB-binding protein (CBP), which is activated by proinflammatory transcription factors, such as nuclear factor-κb (NF-κb), thus switching off the inflammatory genes that are activated by these transcription factors. SLPI, secretory leukoprotease inhibitor; MKP-1, mitogen-activated kinase phosphatase-1; IκB-α, inhibitor of NF-κB; GILZ, glucocorticoid-induced leucine zipper protein; POMC, proopiomelanocortin; CRF, corticotrophin-releasing factor.

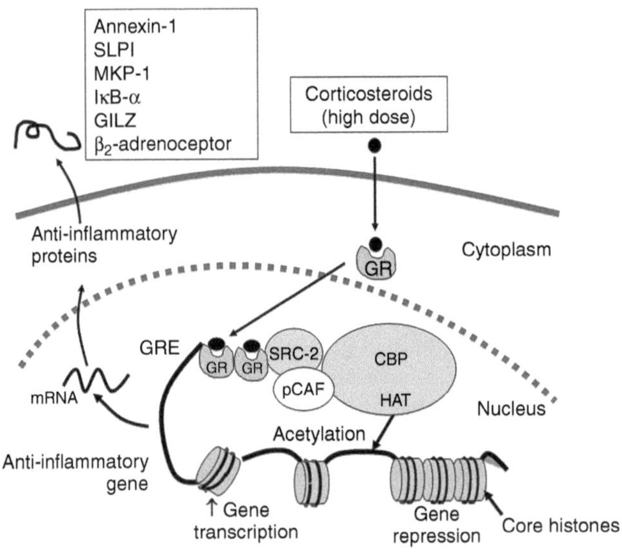


Figure 4. Glucocorticoids activation of anti-inflammatory gene expression. Glucocorticoids bind to cytoplasmic glucocorticoid receptors (GRs) that translocate to the nucleus where they bind to glucocorticoid response elements (GREs) in the promoter region of steroid-sensitive genes and also directly or indirectly to coactivator molecules such as CREB-binding protein (CBP), p300/CBP-activating factor (pCAF), or steroid receptor coactivator-2 (SRC-2), which have intrinsic histone acetyltransferase (HAT) activity, causing acetylation of lysines on histone H4, which leads to activation of genes encoding anti-inflammatory proteins, such as secretory leukoprotease inhibitor (SLPI), mitogen-activated kinase phosphatase-1 (MKP-1), inhibitor of NF-κb (iκb-α), and glucocorticoid-induced leucine zipper protein (GILZ).

## 2.3. Switching off inflammation

The major action of glucocorticoids is to switch off multiple activated inflammatory genes that encode for cytokines, chemokines, adhesion molecules inflammatory enzymes, and receptors [10]. These genes are switched on in the airways by proinflammatory transcription factors, such as nuclear factor-κB (NF-κB) and activator protein-1, both of which are activated in asthmatic airways and switch on inflammatory genes by interacting with coactivator molecules, such as CREB-binding protein, that have intrinsic histone acetyltransferase activity, resulting in acetylation of core histones, which opens up the chromatin structure so that gene transcription is facilitated [11]. In artificial overexpression systems, activated GR may directly interact with NF-κB and AP-1 to inhibit their activity, but this does not appear to occur in asthmatic patients treated with inhaled glucocorticoids [12]. Glucocorticoid-activated GRs also interact with coactivator molecules, and this inhibits the interaction of NF-κB with coactivators, thus reducing histone acetylation [1,13]. Reduction of histone acetylation also occurs through the recruitment of histone deacetylase-2 (HDAC2) to the activated inflammatory gene complex by activated GR, thereby resulting in effective suppression of all activated inflammatory genes within the nucleus (Fig. 5). This accounts for why glucocorticoids are so effective in the control of asthmatic inflammation but also why they are safe, since other activated genes are not affected.

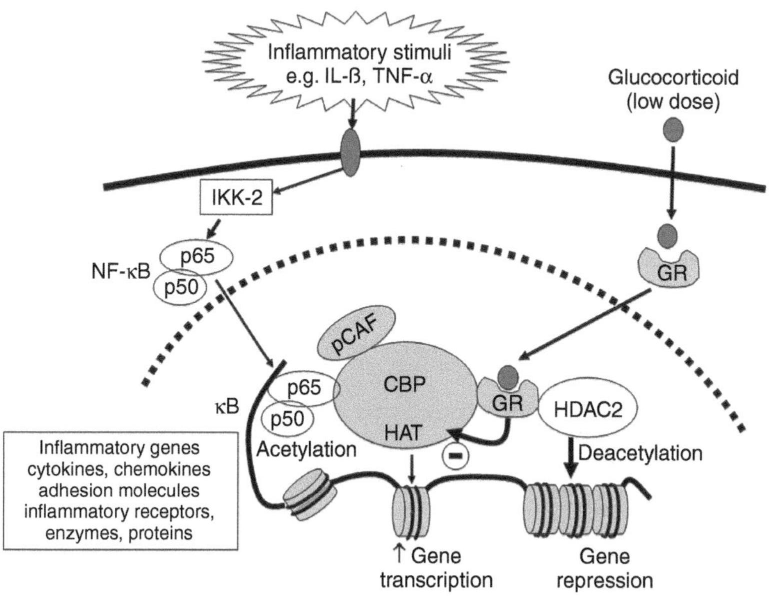


Figure 5. Glucocorticoids suppression of activated inflammatory genes. Inflammatory genes are activated by inflammatory stimuli, such as interleukin-1β (IL-1β) or tumour necrosis factor alpha (TNF-α), resulting in activation of IKK-2 (inhibitor of I-κB kinase-2), which activates the transcription factor, nuclear factor-κB (NF-κB). A dimer of p50 and p65 NF-κB proteins translocates to the nucleus and binds to specific κB recognition sites and also to coactivators, such as CREB-binding protein (CBP) or p300/CBP-activating factor (pCAF), which have intrinsic histone acetyltransferase (HAT) activity. This results in acetylation of core histone H4, resulting in increased expression of genes encoding multiple inflammatory proteins. Glucocorticoid receptors (GRs), after activation by glucocorticoids, translocate to the nucleus and bind to coactivators to inhibit HAT activity directly and recruiting histone deacetylase-2 (HDAC2), which reverses histone acetylation leading in suppression of these activated inflammatory genes.

There may be additional mechanisms that are also important in the anti-inflammatory actions of glucocorticoids. Glucocorticoids have potent inhibitory effects on mitogen-activated kinase signaling pathways through the induction of MKP-1 and this may inhibit the expression of multiple inflammatory genes [7,8]. Some inflammatory genes, for example granulocyte–macrophage colony-stimulating factor, have an unstable messenger RNA that is rapidly degraded by certain RNAses but stabilized when cells are stimulated by inflammatory mediators. Glucocorticoids reverse this effect, resulting in rapid degradation of mRNA and reduced inflammatory protein secretion [14]. This may be through the inhibition of proteins that stabilize mRNAs of inflammatory proteins, such as tristretraprolin [15].

## 3. INTERACTION WITH $\beta_2$-ADRENERGIC RECEPTORS

Inhaled $\beta_2$-agonists and glucocorticoids are frequently used together in the control of asthma, and it is now recognized that there are important molecular interactions between these two classes of drug [16]. As discussed above, glucocorticoids increase the gene transcription of $\beta_2$-receptors, resulting in increased expression of cell surface receptors. This has been demonstrated in human lung in vitro [17] and nasal mucosa in vivo after topical application of a glucocorticoid [18]. In this way, glucocorticoids protect against the down-regulation of $\beta_2$-receptors after long-term administration [19]. This may be important for the non-bronchodilator effects of $\beta_2$-agonists, such as mast cell stabilization. Glucocorticoids may also enhance the coupling of $\beta_2$-receptors to G-proteins, this enhancing $\beta_2$-agonist effects and reversing the uncoupling of $\beta_2$-receptors that may occur in response to inflammatory mediators, such as interleukin-1$\beta$ through a stimulatory effect on a G-protein-coupled receptor kinase [20].

There is also evidence that $\beta_2$-agonists may affect GR and thus enhance the anti-inflammatory effects of glucocorticoids. $\beta_2$-agonists increase the translocation of GR from cytoplasm to the nucleus after activation by glucocorticoids [21]. This effect has now been demonstrated in sputum macrophages of asthmatic patients after an inhaled glucocorticoid and inhaled long-acting $\beta_2$-agonist [22]. This suggests that $\beta_2$-agonists and glucocorticoid enhance each others beneficial effects in asthma therapy.

## 4. PHARMACOKINETICS

Prednisone is readily and consistently absorbed after oral administration with little inter-individual variation. Prednisone is converted in the liver to active prednisolone. Enteric coatings to reduce the incidence of dyspepsia delay absorption but not the total amount of drug absorbed. Prednisolone is metabolized in the liver, and drugs such as rifampicin, phenobarbitone, or phenytoin, which induce CYP450 enzymes, lower the plasma half-life of prednisolone [23]. The plasma half-life is 2–3 h although its biological half-life is approximately 24 h, so that it is suitable for daily dosing. There is no evidence that previous exposure to steroids changes their subsequent metabolism. Prednisolone is approximately 92% protein bound, the majority to a specific protein, transcortin, and the remainder to albumin; it is the unbound fraction that is biologically active. Some patients, usually with severe asthma, apparently fail to respond to glucocorticoids. "Steroid-resistant" asthma is not due to impaired absorption or metabolism of steroids but is due to reduced anti-inflammatory actions of glucocorticoids. Measurement of plasma concentrations of prednisolone are useful in monitoring compliance with inhaled glucocorticoids and in assessing whether a poor therapeutic response to glucocorticoids is due to poor absorption or increased metabolism.

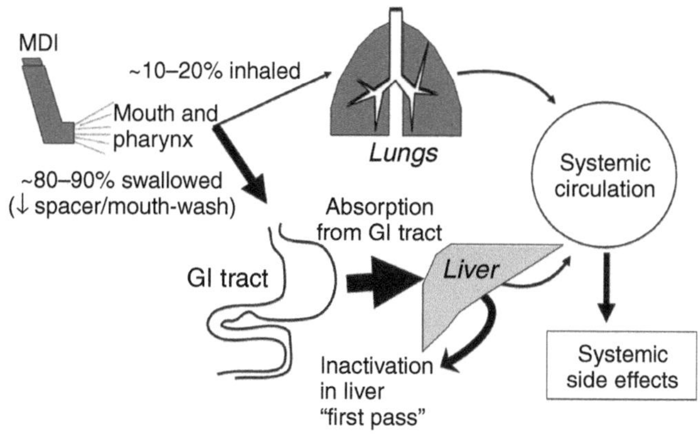


Figure 6. Pharmacokinetics of inhaled glucocorticoids. GI, gastrointestinal.

## 4.1. Inhaled delivery

The pharmacokinetics of inhaled glucocorticoids is important in relation to systemic effects [24–26]. The fraction of steroid which is inhaled into the lungs acts locally on the airway mucosa but may be absorbed from the airway and alveolar surface. This fraction therefore reaches the systemic circulation (Fig. 6). The fraction of inhaled steroid which is deposited in the oropharynx is swallowed and absorbed from the gut. The absorbed fraction may be metabolized in the liver before reaching the systemic circulation (first-pass metabolism). Budesonide and fluticasone propionate (FP) have a greater first-pass metabolism than beclomethasone dipropionate (BDP) and are therefore less likely to produce systemic effects at high inhaled doses. The use of a large volume spacer chamber reduces oropharyngeal deposition and therefore reduces systemic absorption of glucocorticoids although this effect is minimal in glucocorticoids with a high first-pass metabolism [27]. Mouth rinsing and discarding the rinse has a similar effect, and this procedure should be used with high-dose dry powder steroid inhalers, since spacer chambers cannot be used with these devices. The ideal inhaled corticosteroid with optimal therapeutic index should have high lung bioavailability, negligible oral bioavailability, low systemic absorption, high systemic clearance, and high protein binding [28].

A recently introduced glucocorticoid ciclesonide is an inactive prodrug that is activated by esterases in the lung to the active metabolite des-ciclesonide [29]. This may reduce oropharyngeal side effects as esterases appear to be less active in this site than in the lower airways. Ciclesonide is also claimed to be effective as a once daily therapy.

# 5. SYSTEMIC STEROIDS

Hydrocortisone is given intravenously in acute severe asthma. Although the value of glucocorticoids in acute severe asthma has been questioned, others have found that they speed the resolution of attacks. There is no apparent advantage in giving very high doses of intravenous steroids (such as methylprednisolone 1 g) as this only increases the risk of side effects, such as hyperglycemia and increased susceptibility to infections. Intravenous steroids are indicated in patients with acute asthma if lung function is $<30\%$ predicted and in whom there is no significant improvement with a nebulized $\beta_2$-agonist. Intravenous therapy is usually given until a satisfactory response is obtained and then oral prednisolone may be substituted. Oral prednisolone (40–60 mg) has a similar effect to intravenous hydrocortisone and is easier to administer [30,31]. High doses of inhaled glucocorticoids may also substitute for a course of oral steroids in controlling acute exacerbations of asthma. High dose of FP (2000 μg daily) was as effective as a course of oral prednisolone in controlling acute exacerbations of asthma in a family practice setting and in children in an emergency room setting although this route of delivery is more expensive [32,33]. Although doubling the dose of inhaled glucocorticoids was recommended for mild exacerbations of asthma, this does not appear to be useful [34,35], but a fourfold increase in dose appears to be effective [36]. Inhaled steroids have no proven effect in the management of severe acute asthma in a hospital setting [37], but trials with nebulized steroids, which can deliver large doses, are underway.

Maintenance treatment with oral steroids is reserved for patients who cannot be controlled on maximum doses of other therapy, the dose being titrated to the lowest which provides acceptable control of symptoms. For any patient taking regular oral steroids, objective evidence of steroid responsiveness should be obtained before maintenance therapy is instituted. Short courses of oral steroids (30–40 mg prednisolone daily for 1–2 weeks) are indicated for exacerbations of asthma, and the dose may be tailed off over 1 week once the exacerbation is resolved (although the tail-off period is not strictly necessary, patients often find it reassuring).

## 6. INHALED GLUCOCORTICOIDS

There is no doubt that the early use of inhaled glucocorticoids has revolutionized the management of asthma, with marked reductions in asthma morbidity and improvement in health status. Inhaled steroids are now recommended as first-line therapy for all patients with persistent asthma [38]. Inhaled glucocorticoids are very effective in controlling asthma symptoms in asthmatic patients of all ages and severity. Inhaled glucocorticoids improve the quality of life of patients with asthma and allow many patients to lead normal lives, improve lung function, reduce the frequency of exacerbations and may prevent irreversible airway changes. They were first introduced to reduce the requirement for oral glucocorticoids in patients with severe asthma, and many studies have confirmed that the majority of patients can be weaned off oral glucocorticoids [24].

### 6.1. Studies in adults

As experience has been gained with inhaled glucocorticoids, they have been introduced in patients with milder asthma, with the recognition that inflammation is present even in patients with mild asthma. Inhaled anti-inflammatory drugs have now become first-line therapy in any patient who needs to use a $\beta_2$-agonist inhaler more than two to three times a week, and this is reflected in national and international guidelines for the management of chronic asthma. In patients with newly diagnosed asthma, an inhaled corticosteroid (budesonide 600 μg twice daily) reduced symptoms and $\beta_2$-agonist inhaler usage and improved lung function. These effects persisted over the 2 years of the study, whereas in a parallel group treated with inhaled $\beta_2$-agonists alone, there was no significant change in symptoms or lung function [39]. In another study, patients with mild asthma treated with a low dose of inhaled corticosteroid (budesonide 400 μg daily) showed less symptoms and a progressive improvement in lung function over several months, and many patients became completely asymptomatic [40]. There was also a significant reduction in the number of exacerbations; in patients with mild asthma, a low dose of glucocorticoids (budesonide 400 μg daily) significantly reduces exacerbation by around 40% over a 3-year period [41]. Although the effects of inhaled glucocorticoids on AHR may take several months to reach a plateau, the reduction in asthma symptoms occurs much more rapidly, and reduced inflammation is seen within hours [3,4].

High-dose inhaled glucocorticoids may be used for the control of more severe asthma. This markedly reduces the need for maintenance of oral glucocorticoids [42]. With the use of add-on therapies, particularly long-acting $\beta_2$-agonists, most patients can now be controlled on much lower doses of inhaled glucocorticoids so that high doses are needed in only a few patients with severe disease. Inhaled glucocorticoids are the treatment of choice in nocturnal asthma, which is a manifestation of inflamed airways, reducing nocturnal awakening and reducing the diurnal variation in airway function.

Inhaled glucocorticoids effectively control asthmatic inflammation but must be taken regularly. When inhaled glucocorticoids are discontinued, there is usually a gradual increase in symptoms and airway responsiveness back to pretreatment values [43]. Reduction in the dose of inhaled glucocorticoids is associated with an increase in symptoms, and this is preceded by an increase in exhaled NO and sputum eosinophils [44,45].

### 6.2. Studies in children

Inhaled glucocorticoids are equally effective in children. In an extensive study of children aged 7–17 years, there was a significant improvement in symptoms, peak flow variability, and lung function compared to a regular inhaled $\beta_2$-agonist that was maintained over the 22 months of the study [46], but asthma deteriorated when the inhaled glucocorticoids were withdrawn [47]. There was a high proportion of drop-outs (45%) in the group treated with inhaled $\beta_2$-agonist alone. Inhaled glucocorticoids are also effective in younger children. Nebulized budesonide reduces the need for oral glucocorticoids and also improved lung function in children under the age of 3 [48]. Inhaled glucocorticoids given via a large volume spacer improve asthma symptoms and reduce the number of exacerbations in preschool children and in infants.

### 6.3. Dose–response studies

Surprisingly, the dose–response curve for the clinical efficacy of inhaled glucocorticoids is relatively flat, and while all studies have demonstrated a clinical benefit of inhaled glucocorticoids, it has been difficult to demonstrate differences between doses, with most benefit obtained at the lowest doses used [49,50]. This is in contrast to the steeper dose–response for systemic effects, implying that while there is little clinical benefit from increasing doses of inhaled glucocorticoids, the risk of adverse effects is increased. However, the dose–response effect of inhaled glucocorticoids may depend on the parameters measured, and while it is difficult to discern a dose–response when traditional lung function parameters are measured, there may be a dose–response effect in prevention of asthma exacerbations. Thus, there is a significantly greater effect of budesonide 800 μg daily compared to 200 μg daily in preventing severe and mild asthma exacerbations [51]. Normally, a fourfold or greater difference in dose has been required to detect a statistically significant (but often small) difference in effect on commonly measured outcomes such as symptoms, PEF, use of rescue $\beta_2$-agonist, and lung function, and even such large differences in dose are not always associated with significant differences in response. These findings suggest that pulmonary function tests or symptoms may have a rather low sensitivity in the assessment of the effects of inhaled glucocorticoids. This is obviously important for the interpretation of clinical comparisons between different inhaled glucocorticoids or inhalers. It is also important to consider the type of patient included in clinical studies. Patients with relatively mild asthma may have relatively little room for improvement with inhaled glucocorticoids, so that maximal improvement is obtained with relatively low doses. Patients with more severe asthma or with unstable asthma may have more room for improvement and may therefore show a greater response to increasing doses, but it is often difficult to include such patients in controlled clinical trials.

More studies are needed to assess whether other outcome measures such as AHR or more direct measurements of inflammation, such as sputum eosinophils or exhaled NO, may be more sensitive than traditional outcome measures such as symptoms or lung function tests [52–55]. Higher doses of inhaled glucocorticoids are needed to control AHR than to improve symptoms and lung function, and this may have a better long-term outcome in terms of reduction in structural changes of the airways [56]. Measurement of sputum eosinophils to adjust the dose of inhaled glucocorticoids may reduce the overall dose requirement for inhaled glucocorticoids and exacerbations [57,58]. Monitoring of exhaled nitric oxide also reduces the requirement for glucocorticoids but is not yet practical in clinical practice [59].

### 6.4. Prevention of irreversible airway changes

Some patients with asthma develop an element of irreversible airflow obstruction, but the pathophysiological basis of these changes is not yet understood. It is likely that they are the result of chronic airway inflammation and that they may be prevented by treatment with inhaled glucocorticoids. There is some evidence that the annual decline in lung function may be slowed by the introduction of inhaled glucocorticoids [60], and this is supported by a 5-year study of low-dose budesonide in patients with mild asthma [61,62]. Increasing evidence also suggests that delay in starting inhaled glucocorticoids may result in less overall improvement in lung function in both adults and children [63–65]. These studies suggest that introduction of inhaled glucocorticoids at the time of diagnosis is likely to have the greatest impact [64,65]. So far there is no evidence that early use of inhaled glucocorticoids is curative, and even when inhaled glucocorticoids are introduced at the time of diagnosis, symptoms and lung function revert to pretreatment levels when glucocorticoids are withdrawn [63].

### 6.5. Reduction in mortality

In a retrospective review of the risk of mortality and prescribed anti-asthma medication, there was a significant protection provided by regular inhaled corticosteroid therapy [66]. By contrast, asthma mortality appears to increase with increasing usage of short-acting $\beta_2$-agonists, reflecting the fact that increased rescue therapy is a marker of poor asthma control [67]. The increase in use of rescue therapy should result in an increase in the maintenance dose of inhaled glucocorticoids. The long-acting inhaled $\beta_2$-agonist salmeterol is associated with a small increase in asthma mortality, but the excess deaths appear to be related to underuse of inhaled glucocorticoids [68].

### 6.6. Comparison between inhaled glucocorticoids

Several inhaled glucocorticoids are currently prescribable in asthma although their availability varies between countries. There have been relatively few studies comparing efficacy of the different inhaled glucocorticoids, and it is important to take into account the delivery system and the type of patient under investigation when such comparisons are made. Because of the relatively flat dose–response curve for the clinical parameters normally used in comparing doses of inhaled glucocorticoids, it may be difficult to see differences in efficacy of inhaled glucocorticoids. Most comparisons have concentrated on differences in systemic effects at equally efficacious doses although it has often proven difficult to establish dose-equivalence. There are few studies comparing different doses of inhaled glucocorticoids in asthmatic patients. Budesonide has been compared with BDP, and in adults and children it appears to have comparable anti-asthma effects at equal doses, whereas FP appears to be approximately twice as potent as BDP and budesonide [50]. Studies have consistently shown that FP and budesonide have less systemic effects than BDP, triamcinolone, and flunisolide [25]. The new inhaled glucocorticoids mometasone and ciclesonide are claimed to have less systemic effects [29,69].

## 7. CLINICAL USE OF GLUCOCORTRICOIDS IN ASTHMA

Inhaled glucocorticoids are now recommended as first-line therapy for all patients with persistent symptoms. Inhaled glucocorticoids should be started in any patient who needs to use a $\beta_2$-agonist inhaler for symptom control more than three times weekly. It is conventional to start with a low dose of inhaled corticosteroid and to increase the dose until asthma control is achieved. However, this may take time and a preferable approach is to start with a dose of glucocorticoids in the middle of the dose range (400 μg twice daily) to establish control of asthma more rapidly. Once control is achieved (defined as normal or best possible lung function and infrequent need to use an inhaled $\beta_2$-agonist), the dose of inhaled corticosteroid should be reduced in a step-wise manner to the lowest dose needed for optimal control. It may take as long as 3 months to reach a plateau in response, and any changes in dose should be made at intervals of 3 months or more. When daily doses of $\geq$800 μg daily are needed, a large volume spacer device should be used with a metered dose inhaler (MDI) and mouth washing with a dry powder inhaler in order to reduce local and systemic side effects. Inhaled glucocorticoids are usually given as a twice daily dose in order to increase compliance. When asthma is unstable, four times daily dosage is preferable [70].

The dose of inhaled corticosteroid should be increased to 2000 μg daily if necessary, but higher doses may result in systemic effects. It may be preferable to add a low dose of oral corticosteroid, since higher doses of inhaled glucocorticoids are expensive and have a high incidence of local side effects. Nebulized budesonide has been advocated in order to give an increased dose of inhaled corticosteroid and to reduce the requirement for oral glucocorticoids [71], but this treatment is expensive and may achieve its effects largely via systemic absorption. The dose of inhaled corticosteroid should be the minimal dose that controls asthma and once control is achieved, the dose should be slowly reduced [72].

### 7.1. Add-on therapy

Previously, it was recommended to increase the dose of inhaled glucocorticoids if asthma was not controlled, on the assumption that there was residual inflammation of the airways. However, the dose–response effect of inhaled glucocorticoids is relatively flat, so that there is little improvement in lung function after increasing the dose of inhaled glucocorticoids. An alternative strategy is to add some other class of controller drug, and this is more effective than increasing the dose of inhaled glucocorticoids for most patients [73].

In patients in general practice who are not controlled on BDP 200 μg twice daily, addition of salmeterol 50 μg twice daily was more effective than increasing the dose of inhaled corticosteroid to 500 μg twice daily, in terms of lung function improvement, use of rescue $\beta_2$-agonist, and symptom control [74]. This has been confirmed in several other studies [75]. Similar results have been found with another long-acting inhaled $\beta_2$-agonist formoterol, which in addition reduced the frequency of mild and severe asthma exacerbations in patients with mild, moderate, and severe persistent asthma [51,76]. These studies showing the great efficacy of combined glucocorticoids and LABA compared to increased doses of LABA have led to the development of fixed combinations of glucocorticoids and long-acting $\beta_2$-agonists, such as FP/salmeterol and budesonide/formoterol, which may be more convenient for patients [77,78]. These fixed combination, inhalers also ensure that patients do not discontinue their inhaled glucocorticoids when a long-acting bronchodilator is used. For patients with mild persistent asthma combination inhalers are no more effective than the inhaled glucocorticoids alone in controlled trials [79] but may have an advantage in the real world where adherence to regular inhaled glucocorticoids is very low.

Recently, studies have demonstrated that when formoterol combined with budesonide is used as a reliever therapy, this gives better control of asthma compared to the normally used short-acting $\beta_2$-agonist as a rescue therapy with either the same dose of combination inhaler or a high dose of inhaled glucocorticoids as maintenance treatment [80,81]. This advantage is particularly striking in terms of reducing the number of severe exacerbations. When formoterol was used as the reliever therapy, this reduced exacerbations to a greater extent than the short-acting $\beta_2$-agonist terbutaline but the combination was even more effective [82]. This suggests that the "as required" use of inhaled glucocorticoids contributes to the marked reduction in acute exacerbations. The mechanisms by which glucocorticoids as required improve asthma control and reduce exacerbations are not completely understood, but exacerbations of asthma evolve over several days when patients take increasing amounts of rescue medication [83]. During this time, there is increasing inflammation of the airways, as may be measured by exhaled nitric oxide and sputum eosinophils [44]. Taking the inhaled corticosteroid at the same time as the formoterol to relieve symptoms may suppress this evolving inflammation, particularly since glucocorticoids appear to have a relatively rapid onset of effect in suppressing airway inflammation [84].

Addition of low doses of theophylline (giving plasma concentrations of $<10$ mg/L) are more effective than doubling the dose of inhaled budesonide, either in mild or in severe asthma [85–87]. However, this is less effective than using a long-acting inhaled $\beta_2$-agonist as add-on therapy [88].

Anti-leukotrienes have also been used as an add-on therapy [89,90] although this is less effective than addition of long-acting $\beta_2$-agonists [91,92].

## 8. SIDE EFFECTS

The efficacy of inhaled glucocorticoids is now established in short- and long-term studies in adults and children, but there are still concerns about side effects, particularly in children and when high inhaled doses are used. Several side effects have been recognized (Table 2).

Table 2 Side effects of inhaled glucocorticoids

| |
|---|
| *Local side effects* |
| Dysphonia |
| Oropharyngeal candidiasis |
| Cough |
| *Systemic side effects* |
| Adrenal suppression |
| Growth suppression |
| Bruising |
| Osteoporosis |
| Cataracts |
| Glaucoma |
| Metabolic abnormalities (glucose, insulin, triglycerides) |
| Psychiatric disturbances |

### 8.1. Local side efects

Side effects due to the local deposition of the inhaled corticosteroid in the oropharynx may occur with inhaled glucocorticoids, but the frequency of complaints depends on the dose and frequency of administration and on the delivery system used.

The commonest complaint is of hoarseness of the voice (dysphonia) and may occur in over 50% of patients using MDI. Dysphonia is not appreciably reduced by using spacers but may be less with dry powder devices. Dysphonia may be due to myopathy of laryngeal muscles and is reversible when treatment is withdrawn [93]. For most patients, it is not troublesome but may be disabling in singers and lecturers.

Oropharyngeal candidiasis (thrush) may be a problem in some patients, particularly in the elderly, with concomitant oral glucocorticoids and more than twice daily administration [94]. Large volume spacer devices protect against this local side effect by reducing the dose of inhaled corticosteroid that deposits in the oropharynx.

There is no evidence that inhaled corticosteroid, even in high doses, increases the frequency of infections, including tuberculosis, in the lower respiratory tract. There is no evidence for atrophy of the airway epithelium, and even after 10 years of treatment with inhaled glucocorticoids, there is no evidence for any structural changes in the epithelium. Cough and throat irritation, sometimes accompanied by reflex bronchoconstriction, may occur when inhaled glucocorticoids are given via a MDI. These symptoms are likely to be due to surfactants in pressurized aerosols as they disappear after switching to a dry powder corticosteroid inhaler device.

### 8.2. Systemic side effects

The efficacy of inhaled glucocorticoids in the control of asthma is undisputed, but there are concerns about systemic effects of inhaled glucocorticoids, particularly as they are likely to be used over long periods and in children of all ages [25,95]. The safety of inhaled glucocorticoids has been extensively investigated since their introduction 30 years ago [24]. One of the major problems is to decide whether a measurable systemic effect has any significant clinical consequence, and this necessitates careful long-term follow-up studies. As biochemical markers of systemic corticosteroid effects become more sensitive, then systemic effects may be seen more often, but this does not mean that these effects are clinically relevant. There are several case reports of adverse systemic effects of inhaled glucocorticoids, and these may be idiosyncratic reactions, which may be due to abnormal pharmacokinetic handing of the inhaled corticosteroid. The systemic effect of an inhaled corticosteroid will depend on several factors, including the dose delivered to the patient, the site of delivery (gastrointestinal tract and lung), the delivery system used, and individual differences in the patient's response to the corticosteroid. Recent studies suggest that systemic effects of inhaled corticosteroid are less in patients with more severe asthma, presumably as less drug reaches the lung periphery [96,97].

The systemic effect of an inhaled corticosteroid is dependent on the amount of drug absorbed into the systemic circulation. Approximately 90% of the inhaled dose from an MDI deposits in the oropharynx and is swallowed and subsequently absorbed from the gastrointestinal tract. Use of a large volume spacer device markedly reduces the oropharyngeal deposition, and therefore the systemic effects of inhaled glucocorticoids, although this is less important when oral bioavailability is minimal, as with FP. For dry powder inhalers, similar reductions in systemic effects may be achieved with mouth washing and discarding the fluid. All patients using a daily dose of $\geq$800 μg of an inhaled corticosteroid should therefore use either a spacer or mouth washing to reduce systemic absorption. Approximately, 10% of an MDI enters the lung and this fraction (which presumably exerts the therapeutic effect) may be absorbed into the systemic circulation. As the fraction of inhaled corticosteroid deposited in the oropharynx is reduced, the proportion of the inhaled dose entering the lungs is increased. More efficient delivery to the lungs is therefore accompanied by increased systemic absorption, but this is offset by a reduction in the dose needed for optimal control of airway inflammation. For example, a multiple dry powder delivery system, the Turbuhaler, delivers approximately twice as much corticosteroid to the lungs as other devices and therefore has increased systemic effects. However, this is compensated for by the fact that only half the dose is required.

### *8.2.1. Adrenal suppression*

Glucocorticoids may cause hypothalamic–pituitary–adrenal (HPA) axis suppression by reducing corticotrophin (ACTH) production, which reduces cortisol secretion by the adrenal gland. The degree of HPA suppression is dependent on dose, duration, frequency, and timing of corticosteroid administration. Measurement of HPA axis function provides evidence for systemic effects of an inhaled corticosteroid. Basal adrenal cortisol secretion may be measured by a morning plasma cortisol, by 24-h urinary cortisol, or by plasma cortisol profile over 24 h. Other tests measure the HPA response following stimulation with tetracosactrin (which measures adrenal reserve) or stimulation with metyrapone and insulin (which measure the response to stress). There are many studies of HPA axis function in asthmatic patients with inhaled glucocorticoids, but the results are inconsistent as they have often been uncontrolled and patients have also been taking courses of oral glucocorticoids (which may affect the HPA axis for weeks) [24]. BDP, budesonide, and FP at high doses by conventional MDI ($>$1600 μg daily) give a dose-related decrease in morning serum cortisol levels and 24 h urinary cortisol although values still lie well within the normal range. However, when a large volume spacer is used, doses of 2000 μg daily of BDP or budesonide have little effect on 24-h urinary cortisol excretion. Stimulation tests of HPA axis function similarly show no consistent effects of doses of 1500 μg or less of inhaled corticosteroid. At high doses (>1500 μg daily), budesonide and FP have less effect than BDP on HPA axis function. In children, no suppression of urinary cortisol is seen with doses of BDP of 800 μg or less. In studies where plasma cortisol has been measured at frequent intervals there was a significant reduction in cortisol peaks with doses of inhaled BDP as low as 400 μg daily although this does not appear to be dose-related in the range 400–1000 μg. The clinical significance of these effects is not certain, however.

### *8.2.2. Bone metabolism*

Glucocorticoids lead to a reduction in bone mass by direct effects on bone formation and resorption and indirectly by suppression of the pituitary–gonadal and HPA axes, effects on intestinal calcium absorption, renal tubular calcium reabsorption, and secondary hyperparathyroidism [98]. The effects of oral glucocorticoids on osteoporosis and increased risk of vertebral and rib fractures are well known, but there are no reports suggesting that long-term treatment with inhaled glucocorticoids is associated with an increased risk of fractures. Bone densitometry has been used to assess the effect of inhaled glucocorticoids on bone mass. Although there is evidence that bone density is less in patients taking high-dose inhaled glucocorticoids, interpretation is confounded by the fact that these patients are also taking intermittent courses of oral glucocorticoids. Changes in bone mass occur very slowly, and several biochemical indices have been used to assess the short-term effects of inhaled glucocorticoids on bone metabolism. Bone formation has been measured by plasma concentrations of bone-specific alkaline phosphatase, serum osteocalcin, or procollagen peptides. Bone resorption may be assessed by urinary hydroxyproline after a 12-h fast, urinary calcium excretion, and pyridinium cross-link excretion. Inhaled glucocorticoids, even at doses up to 2000 μg daily, have no significant effect on calcium excretion, but acute and reversible dose-related suppression of serum osteocalcin has been reported with BDP and budesonide when given by conventional MDI in several studies. Budesonide consistently has less effect than BDP at equivalent doses and only BDP increases urinary hydroxyproline at high doses. With a large volume spacer, even doses of 2000 μg daily of either BDP or budesonide are without effect on plasma osteocalcin concentrations, however. Urinary pyridinium and deoxypyridinoline cross-links, which are a more accurate and stable measurement of bone and collagen degradation, are not increased with inhaled glucocorticoids (BDP >1000 μg daily), even with intermittent courses of oral glucocorticoids. It is important to monitor changes in markers of bone formation as well as bone degradation, as the net effect on bone turnover is important. There is no evidence that inhaled glucocorticoids increase the frequency of fractures. Long-term treatment with high-dose inhaled glucocorticoids has not been associated with any consistent change in bone density. Indeed, in elderly patients, there may be an increase in bone density due to increased mobility.

### *8.2.3. Connective tissue effects*

Oral and topical glucocorticoids cause thinning of the skin, telangiectasia, and easy bruising, probably as a result of loss of extracellular ground substance within the dermis, due to an inhibitory effect on dermal fibroblasts. There are reports of increased skin bruising and purpura in patients using high doses of inhaled BDP, but the amount of intermittent oral glucocorticoids in these patients is not known. Easy bruising in association with inhaled glucocorticoids is more frequent in elderly patients [99], and there are no reports of this problem in children. Long-term prospective studies with objective measurements of skin thickness are needed with different inhaled glucocorticoids.

### 8.2.4. Cataracts

Long-term treatment with oral glucocorticoids increases the risk of posterior subcapsular cataracts, and there are several case reports describing cataracts in individual patients taking inhaled glucocorticoids [24]. In a recent cross-sectional study in patients aged 5–25 years taking either inhaled BDP or budesonide, no cataracts were found on slit-lamp examination, even in patients taking 2000 μg daily for over 10 years [100]. A slight increase in the risk of glaucoma in patients taking very high does of inhaled glucocorticoids has also been identified [101].

### 8.2.5. Growth

There has been particular concern that inhaled glucocorticoids may cause stunting of growth and several studies have addressed this issue. Asthma itself (as with other chronic diseases) may have an effect on the growth pattern and has been associated with delayed onset of puberty and deceleration of growth velocity that is more pronounced with more severe disease [102]. However, asthmatic children appear to grow for longer, so that their final height is normal. The effect of asthma on growth makes it difficult to assess the effects of inhaled glucocorticoids on growth in cross-sectional studies, particularly as courses of oral glucocorticoids is a confounding factor. Longitudinal studies have demonstrated that there is no significant effect of inhaled glucocorticoids on statural growth in doses of up to 800 μg daily and for up to 5 years of treatment [24]. A meta-analysis of 21 studies, including over 800 children, showed no effect of inhaled BDP on statural height, even with higher doses and long duration of therapy [103], and in a large study of asthmatics treated with inhaled glucocorticoids during childhood, there was no difference in statural height compared to normal children [104]. Another long-term follow-up study showed no effect of glucocorticoids on final height in children treated over several years [105]. Short-term growth measurements (knemometry) have demonstrated that even a low dose of an oral corticosteroid (prednisolone 2.5 mg) is sufficient to give complete suppression of lower leg growth. However, inhaled budesonide up to 400 μg is without effect although some suppression is seen with 800 and with 400 μg of BDP. The relationship between knemometry measurements and final height is uncertain since low doses of oral corticosteroid that have no effect on final height cause profound suppression.

### 8.2.6. Metabolic effects

Several metabolic effects have been reported after inhaled glucocorticoids, but there is no evidence that these are clinically relevant at therapeutic doses. In adults, fasting glucose and insulin are unchanged after doses of BDP up to 2000 μg daily and in children with inhaled budesonide up to 800 μg daily. In normal individuals, high-dose inhaled BDP may slightly increase resistance to insulin. However, in patients with poorly controlled asthma, high doses of BDP and budesonide paradoxically decrease insulin resistance and improve glucose tolerance, suggesting that the disease itself may lead to abnormalities in carbohydrate metabolism. Neither BDP 2000 μg daily in adults nor budesonide 800 μg daily in children have any effect on plasma cholesterol or triglycerides.

### 8.2.7. Psychiatric effects

There are various reports of psychiatric disturbance, including emotional lability, euphoria, depression, aggressiveness, and insomnia, after inhaled glucocorticoids. Only eight such patients have so far been reported, suggesting that this is very infrequent, and a causal link with inhaled glucocorticoids has usually not been established.

### *8.2.8. Pregnancy*

Based on extensive clinical experience, inhaled glucocorticoids appear to be safe in pregnancy although no controlled studies have been performed. There is no evidence for any adverse effects of inhaled glucocorticoids on the pregnancy, the delivery, or on the fetus [106]. It is important to recognize that poorly controlled asthma may increase the incidence of perinatal mortality and retard intra-uterine growth, so that more effective control of asthma with inhaled glucocorticoids may reduce these problems.

## 9. SYSTEMIC GLUCOCORTICOIDS

Oral or intravenous glucocorticoids may be indicated in several situations. Prednisone is converted in the liver to the active prednisolone. In pregnant patients, prednisone is preferable as it is not converted to prednisolone in the fetal liver, thus diminishing the exposure of the fetus to glucocorticoids. Enteric-coated preparations of prednisolone are used to reduce side effects (particularly gastric side effects) and give delayed and reduced peak plasma concentrations although the bioavailability and therapeutic efficacy of these preparations is similar to uncoated tablets. Prednisolone and prednisone are preferable to dexamethasone, betamethasone, or triamcinolone, which have longer plasma half-lives and therefore an increased frequency of adverse effects.

### 9.1. Short courses

Short courses of oral glucocorticoids (30–40 mg prednisolone daily for 1–2 weeks or until the peak flow values return to best attainable) are indicated for exacerbations of asthma, and the dose may be tailed off over 1 week once the exacerbation is resolved. The tail-off period is not strictly necessary, but some patients find it reassuring.

### 9.2. Maintenance glucocorticoids

Maintenance oral glucocorticoids are only needed in a small proportion of asthmatic patients (approximately 1%) with the most severe asthma that cannot be controlled with maximal doses of inhaled glucocorticoids (2000 μg daily) and additional bronchodilators. The minimal dose of oral corticosteroid needed for control should be used, and reductions in the dose should be made slowly in patients who have been on oral glucocorticoids for long periods (e.g., for prednisolone reduction by 2.5 mg per month for doses down to 10 mg daily and thereafter by 1 mg per month). Oral glucocorticoids are usually given as a single morning dose, as this reduces the risk of adverse effects as it coincides with the peak diurnal concentrations. There is some evidence that administration in the afternoon may be optimal for some patients who have severe nocturnal asthma [107]. Alternate day administration may also reduce adverse effects, but control of asthma may not be as good on the day when the oral dose is omitted in some patients.

Intramuscular triamcinolone acetonide (80 mg monthly) has been advocated in patients with severe asthma as an alternative to oral glucocorticoids [108,109]. This may be considered in patients in whom compliance is a particular problem, but the major concern is the high frequency of proximal myopathy associated with this fluorinated corticosteroid. Some patients who do not respond well to prednisolone are reported to respond to oral betamethasone, presumably because of pharmacokinetic handling problems with prednisolone.

### 9.3. Acute severe asthma

Intravenous hydrocortisone is given in acute severe asthma, with a recommended dose of 200 mg intravenously. While the value of glucocorticoids in acute severe asthma has been questioned, others have found that they speed the resolution of attacks [110]. There is no apparent advantage in giving very high doses of intravenous glucocorticoids (such as methylprednisolone 1 g). Indeed, intravenous glucocorticoids have occasionally been associated with an acute severe myopathy [111]. No difference in recovery from acute severe asthma was seen whether i.v. hydrocortisone in doses of 50, 200, or 500 mg 6 hourly was used [112] and another placebo-controlled study showed no beneficial effect of i.v. glucocorticoids [113]. Intravenous glucocorticoids are indicated in acute asthma if lung function is <30% predicted and in whom there is no significant improvement with nebulized $\beta_2$-agonist. Intravenous therapy is usually given until a satisfactory response is obtained and then oral prednisolone may be substituted. Oral prednisolone (40–60 mg) has a similar effect to intravenous hydrocortisone and is easier to administer [30,110]. Oral prednisolone is the preferred treatment for acute severe asthma, providing there are no contraindications to oral therapy [114]. There is some evidence that high doses of nebulized glucocorticoids may also be effective in acute exacerbations of asthma, with a more rapid onset of action [115].

## REFERENCES


1. Barnes PJ. How corticosteroids control inflammation. Br J Pharmacol 2006;148:245–54.
2. Rhen T, Cidlowski JA. Antiinflammatory action of glucocorticoids – new mechanisms for old drugs. New Engl J Med 2005;353:1711–23.
3. Gibson PG, Saltos N, Fakes K. Acute anti-inflammatory effects of inhaled budesonide in asthma: a randomized controlled trial. Am J Respir Crit Care Med 2001;163:32–6.
4. Ketchell RI, Jensen MW, Lumley P, Wright AM, Allenby MI, O'Connor BJ. Rapid effect of inhaled fluticasone propionate on airway responsiveness to adenosine 5'-monophosphate in mild asthma. J Allergy Clin Immunol 2002;110:603–6.
5. Juniper EF, Kline PA, Yan Zieleshem MA, Ramsdale EH, O'Byrne PM, Hargreave FE. Long-term effects of budesonide on airway responsiveness and clinical asthma severity in inhaled steroid-dependent asthmatics. Eur Respir J 1990;3:1122–7.
6. Lewis-Tuffin LJ, Cidlowski JA. The physiology of human glucocorticoid receptor beta (hGRbeta) and glucocorticoid resistance. Ann N Y Acad Sci 2006;1069(1):1–9.
7. Barnes PJ. Corticosteroid effects on cell signalling. Eur Respir J 2006;27:413–26.
8. Clark AR. MAP kinase phosphatase 1: a novel mediator of biological effects of glucocorticoids. J Endocrinol 2003;178:5–12.
9. Dostert A, Heinzel T. Negative glucocorticoid receptor response elements and their role in glucocorticoid action. Curr Pharm Des 2004;10:2807–16.
10. Barnes PJ, Adcock IM. How do corticosteroids work in asthma? Ann Intern Med 2003;139: 359–70.

11. Barnes PJ, Adcock IM, Ito K. Histone acetylation and deacetylation: importance in inflammatory lung diseases. Eur Respir J 2005;25:552–63.
12. Hart L, Lim S, Adcock I, Barnes PJ, Chung KF. Effects of inhaled corticosteroid therapy on expression and DNA-binding activity of nuclear factor-kB in asthma. Am J Respir Crit Care Med 2000;161:224–31.
13. Ito K, Barnes PJ, Adcock IM. Glucocorticoid receptor recruitment of histone deacetylase 2 inhibits IL-1b-induced histone H4 acetylation on lysines 8 and 12. Mol Cell Biol 2000;20:6891–903.
14. Bergmann MW, Staples KJ, Smith SJ, Barnes PJ, Newton R. Glucocorticoid inhibition of GM-CSF from T cells is independent of control by NF-kB and CLE0. Am J Respir Cell Mol Biol 2004;30: 555–63.
15. Brook M, Tchen CR, Santalucia T, McIlrath J, Arthur JS, Saklatvala J, Clark AR. Posttranslational regulation of tristetraprolin subcellular localization and protein stability by p38 mitogen-activated protein kinase and extracellular signal-regulated kinase pathways. Mol Cell Biol 2006;26:2408–18.
16. Barnes PJ. Scientific rationale for combination inhalers with a long-acting b2-agonists and corticosteroids. Eur Respir J 2002;19:182–91.
17. Mak JCW, Nishikawa M, Barnes PJ. Glucocorticosteroids increase b2-adrenergic receptor transcription in human lung. Am J Physiol 1995;12:L41–6.
18. Baraniuk JN, Ali M, Brody D, Maniscalco J, Gaumond E, Fitzgerald T, Wonk G, Mak JCW, Bascom R, Barnes PJ,Troost T. Glucocorticoids induce b2-adrenergic receptor function in human nasal mucosa. Am J Respir Crit Care Med 1997;155:704–10.
19. Mak JCW, Nishikawa M, Shirasaki H, Miyayasu K, Barnes PJ. Protective effects of a glucocorticoid on down-regulation of pulmonary b2-adrenergic receptors *in vivo*. J Clin Invest 1995;96:99–106.
20. Mak JC, Chuang TT, Harris CA, Barnes PJ. Increased expression of G protein-coupled receptor kinases in cystic fibrosis lung. Eur J Pharmacol 2002;436:165–72.
21. Roth M, Johnson PR, Rudiger JJ, King GG, Ge Q, Burgess JK, Anderson G, Tamm M, Black JL. Interaction between glucocorticoids and b2 agonists on bronchial airway smooth muscle cells through synchronised cellular signalling. Lancet 2002;360:1293–9.
22. Usmani OS, Ito K, Maneechotesuwan K, Ito M, Johnson M, Barnes PJ, Adcock IM. Glucocorticoid receptor nuclear translocation in airway cells following inhaled combination therapy. Am J Respir Crit Care Med 2005;172:704–12.
23. Gambertoglio JG, Amend WJC, Benet LZ. Pharmacokinetics and bioavailability of prednisone and prednisolone in healthy volunteers and patients: a review. J Pharmacokin Biopharm 1980;8:1–52.
24. Barnes PJ, Pedersen S, Busse WW. Efficacy and safety of inhaled corticosteroids: an update. Am J Respir Crit Care Med 1998;157:S1–53.
25. Lipworth BJ. Systemic adverse effects of inhaled corticosteroid therapy: a systematic review and meta-analysis [see comments]. Arch Intern Med 1999;159:941–55.
26. Derendorf H, Hochhaus G, Meibohm B, Mollmann H, Barth J. Pharmacokinetics and pharmacodynamics of inhaled corticosteroids. J Allergy Clin Immunol 1998;101:S440–6.
27. Brown PH, Greening AP, Crompton GK. Large volume spacer devices and the influence of high dose beclomethasone diproprionate on hypothalamo-pituitary-adrenal axis function. Thorax 1993;48: 233–8.
28. Derendorf H. Corticosteroid pharmacokinetic/pharmacodynamic parameters and their relationship to safety and efficacy. Allergy Asthma Proc 2005;26:327–35.
29. Reynolds NA, Scott LJ. Ciclesonide. Drugs 2004;64:511–19.
30. Harrison BDN, Stokes TC, Hart GJ, Vaughan DA, Ali NJ, Robinson AA. Need for intravenous hydrocortisone in addition to oral prednisolone in patients admitted to hospital with severe asthma without ventilatory failure. Lancet 1986;i:181–4.
31. Storr J, Barrell E, Barry W, Lenney W, Hatcher G. Effect of a single oral dose of prednisolone in acute childhood asthma. Lancet 1987;1:879–82.
32. Levy ML, Stevenson C, Maslen T. Comparison of short courses of oral prednisolone and fluticasone propionate in the treatment of adults with acute exacerbations of asthma in primary care. Thorax 1996;51:1087–92.

33. Manjra AI, Price J, Lenney W, Hughes S, Barnacle H. Efficacy of nebulized fluticasone propionate compared with oral prednisolone in children with an acute exacerbation of asthma. Respir Med 2000;94:1206–14.
34. Harrison TW, Oborne J, Newton S, Tattersfield AE. Doubling the dose of inhaled corticosteroid to prevent asthma exacerbations: randomised controlled trial. Lancet 2004;363:271–5.
35. FitzGerald JM, Becker A, Sears MR, Mink S, Chung K, Lee J. Doubling the dose of budesonide versus maintenance treatment in asthma exacerbations. Thorax 2004;59:550–6.
36. Foresi A, Morelli MC, Catena E. Low-dose budesonide with the addition of an increased dose during exacerbations is effective in long-term asthma control. On behalf of the Italian Study Group. Chest 2000;117:440–6.
37. Edmonds ML, Camargo CA Jr, Pollack CV, Rowe BH. Early use of inhaled corticosteroids in the emergency department treatment of acute asthma. Cochrane Database Syst Rev 2003;CD002308.
38. Global Initiative for Asthma Global strategy for asthma management and prevention. NHLBI/WHO Workshop Report. www.ginasthma.com, 2006
39. Haahtela T, Jarvinen M, Kava T, Kiviranta K, Koskinen S, Lehtonen K, Nivander K, Persson T, Reinikainen R, Selroos O, Sovijarvi A, Stenius-Aarniala B, Svahn T, Tammivaara R, Laitinen LA. Comparison of a b2-agonist terbutaline with an inhaled steroid in newly detected asthma. New Engl J Med 1991;325:388–92.
40. Juniper EF, Kline PA, Vanzieleghem MA, Ramsdale EH, O'Byrne PM, Hargreave FE. Effect of long-term treatment with an inhaled corticosteroid (budesonide) on airway hyperresponsiveness and clinical asthma in nonsteroid-dependent asthmatics. Am Rev Respir Dis 1990;142:832–6.
41. Pauwels RA, Pedersen S, Busse WW, Tan WC, Chen YZ, Ohlsson SV, Ullman A, Lamm CJ, O'Byrne PM. Early intervention with budesonide in mild persistent asthma: a randomised, double-blind trial. Lancet 2003;361:1071–6.
42. Mash B, Bheekie A, Jones PW. Inhaled vs oral steroids for adults with chronic asthma. Cochrane Database Syst Rev 2000;2:CD002160.
43. Vathenen AS, Knox AJ, Wisniewski A, Tattersfield AE. Time course of change in bronchial reactivity with an inhaled corticosteroid in asthma. Am Rev Respir Dis 1991;143:1317–21.
44. Jatakanon A, Lim S, Barnes PJ. Changes in sputum eosinophils predict loss of asthma control. Am J Respir Crit Care Med 2000;161:64–72.
45. Leuppi JD, Salome CM, Jenkins CR, Anderson SD, Xuan W, Marks GB, Koskela H, Brannan JD, Freed R, Andersson M, Chan HK, Woolcock AJ. Predictive markers of asthma exacerbation during stepwise dose reduction of inhaled corticosteroids. Am J Respir Crit Care Med 2001;163:406–12.
46. van Essen-Zandvliet EE, Hughes MD, Waalkens HJ, Duiverman EJ, Pocock SJ, Kerrebijn KF. Effects of 22 months of treatment with inhaled corticosteroids and/or b2-agonists on lung function, airway responsiveness and symptoms in children with asthma. Am Rev Respir Dis 1992;146:547–54.
47. Waalkens HJ, van Essen-Zandvliet EE, Hughes MD, Gerritsen J, Duiverman EJ, Knol K, Kerrebijn KF. Cessation of long-term treatment with inhaled corticosteroids (budesonide) in children with asthma results in deterioration. Am Rev Respir Dis 1993;148:1252–7.
48. Berger WE. Budesonide inhalation suspension for the treatment of asthma in infants and children. Drugs 2005;65:1973–89.
49. Busse WW, Chervinsky P, Condemi J, Lumry WR, Petty TL, Rennard S, Townley RG. Budesonide delivered by Turbuhaler is effective in a dose-dependent fashion when used in the treatment of adult patients with chronic asthma. J Allergy Clin Immunol 1998;101:457–63.
50. Adams NP, Jones PW. The dose-response characteristics of inhaled corticosteroids when used to treat asthma: an overview of Cochrane systematic reviews. Respir Med 2006;100:1297–306.
51. Pauwels RA, Lofdahl C-G, Postma DS, Tattersfield AE, O'Byrne PM, Barnes PJ, Ullman A. Effect of inhaled formoterol and budesonide on exacerbations of asthma. New Engl J Med 1997;337: 1412–18.
52. Lim S, Jatakanon A, John M, Gilbey T, O'Connor BJ, Chung KF, Barnes PJ. Effect of inhaled budesonide on lung function and airway inflammation. Assessment by various inflammatory markers in mild asthma. Am J Respir Crit Care Med 1999;159:22–30.

53. Jatakanon A, Lim S, Chung KF, Barnes PJ. An inhaled steroid improves markers of inflammation in asymptomatic steroid-naive asthmatic patients. Eur Respir J 1998;12:1084–8.
54. Jatakanon A, Kharitonov S, Lim S, Barnes PJ. Effect of differing doses of inhaled budesonide on markers of airway inflammation in patients with mild asthma. Thorax 1999;54:108–14.
55. Kharitonov SA, Donnelly LE, Montuschi P, Corradi M, Collins JV, Barnes PJ. Dose-dependent onset and cessation of action of inhaled budesonide on exhaled nitric oxide and symptoms in mild asthma. Thorax 2002;57:889–96.
56. Sont JK, Willems LN, Bel EH, van Krieken JH, Vandenbroucke JP, Sterk PJ. Clinical control and histopathologic outcome of asthma when using airway hyperresponsiveness as an additional guide to long- term treatment. The AMPUL Study Group. Am J Respir Crit Care Med 1999;159:1043–51.
57. Green RH, Brightling CE, McKenna S, Hargadon B, Parker D, Bradding P, Wardlaw AJ, Pavord ID. Asthma exacerbations and sputum eosinophil counts: a randomised controlled trial. Lancet 2002;360:1715–21.
58. Jayaram L, Pizzichini MM, Cook RJ, Boulet LP, Lemiere C, Pizzichini E, Cartier A, Hussack P, Goldsmith CH, Laviolette M, Parameswaran K, Hargreave FE. Determining asthma treatment by monitoring sputum cell counts: effect on exacerbations. Eur Respir J 2006;27:483–94.
59. Smith AD, Cowan JO, Brassett KP, Herbison GP, Taylor DR. Use of exhaled nitric oxide measurements to guide treatment in chronic asthma. New Engl J Med 2005;352:2163–73.
60. Dompeling E, Van Schayck CP, Molema J, Folgering H, van Grusven PM, van Weel C. Inhaled beclomethasone improves the course of asthma and COPD. Eur Resp J 1992;5:945–52.
61. O'Byrne PM, Pedersen S, Busse WW, Tan WC, Chen YZ, Ohlsson SV, Ullman A, Lamm CJ, Pauwels RA. Effects of early intervention with inhaled budesonide on lung function in newly diagnosed asthma. Chest 2006;129:1478–85.
62. Lange P, Scharling H, Ulrik CS, Vestbo J. Inhaled corticosteroids and decline of lung function in community residents with asthma. Thorax 2006;61:100–4.
63. Haahtela T, Järvinsen M, Kava T, Kiviranta K, Koskinen S, Lemtonen K, Nikander K, Person T, Selroos O, Sovijäri A, Stenius-Aarniala B, Svahn T, Tammivaara R, Laitinen LA. Effects of reducing or discontinuing inhaled budesonide in patients with mild asthma. New Engl J Med 1994;331:700–5.
64. Agertoft L, Pedersen S. Effects of long-term treatment with an inhaled corticosteroid on growth and pulmonary function in asthmatic children. Resp Med 1994;5:369–72.
65. Selroos O, Pietinalcho A, Lofroos A-B, Riska A. Effect of early and late intervention with inhaled corticosteroids in asthma. Chest 1995;108:1228–34.
66. Suissa S, Ernst P, Benayoun S, Baltzan M, Cai B. Low-dose inhaled corticosteroids and the prevention of death from asthma. N Engl J Med 2000;343:332–6.
67. Ernst P, Habbick B, Suissa S, Hemmelgarn B, Cockcroft D, Buist AS, Horwitz RI, McNutt M, Spitzer WO. Is the association between inhaled beta-agonist use and life-threatening asthma because of confounding by severity. Am Rev Respir Dis 1993;148:75–9.
68. Nelson HS, Weiss ST, Bleecker ER, Yancey SW, Dorinsky PM. The Salmeterol Multicenter Asthma Research Trial: a comparison of usual pharmacotherapy for asthma or usual pharmacotherapy plus salmeterol. Chest 2006;129:15–26.
69. Nathan RA, Nayak AS, Graft DF, Lawrence M, Picone FJ, Ahmed T, Wolfe J, Vanderwalker ML, Nolop KB, Harrison JE. Mometasone furoate: efficacy and safety in moderate asthma compared with beclomethasone dipropionate. Ann Allergy Asthma Immunol 2001;86:203–10.
70. Malo J-L, Cartier A, Merland N, Ghezzo H, Burke A, Morris J, Jennings BH. Four-times-a-day dosing frequency is better than twice-a-day regimen in subjects requiring a high-dose inhaled steroid, budesonide, to control moderate to severe asthma. Am Rev Respir Dis 1989;140:624–8.
71. Otulana BA, Varma N, Bullock A, Higenbottam T. High dose nebulized steroid in the treatment of chronic steroid-dependent asthma. Resp Med 1992;86:105–8.
72. Hawkins G, McMahon AD, Twaddle S, Wood SF, Ford I, Thomson NC. Stepping down inhaled corticosteroids in asthma: randomised controlled trial. Br Med J 2003;326:1115.
73. Kankaanranta H, Lahdensuo A, Moilanen E, Barnes PJ. Add-on therapy options in asthma not adequately controlled by inhaled corticosteroids: a comprehensive review. Respir Res 2004;5:17.

74. Greening AP, Ind PW, Northfield M, Shaw G. Added salmeterol versus higher-dose corticosteroid in asthma patients with symptoms on existing inhaled corticosteroid. Lancet 1994;344:219–24.
75. Shrewsbury S, Pyke S, Britton M. Meta-analysis of increased dose of inhaled steroid or addition of salmeterol in symptomatic asthma (MIASMA). Br Med J 2000;320:1368–73.
76. O'Byrne PM, Barnes PJ, Rodriguez-Roisin R, Runnerstrom E, Sandstrom T, Svensson K, Tattersfield A. Low dose inhaled budesonide and formoterol in mild persistent asthma: the OPTIMA randomized trial. Am J Respir Crit Care Med 2001;164:1392–7.
77. Chapman KR, Ringdal N, Backer V, Palmqvist M, Saarelainen S, Briggs M. Salmeterol and fluticasone propionate (50/250 mg) administered via combination diskus inhaler: as effective as when given via separate diskus inhalers [In Process Citation]. Can Respir J 1999;6:45–51.
78. Shapiro G, Lumry W, Wolfe J, Given J, White MV, Woodring A, Baitinger L, House K, Prillaman B, Shah T. Combined salmeterol 50 mg and fluticasone propionate 250 mg in the Diskus device for the treatment of asthma. Am J Respir Crit Care Med 2000;161:527–34.
79. Ni CM, Greenstone IR, Ducharme FM. Addition of inhaled long-acting beta2-agonists to inhaled steroids as first line therapy for persistent asthma in steroid-naive adults. Cochrane Database Syst Rev 2005;CD005307.
80. O'Byrne PM, Bisgaard H, Godard PP, Pistolesi M, Palmqvist M, Zhu Y, Ekstrom T, Bateman ED. Budesonide/formoterol combination therapy as both maintenance and reliever medication in asthma. Am J Resp Crit Care Med 2005;171:129–36.
81. Rabe KF, Pizzichini E, Stallberg B, Romero S, Balanzat AM, Atienza T, Lier PA, Jorup C. Budesonide/formoterol in a single inhaler for maintenance and relief in mild-to-moderate asthma: a randomized, double-blind trial. Chest 2006;129:246–56.
82. Rabe KF, Atienza T, Magyar P, Larsson P, Jorup C, Lalloo UG. Effect of budesonide in combination with formoterol for reliever therapy in asthma exacerbations: a randomised controlled, double-blind study. Lancet 2006;368:744–53.
83. Tattersfield AE, Postma DS, Barnes PJ, Svensson K, Bauer CA, O'Byrne PM, Lofdahl CG, Pauwels RA, Ullman A. Exacerbations of asthma. A descriptive study of 425 severe exacerbations. Am J Respir Crit Care Med 1999;160:594–9.
84. Barnes PJ. New therapies for asthma. Trends Mol Med 2006;12:515–20.
85. Evans DJ, Taylor DA, Zetterstrom O, Chung KF, O'Connor BJ, Barnes PJ. A comparison of low-dose inhaled budesonide plus theophylline and high-dose inhaled budesonide for moderate asthma. New Engl J Med 1997;337:1412–18.
86. Ukena D, Harnest U, Sakalauskas R, Magyar P, Vetter N, Steffen H, Leichtl S, Rathgeb F, Keller A, Steinijans VW. Comparison of addition of theophylline to inhaled steroid with doubling of the dose of inhaled steroid in asthma. Eur Respir J 1997;10:2754–60.
87. Lim S, Jatakanon A, Gordon D, Macdonald C, Chung KF, Barnes PJ. Comparison of high dose inhaled steroids, low dose inhaled steroids plus low dose theophylline, and low dose inhaled steroids alone in chronic asthma in general practice. Thorax 2000;55:837–41.
88. Wilson AJ, Gibson PG, Coughlan J. Long acting beta-agonists versus theophylline for maintenance treatment of asthma. Cochrane Database Syst Rev 2000;CD001281.
89. Laviolette M, Malmstrom K, Lu S, Chervinsky P, Pujet JC, Peszek I, Zhang J, Reiss TF. Montelukast added to inhaled beclomethasone in treatment of asthma. Am J Respir Crit Care Med 1999;160:1862–8.
90. Price DB, Hernandez D, Magyar P, Fiterman J, Beeh KM, James IG, Konstantopoulos S, Rojas R, van Noord JA, Pons M, Gilles L, Leff JA. Randomised controlled trial of montelukast plus inhaled budesonide versus double dose inhaled budesonide in adult patients with asthma. Thorax 2003;58:211–16.
91. Nelson HS, Busse WW, Kerwin E, Church N, Emmett A, Rickard K, Knobil K. Fluticasone propionate/salmeterol combination provides more effective asthma control than low-dose inhaled corticosteroid plus montelukast. J Allergy Clin Immunol 2000;106:1088–95.
92. Ducharme F, Schwartz Z, Hicks G, Kakuma R. Addition of anti-leukotriene agents to inhaled corticosteroids for chronic asthma. Cochrane Database Syst Rev 2004;CD003133.

93. Williamson IJ, Matusiewicz SP, Brown PH, Greening AP, Crompton GK. Frequency of voice problems and cough in patients using pressurised aersosol inhaled steroid preparations. Eur Resp J 1995;8:590–2.
94. Toogood JA, Jennings B, Greenway RW, Chung L. Candidiasis and dysphonia complicating beclomethasone treatment of asthma. J Allergy Clin Immunol 1980;65:145–53.
95. Kamada AK, Szefler SJ, Martin RJ, Boushey HA, Chinchilli VM, Drazen JM, Fish JE, Israel E, Lazarus SC, Lemanske RF. Issues in the use of inhaled steroids. Am J Respir Crit Care Med 1996;153:1739–48.
96. Brutsche MH, Brutsche IC, Munawar M, Langley SJ, Masterson CM, Daley-Yates PT, Brown R, Custovic A, Woodcock A. Comparison of pharmacokinetics and systemic effects of inhaled fluticasone propionate in patients with asthma and healthy volunteers: a randomised crossover study. Lancet 2000;356:556–61.
97. Harrison TW, Wisniewski A, Honour J, Tattersfield AE. Comparison of the systemic effects of fluticasone propionate and budesonide given by dry powder inhaler in healthy and asthmatic subjects. Thorax 2001;56:186–91.
98. Efthimou J, Barnes PJ. Effect of inhaled corticosteroids on bone and growth. Eur Respir J 1998; 11:1167–77.
99. Roy A, Leblanc C, Paquette L, Ghezzo H, Cote J, Cartier A, Malo J-L. Skin bruising in asthmatic subjects treated with high does of inhaled steroids: frequency and association with adrenal function. Eur Respir J 1996;9:226–31.
100. Simons FER, Persaud MP, Gillespie CA, Cheang M, Shuckett EP. Absence of posterior subcapsular cataracts in young patients treated with inhaled glucocorticoids. Lancet 1993;342:736–8.
101. Garbe E, LeLorier J, Boivin J-F, Suissa S. Inhaled and nasal glucocorticoids and the risks of ocular hypertension or open-angel glaucoma. JAMA 1997;227:722–7.
102. Pedersen S. Do inhaled corticosteroids inhibit growth in children. Am J Respir Crit Care Med 2001;164:521–35.
103. Allen DB, Mullen M, Mullen B. A meta-analysis of the effects of oral and inhaled corticosteroids on growth. J Allergy Clin Immunol 1994;93:967–76.
104. Silverstein MD, Yunginger JW, Reed CE, Petterson T, Zimmerman D, Li JT, O'Fallon WM. Attained adult height after childhood asthma: effect of glucocorticoid therapy. J Allergy Clin Immunol 1997;99:466–74.
105. Agertoft L, Pedersen S. Effect of long-term treatment with inhaled budesonide on adult height in children with asthma. New Engl J Med 2000;343:1064–9.
106. Schatz M. Asthma and pregnancy. Lancet 1999;353:1202–4.
107. Beam WR, Ballard RD, Martin RJ. Spectrum of corticosteroid sensitivity in nocturnal asthma. Am Rev Respir Dis 1992;145:1082–6.
108. McLeod DT, Capewell SJ, Law J, MacLaren W, Seaton A. Intramuscular triamcinolone acetamide in chronic severe asthma. Thorax 1985;40:840–5.
109. Ogirala RG, Aldrich TK, Prezant DJ, Sinnett MJ, Enden JB, Williams MH. High dose intramuscular triamcinolone in severe life-threatening asthma. New Engl J Med 1991;329:585–9.
110. Engel T, Heinig JH. Glucocorticoid therapy in acute severe asthma - a critical review. Eur Respir J 1991;4:881–9.
111. Decramer M, Lacquet LM, Fagard R, Rogiers P. Corticosteroids contribute to muscle weakness in chronic airflow obstruction. Am J Respir Crit Care Med 1995;150:11–16.
112. Bowler SD, Mitchell CA, Armstrong JG. Corticosteroids in acute severe asthma: effectiveness of low doses. Thorax 1992;47:584–7.
113. Morell F, Orkiols R, de Gracia J, Curul V, Pujol A. Controlled trial of intravenous corticosteroids in severe acute asthma. Thorax 1992;47:588–91.
114. British Thoracic Society. The British guidelines on asthma management. Thorax 1997;52 (Suppl 1):S1–21.
115. Devidayal Singhi S, Kumar L, Jayshree M. Efficacy of nebulized budesonide compared to oral prednisolone in acute bronchial asthma. Acta Paediatr 1999;88:835–40.

# *Index*

FSC
www.fsc.org
MIX
Papier | Fördert gute Waldnutzung
FSC® C083411

Kolibri 360 GmbH
Ferdinand-Jühlke-Straße 7
99095 Erfurt, Deutschland
produktsicherheit@kolibri360.de